LATIN SQUARES

New Developments
in the Theory and Applications

ANNALS OF DISCRETE MATHEMATICS 46

General Editor: Peter L. HAMMER
Rutgers University, New Brunswick, NJ, U.S.A.

Advisory Editors:

C. BERGE, Université de Paris, France
R.L. GRAHAM, AT&T Bell Laboratories, NJ, U.S.A.
M.A. HARRISON, University of California, Berkeley, CA, U.S.A.
V. KLEE, University of Washington, Seattle, WA, U.S.A.
J.H. VAN LINT, California Institute of Technology, Pasadena, CA, U.S.A.
G.C. ROTA, Massachusetts Institute of Technology, Cambridge, MA, U.S.A.
T. TROTTER, Arizona State University, Tempe, AZ, U.S.A.

LATIN SQUARES

New Developments in the Theory and Applications

J. DÉNES

*Industrial and Scientific Consultant
Formerly Head of Mathematics
Institute for Research and
  Co-ordination of Computing Techniques (SZKI)
Budapest, Hungary*

and

A.D. KEEDWELL

*Department of Mathematical
  and Computing Sciences
University of Surrey
Guildford, United Kingdom*

With specialist contributions by

G.B. BELYAVSKAYA
A.E. BROUWER
T. EVANS
K. HEINRICH
C.C. LINDNER
D.A. PREECE

1991

NORTH-HOLLAND – AMSTERDAM • NEW YORK • OXFORD • TOKYO

ELSEVIER SCIENCE PUBLISHERS B.V.
Sara Burgerhartstraat 25
P.O. Box 211, 1000 AE Amsterdam, The Netherlands

Distributors for the United States and Canada:

ELSEVIER SCIENCE PUBLISHING COMPANY, INC.
655 Avenue of the Americas
New York, N.Y. 10010, U.S.A.

Library of Congress Cataloging-in-Publication Data

```
Latin squares   new developments in the theory and applications /
   edited by J. Dénes and A.D. Keedwell ; with specialist contributions
   by G.B. Belyavskaya ... [et al.].
       p.   cm. -- (Annals of discrete mathematics ; 46)
   Includes bibliographical references and index.
   ISBN 0-444-88899-3
   1. Magic squares.   I. Dénes, J. (József).   II. Keedwell, A. D.
III. Beliavskaia, G. B.   IV. Series.
QA165.L38   1991
512.9'25--dc20                                              90-21564
                                                                 CIP
```

ISBN: 0 444 88899 3

© ELSEVIER SCIENCE PUBLISHERS B.V., 1991

All rights reserved. No part of this publication may be reproduced, stored in a retrieval system, or transmitted, in any form or by any means, electronic, mechanical, photocopying, recording or otherwise, without the prior written permission of the publisher, Elsevier Science Publishers B.V. / Physical Sciences and Engineering Division, P.O. Box 103, 1000 AC Amsterdam, The Netherlands.

Special regulations for readers in the U.S.A. – This publication has been registered with the Copyright Clearance Center Inc. (CCC), Salem, Massachusetts. Information can be obtained from the CCC about conditions under which photocopies of parts of this publication may be made in the U.S.A. All other copyright questions, including photocopying outside of the U.S.A., should be referred to the copyright owner, Elsevier Science Publishers B.V., unless otherwise specified.

No responsibility is assumed by the publisher for any injury and/or damage to persons or property as a matter of products liability, negligence or otherwise, or from any use or operation of any methods, products, instructions or ideas contained in the material herein.

Transferred to digital printing 2005

FOREWORD

Sixteen years ago the editors of the present volume published a book entitled "Latin Squares and their Applications". That book included a list of 73 unsolved problems of which about 20 have been completely solved in the intervening period and about 10 more have been partially solved. Thus, the hope which I expressed in my foreword to the former book that many of the problems would become theorems of Mr. or Ms. So and So has become a reality.

The present work comprises six contributed chapters and also six further chapters written by the editors themselves. As well as discussing the advances which have been made in the subject matter of most of the chapters of the earlier book, this new book contains at least one chapter which deals with a subject (that of r-orthogonal latin squares) which did not exist when the earlier book was written.

The success of the former book is shown by the two or three hundred published papers which deal with questions raised by it. I hope that this new book will be just as successful and that there will again be many unsolved problems some of which I hope will not stay unsolved for long.

PAUL ERDÖS

CONTENTS

Preface	xi
Acknowledgements	xiii

CHAPTER 1. INTRODUCTION. (J. Dénes and A. D. Keedwell)

(1) Basic definitions.	1
(2) Orthogonal latin squares.	2
(3) Isotopy and parastrophy.	4

CHAPTER 2. TRANSVERSALS AND COMPLETE MAPPINGS. (J. Dénes and A. D. Keedwell)

(1) Basic facts and definitions.	7
(2) Partial transversals.	9
(3) Number of transversals in a latin square.	14
(4) Sets of mutually orthogonal latin squares with no common transversal.	23
(5) Sets of mutually orthogonal latin squares which are not extendible.	28
(6) Generalizations of the concepts of transversal and complete mapping.	33
ADDITIONAL REMARKS.	39

CHAPTER 3. SEQUENCEABLE AND R-SEQUENCEABLE GROUPS: ROW COMPLETE LATIN SQUARES. (J. Dénes and A. D. Keedwell)

(1) Row-complete latin squares and sequenceable groups.	43
(2) Quasi-complete latin squares, terraces and quasi-sequenceable groups.	58
(3) R-sequenceable and R_h-sequenceable groups.	67
(4) Super P-groups.	75
(5) Tuscan squares and a graph decomposition problem.	79
(6) More results on the sequencing and 2-sequencing of groups.	84
ADDITIONAL REMARKS.	99

CHAPTER 4. LATIN SQUARES WITH AND WITHOUT SUBSQUARES OF PRESCRIBED
TYPE. (K. Heinrich) 101
   (1) Introduction. 102
   (2) Without subsquares. 113
   (3) With subsquares. 119
   (4) With subsquares and orthogonal. 133
   (5) Acknowledgement. 147
   ADDITIONAL REMARKS BY THE EDITORS. 147

CHAPTER 5. RECURSIVE CONSTRUCTIONS OF MUTUALLY ORTHOGONAL LATIN
SQUARES. (A. E. Brouwer) 149
   (1) Introductory definitions. 150
   (2) Pairwise balanced designs - definitions. 151
   (3) Simple constructions for transversal designs. 152
   (3)* Examples. 156
   (4) Wilson's construction. 159
   (4)* Examples. 161
   (5) Weighting and holes. 162
   (5)* Examples. 164
   (6) Asymptotic results. 165
   (7) Table of values of N(v) up to v=200. 166
   ADDITIONAL REMARKS BY THE EDITORS. 166

CHAPTER 6. r-ORTHOGONAL LATIN SQUARES. (G. B. Belyavskaya)
   (1) Some weaker modifications of the concept of orthogonality. 169
   (2) r-Orthogonal latin squares and quasigroups. 171
   (3) Partial admissibility of quasigroups, its connection with
r-orthogonality. 177
   (4) Spectra of partial orthogonality of latin squares (quasigroups). 186
   (5) Near-orthogonal and perpendicular latin squares. 190
   (6) r-Orthogonal sets of latin squares. 195
   (7) Applications of r-orthogonal latin squares and problems
raised thereby. 200

CHAPTER 7. LATIN SQUARES AND UNIVERSAL ALGEBRA. (T. Evans) 203
   (1) Universal algebra preliminaries. 204
   (2) Varieties of latin squares. 206

Contents

(3)	Varieties of orthogonal latin squares.	208
(4)	Euler's conjecture.	211
(5)	Free algebras and orthogonal latin squares.	212

CHAPTER 8. EMBEDDING THEOREMS FOR PARTIAL LATIN SQUARES. (C. C. Lindner) 217

    (1) Introduction. 218
    (2) Systems of distinct representatives. 219
    (3) The theorems of Ryser and Evans (on latin rectangles and squares). 222
    (4) Cruse's theorems (on commutative latin rectangles and squares). 225
    (5) Embedding idempotent latin squares. 229
    (6) Conjugate quasigroups and identities. 236
    (7) Embedding semisymmetric and totally symmetric quasigroups. 240
    (8) Embedding Mendelsohn and Steiner triple systems. 243
    (9) Summary of embedding theorems. 253
    (10) The Evans' conjecture. (Smetaniuk's proof.) 254
    APPENDIX (1). Alternative description of Smetaniuk's proof of the Evans' conjecture. 261
    APPENDIX (2). Additional Bibliography. 265

CHAPTER 9. LATIN SQUARES AND CODES. (J. Dénes and A. D. Keedwell) 267

    (1) Basic facts about error-detecting and correcting codes. 268
    (2) Codes based on orthogonal latin squares and their generalizations. 272
    (3) Row and column complete latin squares in coding theory. 283
    (4) Two-dimensional coding problems. 290
    (5) Secret-sharing systems. 303
    (6) Miscellaneous results. 308
    ADDITIONAL REMARKS. 314

CHAPTER 10. LATIN SQUARES AS EXPERIMENTAL DESIGNS. (D. A. Preece)

    (1) Introduction. 317
    (2) The design and and analysis of experiments. 318

(3) Some practical examples of latin squares used as row-and-column designs. 322
(4) Some other uses of latin squares in experimental design. 324
(5) The use of latin squares in experiments with changing treatments. 327
(6) Other "latin" experimental designs. 329
(7) Statistical analysis of latin square designs. 331
(8) Randomization of latin square designs. 338
(9) Polycross designs. 341

CHAPTER 11. LATIN SQUARES AND GEOMETRY. (J. Dénes and A. D. Keedwell)
(1) Complete sets of mutually orthogonal latin squares and projective planes. 343
(2) Projective planes of orders 9, 10, 12 and 15. 346
(3) Non-desarguesian projective planes of prime order. 351
(4) Digraph complete sets of latin squares and incidence matrices. 352
(5) Complete sets of column orthogonal latin squares and affine planes. 358
(6) The Paige-Wexler latin squares. 360
(7) Miscellanea. 373
ADDENDUM. 377

CHAPTER 12. FREQUENCY SQUARES. (J. Dénes and A. D. Keedwell)
(1) F-squares and orthogonal F-squares. 381
(2) Enumeration and classification of F-squares. 388
(3) Completion of partial F-squares. 389
(4) F-rectangles and other generalizations. 392
(5) A generalized Bose construction for orthogonal F-squares. 396
ADDITIONAL REMARKS. 398

Bibliography 399
Subject Index 444

PREFACE

In 1974, the authors of the present book published a book "Latin Squares and their Applications" in which they tried to collect and collate all the literature on the subject which then existed. They also published what they hoped would prove to be a comprehensive bibliography.

Perhaps partly as a result of this attempt, interest in the subject has exploded in the last fifteen years with the result that as many new papers on it have been published in those years as in the previous two hundred years. By the early 1980's, it appeared to us, therefore, that it was time that we made an attempt to repeat the task and we have been encouraged to do so by many requests from all parts of the World. However, it soon became clear that a book of about 800 pages would be required with a further 200 pages of bibliography. That being so, we decided in consultation with the publishers that our best strategy would be to try to be comprehensive on a limited range of topics, which should so far as possible include the core topics, and to defer remaining topics to a subsequent volume. The latter is referred to as "Part II" in the present book and its eventual appearance will depend very much on whether the present work is regarded by its publishers as "successful".

Most of the material in the present book is new and it can be read independently of the former one in the sense that, for almost all of the results which are not proved, reference is made to their original source unless they have become "folklore" in the subject. However, the earlier book (which is referred to throughout as [DK]) is frequently cited as an alternative reference source and for background information which space does not allow us to repeat in the present one.

Because developments in the subject of latin squares have been so rapid and have diverged in so many different directions, we decided that the only way in which we could complete our task within a reasonable time was by inviting a number of acknowledged experts on particular aspects of the subject to contribute chapters to the present work. Ten mathematicians were approached and all agreed to help us. It is with very much regret that, because the total number of pages is limited, we are not able to include all the contributed Chapters within the present Volume. Those that do not appear herein will, we hope, be included in Part II.

We have a further apology to make to our contributing authors. Because they completed their Chapters at different times, not all are equally up-to-date. We take full responsibility for the long time which has elapsed before the complete manuscript was ready for publication. This is partly due to our under-estimating the time that writing would take and partly due to problems encountered in producing error-free camera-ready copy with the resources available to us.

All the Chapters were prepared on an "Advent" mathematical word-processor which, though the best available to us in the early 1980's and one of the first full-page-screen processors to appear on the market, has somewhat limited facilities by to-day's standards. Readers will notice that formulae are sometimes unavoidably split, that some symbols are of unsuitable shape or size and that some which were not available on the processor have had to be written-in by hand. This word-processor is no longer in satisfactory working order and one unfortunate consequence is that hardly any alterations of text could be made in the final months before completion of the work. We have tried to compensate for the latter fact and also for the fact that not all parts of the text could be equally up-to-date by adding "Additional Remarks" at the ends of several of the Chapters. It was also necessary to use a different word-processor for the typing of this Preface, the Contents pages, the Bibliography and Subject Index.

Our original intention of including a comprehensive bibliography of all papers on latin squares has had to be abandoned owing to lack of space. (We apologize therefore to the many mathematicians who sent us papers which we have been unable to include.) The present Bibliography contains only papers which are actually mentioned in the text. Also, for ease of reference, where papers cited have joint authorships, the order of authors has been made alphabetical. In some (but not all) of the Chapters, a similar transposition of names of joint authorships has been made. The present authors apologize for their lack of consistency in this regard.

As regards references to the Bibliography, with two exceptions these are in the form of an author reference followed by a date. For example, the statements "In (1963), R.H.Bruck has shown ... " or "R.H.Bruck(1963) has shown ... " are both references to the paper of R. H. Bruck which is listed under the publication date of 1963. Exceptionally, the earlier book of the present authors is referred to as [DK] throughout the present Volume and, in

Chapter 6, two papers of G. B. Belyavskaya which are very frequently cited are abbreviated to B(1976) and B(1977).

For the numbering of Theorems, Figures and Examples within each Chapter, decimal notation has been used and cross-references are in the style "See Theorem 2.6 of Chapter 10" which is a reference to the sixth Theorem of the second Section of that Chapter. In the Subject Index, page references are to within-chapter page numbers and take the form "3:36,55, 9:16" which indicates that the topic is discussed on pages 36 and 55 of Chapter 3 and on page 16 of Chapter 9.

Some editorial changes to the Contributed Chapters have been made so as to bring their layout style and format into consistency with that of the rest of the book.

We should like to call attention to three other publications on the subject of latin squares: the first, T.Evans and C.C.Lindner(1977), is devoted to problems of embedding but is now very much out-of-date; the other two are I. Bosák(1976), which will be of special interest to Slovak readers, and a recent work of V.D.Belousov and G.B.Belyavskaya(1989) which provides a succinct summary of the main results in [DK] for the benefit of Russian readers.

Also, we call attention to two historical surveys of non-associative binary systems (especially quasigroups and loops), one in Russian and one in English, which are not mentioned elsewhere in this book: namely, V.D. Belousov(1967a) and R.H.Bruck(1958). Owing to lack of space, recent results on the rôle of a latin square as the multiplication table of a quasigroup have had to be deferred to the planned Part II of this book.

Finally, we urge readers to look at the Additional Remarks at the ends of Chapters for the latest information before reading the Chapters in detail.

ACKNOWLEDGEMENTS

Firstly, we wish to express our thanks to the many mathematicians who have sent us copies of their papers, both published and forthcoming, to the many who offered us encouragement by their interest and enquiries and, above all, to our contributing authors, to Paul Erdős who provided the Foreword, and to the following mathematicians who kindly acted as referees for individual Chapters: G.B.Belyavskaya, D.A.Drake, S.W.Golomb, J.C.Gower, K.Heinrich, A.J.W.Hilton, D.F.Hsu, G.L.Mullen, J.M.Nowlin-Brown. We are also very grateful to Amanda Woods and, latterly, Sally Fenwick, of the University of Surrey Mathematics Office, for their careful typing and for the many hours they spent in producing accurate and well-laid-out camera-ready copy.

We owe thanks also to the Soros Foundation for providing financial assistance to enable the first author to spend an extended period in England during the later stages of the writing; to Dr. Paul Fisher, a former colleague of the second author, for helping with compilation and editorial work on the Bibliography and to Dr. Donald Preece for checking it for internal consistency; to Dr. S. Lajos who assisted us in making translations from Russian to English; and to Prof. S. Csibi of the Institute for Communications and Electronics, Technical University of Budapest, for helping behind-the-scenes in several ways.

The first author also wishes to thank Dr. J. Merza and his colleagues from the library of the Mathematical Institute of the Hungarian Academy of Sciences for the care and trouble which they took to provide him with copies of research papers, often from difficult-to-obtain journals. The second author wishes to express similar thanks to the librarians of the University of Surrey library for their helpfulness at all times.

1st JUNE, 1990. J. DÉNES
  A. D. KEEDWELL

CHAPTER 1

INTRODUCTION (J.Dénes and A.D.Keedwell)

In this preliminary chapter, we provide a number of basic definitions and introduce some of the concepts which will be used throughout the book. Our intention is to ensure that the reader who does not have easy access to a copy of [DK] will be able to read the present book without difficulty.

(1) Basic Definitions.

A latin square of order n is an $n \times n$ matrix L whose entries are taken from a set S of n symbols and which has the property that each symbol from S occurs exactly once in each row and exactly once in each column of L.

A latin square based on the natural numbers as symbol set is reduced or in standard form if the elements of its first row and column are in natural order.

A set S on which a binary operation (.) is defined forms a quasigroup with respect to that operation if, when any two elements a,b of S are given, the element $a.b$ is in S and the equations $a.x = b$ and $y.a = b$ each have exactly one solution in S.

A loop H is a quasigroup which has an identity element: that is, a quasigroup in which there exists an element e of H with the property that $e.x = x.e = e$ for all x of H.

It is easy to see that the multiplication table (often called the Cayley table) of a quasigroup is a latin square (Theorem 1.1.1 of [DK]). Moreover, if the quasigroup is commutative, then the latin square is symmetric. If the quasigroup is a loop and the identity element e is taken as first element of the row border and the column border of its multiplication table, then the first row and the first column of the latin square coincide with its borders. As an illustrative example, we exhibit in

Figure 1.1 the multiplication table of a loop of order five whose identity element is denoted by e. This is the smallest order for which there exists a loop which is not a group.

A quasigroup $(Q,.)$ for which $a.a = a$ for all a in Q is called <u>idempotent</u>. One for which $a.a = b.b$ is called <u>unipotent</u>. These same adjectives are applied to the corresponding latin squares. Thus, the latin square exhibited in Figure 1.1 is unipotent.

The Cayley table of a group (associative quasigroup) has a property additional to that of being a latin square. Namely, it satisfies the so-called <u>quadrangle criterion</u>. If the entry in the i-th row and j-th

	e	a	b	c	d
e	e	a	b	c	d
a	a	e	c	d	b
b	b	d	e	a	c
c	c	b	d	e	a
d	d	c	a	b	a

Figure 1.1

column is denoted by a_{ij} and if $h,i,j,k,\ldots$, are any indices, it follows from the equations $a_{ij} = a_{i_1 j_1}$, $a_{hj} = a_{h_1 j_1}$, $a_{hk} = a_{h_1 k_1}$ that $a_{ik} = a_{i_1 k_1}$. (For a proof, see Theorem 1.2.1 of [DK]).

(2) <u>Orthogonal latin squares</u>.

Two latin squares $L_1 = ||a_{ij}||$ and $L_2 = ||b_{ij}||$ on n symbols, say $1,2,\ldots,n$, are said to be <u>orthogonal</u> if every ordered pair of symbols occurs exactly once among the n^2 pairs (a_{ij}, b_{ij}), $i = 1,2,\ldots,n$; $j = 1,2,\ldots,n$. A pair of orthogonal latin squares A_1 and A_2 of the smallest possible order 3 is exhibited in Figure 5.1 of Chapter 2. (In that Figure, the symbol set used is $0,1,2$.)

A pair of orthogonal latin squares is also called a <u>graeco-latin square</u>, especially in statistical applications, because L. Euler (1779) used Greek letters for one square of the pair and latin letters for the other. See, for example, A.D.Keedwell(1983a).

If we consider the (exactly) n cells of the latin square L_2 all of which contain the same fixed entry h say ($1 \leq h \leq n$), then the entries in the corresponding cells of the latin square L_1 must be all different, otherwise the squares would not be orthogonal. Since the symbol h occurs exactly once in each row and once in each column of the latin square L_2, we see that the n entries of L_1 corresponding to the entry h in L_2 have the property that they are all different and also occur one in each row and one in each column of the square L_1. A set of n different symbols having this property in a latin square of order n is usually called a transversal of the latin square. (Other names have sometimes been used, more particularly the term directrix and occasionally the confusing term diagonal.)

It will be immediately obvious from the above remarks that a given latin square possesses an orthogonal mate (latin square orthogonal to it) if and only if it has n disjoint transversals. (Theorem 5.1.1 of [DK]).

The concepts of transversal and orthogonal latin squares date from the time of L.Euler(1779), two hundred years ago. Euler had been able to construct pairs of orthogonal latin squares of every odd order and of every order divisible by four but not of any order congruent to 2 modulo 4. He conjectured that pairs of orthogonal latin squares do not exist for such orders. The conjecture remained unresolved until the present century. In 1900, G.Tarry(1900a) proved that the conjecture was true for the order six by an exhaustive enumeration of cases. Subsequently, a number of much shorter proofs of this result have been obtained. In particular, we refer the reader to R.A.Fisher and F.Yates(1934), K.Yamamoto(1954), A.K.Rybnikov and N.M.Rybnikova(1966), and for two really short proofs to D.R.Stinson(1984) and to D.Betten(1983,1984).

Much later, R.C.Bose, S.S.Shrikhande and E.T.Parker(1960) proved that the Euler conjecture was false for all orders n of the form 4k+2 except n = 2 or 6 by providing a constructive method of obtaining pairs of orthogonal latin squares of all of these orders. Their proof was lengthy and involved concepts from the theory of experimental design. However, it undoubtedly stimulated much new research on the structure and properties of latin squares. Another stimulus was the increasing interest in the theory of finite projective planes, since such a plane can be viewed as a so-called <u>complete set</u> of mutually orthogonal latin squares (Theorem 5.2.2 of [DK]). A set of n-1 pairwise orthogonal latin squares of order n is called a <u>complete set</u> because it is easily seen that no larger set can exist (Theorem 5.1.5 of [DK]).

Since 1960, much shorter disproofs of Euler's conjecture have been obtained. The first of these was that of D.J.Crampin and A.J.W.Hilton (1975) but this involved the use of a computer. A short direct and elegant proof was obtained by L.Zhu(1977) and published in Chinese in that year. Later, an English version was published in Ars Combinatoria, L.Zhu (1982), with editorial comment by one of the present authors. The reader interested in this subject is strongly recommended to look at Zhu's proof. However, he should take cognizance of the following unfortunate misprints: On page 49 of the Ars Combinatoria paper, the last line should read "(Note that we have proved that there exist only two distinct k-th power residues other than unity when $p = 3k+1$). In the case when $p = 3k-1$, we show that the cubes of..."

Page 51, line 4, should read "... then $\alpha^{3x} = \alpha^t$ mod p or ..."

Page 53, last line, should read "If v is divisible by an odd prime p other than 3, then $v = 2(2s+1)3^h = u.3^h$ where $p|(2s+1)$ and ...".

Page 54, line 9, should read "Sets of four MOLS...".

(3) <u>Isotopy and parastrophy</u>

Latin squares which can be obtained one from the other by a permutation of rows, a permutation of columns or a renaming of the symbols are in some sense not mathematically different. We call them <u>isotopic</u> or, sometimes, <u>equivalent</u>.

Similarly, two sets of mutually orthogonal latin squares of the same order are often called <u>equivalent</u> if the numbers of squares in the two sets are the same and if the squares of the two sets can be put into one-to-one correspondence in such a way that each pair is equivalent relative to the same renaming of symbols and reordering of rows and columns.

There is an analogous concept of isotopy between quasigroups. Let $(G, \cdot)$ and $(H, *)$ be two quasigroups of the same order. An ordered triple (α,β,γ) of one-to-one mappings α,β,γ of the set G onto the set H is called an <u>isotopism</u> of $(G,.)$ upon $(H,*)$ if $(x\alpha)*(y\beta) = (x.y)\gamma$ for all x,y in G. The quasigroups $(G,.)$ and $(H,*)$ are then said to be <u>isotopic</u>.

With any given quasigroup (Q,θ), there are associated, generally speaking, five <u>conjugate</u> or <u>parastropic</u> quasigroups defined on the same set G by the five operations θ^*, $^{-1}\theta$, θ^{-1}, $^{-1}(\theta^{-1}) \equiv (^{-1}\theta)^*$, $(^{-1}\theta)^{-1} \equiv (\theta^{-1})^*$ where, if $a\theta b = c$ for $a,b,c \in Q$, we write $b(\theta^*)a = c$, $c(^{-1}\theta)b = a$, $a(\theta^{-1})c = b$, $b\{^{-1}(\theta^{-1})\}c = a$, $c\{(^{-1}\theta)^{-1}\}a = b$.
It is immediately clear from the definitions that the operations $(^{-1}\theta)^*$ and $(\theta^{-1})^*$ are respectively the same as the operations $^{-1}(\theta^{-1})$ and $(^{-1}\theta)^{-1}$ as we have indicated. In S.K.Stein(1956) and (1957) and V.D.Belousov (1965), these five operations associated with a given opeation θ have been called conjugates of θ, while in A.Sade(1959a) they have been called parastrophes of θ. The significance of these operations from the point of view of latin squares is that, if the multiplication table of the quasigroup (Q,θ) is given by the bordered latin square L, then (Q,θ^*) has multiplication table given by the transpose of L, while in $(Q,^{-1}\theta)$, (Q,θ^{-1}) the roles of row and element number, column and element number, respectively, are interchanged. In the remaining two parastrophic quasigroups, the roles of row, column, and element number are all three permuted.

If the quasigroup (Q,θ) satisfies a given identity, for example the associative law, then in general each of its conjugates will satisfy a different <u>conjugate identity</u>. Thus, for example, the quasigroup (Q,θ^{-1}) satisfies the identity $(x\theta^{-1}y)\theta^{-1}(x\theta^{-1}z) = y\theta^{-1}z$ if (Q,θ) is associative. A table listing the identities conjugate to a number of well-known identities appears as Figure 2.1.8 of [DK].

A set of latin squares which comprises all the latin squares isotopic to a given latin square (an <u>isotopy class</u>) together with all their conjugates is called a <u>main class</u> of latin squares (sometimes a <u>transformation set</u> or an <u>adjugate class</u>). The latin squares of a given order can be classified by separating them first into their main classes and then into their isotopy classes. (See chapter 4 of [DK].) In addition, we say that two latin squares are <u>isomorphic</u> if their corresponding quasigroups are isomorphic:

that is, if one can transform one square into the other by making the same permutation of rows, columns and symbols simultaneously.

In Part II of the present book, the question of when a latin square is orthogonal to all its parastrophes is discussed.

We end this section and this introductory chapter by calling the reader's attention to an important theorem due to V.D. Belousov(1966) of which a new short proof has been published since [DK] was written. Belousov proved that a quasigroup which satisfies any irreducible balanced identity is isotopic to a group. (An identity $W_1 = W_2$ defined on a quasigroup $(Q,.)$ is <u>balanced</u> if each variable x which occurs on one side W_1 of the identity occurs on the other side W_2 too and if no variable occurs in W_1 or W_2 more than once. The identity is <u>reducible</u> if either (i) each of W_1 and W_2 contains a "free element" x so that W_1 is of the form $U_1.x$ or $x.V_1$ and W_2 is likewise of the form $U_2.x$ or $x.V_2$ (where U_i or V_i represents a subword of W_i) or (ii) W_1 has the product xy of two free elements as a subword and W_2 has one of the products xy or yx as a subword, or the dual of this statement. For examples, we refer the reader to [DK], pages 68 and 69). In M.A.Taylor(1978) a short and quite elementary proof of the following generalized version of Belousov's result has been given: Let $W_1 = W_2$ be a balanced identity with the property that W_1 contains a subword xy and that x and y are separated in W_2. Then every quasigroup $(Q,.)$ which satisfies this identity is isotopic to a group. (We say that x and y are <u>separated</u> in W_2 if neither xy nor yx occurs in W_2: that is, at least one bracket lies between x and y. Taylor's proof involves showing that the quadrangle criterion is satisfied by such a quasigroup $(Q,.).$)

CHAPTER 2

TRANSVERSALS AND COMPLETE MAPPINGS (J. Dénes and A.D. Keedwell)

The concept of transversal and the closely related one of complete mapping are fundamental to the theory of latin squares. In this chapter, we summarize the basic facts about these concepts and give an account of more recent generalizations and results.

(1) Basic facts and definitions.

DEFINITIONS. A transversal of a latin square of order n is a set of n cells, one in each row, one in each column, such that no two of the cells contain the same symbol. [L. Euler (1779)]. A complete mapping of a group, loop, or quasigroup $(G, \cdot)$ is a bijection $x \to \theta(x)$ of G onto G such that the mapping $x \to \phi(x)$ defined by $\phi(x) = x.\theta(x)$ is again a bijection of G. [H.B. Mann (1942)].

We shall assume the following facts to be well-known:
(1) If Q is a quasigroup which possesses a complete mapping then its multiplication table is a latin square with a transversal; and conversely.
(2) If L is a latin square which satisfies the quadrangle criterion (that is, is isotopic to the multiplication table of a group) and which possesses at least one transversal, then L has a decomposition into n disjoint transversals and consequently has an orthogonal mate.
(3) For a commutative quasigroup of odd order, the identity mapping is a complete mapping [A. Sade (1960)] and the same result is true for Bol loops of odd order [D.A. Robinson (1966)].

A loop is called a Bol loop if, for all elements x, y, z, $[(xy)z]y = x[(yz)y]$. The class of Bol loops includes all Moufang loops, all Bruck loops and, of course, all groups.

(4) An abelian group has a complete mapping if and only if the product of all its elements is equal to the identity; that is, if and only if it does not have a unique element of order 2. [G.A.Miller(1903), K.G.Ramanathan(1947), L.J.Paige(1947) and (1951). See also Remark (2) below.]

(5) For a non-abelian group to have a complete mapping it is necessary that the product of all its elements in some order be equal to the identity and sufficient (but certainly not necessary) that it be R-sequenceable. [L..J.Paige(1951).]

(6) A finite group which has a cyclic Sylow 2-subgroup does not possess a complete mapping. [M.Hall and L.J.Paige(1955).]

(7) A finite soluble group whose Sylow 2-subgroups are non-cyclic has a complete mapping [M.Hall and L.J.Paige(1955).]

(8) All finite alternating groups and all finite symmetric groups except S_2 and S_3 possess complete mappings [M.Hall and L.J.Paige(1955).]

Closely related to (5) above is the concept of a P-group, first conceived by L.Fuchs. Let G be a finite group of order n and G' its commutator subgroup. Since the elements of G commute mod G', every product of the n elements of G is in the same coset of G'. (If G has a complete mapping, this coset is G' itself.) L. Fuchs proposed the study of groups G for which it is true, conversely, that every element of the appropriate coset of G' in G is expressible as the product of the n elements of G in some suitable order. J.Denes later christened such groups as P-groups. A.Rhemtulla(1969) showed that every finite soluble group is a P-group and thirteen years later J.Denes and P.Hermann(1982) proved that every finite group is a P-group. Recently, A.D.Keedwell(1983c) and (1984) has introduced a generalization of this concept under the name of super P-group (see Chapter 3 of the present book).

REMARK (1). More details concerning all the above facts and proofs of most of them will be found in section 1.4 of the present authors' earlier book.

REMARK (2). In their earlier book [DK], the present authors had attributed the realization that the product of all the elements of an abelian group is equal to the identity unless the group has a unique element of order two to L.J.Paige. However, see J.Dénes(1984), it turns out that this result was originally proved by G.A.Miller(1903). Also, in addition to the proofs given by L.J.Paige(1947) and M.Hall(1948), there was also one given by K.G.Ramanathan(1947).

REMARK (3). Regarding statements (5) and (7) above, the present authors have very recently shown that all finite non-soluble groups have non-cyclic Sylow 2-subgroups and, moreover, that there exists an ordering of the elements of any finite non-soluble group for which their product is equal to the identity element. This has led them to propose the new conjecture that every finite non-soluble group has a complete mapping. For details, see J.Dénes and A.D.Keedwell(1989) and (1990c).

(2) Partial transversals.

From (1) and (6) above, it is already clear that many latin squares do not have any transversals. In particular, a latin square of order $4n+2$ which has a latin subsquare of order $2n+1$ does not possess any transversals. [H.B.Mann(1944). See also theorem 5.1.4 of [DK].] More generally, a latin square of order $n = mq$ and of q-step type has no transversals if m is even and q is odd. (See the next section.)

This fact has led several authors to try to determine a lower bound for the length of a <u>partial transversal</u> : that is, a maximal set of cells, each in a different row and each in a different column and such that no two contain the same symbol. K.K.Koksma(1969) showed that an arbitrary latin square of order n ($n \geq 7$) has at least one partial transversal of length t with $t \geq (2n+1)/3$. (His method is described in Section 3.1 of [DK].) Eight years later, D.A.Drake(1977) was able to improve this to $t \geq 3n/4$ using a rather simpler method that of Koksma. In the following year, S.M.P.Wang(1978) obtained $t \geq (9n-15)/11$. However, in the same year, A.E.Brouwer, A.J.de Vries and R.M.A.Wieringa(1978) and, independently, D.E.Woolbright(1978) obtained $t \geq n-\sqrt{n}$. More precisely, they obtained $(n-t)^2 \leq t$ whence $t \geq n+\frac{1}{2} - \sqrt{(n+\frac{1}{4})}$ which is one integer larger than $n-\sqrt{n}$ for some values of n. Their arguments are essentially the same though differently expressed. Both arguments use graphs as a descriptive tool but, in the opinion of the present authors, the introduction of a graph in this particular case does not lead to any real simplification so we offer a graph-free version of the proof in theorem 2.2 below. Also, despite the fact that this theorem supercedes Drake's earlier result, the ideas used by Drake seem to us sufficiently interesting to merit us including a proof of his result also (in theorem 2.1).

More recently, P.W.Shor[1] has proved that $t \geq n - 5 \cdot 3 (\log n)^2$. The latter bound is better than the bound $n - \sqrt{n}$ for all $n > 2,000,000$. Shor's argument is novel and does not make use of graphs. However, because it is relatively lengthy, we refer the reader to the original paper for the proof. (The authors hope that, by employing a combination of the various methods, one of their readers may be able to improve the bound yet further.)

THEOREM 2.1 (D.A.Drake). Let L be a latin square of order n whose longest partial transversal has length t. Then $t \geq \min(3n/4, n-2)$.

Proof. Let t be the length of a maximal partial transversal of the n×n latin square L. By means of row, column and symbol permutations, we may bring L to the form $L = [\ell_{ij}] = \begin{vmatrix} A & B \\ C & D \end{vmatrix}$

where A is of size t×t and has $\ell_{ii} = i$ for $i = 1, 2, \ldots, t$. We shall call the integers $1, 2, \ldots, t$ minor numbers and the integers $t+1, t+2, \ldots, n$ major numbers. We partition the minor numbers into three types. The number h is of type 1 if neither the h-th row nor the h-th column of L contains any major numbers in the last n−t cells. The number h is of type 3 if either the h-th row or the h-th column of L has two or more major numbers in its last n−t cells. All remaining minor numbers are of type 2. Let there be i,j,k numbers of types 1,2,3 respectively. Then $i+j+k = t$.

Let T be the partial transversal formed by the cells (i,i), $i \leq t$. Since T is maximal, D cannot contain any major numbers.

Suppose now that there is a minor number h such that two cells in the h-th row of B (say ℓ_{hu} and ℓ_{hv}, $u,v > t$) and one cell in the h-th column of C (say ℓ_{wh}, $w > t$) are all major numbers. Since ℓ_{wh} cannot be equal to both ℓ_{hu} and ℓ_{hv} we can replace the cell (h,h) of T by the cells (w,h) and either (h,u) or (h,v) and thus extend T. This contradiction shows that no such minor number h exists and consequently, for every type 3 number h, the entries in the n−t cells of row h of B together with the entries in the n−t cells of column h of C include at most n−t major numbers (which are either all in B or all in C). Since, for a type 1 number h, there are no major numbers among the entries in the h-th row of B or the h-th column of C, we can deduce that the total number of major numbers which occur in B and C together cannot exceed $2j+(n-t)k$.

Next, let h be any minor number of type 3 and suppose that the cells (h,u) and (h,v) of B contain major numbers, say ℓ_{hu} and ℓ_{hv}. If h occurs in row w of D then it occurs in at most one of the cells (w,u) and (w,v). Suppose for definiteness that $\ell_{wu} \neq h$. Then T can be extended by replacing cell (h,h) of T by the cells (h,u) and (w,x), where $\ell_{wx} = h$. This is again a contradiction and shows that h cannot occur in any row of D. That is, no type 3 minor numbers occur in D. If h is a minor number of type 2 then at most one cell of the h-th row of B contains a major number. We may suppose that ℓ_{hu} is a major number. If h occurs in any column of D other than the u-th column: for example, if $\ell_{wx} = h$, $x \neq u$, then T can be extended as before by replacing cell (h,h) of T by the cells (h,u) and (w,x). From this, we deduce that a type 2 minor number occurs at most once in D.

Since all of the $(n-t)^2$ cells of D contain minor numbers, we deduce that $(n-t)i+j \geq (n-t)^2$ because the left hand side of this inequality represents the largest possible number of minor numbers which can occur in D. Now the total number of major numbers which occur in B and C together is $2(n-t)^2$ because each occurs once in each column of B and once in each row of C. Combining this with our earlier result, we get the inequality $2j+(n-t)k \geq 2(n-t)^2$. Addition of the two inequalities gives $(n-t)i+3j+(n-t)k \geq 3(n-t)^2$ or, equivalently, $i+(3/(n-t))j+k \geq 3(n-t)$. So whenever $t < n-2$ (implying $3 \leq n-t$) we have $t = i+j+k \geq i+(3/(n-t))j+k \geq 3(n-t)$. Thus, if $t < n-2$ then $t \geq 3n/4$. □

THEOREM 2.2 (D.E.Woolbright/A.E.Brouwer et al) Every latin square of order n has a partial transversal of length t for some t such that $(n-t)^2 \leq t$.

Proof. Let t be the length of a maximal partial transversal of the n×n latin square L. By means of row, column and symbol permutations, we may bring L to the form $L = [\ell(i,j)] = \begin{vmatrix} A & B \\ C & D \end{vmatrix}$ where A is of size t×t and has $\ell(i,i) = i$ for $i = 1,2,\ldots,t$ as in Drake's theorem. We shall again call the integers $1,2,\ldots,t$ <u>minor numbers</u> and the integers $t+1, t+2, \ldots, n$ major numbers. Because t is maximal, D contains only minor numbers.

Define sets A_i ($i = 0,1,\ldots,n-t$) recursively by $A_0 = \phi$, $A_i = \{j: \ell(j,t+i) \in A_{i-1} \cup M\}$, where M is the set of major numbers.

Then A_1 contains the row-numbers of all those rows of L which contain major numbers in the (t+1)-th column, A_2 contains the row-numbers of all those rows of L which contain in the (t+2)-th column either major numbers or else entries from A_1, and so on. For example, in Figure 2.1, the numbers $1,2,\ldots,9$ are minor numbers and the sets just defined are as follows:

$A_1 = \{2,5,9\}$, $A_2 = \{1,3,4,6,7,8\}$, $A_3 = \{1,2,3,4,5,7,9,11,12\}$.

1	5	2	8
. 2	10	8	11
. . 3	4	11	10
. . . 4	7	10	12
. . . . 5	11	3	4
. . . . . 6	3	5	2
. . . . . . 7 . . .	1	9	3
. . . . 8 . .	6	12	9
. . . . 9 .	12	6	7
. . . .	2	7	5
. . . .	8	4	6
. . . .	.	.	1

Figure 2.1

We claim that every A_i consists of minor numbers. If the claim is true, then $|A_i| = i(n-t)$ for each i, since all n-t major numbers occur once each in the (t+i)-th column, $i = 1, 2, \ldots, n-t$, and are obviously distinct from all minor numbers in that column. In particular, $(n-t)^2 = |A_{n-t}| \leq t$, so the theorem holds if our claim is true.

The claim is obviously true for $i = 0$ (vacuously) and for $i = 1$.

Consider next the set A_2. It contains the row-numbers of rows of B whose second column entries are major numbers and also of those rows of L for which the entry in the (t+2)th column is a member of the set A_1. Suppose that one of the rows of L which has an element from A_1 as entry in the (t+2)th column is a row of D and so corresponds to a major number. If this element from A_1 is r then we are postulating that r occurs in the second column of D and also, by definition of A_1, that the entry in cell (r,t+1) of L is a major number. However, that being so, we can replace cell (r,r) of L by the cell (r,t+1) together with the cell of the second column of D which contains r to give a partial transversal of L of length t+1, contrary to hypothesis. We conclude that all the elements of A_2 must be minor numbers. We now take as induction hypothesis that we have already proved that all elements of A_i are minor numbers. We wish to show that no element of A_i can occur in the (i+1)th column of D. Suppose on the contrary that cell (z,t+i+1) contains an entry $y \in A_i$ (and so $y \leq t$). Then there is a cell (y,t+i) which contains an entry $x \in A_{i-1}$ or else a major number. In the former case, there is a cell (x,t+i-1) which contains an entry $w \in A_{i-2}$ or else a major number, and so on. Thus, we require to show that there does not exist a sequence of cells $(a_0, t+j)$, $(a_1, t+j+1)$, $(a_2, t+j+2)$, $\ldots$, $(a_k, t+j+k)$ where $a_{h-1} = \ell(a_h, t+j+h)$ for $h = 1, 2, \ldots, k$, and with the properties (i) that cell $(a_0, t+j)$ contains a major number, and (ii) that $a_k > t$, but $a_{k-1} \leq t$.

We first note that if such a sequence exists then all its cells are in distinct columns. If two cells of the sequence are in the same row, say $a_u = a_v$, we consider the shorter sequence obtained by omitting the cells $(a_{u+1}, t+j+u+1)$, $(a_{u+2}, t+j+u+2)$, $\ldots$, $(a_v, t+j+v)$. Since $a_u = a_v = \ell(a_{v+1}, t+j+v+1)$, this shortened sequence has the same properties as the original sequence. Thus, the shortest possible sequence with the properties postulated has all its cells in different rows and also in different columns. Moreover, no two cell entries are equal since $\ell(a_w, t+j+w) = \ell(a_x, t+j+x)$ implies $a_{w-1} = a_{x-1}$, a contradiction. Since $(a_0, t+j)$ is a major number and $a_k > t$, it now follows that if we use the above set of cells as replacements for the cells $(a_0,a_0), (a_1,a_1), \ldots, (a_{k-1},a_{k-1})$ of our maximal partial transversal of L, we shall obtain a new partial transversal of length $t+1$. This contradiction shows that all elements of A_i occur in the $(i+1)$th column of B rather than D. Hence, A_{i+1} consists entirely of minor numbers and, moreover, $|A_{i+1}| = (i+1)(n-t)$. By induction on i, $|A_{n-t}| = (n-t)^2 \leq t$, as required. []

From $(n-t)^2 \leq t$, we have $(n-t)^2+(n-t)+\frac{1}{4} \leq n+\frac{1}{4}$. Hence, $n-t+\frac{1}{2} \leq \sqrt{(n+\frac{1}{4})}$ and so $t \geq n+\frac{1}{2} -\sqrt{(n+\frac{1}{4})}$.

We conclude our discussion of partial transversals by observing that none of the bounds so far obtained is anywhere near the bound $t \geq n-1$ conjectured by A.Brualdi (see [DK], page 103) and independently by S.K. Stein(1975) (see also Section 6 below) or the conjecture of H.J.Ryser (1967) that every odd order latin square has a complete transversal. No counterexample to either of these conjectures has so far been obtained so far as the present authors are aware.

Also, we draw the reader's attention at this point to the closely related concept of partial admissibility developed by G.B.Belyavskaya and A.F.Russu(1976) and discussed in detail in Chapter 6.

One of the present authors believes that the following may be true: For every integer k relatively prime to n, there exists a quasigroup in each isotopy class of n for which the k-th powers of its elements are all distinct. If so, the truth of Ryser's conjecture would follow from the case $k = 2$.

For an analogous generalization of the Brualdi-Stein conjecture, see Section 6 below.

(3) <u>Number of transversals in a latin square</u>.

Suppose now that L is a latin square which has an orthogonal mate. Then L must have at least n transversals. Indeed, every latin square orthogonal to L defines a set of n transversals of L. If L is a member of m mutually orthogonal latin squares, then L has at least $(m-1)n$

transversals. It is therefore of interest in connection with the investigation of orthogonal latin squares to count the transversals of a given latin square L.

Such a count for latin squares of order 6 has been carried out by D. McCarthy(1976) in connection with his investigation of the existence of (7,1)-designs. [An (r,1) design comprises a set of v elements arranged into subsets called blocks such that each element occurs in r of the blocks and each pair of elements in exactly one block. The blocks need not be of equal size.] McCarthy noted that, of the 22 isotopy classes of 6×6 latin squares (see page 141 of [DK]), only 17 have distinct transversal structures and he found by computer that 8 of these classes consist of squares which have no transversals, 6 comprise squares which have 8 transversals, 2 comprise squares with 24 transversals and 1 comprises squares with 32 transversals. However, he found that a latin square of order 6 contains at most 6 transversals with the property that, when taken in pairs, they have at most one cell in common and of course, because no latin square of order 6 has an orthogonal mate, none has as many as 6 cell-disjoint transversals.

In his investigation of the existence of pairs and triples of mutually orthogonal latin squares of order 10, E.T.Parker(1978) carried out a similar computer search with the aim of finding 10×10 latin squares with a large number of transversals. In the course of his investigation, Parker came across the following remarkable latin square

$$C = \begin{vmatrix} A & B \\ B & A \end{vmatrix}, \text{ where } A = \begin{vmatrix} 5 & 9 & 3 & 2 & 1 \\ 9 & 3 & 2 & 1 & 0 \\ 3 & 2 & 1 & 0 & 4 \\ 2 & 1 & 0 & 4 & 3 \\ 1 & 0 & 4 & 3 & 2 \end{vmatrix}, B = A+5J$$

and J is the all-ones matrix, which has 5504 transversals and about one million orthogonal mates but yet is not a member of any triad of mutually orthogonal 10×10 latin squares. The existence or not of such a triad remains an open question. (For more details of investigations into the latter question, see Chapters 5 and 11).

Some more general results concerning the number t_n of transversals possessed by a latin square of order n have been obtained by G.B. Belyavskaya and A.F.Russu(1975). These authors have proved among other things that, for a latin square in standard form (first row and column in natural order if $0, 1, \ldots, n-1$ are used as symbols) t_n is congruent to zero modulo the number of elements in the left nucleus N_λ (or the right nucleus N_μ) of the loop G whose multiplication table is given by the latin square. We discuss some of their results below. As an illustration, we can deduce immediately from the results just mentioned that the six classes of 6×6 squares which McCarthy found to have 8 transversals cannot correspond to either of the group squares because $8 \neq 0 \bmod 6$. (In fact, neither group square has any transversals by fact (6) in section 1 of this chapter.)

We start with some definitions:

DEFINITIONS. Let $(Q, \cdot)$ be a quasigroup. A permutation λ of Q is a <u>left-regular permutation</u> of the quasigroup if $(xy)\lambda = x\lambda \cdot y$ for all $x, y \in Q$. A permutation ρ is a <u>right regular permutation</u> if $(xy)\rho = x \cdot y\rho$ for all $x, y \in Q$, and a permutation ϕ of Q is a <u>middle regular permutation</u> of the quasigroup if there exists a (conjugate) permutation ϕ^* of Q such that $x\phi \cdot y = x \cdot y\phi^*$ for all $x, y \in Q$.

The set of all left regular permutations of $(Q, \cdot)$ forms a group L since $\lambda_1, \lambda_2 \in L \Rightarrow (xy)\lambda_1\lambda_2 = (x\lambda_1 \cdot y)\lambda_2 = x\lambda_1\lambda_2 \cdot y$, whence $\lambda_1\lambda_2 \in L$; and if $x\lambda = z$, we have $(zy)\lambda^{-1} = (x\lambda \cdot y)\lambda^{-1} = (xy)\lambda\lambda^{-1} = xy = z\lambda^{-1} \cdot y$. So, $\lambda^{-1} \in L$.

The middle and right regular permutations also form groups R and $\bar{\Phi}$. The set of all permutations conjugate to the permutations of $\bar{\Phi}$ form a group $\bar{\Phi}^*$.

<u>LEMMA 3.1</u>. In the special case that $(Q, \cdot)$ is a group, $L = \{L_a : a \in Q\}$, $R = \{R_a : a \in Q\}$ and $\bar{\Phi} = R$, $\bar{\Phi}^* = L$, where $xL_a = ax$, $xR_a = xa$.

<u>Proof</u>: Suppose that $\lambda \in L$ and that $x\lambda = z$. There exists an element $a \in Q$ such that $ax = z$. Then, for each $b \in Q$, $ab = a[x(x^{-1}b)] = (ax)(x^{-1}b) = z(x^{-1}b)$. Also, $b\lambda = (x \cdot x^{-1}b)\lambda = x\lambda \cdot x^{-1}b = z(x^{-1}b)$ and so $b\lambda = ab = bL_a$ for all $b \in Q$. Thus $\lambda = L_a$ and $L = \{L_a : a \in Q\}$.

Suppose that $\phi \in \bar{\Phi}$ and that $x\phi = z$. There exists an element $a \in Q$ such that $xa = z$. Then $x = za^{-1} = x\phi \cdot a^{-1} = x \cdot a^{-1}\phi^*$, where ϕ^* is the conjugate of ϕ. Hence, $a^{-1}\phi^* = e$. It follows that $b\phi \cdot a^{-1} = b \cdot a^{-1}\phi^* = be = b$. That is, $b\phi = ba = bR_a$ for all $b \in Q$. So, $\phi = R_a$ and $\bar{\Phi} = \{R_a : a \in Q\}$.

The other proofs are similar. []

DEFINITIONS. The <u>left</u>, <u>middle</u> and <u>right nuclei relative to an element h</u> of the quasigroup $(Q, \cdot)$ are defined by $N_\ell(h) = \{h\lambda : \lambda \in L\}$, $N_m(h) = \{h\phi : \phi \in \bar{\Phi}\}$, and $N_r(h) = \{h\rho : \rho \in R\}$.

Suppose that two left regular permutations $\lambda_1, \lambda_2 \in L$ were to intersect: that is, for some $a \in Q$, $a\lambda_1 = a\lambda_2$. Let $b \in Q$. Then there exists $x \in Q$ such that $b = ax$ and so $b\lambda_1 = (ax)\lambda_1 = a\lambda_1 \cdot x = a\lambda_2 \cdot x = (ax)\lambda_2 = b\lambda_2$. It follows that λ_1 and λ_2 do not intersect unless they coincide and so $|N_\ell(h)| = |L|$.

Similarly, it is easy to show that $|N_m(h)| = |\bar{\Phi}|$ and $|N_r(h)| = |R|$. It follows also that the left nuclei relative to different elements h_1 and h_2 have equal cardinals.

Let us define an equivalence relation ~ on the set Q by saying that h' ~ h if there is a left regular permutation λ of the quasigroup $(Q, \cdot)$ such that $h' = h\lambda$. (Because the left regular permutations form a group L, it is easy to see that ~ is an equivalence relation.) The equivalence classes into which ~ separates Q are the distinct left nuclei $N_\ell(h) = \{h\lambda_1, h\lambda_2, \ldots, h\lambda_t\}$, $N_\ell(k) = \{k\lambda_1, k\lambda_2, \ldots, k\lambda_t\}$, ..., of $(Q, \cdot)$ relative to the elements $h, k, \ldots$, where $t = |L|$. Since each of these has cardinal $|L|$, we deduce that the order of the group of left regular permutations of the quasigroup $(Q, \cdot)$ divides the order of the quasigroup. By similar arguments we can show that the orders of the groups $\bar{\Phi}$ and R divide the order of the quasigroup.

DEFINITION. An ordered triple (α, β, γ) of permutations of the set Q is called an <u>autotopism</u> of the quasigroup $(Q, \cdot)$ if $(x\alpha \cdot y\beta)\gamma^{-1} = x \cdot y$ for all x, y in Q.

The autotopisms $(\alpha_1,\beta_1,\gamma_1), (\alpha_2,\beta_2,\gamma_2), \ldots$, of $(Q,\cdot)$ form a group $(A,\cdot)$ relative to the composition $(\alpha_1,\beta_1,\gamma_1) \cdot (\alpha_2,\beta_2,\gamma_2) = (\alpha_1\alpha_2,\beta_1\beta_2,\gamma_1\gamma_2)$.

LEMMA 3.2. If $\lambda \in L$, $\rho \in R$ and $\phi \in \bar{\Phi}$, then $(\lambda,1,\lambda), (1,\rho,\rho), (\phi,(\phi^*)^{-1},1)$ and $(\phi^{-1},\phi^*,1)$ are autotopisms of $(Q,\cdot)$.

Proof $(x\lambda \cdot y)\lambda^{-1} = x \cdot y$ because $\lambda^{-1} \in L$.
$(x,y\rho)\rho^{-1} = x \cdot y$ because $\rho^{-1} \in R$.
$x\phi \cdot z = x \cdot z\phi^*$ and so $x\phi \cdot y(\phi^*)^{-1} = x \cdot y$, where $y = z\phi^*$.
$z \cdot y\phi^* = z\phi \cdot y$ and so $x\phi^{-1} \cdot y\phi^* = x \cdot y$, where $x = z\phi$. □

DEFINITION. A quasigroup which has at least one complete mapping is called <u>admissible</u>.

LEMMA 3.3. If θ is a complete mapping of the admissible quasigroup $(Q,\cdot)$ so that $x \cdot x\theta = x\theta'$ for all $x \in Q$ and if (α,β,γ) is an autotopism of $(Q,\cdot)$, then the mapping $\bar{\theta} = \alpha^{-1}\theta\beta$ is a complete mapping of $(Q,\cdot)$ and $\bar{\theta}' = \alpha^{-1}\theta'\gamma$.

Proof Since $x \cdot x\theta = x\theta'$, we have $x\alpha \cdot x\theta\beta = x\theta'\gamma$ and so $y \cdot y\alpha^{-1}\theta\beta = y\alpha^{-1}\theta'\gamma$, where $y = x\alpha$. □

COROLLARY. If θ is a complete mapping of $(Q,\cdot)$ and $\lambda \in L$, $\rho \in R$, $\phi \in \bar{\Phi}$ then the permutations $\lambda^{-1}\theta, \theta\rho, \phi\theta\phi^*$ are complete mappings.

Proof We use the fact that $(\lambda,1,\lambda), (1,\rho,\rho)$ and $(\phi^{-1},\phi^*,1)$ are autotopisms of $(Q,\cdot)$. □

Let $(Q,\cdot)$ be an admissible quasigroup with left-regular permutation group L. Then the relation $\theta_1 \underset{L}{\sim} \theta_2$ if there is a permuation $\lambda \in L$ such that $\theta_2 = \lambda^{-1}\theta_1$ is an equivalence relation on the set Δ of complete mappings of $(Q,\cdot)$ which, by virtue of the corollary above, separates the complete mappings of $(Q,\cdot)$ into classes of the form $K_L(\theta) = \{\lambda^{-1}\theta : \lambda \in L\}$ each of cardinal $|L|$.

Similarly, the relation $\theta_1 \underset{R}{\sim} \theta_2$ if there is a permutation $\rho \in R$ such that $\theta_2 = \theta_1\rho$ is an equivalence relation on Δ which separates Δ into equivalence classses of the form $K_R(\theta) = \{\theta\rho : \rho \in R\}$ each of cardinal $|R|$. The relation $\theta_1 \underset{\bar{\Phi}}{\sim} \theta_2$ if there exists a permutation $\phi \in \bar{\Phi}$ with conjugate ϕ^* such that

2:13 Transversals and complete mappings

$\theta_2 = \phi\theta_1\phi^*$ is an equivalence relation on Δ with classes of the form $K_\phi(\theta) = \{\phi\theta\phi^* : \phi \in \tilde{\Phi}\}$. These classes each have cardinal $|\tilde{\Phi}|$: for, if not, there would exist middle regular mappings ϕ, ψ such that $\phi\theta\phi^* = \psi\theta\psi^* = \bar{\theta}$ say. Then, because $(\phi^{-1}, \phi^*, 1)$ and $(\psi^{-1}, \psi^*, 1)$ are autotopisms of $(Q, \cdot)$ we can deduce from lemma 3.3 that $\bar{\theta}' = \phi\theta' = \psi\theta'$, where $x \cdot x\theta = x\theta'$ and $x \cdot x\bar{\theta} = x\bar{\theta}'$ for all $x \in Q$. It follows that $\phi = \psi$. Consequently, all mappings in $K_\phi(\theta)$ are distinct and so the cardinal of this set is $|\tilde{\Phi}|$.

If θ is a complete mapping of the quasigroup $(Q, \cdot)$ and $x \cdot x\theta = x\theta'$ for all $x \in Q$, then the entries in the cells of the transversal defined by the complete mapping θ are the elements $x\theta'$, $x \in Q$. Let Δ' denote the set of all such mappings of $(Q, \cdot)$. We shall call Δ' the <u>set of transversals</u> of $(Q, \cdot)$.

To each of the equivalence relations $\underset{L}{\tilde{}}$, $\underset{R}{\tilde{}}$, $\underset{\Phi}{\tilde{}}$ which we defined above on the set Δ, there corresponds an equivalence relation on the set Δ'.

First suppose that $\theta_2 = \lambda^{-1}\theta_1$. Then, for $x \in Q$, $x\theta_2' = x \cdot x\theta_2 = x \cdot x\lambda^{-1}\theta_1 = y\lambda \cdot y\theta_1 = (y \cdot y\theta_1)\lambda = y\theta_1'\lambda = x\lambda^{-1}\theta_1'\lambda$. So $\theta_2' = \lambda^{-1}\theta_1'\lambda$. The corresponding equivalence classes take the form $K_L'(\theta') = \{\lambda^{-1}\theta'\lambda : \lambda \in L\}$.

Similarly, if $\theta_2 \underset{R}{\tilde{}} \theta_1$ then $\theta_2' = \theta_1'\rho$; if $\theta_2 \underset{\Phi}{\tilde{}} \theta_1$ then $\theta_2' = \phi\theta_1'$. For the latter case, we have $x\theta_2' = x \cdot x\theta_2 = x \cdot x\phi\theta_1\phi^* = x\phi \cdot x\phi\theta_1 = x\phi\theta_1'$.

The corresponding equivalence classes of Δ' are
$K_R'(\theta') = \{\theta'\rho : \rho \in R\}$ and $K_\Phi'(\theta') = \{\phi\theta' : \phi \in \tilde{\Phi}\}$.

Next we show that complete mappings from the same equivalence class are disjoint. Consider first $\theta_1, \theta_2 \in K_L$. There exists $\lambda \in L$ such that $\theta_2 = \lambda^{-1}\theta_1$. Then $y\theta_2 = y\theta_1$ for some $y \in Q$ implies that $y\lambda^{-1}\theta_1 = y\theta_1$, whence $y\lambda^{-1} = y\varepsilon$ where ε is the identity of L. But two left regular permutations do not intersect unless they coincide, so $y\theta_2 \neq y\theta_1$ unless $\theta_2 = \theta_1$. A similar proof holds for $\theta_1, \theta_2 \in K_R$. Finally, consider $\theta_1, \theta_2 \in K_\Phi$. There exists $\phi \in \tilde{\Phi}$ with conjugate ϕ^* such that $\theta_2 = \phi\theta_1\phi^*$. Then $y\theta_2 = y\theta_1$ for some $y \in Q$ implies $y\phi\theta_1\phi^* = y\theta_1$. From this we have $y\theta_1' = y \cdot y\theta_1 = y \cdot y\phi\theta_1\phi^* = y\phi \cdot y\phi\theta_1 = y\phi\theta_1'$. Hence, $y\varepsilon = y\phi$. Two middle regular permutations do not intersect unless they coincide so $y\theta_2 \neq y\theta_1$ unless $\theta_2 = \theta_1$.

It follows also that transversals from the same equivalence class K'_L, K'_R or K'_Φ of the set Δ' are disjoint. Hence we get

<u>Main Result of G.B.Belyavskaya and A.F.Russu</u>. Let $(Q,\cdot)$ be an admissible quasigroup and let M be any one of the groups L, R or $\overline{\Phi}$. Then the set Δ of all complete mappings of Q (or set Δ' of all transversals of Q) is broken up by the abovementioned equivalence relation $\underset{L}{\sim}$, $\underset{R}{\sim}$, or $\underset{\overline{q}}{\sim}$ defined by M into disjoint equivalence classes each of which has cardinal $|M|$.

This yields at once the following theorem.

<u>THEOREM 3.4</u>. If $(Q,\cdot)$ is an admissible quasigroup of order n and t_n is the number of its complete mappings (transversals) then
$$t_n = k_1 \cdot |L| = k_2 \cdot |R| = k_3 \cdot |\overline{\Phi}|$$
for some positive integers k_1, k_2 and k_3.

Alternatively, since the left, right and middle nuclei of a quasigroup $(Q,\cdot)$ relative to an element $h \in Q$ have cardinals $|L|$, $|R|$ and $|\overline{\Phi}|$ respectively, we can say that $t_n = k_1 \cdot |N_\ell(h)| = k_2 \cdot |N_r(h)| = k_3 \cdot |N_m(h)|$. □

DEFINITION. An element a of a quasigroup $(Q,\cdot)$ which satisfies the relation $(ax)y = a(xy)$ for all $x, y \in Q$ is said to <u>associate on the left</u>. The set of all elements which associate on the left form the <u>left nucleus of Q.</u>

Suppose that $(Q,\cdot)$ has a left identity element e and that $\lambda \in L$. Let $e\lambda = a$. Then, for $x \in Q$, $x\lambda = (ex)\lambda = e\lambda . x = ax$. So, $\lambda = L_a$. Also, the element a associates on the left since $(ax)y = (e\lambda.x)y = (ex)\lambda.y = x\lambda.y = (xy)\lambda = a(xy)$. Thus, in this case, Q has a left nucleus $N_\ell = N_\ell(e)$ and $L = \{L_a : a \in N_\ell\}$. (cf. Lemma 3.1 above.)

Similarly, if $(Q,\cdot)$ has a right identity e^*, then Q has a right nucleus $N_r = N_r(e^*)$ and $R = \{R_a : a \in N_r\}$.

If $(Q,\cdot)$ is a loop with identity element e, then it has left, right and middle nuclei and each of these sets is a group. Also, $N_\ell = N_\ell(e)$, $N_r = N_r(e)$. Since the multiplication table of a loop is a latin square in standard form and since, conversely, every latin square in standard form defines a loop whose multiplication table is that square, we can deduce that

THEOREM 3.5. The number t_n of transversals of a latin square L in standard form is congruent to zero modulo $|N_\ell|$ or $|N_r|$, where $N_\ell(N_r)$ is the left (right) nucleus of the loop $(Q,\cdot)$ of which L is the multiplication table.

In the special case that $(Q,\cdot)$ is a group of order n, we showed earlier that $L = \{L_a, a \in Q\}$, $R = \{R_a, a \in Q\}$ and that $\bar{\varphi} = R$, $\bar{\varphi}^* = L$. It follows that each of L, R and $\bar{\varphi}$ has cardinal n. That is, n divides t_n.

Also,
$$\left.\begin{array}{l} \theta_2 \underset{L}{\sim} \theta_1 \Longleftrightarrow \theta_2 = L_a \theta_1 \\ \theta_2 \underset{R}{\sim} \theta_1 \Longleftrightarrow \theta_2 = \theta_1 R_a \\ \theta_2 \underset{\bar{\varphi}}{\sim} \theta_1 \Longleftrightarrow \theta_2 = R_a \theta_1 L_a \end{array}\right\} \text{ for some } a \in Q$$

$$\theta_2' \underset{L}{\sim} \theta_1' \Longleftrightarrow \theta_2' = L_{a^{-1}} \theta_1' L_a$$
$$\theta_2' \underset{R}{\sim} \theta_1' \Longleftrightarrow \theta_2' = \theta_1' R_a$$
$$\theta_2' \underset{\bar{\varphi}}{\sim} \theta_1' \Longleftrightarrow \theta_2' = R_a \theta_1'.$$

From this, we get another special case of Belyavskaya and Russu's main result which was originally proved by J.Singer (1960) for the particular case of cyclic groups.

THEOREM 3.6. Let $(Q,\cdot)$ be an admissible group of order n and let t_n be the number of its complete mappings (transversals), then $t_n \equiv 0 \mod n$.

COROLLARY (1). If a quasigroup $(Q,\cdot)$ is isotopic to an admissible group, then the number of its complete mappings (transversals) is a multiple of n.

Proof. A quasigroup $(Q,\circ)$ is isotopic to a quasigroup $(Q,\cdot)$ if there exists a triple of permutations (α,β,γ) such that $x \circ y = (x\alpha \cdot y\beta)\gamma^{-1}$. Thus, different transversals (complete mappings) of a quasigroup correspond to different transversals of a quasigroup isotopic to it. Furthermore, disjoint transversals are transformed under isotopy into disjoint transversals. []

<u>COROLLARY (2)</u> Let $(Q,\cdot)$ be an admissible group of order n with t_n = kn. Then the latin square formed by the Cayley table of $(Q,\cdot)$ has at least k orthogonal mates and exactly k different transversals pass through each cell of the latin square.

Proof. Each equivalence class into which the t_n transversals (complete mappings) of $(Q,\cdot)$ are split consists of n disjoint transversals (since $|L| = |R| = |\bar{\phi}| = n$) and so defines an orthogonal mate for $(Q,\cdot)$. Since $t_n = kn$, there are k equivalence classes and since the transversals of each equivalence class cover all n^2 cells of the multiplication table of $(Q,\cdot)$, exactly one transversal out of each of the k equivalence classes passes through each cell of this multiplication table. []

By developing their ideas further, G.B.Belyavskaya and A.F.Russu (1975) have also obtained the following refinements of the last result.

THEOREM 3.7 Let $(Q,\cdot)$ be an admissible non-abelian group and k be the number of complete mappings which fix the identity element of the group. Then the group has at least 2k orthogonal mates.

THEOREM 3.8 Let $(Q,\cdot)$ be an admissible abelian group, k be the number of complete mappings which fix the identity element of the group and h be the number of complete automorphisms of the group (that is, complete mappings which are also automorphisms). Then the group has at least 3k-2h orthogonal mates.

We refer the reader to Belyavskaya and Russu's original paper for the proofs of these two theorems.

Belyavskaya and Russu have also shown that, for a group of prime order p which has h complete automorphisms and k complete mappings which fix the identity element of the group, $k > h$ implies $k \geq p+h$. Combining this with the last theorem above we deduce that, if a group of prime order p has h complete automorphisms and k complete mappings which fix the identity element of the group then, provided that $k > h$, the

Cayley table of the group has at least 3p+h orthogonal mates. In particular, for the cyclic group of order 7 in which $k > h$ and in which five of the six automorphisms are complete mappings, we deduce the existence of at least $3.7+5 = 26$ orthogonal mates. On the other hand, for the cyclic group of order 5 we have $k \leq h$ (since all complete mappings which fix the identity are automorphisms) and so no similar deduction can be made.

(4) <u>Sets of mutually orthogonal latin squares with no common transversal</u>.

From a consideration of the number of transversals possessed by a single latin square, the next step is an investigation of the number of common transversals possessed by a set of mutually orthogonal latin squares. This is a first step in determining whether or not such a set is extendible. As on page 453 of [DK], we shall define a <u>k-transversal</u> of a set of k mutually orthogonal latin squares of order n as a set of n cells, common to all the squares and taken one from each row and one from each column, whose entries in each square form a transversal of that square.

D.A.Drake(1977) has given a very useful general criterion for the existence of such a k-transversal which we shall now explain.

Drake's criterion applies to any set $C = \{C^0, C^1, \ldots, C^{k-1}\}$ of k mutually orthogonal latin squares of order mq having structure of the following form. Let $M = \{A^0, A^1, \ldots, A^{k-1}\}$ where $A^t = (a_{rs}^t)$, $1 \leq r, s \leq m$, be a set of k mutually orthogonal latin squares of order m defined on the symbols $0, 1, \ldots, m-1$ and let $Q_{ij} = \{B^{ij0}, B^{ij1}, \ldots, B^{ij(k-1)}\}$ be a set of k mutually orthogonal latin squares of order q defined on the symbols $0, 1, \ldots, q-1$. Here i,j take values from 1 to m and the sets Q_{ij} may be different or all the same. Then C_{ij}^t is the latin square of order q whose (r,s)th entry is (a_{ij}^t, b_{rs}^{ijt}), where b_{rs}^{ijt} is the entry in the (r,s)th cell of square B^{ijt}, and

$$C^t = \begin{vmatrix} C^t_{11} & C^t_{12} & \cdots & C^t_{1m} \\ C^t_{21} & C^t_{22} & \cdots & C^t_{2m} \\ \vdots & \vdots & \vdots & \vdots \\ C^t_{m1} & C^t_{m2} & \cdots & C^t_{mm} \end{vmatrix} = \begin{vmatrix} (a^t_{11}, B^{11t}) & (a^t_{12}, B^{12t}) & \cdots & (a^t_{1m}, B^{1mt}) \\ (a^t_{21}, B^{21t}) & (a^t_{22}, B^{22t}) & \cdots & (a^t_{2m}, B^{2mt}) \\ \vdots & \vdots & \vdots & \vdots \\ (a^t_{m1}, B^{m1t}) & (a^t_{m2}, B^{m2t}) & \cdots & (a^t_{mm}, B^{mmt}) \end{vmatrix}$$

In particular, we observe that if C^t is cyclic with respect to the subsquares C^t_{ij} then it is of <u>q-step type</u> as defined by E. Maillet (1894) (see page 445 of [DK]).

In order to obtain Drake's criterion, we represent the set M of k mutually orthogonal squares by means of the corresponding (k+2)-net (see page 270 of [DK]) comprising (k+2) pencils of m lines each, with vertices $R, S, L^0, L^1, \ldots, L^{k-1}$ on ℓ_∞, where lines through R are indexed by rows, lines through S by columns and lines through L^t by entries of the latin square A^t. The point (i,j) of the net is the intersection of the row line i with the column line j. To each point (i,j) we assign a non-negative integer f(i,j). If the map f(i,j) from points to non-negative integers has the property that the sum of the integers assigned to the points of a line is a constant d for all lines of the net M, we say that d is <u>positively represented on M</u>. This concept is due to R. H. Bruck (1951). (See also section 9.2 of [DK]). We then have

<u>THEOREM 4.1 (D.A.Drake)</u>: If the set C of k mutually orthogonal mq×mq latin squares has a k-transversal T then q is positively represented on the net M.

<u>Proof</u>. We define f(i,j) to be the number of cells of the k-transversal T which lie in the latin subsquare

$$C^0_{ij} = (a^0_{ij}, B^{ij0}) = \begin{bmatrix} (a^0_{ij}, b^{ij0}_{11}) & \cdots & (a^0_{ij}, b^{ij0}_{1q}) \\ & \vdots & \\ (a^0_{ij}, b^{ij0}_{q1}) & \cdots & (a^0_{ij}, b^{ij0}_{qq}) \end{bmatrix}$$

Then $\sum_{j=0}^{m-1} f(i,j) = q$ because T contains one cell from each row of C^0 and $(C^0_{i1} \ C^0_{i2} \ \cdots \ C^0_{iq})$ is a complete set of q rows of C^0. Thus $\Sigma f(i,j) = q$ for the points of each row line of M. Similarly, $\Sigma f(i,j) = q$ for the points of each column line of M. Finally, the points of the line L say with label h of the pencil with vertex L^v are all those which correspond to cells (i,j) which in the latin square A^v have entry h. So the sum of the values of f(i,j) over the points of L is equal to the number of cells of T which in C^v are occupied by pairs (a_{rs}^v, b_{xy}^{rsv}) for which $a_{rs}^v = h$. This number is again q because there are q ordered pairs of symbols of form (h,z) among the entries of the square C^v. (z can be any one of the symbols $0, 1, \ldots, q-1$). []

COROLLARY (1). If $k = m-1$, the set C of k mutually orthogonal mq×mq latin squares has no common transversal unless m divides q.

Proof. Suppose that a common transversal exists. Then q is positively represented on the net M and so $\sum_{Q \in M} f(Q) = mq$ because all points (i,j) lie on one of the m lines of the pencil with vertex R (or on the m lines of any other parallel class).

Since each line of M has m points, the average value of f(Q) is q/m. Suppose that for some point P, $f(P) = (q/m)+x$, $x > 0$. Then for each line L_i through P, $\sum_{L_i \setminus P} f(Q) = (m-1)(q/m)-x$. However, since $k = m-1$, M is an affine plane (Theorem 8.1.1 of [DK]) and so every point Q of M\P lies on one of the m+1 lines through P. Consequently,
$$mq = \sum_{Q \in M} f(Q) = f(P) + \sum_{Q \in M \setminus P} f(Q) = (q/m)+x+(m+1)[(m-1)(q/m)-x]$$
$$= m^2(q/m)-mx < mq.$$
This contradiction shows that in fact $f(Q) = q/m$ for every point $Q \in M$ and so q/m must be an integer. That is, m divides q. []

COROLLARY (2). Let s be a positive integer with prime factorization $s = p_1^{t_1} p_2^{t_2} \ldots p_v^{t_v}$ and let $m = \min(p_j^{t_j})$. Then there exists a set of m-1 mutually orthogonal latin squares of order s without a common transversal.

Proof. Since m is a prime power, we can construct a set $M = \{A^0, A^1, \ldots, A^{m-2}\}$ of m-1 mutually orthogonal latin squares of order m and, since $s/m = \prod_{j \neq h} p_j^{t_j}$ (where $m = p_h^{t_h}$) is a product of prime powers, there exists a set Q^* of at least $\min_{j \neq h} (p_j^{t_j}-1) \geq m-1$ mutually orthogonal latin squares of order $q = {}^s/m$ by MacNeish's theorem. (See [DK], page 390.) From these we can select a subset $Q = \{B^0, B^1, \ldots, B^{m-2}\}$ of $m-1$ squares and thence a set $C = [C_{ij}^t] = [a_{rs}^t, B^t]$ of $m-1$ mutually orthogonal latin squares of order $mq = s$ as in Drake's theorem above. Since m does not divide q, the set C has no common transversal by Corollary (1). []

COROLLARY (3) (a) Let s be a positive integer not equal to 6 which is divisible by 3 but not by 9. Then there is a pair of orthogonal latin squares of order s without a common transversal.

(b) Let $s \neq 8, 12, 24$ or 40 be a positive integer which is divisible by 4 but not by 16. Then there is a triple of mutually orthogonal latin squares of order s without a common transversal.

Proof. For (a), let $m = 3$ and $q = {}^s/3$. then $q \neq 2, 3, 6$ otherwise $s = 6$ or s is divisible by 9. Consequently there exist pairs A, Q of mutually orthogonal latin squares of orders m, q respectively from which we can construct a pair C of orthogonal latin squares of order $mq = s$ to which Corollary 1 applies.

For (b), let $m = 4$ and $q = {}^s/4 \neq 2, 3, 6, 10$. Consequently there exist triads A, Q of mutually orthogonal latin squares of orders m, q respectively from which we can construct a triple C of mutually orthogonal latin squares of order $mq = s$ to which Corollary 1 applies. (The fact that there exist triples of mutually orthogonal latin squares of all orders except $2, 3, 6$ and possibly $10, 14$ was shown by S.M.P.Wang(1978) and, recently, a triple of mutually orthogonal latin squares of order 14 has been obtained by D.T.Todorov(1985). We discuss these results in more detail later in the present book.) []

COROLLARY (4). [E.Maillet(1894)] A latin square C of order $s = mq$ and of q-step type has no transversals if m is even and q is odd.

Proof. E.Maillet's original proof was described on page 447 of [DK]. The following much shorter proof is due to D.A.Drake(1977). Let

2:21   *Transversals and complete mappings*   27

$M = \{A\}$ be the cyclic latin square $\begin{vmatrix} 0 & 1 & 2 & \ldots & m-1 \\ 1 & 2 & 3 & \ldots & 0 \\ . & . & . & \ldots & . \end{vmatrix}$ and let $Q_{ij} = \{B^{ij}\}$ be

a single latin square of order q defined on the symbols $0, 1, \ldots, q-1$. Then

$$C = \begin{vmatrix} (0, B^{00}) & (1, B^{01}) & \ldots & (m-1, B^{0,m-1}) \\ (1, B^{10}) & (2, B^{11}) & \ldots & (0, B^{1,m-1}) \\ . & . & \ldots & . \end{vmatrix} ,$$

where the squares B^{ij} may be different or all the same, is a latin square of q-step type as defined by Maillet. If it has a transversal then q is positively represented on the 3-net M which represents A. Let R,S,L be the vertices of the three parallel classes of this net which correspond to rows, columns and symbols respectively. We index the rows and columns of A by the symbols $0, 1, \ldots, m-1$. Then the (i,j)th cell of A contains the symbol i+j, mod m. If i+j < m, this symbol is i+j. If i+j ⩾ m, this symbol is i+j-m.

Since q is positively represented (by f) on M, the sum of the values of f(i,j) taken over the points of the line through L which corresponds to the symbol d is q. That is, $\underset{\substack{i+j=d \\ (\mod m)}}{\Sigma} f(i,j) = q$. Likewise, because q is positively represented on M, we have

$\underset{i=0}{\overset{m-1}{\Sigma}} f(i,j) = q$ and $\underset{j=0}{\overset{m-1}{\Sigma}} f(i,j) = q$. It follows that each of the double sums

$\underset{d=0}{\overset{m-1}{\Sigma}} d \underset{\substack{i+j=d \\ (\mod m)}}{\Sigma} f(i,j),\ \underset{j=0}{\overset{m-1}{\Sigma}} j \underset{i=0}{\overset{m-1}{\Sigma}} f(i,j),\ \underset{i=0}{\overset{m-1}{\Sigma}} i \underset{j=0}{\overset{m-1}{\Sigma}} f(i,j)$ is equal to $\tfrac{1}{2} m(m-1)q$. So,

$\underset{i=0}{\overset{m-1}{\Sigma}} i \underset{j=0}{\overset{m-1}{\Sigma}} f(i,j) + \underset{j=0}{\overset{m-1}{\Sigma}} j \underset{i=0}{\overset{m-1}{\Sigma}} f(i,j) - \underset{d=0}{\overset{m-1}{\Sigma}} d \underset{\substack{i+j=d \\ (\mod m)}}{\Sigma} f(i,j) = \tfrac{1}{2} m(m-1) q.$

However, the left-hand side of this equality can be computed in another way by first grouping together those terms which involve $[f(i,j)]_{i+j=d}$ to give the summation $[\underset{i}{\Sigma} \underset{j}{\Sigma} (i+j-d) f(i,j)]_{\substack{i+j=d \\ (\mod m)}}$. Since $(i+j-d)_{i+j=d} = 0$ if

$i+j < m$ and $(i+j-d)_{i+j=d} = m$ if $i+j \geq m$, m is a factor of the double sum. It follows that m divides $\frac{1}{2}m(m-1)q$ and so $\frac{1}{2}(m-1)q$ is an integer. But this is impossible if m is even and q is odd. The non-existence of a transversal in this case follows. □

(5) <u>Sets of mutually orthogonal latin squares which are not extendible</u>.

There exist many sets of mutually orthogonal latin squares of a given order n which possess common transversals but which, despite this, are still not extendible: that is, do not possess a common orthogonal mate. On the other hand, S.S.Shrikhande(1961) showed that a set of $k = n-3$ mutually orthogonal latin squares of order n can always be extended to a complete set except when $n = 4$ and R.H.Bruck(1963) improved this result to show that a set of $k \geq n-1-(2n)^{1/4}$ squares can likewise be extended. He also showed that a set of $k > n-\sqrt{n}-2$ squares has at most $n(n-1-k)$ k-transversals and that, if it can be extended at all, then the extended set of squares is unique. Later, T.G.Ostrom(1964) obtained further results of the same kind. (For the details, see Chapter 9 of [DK].)

Recently, a number of authors have obtained results on this subject: more precisely, they have worked on the related question of determining the minimum number of lines which a partial projective plane of order n must possess if it is to be extendible to a complete plane. A <u>partial projective plane of order n</u> is an incidence structure comprising a set of n^2+n+1 points arranged into subsets called lines such that (i) each line contains exactly n+1 points and (ii) any two lines meet in a unique point. It follows that the number of lines which are incident with a particular point P (called the <u>valence</u> of P) is at most n+1. We denote by b the total number of lines of the partial plane. In particular, a set of k mutually orthogonal latin squares of order n is equivalent to a partial projective plane of order n with $b = (k+2)n+1$ lines. To see this, we first represent the latin squares as a (k+2)-net of order n (as, for example, on page 270 of [DK]) and we then join the vertices of the k+2 line-pencils of this net by an additional "line at infinity". We say that a partial projective plane Σ of order n with b lines is <u>line extendible</u> if there exists a set of n+1 points $P_1, P_2, \ldots, P_{n+1}$ in Σ with the property that every line of Σ is incident with exactly one member of the set $\{P_1, P_2, \ldots, P_{n+1}\}$. For, in that case, the set of points $P_1, P_2, \ldots, P_{n+1}$ can be adjoined as a line to Σ. We easily see that, in the case when the partial projective plane arises from a set of k mutually orthogonal latin squares, these squares have a k-transversal if and only if the partial plane is line-extendible.

It is easy to see (i) that $b \leq n^2+n+1$ for any partial projective plane of order n (with equality if and only if the plane is complete) and (ii) that, if P is any point of a partial projective plane Σ of order n and S is the set of points of Σ which are not joined to P, then $|S| = n[(n+1)-(\text{valence of P})] = nd_P$ say and each line not incident with P is incident with $d_P = n+1-(\text{valence of P})$ points of S. The integer d_P is called the <u>deficiency</u> of P. It follows at once that, if any point P of Σ has valence n, then Σ is line-extendible. In D.McCarthy and S.A.Vanstone (1977a), it was shown that

(1) if $b < n^2+n+1$ and no point has valence n, then $b < n^2$ and every point P has valence $\geq b-(n^2-n)$;

(2) if Σ is a partial projective plane of order n with $b = n^2-\alpha$ ($\alpha > 2$) and with $n > 2\alpha^2+3\alpha+2$ or if $n^2-2 \leq b \leq n^2+n$ then Σ is line-extendible. D.McCarthy and S.A.Vanstone's arguments were expressed in terms of (r,1)-designs, which we defined in section (3) of this chapter.

S.Dow(1983a) has improved the latter result to the following:

THEOREM 5.1. Let Σ be a partial projective plane of order n. If $b = n^2-\alpha$ ($\alpha > 0$) and $n > \frac{1}{4}\alpha^2+3\alpha+6$, then Σ is line-extendible.

COROLLARY (1). Any partial projective plane of order n with more than $n^2-2(n+3)^{1/2}+6$ lines can be embedded in a projective plane of order n.

In the same paper, Dow has shown that if $b > n^2-n+1$, then the embedding of the partial projective plane in a complete plane is unique.

We may compare these results of Dow with the earlier results of Bruck. According to Bruck, a (k+2)-net comprising at least $n^2+n+1-n(2n)^{1/4}$ lines can be extended to a complete plane and the embedding is unique whenever the number of lines is greater than $n^2+1-n^{3/2}$. For this special case, these results are considerably better than those of Dow.

In antithesis to these results, A.A.Bruen(1972) has exhibited an unimbeddable set of mutually orthogonal latin squares having deficiency $n^{1/2}$ for the special case when n is the square of an odd prime and has shown that this particular system is maximal as regards the property of being unimbeddable.

A basic result obtained much ealier by H.B.Mann(1944) is that a latin square of order 4n+1 or 4n+2 which has a latin subsquare of the maximum possible order 2n or 2n+1 respectively has no orthogonal mate. (For a more precise statement of Mann's theorem, see Theorem 12.3.2 of [DK].) An analogous result for squares of order 4n+3 due to D.A.Drake (1977) is the following:

THEOREM 5.2 (D.A.Drake). Let $M = \{L_0, L_1, \ldots, L_{k-1}\}$ be a set of k mutually orthogonal latin squares of order 4n+3 and suppose that L_0 has a latin subsquare A of maximal order 2n+1, then $k \leq 2n+2$.

Proof. Let $L_0 = \begin{bmatrix} A & B \\ (2n+1)\times(2n+1) & (2n+1)\times(2n+2) \\ \hline C & D \\ (2n+2)\times(2n+1) & (2n+2)\times(2n+2) \end{bmatrix}$

Consider a fixed cell (i,j) in the subsquare D. Suppose that this cell of the latin square L_u contains the symbol u. Then the set of 4n+3 cells of the square L_0 which in L_u all contain the symbol u will be called a "transal" of the square L_0 through the cell (i,j). In each latin square L_v, $v > 0$, $v \neq u$, this set of cells form a transversal. Consequently, the transals through the cell (i,j) defined by the latin squares L_u and L_v are distinct from each other. If L_0 belongs to a set of k mutually orthogonal latin squares, there exist at least k-1 transals of L_0 through any cell (i,j) of D.

Let $S = \{1, 2, \ldots, 4n+3\}$ be the symbols of L_0 and $H = \{1, 2, \ldots, 2n+1\}$ be those of the latin subsquare A. Each $h \in H$ occurs 2n+1 times in A, not in B or C, and hence 2n+2 times in D. The elements of S\H occupy the remaining $(2n+2)^2 - (2n+2)(2n+1) = 2n+2$ places in D. Let Q denote these cells in D. Let T be a transal of L_0 which occupies a cells of A and q cells of Q. T necessarily contains $(2n+1)-a$ cells of B and of C and the elements of these cells are elements of $S\backslash H$. Consequently, T includes all together $2(2n+1-a)+q$ cells of L_0 which contain elements of $S\backslash H$. But the cells of T in L_0 form a transversal of L_0, so just $2n+2$ of them contain elements of $S\backslash H$. Hence, $2(2n+1-a)+q = 2n+2$; so $q = 2(a-n)$ and is even.

Consider now a particular cell (i,j) of Q in D. If $q = 0$, no transals of L_0 pass through this cell and so L_0 has no orthogonal mate. If $q \geq 2$, at most $2n+1$ transals pass through it since each transal occupies at least one further cell of the $2n+2$ cells of Q. It follows that L_0 belongs to a set of at most $2n+2$ mutually orthogonal latin squares. []

As Drake has remarked, no corresponding theorem for the case of a latin square of order $4n+4$ with a latin subsquare of the maximum order $2n+2$ exists, since, in the particular case $4n+4 = 2^m$, a complete set of mutually orthogonal latin squares exists all of which have latin subsquares of order 2^{m-1}. This set of squares arises from the desarguesian projective plane of order 2^m.

In his paper (1977), D.A.Drake used the above theorem to help him in his attempt to determine the spectrum of pairs of integers (n,k) for which there exists a maximal set of k mutually orthogonal latin squares: that is a set of k mutually orthogonal latin squares which cannot be embedded in a set of $k+1$ such squares. Using this theorem in conjunction with theorem 4.1 of the preceding section and also the earlier results of R.H.Bruck and E.Maillet mentioned there, he was able to handle all pairs (n,k) with $n \leq 8$ except the pairs $(8,4)$ and $(7,3)$. For $(7,3)$, he was able to show that no maximal set exists provided there are no omissions in the list of pairs of orthogonal latin squares of order 7 given in H.W.Norton (1939). D.Jungnickel and G.Grams(198x) have more recently shown that maximal sets of mutually orthogonal latin squares exist for $n = 9$ and $k = 1,2,3,5,8$ and that no maximal sets exist for $n = 9$ and $k = 6,7$ (the latter fact follows from R.H.Bruck's theorem). They were unable to resolve the case $n = 9$, $k = 4$.

The maximal sets of squares obtained by Jungnickel and Grams are regular in the sense of the following definition.

DEFINITION. A <u>regular</u> set of pairwise orthogonal latin squares is a set admitting a regular abelian automorphism group on the symbols such that each automorphism induces the same permutation of the rows in all the squares simultaneously.

D.Jungnickel(1978) and (1980) had earlier shown how to construct such sets from an $(s,r:G)$-difference matrix. An <u>$(s,r;G)$-difference matrix</u> is an $r \times s$ matrix $D = ||d_{ij}||$ with entries from an additively written group G of order s such that the set $\{d_{hj}-d_{ij} : j = 1,2,\ldots,s\}$ of differences is the entire set G for each choice of h and i, $h \neq i$. It gives rise to a regular set of $r-1$ mutually orthogonal latin squares, $A_1, A_2, \ldots, A_{r-1}$, such that the ℓth square $A_\ell = ||a^\ell_{ij}||$ is given by $a^\ell_{ij} = d_{\ell j}+g_i$, where $G = \{g_1=0,g_2,g_3,\ldots,g_s\}$, provided that the matrix D is in <u>normal form</u> with every entry of the last row (and column) equal to the identity element 0 of G. For an illustrative example, see Figure 5.1. Jungnickel and Grams remark that if D is maximal in the sense that no further row can be adjoined to D to give an $(s,r+1;G)$-difference matrix, then the mutually orthogonal latin squares which can be constructed from it form a maximal set. They appeal to T.G.Ostrom(1966) for a proof.

$$D = \begin{bmatrix} 2 & 1 & 0 \\ 1 & 2 & 0 \\ 0 & 0 & 0 \end{bmatrix} \quad A_1 = \begin{vmatrix} 2 & 1 & 0 \\ 0 & 2 & 1 \\ 1 & 0 & 2 \end{vmatrix} \quad A_2 = \begin{vmatrix} 1 & 2 & 0 \\ 2 & 0 & 1 \\ 0 & 1 & 2 \end{vmatrix}$$

$$G = \{0,1,2\} \cong (\mathbb{Z}_3,+)$$

Figure 5.1

Some further results relating to the extendibility of sets of mutually orthogonal latin squares (expressed in terms of geometric nets) will be found in A.Bruen and J.C.Fisher(1973) and (1974), S.Dow(1983b), D.Jungnickel(1984a) and, more indirectly, in several of the papers referred to therein which are not included in the bibliography of the present book.

(6) **Generalizations of the concepts of transversal and complete mapping.**

There are at least three distinct ways in which the idea of a transversal can be generalized. In the first, we think of a latin square as a collection of triples (i,j,ℓ) in which i denotes the row number, j denotes the column number and ℓ denotes the entry in the (i,j)th cell. A transversal is then a set of triples such that each symbol occurs exactly once in each of the three places. Instead of triples, we may think of k-tuples. In the latin square, i,j,ℓ are related by the fact that there exists a quasigroup $(Q,\cdot)$ for which $\ell = i.j$. We may replace this 2-ary quasigroup by a (k-1)-ary quasigroup with (k-1)-ary operation $Q^{k-1} \to Q$, as was done, for example by V.D.Belousov(1972), and in earlier papers cited therein, or we may specify our k-tuples in some other way, as was done by P.Lecointe(1970).

A second kind of generalization arises if we replace our latin square by some other kind of n-square: that is, a square n×n matrix whose entries are the symbols $0,1,2,3,\ldots$, as was done by S.K.Stein(1975). Stein made the following further definitions:

An equi-n-square is an n-square in which each of n symbols occurs exactly n times. If m < n, an (m,n)-rectangle is an m×n matrix whose entries are the symbols $0,1,2,3,\ldots$.

A transversal of an n-square or an (m,n)-rectangle is a set of cells, one from each row and no two from the same column. A partial transversal is a subset of a transversal. A transversal is latin if no two cells contain the same symbol. (Thus, all transversals considered in the preceding sections have been latin in the Stein sense.) A row (or column) of an n-square is latin if no two of its cells contain the same symbol.

In his paper (1975), Stein has obtained bounds for the number of transversals in an n-square or an (m,n)-rectangle and for the number of distinct elements in the cells of a transversal. He has also made a number of interesting conjectures which generalize those of Brualdi and Ryser mentioned earlier in this chapter. In particular, (1) that an equi-n-square has a transversal with at least n-1 distinct symbols; (2) that an n-square in which each symbol appears at most n-1 times has a latin transversal; (3) that a row-latin (n-1,n)-rectangle has a latin transversal.

The following conjecture of one of the present authors is connected with the first of the above conjectures: Let n-1 cells of an n×n latin square (or, indeed, of an equi-n-square) be filled in such a way that the conditions for completion are not violated. Then not only can the square be completed (T. Evans' theorem, see chapter 8) but also it can be completed to a member of any chosen isotopy class of squares. If this conjecture were to be proved true then the Brualdi-Stein conjecture of section (2) above would follow as would Stein's more general conjecture concerning equi-n-squares.

A third generalization of the concept of transversal arises in an n×n latin square if, instead of considering a set of n cells, one from each row, one from each column, and including one occurrence of each symbol, we consider a set of nk cells, k from each row, k from each column, and including k occurrences of each symbol. In the case when k = 2 such an object is called a _duplex_. Objects of this kind have been studied by D.J. Finney and G.H. Freeman and have statistical applications. For example, although the cells of a 6×6 cyclic latin square cannot be partitioned into six disjoint transversals, they can be partitioned into three disjoint duplexes. [See G.H. Freeman (1985).]

The related concept of a complete mapping of a group, loop, or quasigroup can also be generalized. Two such generalizations which have proved fruitful are those of near complete mapping and generalized (K, λ) complete mapping which have been introduced and developed by D.F. Hsu and one of the present authors.

We recall that a complete mapping of a finite group, loop or quasigroup G of order n is a permutation $g \to \theta(g)$ of the elements of G such that the mapping $g \to \phi(g)$, where $\phi(g) = g \cdot \theta(g)$, is again a permutation of G. In the case when G is a group or loop, the complete mapping is said to be _in canonical form_ if $\theta(e) = e$, where e is the identity element of G.

The two generalizations which we are about to introduce have so far been developed only for the case when G is a group and we shall assume that this is the case for the rest of this section. We note first that, if θ' is a complete mapping of G which is not in canonical form, then the mapping

$$\theta : g \to \theta'(g) \cdot [\theta'(e)]^{-1}$$

is a complete mapping and is in canonical form.

If θ is a complete mapping of G which is in canonical form then, when the permutation φ is written as a product of cycles

$$\varphi = (e)(g_{11}\ g_{12}\ \cdots\ g_{1k_1})(g_{21}\ g_{22}\ \cdots\ g_{2k_2})\ \cdots\ (g_{s1}\ g_{s2}\ \cdots\ g_{sk_s}),$$

we have that $\theta(g_{ij}) = g_{ij}^{-1}g_{i,j+1}$, where the second suffix is taken modulo k_h in the (h+1)th cycle, h = 1,2,...,s. The special case in which G has a complete mapping for which s = 1 corresponds to the case in which G is R-sequenceable. (See chapter 3) If $e, a_1, a_2, \ldots, a_{n-1}$ is the R-sequencing with partial products e, $b_1 = ea_1$, $b_2 = ea_1a_2, \ldots, b_{n-2} = ea_1a_2\cdots a_{n-2}$, $b_{n-1} = ea_1a_2\cdots a_{n-1} = e$ then the corresponding complete mapping is φ = (c)(e $b_1\ b_2\ \cdots\ b_{n-2}$) where c is the element of G which does not appear among the partial products. In canonical form, this is φ = (e)($c^{-1}\ c^{-1}b_1\ c^{-1}b_2\ \cdots\ c^{-1}b_{n-2}$). (See D.F.Hsu and A.D.Keedwell(1984).)

The concept of near complete mapping is related in an analogous way to sequenceable groups.

DEFINITION. A finite group (G,·) is said to have a <u>near complete mapping</u> θ if its elements can be arranged in such a way as to form a single non-cyclic sequence of length h and s cyclic sequences of lengths $k_1, k_2, \ldots, k_s$, say

$$[g_1'\ g_2'\ \cdots\ g_h'](g_{11}\ g_{12}\ \cdots\ g_{1k_1})(g_{21}\ g_{22}\ \cdots\ g_{2k_2})\cdots(g_{s1}\ g_{s2}\ \cdots\ g_{sk_s})$$

in such a way that the elements $\theta(g_i') = g_i'^{-1}g_{i+1}'$ and $\theta(g_{ij}) = g_{ij}^{-1}g_{i,j+1}$ together with the elements $\theta(g_{ik_i}) = g_{ik_i}^{-1}g_{i1}$ comprise the non-identity elements of G. The mapping θ maps $G\setminus\{g_h'\}$ one-to-one onto $G\setminus\{e\}$ and the mapping φ defined by $\varphi(g) = g \cdot \theta(g)$ for all g∈G maps $G\setminus\{g_h'\}$ one-to-one onto $G\setminus\{g_1'\}$. If $g_1' = e$, the near complete mapping is in <u>canonical form</u>.

We note that a complete mapping in canonical form may be regarded as a special case of a near complete mapping in canonical form : namely one for which the non-cyclic sequence has length one and comprises the identity element alone. On the other hand, a near complete mapping in canonical form for which h = ord G exists if and only if G is <u>sequenceable</u>. In this case, the non-cyclic sequence is the sequence of partial products [e b_1 b_2 ... b_{n-1}], where $e, a_1, a_2, \ldots, a_{n-1}$ is the sequencing and $b_1 = ea_1$, $b_2 = ea_1a_2, \ldots, b_{n-1} = ea_1a_2 \ldots a_{n-1}$. (See chapter 3.)

In D.F.Hsu and A.D.Keedwell(1984), it has been shown that complete mappings and near complete mappings of a finite group G are coextensive with left neofields based on G.

DEFINITION. A set N on which two binary operations (+) and (·) are defined is called a <u>left neofield</u> if

(i) (N,+) is a loop, with identity element 0 say;

(ii) (N\{0}, ·) is a group;

and (iii) a(b+c) = ab+ac for all a,b,c ∈ N.

If also the right distributive law (b+c)a = ba+ca holds for all a,b,c ∈ N, then N is a <u>neofield</u>.

In order to be able to state Hsu and Keedwell's main theorems, we also need the following definition:

DEFINITION. Let θ be a near complete mapping in canonical form of a group (G,·). The element g'_h of G which has no image under θ will be called its <u>exdomain element</u> and we shall usually denote it by η.

We then have

THEOREM 6.1. Let (N,+,.) be a finite left neofield with multiplicative group (G,·) where G = N\{0}. Then, if 1+1 = 0 in N, N defines a complete mapping (in canonical form) of G. If 1+1 ≠ 0 but 1+η = 0, N defines a near complete mapping of G with η as exdomain element.

Conversely, let (G,·) be a finite group with identity element 1 which possesses a complete mapping θ in canonical form. Let 0 be a symbol not in the set G and define N = G∪{0}. Then (N,+,.) is a left neofield, where we define $\psi(w) = 1+w = w\theta(w)$ for all $w \neq 0,1$ and $\psi(1) = 0$. Also $x+y = x(1+x^{-1}y)$ for $x \neq 0$, $0+y = y$ and $0.x = 0 = x.0$ for all $x \in N$.

Alternatively, let (G,·) possess a near complete mapping θ in canonical form. Then, with N defined as before, (N,+,.) is a left neofield where we define $\psi(w) = 1+w = w\theta(w)$ for all $w \neq 0, \eta$, where η is the exdomain element of θ, and $\psi(0) = 1$, $\psi(\eta) = 0$. Also $x+y = x(1+x^{-1}y)$ for $x \neq 0$ as before, $0+y = y$ and $0.x = 0 = x.0$ for all $x \in N$.

<u>THEOREM 6.2</u> A finite left neofield constructed as above from a group (G,·) is a neofield if and only if the mapping θ maps conjugacy classes of G to conjugacy classes and, in the case when θ is a near complete mapping, if and only if we have additionally that the exdomain element η is in the centre of G.

We observe that every finite field and every finite nearfield [see H.Zassenhaus(1958)] is a finite left neofield and consequently defines a complete or near complete mapping of its multiplicative group. Also, in these cases, the exdomain element η is the element which is usually denoted by −1 and is an element of multiplicative order 2. More generally, if η is the exdomain element of a near complete mapping of any finite abelian group, it has multiplicative order 2. This ceases to be true for non-abelian groups and leads to the interesting consequence that there are (infinitely many) finite left neofields for which $(-1)^2 \neq 1$. However, if the additive loop of a left neofield for which $1+1 = 0$ has the left or right inverse property or is commutative or is associative then $(-1)^2 = 1$. The proofs of these assertions are in A.D.Keedwell(1983e).

We turn next to the second generalization of the concept of complete mapping which has been developed by D.F.Hsu and one of the present authors. This second generalization in fact embraces the first and, like the first, has so far been made applicable only to groups.

DEFINITIONS. A (K,λ) complete mapping, where $K = \{k_1, k_2, \ldots, k_s\}$ and the k_i are integers such that $\sum_{i=1}^{s} k_i = \lambda(|G|-1)$, is an arrangement of the non-identity elements of G (each used λ times) into s cyclic sequences of lengths $k_1, k_2, \ldots, k_s$, say

$$(g_{11}\ g_{12} \cdots g_{1k_1})(g_{21}\ g_{22} \cdots g_{2k_2}) \cdots \cdots (g_{s1}\ g_{s2} \cdots g_{sk_s}),$$

such that the elements $g_{ij}^{-1} g_{i,j+1}$ (where $i = 1, 2, \ldots, s$; and the second suffix j is added modulo k_i) comprise the non-identity elements of G each counted λ times.

A (K,λ) near complete mapping, where $K = \{h_1, h_2, \ldots, h_r; k_1, k_2, \ldots, k_s\}$ and the h_i and k_j are integers such that $\sum_{i=1}^{r} h_i + \sum_{j=1}^{s} k_j = \lambda |G|$, is an arrangement of the elements of G (each used λ times) into r sequences with lengths $h_1, h_2, \ldots h_r$ and s cyclic sequences with lengths $k_1, k_2, \ldots, k_s$, say

$$[g'_{11}\ g'_{12} \cdots g'_{1h_1}] \cdots [g'_{r1}\ g'_{r2} \cdots g'_{rh_r}](g_{11}\ g_{12} \cdots g_{1k_1}) \cdots (g_{s1}\ g_{s2} \cdots g_{sk_s})$$

such that the elements $(g'_{ij})^{-1} g'_{i,j+1}$ and $g_{ij}^{-1} g_{i,j+1}$ together with the elements $g_{ik_i}^{-1} g_{i1}$ comprise the non-identity elements of G each counted λ times. (We have $\Sigma(h_i-1) + \Sigma k_j = \lambda(|G|-1)$ so it is immediate from the definition itself that $r = \lambda$.)

We illustrate these two definitions by means of the following two examples taken from D.F.Hsu and A.D.Keedwell(1984).

Example 6.1 $(a\ a^3)(a^2\ a^6)(a^4\ a^5)(a\ a^2\ a^4)(a^3\ a^6\ a^5)$ is a (K,2) complete mapping of the cyclic group $C_7 = gp\{a : a^7 = e\}$, where $K = \{2, 2, 2, 3, 3\}$.

Example 6.2 $[e\ ba][e\ ba^2](a^2\ b\ ba^2\ a)(a\ b\ ba\ a^2)$ is a (K,2) near complete mapping of the dihedral group $D_3 = gp\{a, b : a^3 = b^2 = e, ab = ba^{-1}\}$, where $K = \{2, 2; 4, 4\}$.

In D.F.Hsu and A.D.Keedwell(1985), a detailed account is given of how generalized complete and near complete mappings may be used to construct a wide class of block designs of so-called Mendelsohn type and it is also shown how suitable generalized complete and near complete mappings may be constructed.

We end this section by drawing the reader's attention to several recent papers on complete mappings of finite fields. By a complete mapping of a finite field is meant a complete mapping of its additive group. (Contrast theorem 6.1 above which constructs complete mappings of the multiplicative group of a finite field of characteristic 2.) It is known that any mapping θ of a finite field $GF[q]$ into itself can be represented by a unique polynomial $f(x)$ of degree less than q (called the reduced degree). Such a polynomial f is called a complete mapping polynomial if the induced mapping $x \to \theta(x) = f(x)$ is a permutation of $GF[q]$ and if $x \to x+\theta(x) = \phi(x)$ is also a permutation of $GF[q]$. H.Niedereiter and K.H.Robinson (1982) have determined all complete mapping polynomials of degree less than 6 and have also obtained a necessary and sufficient condition for a binomial of the form $ax^{(q+1-n)/n}+bx$, where $q \equiv 1 \mod n$, to be a complete mapping polynomial of $GF[q]$. In H.Niederreiter and K.H.Robinson(1981), these authors have shown that nonsimple Bol loops of order pr, where p and r are odd primes with $p > r$, can be characterized by pairs of complete mappings of $GF[p]$. G.L.Mullen and H.Niederreiter (1987) have investigated under what circumstances a Dickson polynomial is a complete mapping polynomial.

Some more recent investigations of complete mappings of $GF[q]$ and of related graphs (called orthomorphism graphs) will be found in A.B.Evans(1987a,b), (1988) and (1989a). (See also the additional remarks below.) Evans makes use of the theory of cyclotomy as well as permutation polynomials in his investigations.

ADDITIONAL REMARKS

In Section 1 of this Chapter we discussed Hall and Paige's results concerning the conditions under which a finite group possesses a complete mapping. In addition to the results mentioned there, Hall and Paige proved that "if G is a finite group with a non-cyclic Sylow 2-subgroup S and if G has [G:S] Sylow 2-subgroups and the intersection of every pair of these Sylow 2-subgroups is the identity, then G has a complete mapping. O.M.Di Vincenzo(1989) has recently strengthened this result by omission of the requirement that the number of Sylow 2-subgroups is [G:S].

Also in Section 2, we introduced the concept of a P-group. In J.Dénes and A.D.Keedwell(1989), we gave a more extended history of the origin of this concept. However, it turns out that the latter history was still incomplete. Only very recently did we become aware of the following further facts. On page 664 of the June-July edition of American Mathematical Monthly under the title "Wilsonian products in a group", C.C.Lindner posed the problem whether, for a group of odd order, the set of products of all the elements of G, taken in any order, is in the commutator subgroup. The Editor of the Monthly posed the further question as to whether the set of all such products exhausts the commutator subgroup: that is, is every finite group of odd order a P-group? Also published on the same page was the comment of S.W.Golomb to the effect that he had ben interested in the latter question since 1951 and had finally published it as a problem in S.W.Golomb(1970), as mentioned in J.Dénes and A.D.Keedwell(1989). In response to the question, P.Yff submitted a solution to the Monthly but, according to his recent letter to one of the present Authors, this solution was never published. Quite recently, P.Yff has also provided a solution for the case of groups of even order: namely, every group of even order (as well as every group of odd order) is a P-group. Because Yff's proofs use more elementary concepts than the proof given in J.Dénes and P.Hermann(1982), and, in particular, do not use the Feit-Thompson theorem, the former author has urged Yff to submit them for publication. (Note that the question whether all finite groups are P-groups was Problem 1.4 of [DK].)

With reference to R.A.Brualdi's conjecture discussed in Section 2 of this Chapter, a paper by J.J.Derienko(1988) has been published in Matematicki Issledovania which purports to prove the truth of the conjecture. However, this paper contains a number of errors and, in the opinion of members of the Kishinev school of Combinatorialists, it is not repairable so the conjecture must be regarded as remaining open.

Three further papers related to the subject matter of Sections 4 and 5 are G.H.J.Van Rees(199α), E.T.Parker and L.Somer(1988) and C.F.Woodcock(1986). In his preprint, Van Rees calls a latin square which has no orthogonal mate, a Bachelor square. (Throughout the preprint, Batchelor square is written but the authors suspect that this is an error.) He has discussed the existence and relative frequency of existence of such squares and, among other things, has reproved Theorem 5.2, due to D.A.Drake. Parker and Somer have generalized Theorem 12.3.2(a) of [DK], due to H.B.Mann(1944), to the following: Let $t \geq 2$ and let L be a latin square of order $4t+2$ with a square sub-array of side $2t+1$ in which all entries are from a set of $2t+1$ elements except for less than $t+u$ of the cells, where $u=1$ if $t=2$ and $u=\sqrt{\{(t+1)/8\}} + \frac{1}{2}$ otherwise. Then there does not exist a complete set of $4t$ mutually orthogonal latin squares each orthogonal to L." Woodcock has shown that no complete set of m.o.l.s. all isotopic to the cyclic square exist when $n \equiv 15$ mod 18. This provides a partial solution to Problem 5.2 of [DK]. (Another result relevant to this Problem is that no latin power set containing $n-1$ latin squares and which is based on a group exists unless n is a prime. See J.Dénes, G.L.Mullen and S.J.Suchower(199α) for the proof and the Additional Remarks at the end of Chapter 5 for the definition of a latin power set.)

Related to the subject matter of Section 6 is the following result of P.Erdös and J.Spencer(199α): "Let B be an arbitrary $n \times n$ matrix. Let $k \leq (n-1)/16$ and suppose that no entry of B occurs more than k times. Then B has a latin transversal." Another recent paper which discusses unsolved problems concerning transversals of latin squares and equ-n-squares is P.Erdös, D.R.Hickerson, D.G.Norton and S.K.Stein(1988). In his recent article "Unsolved problems come of age", R.K.Guy(1989) discusses both of the papers just mentioned, the Brualdi conjecture itself (which has also been attributed to other authors) and the related more general conjecture of one of the present authors given in Section 6 (and originally published in J.Dénes(1986)). See page 907 of that article for further comment and for correction of a previously published remark.

In the discussion of generalized complete mappings in Section 6, we asserted that a complete mapping of a finite group or quasigroup G is a mapping $g \to \theta(g)$ of the elements of G such that the mapping $g \to \phi(g)$, where

$\phi(g) = g \cdot \theta(g)$ is again a permutation of G. The mapping ϕ has been called an <u>orthomorphism</u> in D.M.Johnson, A.L.Dulmage and N.S.Mendelsohn(1961) and in a number of recent papers. For consistency with this distinction between the mappings θ and ϕ, the concepts of (K,λ) complete mapping and (K,λ) near complete mapping introduced by D.F.Hsu and one of the present authors ought instead to have been called <u>(K,λ) orthomorphism</u> and <u>(K,λ) near orthomorphism</u>, as Hsu himself has pointed out. However, recently D.F.Hsu(1990) has used the term near orthomorphism in a slightly different way. In several recent papers orthogonal orthomorphisms have been investigated with the aid of the so-called orthomorphism graph. The <u>orthomorphism graph</u> of a group G is the graph with vertex set consisting of all orthomorphisms of G, two orthomorphisms α and β of G being represented by adjacent vertices if and only if the mapping $\alpha^{-1}\beta$ is a permutation of G: that is, if and only if α and β are orthogonal orthomorphisms. We mention, in particular, A.B.Evans(1987a,b,1988,1989a,b,c,d), D.F.Hsu(1990).

An <u>orthogonal latin square graph</u> is a graph whose vertices are latin squares of the same order and with adjacency of vertices corresponding to orthogonality of latin squares. It has been proved that every finite graph can be realised as an orthogonal latin square graph. See C.C.Lindner, E.Mendelsohn, N.S.Mendelsohn and B.Wolk(1979) and, for a shorter proof, P.Erdös and A.B.Evans(1989). Further results on this topic and on more general connections between latin squares and graphs are in the manuscript which the Authors have prepared for a Chapter on this subject which is to appear in Part II of this Book.

CHAPTER 3

SEQUENCEABLE AND R-SEQUENCEABLE GROUPS: ROW COMPLETE LATIN SQUARES (J.Dénes and A.D.Keedwell)

In the early 1950's, it was observed by B.J.Bugelski(1949), E.J.Williams(1949,1950) and others that a particular kind of latin square which subsequently came to be known as a row complete latin square had exactly the properties required of a change-over design in statistical experimentation. Details of this application are given in section 2.3 of [DK] and a more recent account is in the article on Change-over Designs in Volume 1 of the Encyclopaedia of Statistical Sciences [see H.D.Patterson(1982)]. See also Chapter 10 of the present book. A connection with algebra was made when B.Gordon(1961) proved that any finite group which has the property known as sequenceability can be used to construct a row complete latin square of the same order. In this Chapter, we first summarize briefly those results on the subject which have been described in detail in [DK] and then give an account of subsequent developments.

(1) Row-complete latin squares and sequenceable groups.

DEFINITION. An n×n latin square $L = (\ell_{ij})$ is called row complete if the $n(n-1)$ ordered pairs $(\ell_{ij}, \ell_{i,j+1})$ are all distinct. It is column complete if the $n(n-1)$ ordered pairs $(\ell_{ij}, \ell_{i+1,j})$ are all distinct. A latin square is complete if it is both row complete and column complete.

DEFINITION. A finite group $(G, \cdot)$ is called sequenceable if its elements $a_0 = e, a_1, a_2, \ldots, a_{n-1}$ can be ordered in such a way that all the partial products $b_0 = a_0$, $b_1 = a_0 a_1$, $b_2 = a_0 a_1 a_2, \ldots, b_{n-1} = a_0 a_1 a_2 \ldots a_{n-1}$ are different.

B.Gordon(1961) showed that the combinatorial concept of complete latin square and the algebraic concept of sequenceable group are related by the following theorem.

THEOREM 1.1. A sufficient condition for the existence of a complete latin square L of order n is that there exist a sequenceable group of order n.

Proof. If $b_0, b_1, b_2, \ldots, b_{n-1}$ are the partial products defined by the group sequencing then it is easy to check that the latin square L whose (i,j)-th cell contains $b_i^{-1} b_j$, $i,j = 0, 1, \ldots, n-1$, is both row complete and column complete. []

This result raised the questions whether there exist complete (or row complete) latin squares which are not isotopic to any group and whether there exist latin squares which are row complete but cannot be made column complete by any reordering of their rows. P.J.Owens(1976) answered the first question in the affirmative by constructing for every even order greater than six a complete latin square which is not isotopic to a group. Owens also showed that every row complete latin square of even order less than eight has the group property. As it had been shown earlier by several authors that there are no row complete latin squares of orders 3,5 or 7, this was sufficient to prove that eight was the smallest possible order for which a non-group-isotopic complete latin square exists. With regard to the second question, one of the present authors [see A.D.Keedwell(1975,1976c)] gave the answer "No" if the latin square is the Cayley table of a group while P.J.Owens(1976) gave the answer "Yes" in the contrary case. Owens obtained all standard form row complete latin squares of order eight or less and checked that all could be made column complete by a rearrangement of their rows. He then constructed a row complete latin square of the smallest possible even order ten for which this is not the case and described a construction which would provide such squares for an infinite number of further even orders greater than eight. On the other hand, for row complete latin squares which are group isotopic, we have the following theorem:

THEOREM 1.2. Every row complete latin square which represents the multiplication table of a group can be made column complete as well as row complete by suitably reordering its rows.

Proof. Let the given square be the multiplication table of the group G where $h_1, h_2, \ldots, h_n$ and $g_1, g_2, \ldots, g_n$ are two orderings of the elements of G, as in Figure 1.1.

$$
\begin{array}{c|cccccc}
 & h_1 & h_2 & \cdots & h_u & \cdots & h_v & \cdots & h_n \\
\hline
g_1 & g_1 h_1 & g_1 h_2 & & & & & & \\
g_2 & g_2 h_1 & g_2 h_2 & & & & & & \\
\vdots & & & & & & & & \\
g_s & \text{-----------} & g & & & & \\
\vdots & & & & & & & & \\
g_t & \text{--------------------} & g & & \\
\vdots & & & & & & & & \\
g_n & \text{-----------------------------} & g_n h_n \\
\end{array}
$$

Figure 1.1

Since the square is row complete, the elements $h_1^{-1} h_2, h_2^{-1} h_3, \ldots, h_{n-1}^{-1} h_n$ are all distinct and are the non-identity elements of the group in a new order: for suppose that $h_u^{-1} h_{u+1} = h_v^{-1} h_{v+1} = k$ say. Let the arbitrary element g of G occur in the s-th row of column u and in the t-th row of column v. Then $g = g_s h_u = g_t h_v$. The entries in the (u+1)-th column of row s and in the (v+1)-th column of row t are $g_s h_{u+1} = (g_s h_u)(h_u^{-1} h_{u+1}) = gk$ and $g_t h_{v+1} = (g_t h_v)(h_v^{-1} h_{v+1}) = gk$ respectively. Hence, the ordered pair (g, gk) occur as adjacent elements in both the s-th and the t-th rows of the square, contrary to hypothesis.

Now let the rows be reordered according to the permutation

$$\begin{bmatrix} g_1 & g_2 & \cdots & g_n \\ h_1^{-1} & h_2^{-1} & \cdots & h_n^{-1} \end{bmatrix}$$

so that the reordered square takes the form shown in Figure 1.2. This reordering will not affect the row completeness.

Moreover, in the new square each ordered pair of elements will occur at most once as a pair of adjacent elements in the columns : for, suppose that the entries of the (s,u)-th and (t,v)-th cells are the same, equal to g say. Then, $h_s^{-1}h_u = g = h_t^{-1}h_v$. The entries of the $(s+1,u)$-th and $(t+1,v)$-th cells must then be distinct, for $h_{s+1}^{-1}h_u = h_{t+1}^{-1}h_v$ would imply $(h_{s+1}^{-1}h_s)(h_s^{-1}h_u) = (h_{t+1}^{-1}h_t)(h_t^{-1}h_v)$ and so $(h_{s+1}^{-1}h_s)g = (h_{t+1}^{-1}h_t)g$. But then $h_{s+1}^{-1}h_s = h_{t+1}^{-1}h_t$ whence $(h_{s+1}^{-1}h_s)^{-1} = (h_{t+1}^{-1}h_t)^{-1}$. Thus, we would have $h_s^{-1}h_{s+1} = h_t^{-1}h_{t+1}$ which is contrary to hypothesis. This shows that the new square is column complete as well as row complete and so proves the theorem. □

	h_1	h_2	...	h_u	...	h_v	...	h_n
h_1^{-1}	e	$h_1^{-1}h_2$						
h_2^{-1}	$h_2^{-1}h_1$	e						
⋮								
h_s^{-1}	—	—	—	—	g			
⋮								
h_t^{-1}	—	—	—	—	—	—	g	
⋮								
h_n^{-1}	—	—	—	—	—	—	—	e

Figure 1.2

B. Gordon's theorem (Theorem 1.1 above) can be re-expressed in the following stronger form, as was first pointed out by one of the present authors. [See A.D.Keedwell(1976c).]

THEOREM 1.3 The multiplication table of a finite group G can be written in the form of a row complete latin square if and only if the group is sequenceable.

Proof. To see the necessity of the condition, let the row complete latin square $L = (g_{ij})$ be the multiplication table of G so that $g_{ij} = g_i g_j$. In that case, $g_{ij}^{-1} g_{i,j+1} = g_j^{-1} g_i^{-1} g_i g_{j+1} = g_j^{-1} g_{j+1} = h_j$ say, for all values of

i. Suppose that $h_j = h_{j'}$ for $j' \neq j$. Then, because $g_{i'j'} = g_{ij}$ for some value of i' (each element of G occurs exactly once in each column of L), we have $g_{i',j'+1} = g_{i'j'}h_{j'} = g_{ij}h_j = g_{i,j+1}$. However, this contradicts the row completeness of L. Thus $h_j \neq h_{j'}$ unless $j' = j$. Consider now the first row of L. Its j-th element is $g_{1j} = g_{11}(g_{11}^{-1}g_{12})(g_{12}^{-1}g_{13})\cdots(g_{1,j-1}^{-1}g_{1j})$ $= g_{11}h_1h_2\ldots h_{j-1}$. Since the elements of the first row of L are all different, it follows that the partial products $\prod_{k=1}^{j-1} h_k$ for $j = 2,3,\ldots,n$ are all distinct, where the elements $h_1, h_2, \ldots, h_{n-1}$ are the non-identity elements of G. That is, the elements $e, h_1, h_2, \ldots, h_{n-1}$ form a sequencing for G.

To see the sufficiency of the condition, consider the latin square $L = (g_{ij})$, where $g_{ij} = p_i^{-1}p_j$, p_i being one of the partial products defined above. We require to show that the ordered pair of elements (u,v) of G occur consecutively in some row of L. That is, we require to find integers i,j such that $p_i^{-1}p_j = u$ and $p_i^{-1}p_{j+1} = v$. From these two equations, $ug_{j+1} = v$. This determines j. Then $p_i = p_j u^{-1}$ and this fixes i. Thus, every pair of elements of G occurs exactly once and L is row complete.
□

In view of this critical connection between complete (or row complete) latin squares based on groups and the sequenceability property, it becomes important to discover which groups are sequenceable or, from the experimental design point of view, for which orders there exist sequenceable groups. In his paper, B.Gordon(1961) proved that a finite abelian group is sequenceable if and only if it is the direct product of two groups A and B such that A is a cyclic group of order 2^k, $k > 0$, and B is of odd order: that is, if and only if the group has a unique element of order 2. (The proof is in [DK], Section 2.3.) This is precisely the condition that an abelian group does not possess a complete mapping. (See Chapter 2 of the present book). Also, it shows that no sequenceable abelian group of odd order can exist. Since the smallest non-abelian group of odd order has order 21, no row complete latin squares of odd orders less than 21 and which are based on groups can exist. Moreover, since every group of prime order is abelian (and cyclic) no group-based row complete latin square of prime order exists.

The only infinite class of finite non-abelian groups of odd order which are sequenceable so far known is the class of non-abelian groups of order pq, where p and q are distinct odd primes with p < q and such that 2 is a primitive root modulo p. [A.D.Keedwell(1981a)]. It is also known that the two non-abelian groups of order 27 are sequenceable [A.D.Keedwell(1974), B.A.Anderson(1987a)]. A non-abelian group of order pq cannot exist unless p divides q-1 and so there cannot exist row complete latin squares isotopic to groups for either of the orders 15 or 33. The same is true for the orders 9 and 25 since groups of these orders are abelian and do not have any element of order 2. Nevertheless, row complete squares of each of these orders have been obtained. [See A.Hedayat and K.Afsarinejed(1978); D.S.Archdeacon, J.H.Dinitz, D.R.Stinson, T.W.Tillson(1980); J.H.Dinitz and D.R.Stinson(1982); and P.J.Owens(1983).] It seems likely that the methods of construction used by all of these authors are similar. We shall describe a version due mainly to P.J.Owens in the hope that our readers may see how to generalize it to provide a construction for row complete latin squares of all orders p^2 or pq, where p and q are primes. Such a generalization has so far eluded all the abovementioned investigators as well as the present authors.

The construction we describe applies to all orders 3q, where q is an odd prime (including the case q = 3). We assume that the symbols to be used in the square are the integers $1, 2, \ldots, 3q$. The building bricks of the square are the columns $\alpha_1 = (1\ 2\ \ldots\ q)^T$, $\beta_1 = (q+1\ q+2\ \ldots\ 2q)^T$, $\gamma_1 = (2q+1\ 2q+2\ \ldots\ 3q)^T$ and we denote by α_i, β_i, γ_i the i-th cyclic rearrangement of these columns. Thus, $\alpha_i = (i\ i+1\ \ldots\ i-1)^T$, $\beta_i = (q+i\ q+i+1\ \ldots\ q+i-1)^T$, $\gamma_i = (2q+i\ 2q+i+1\ \ldots\ 2q+i-1)^T$. We now construct a row complete latin square of order $3(2r+1)$, where q = 2r+1, with rows as follows:

$$\begin{bmatrix} \alpha & \beta & \gamma & \alpha & \beta & \gamma & \beta & \gamma & \alpha \\ \beta & \alpha & \alpha & \gamma & \gamma & \alpha & \gamma & \beta & \beta \\ \gamma & \gamma & \beta & \beta & \alpha & \beta & \alpha & \alpha & \gamma \end{bmatrix} \begin{bmatrix} \alpha & \beta & \gamma & \gamma & \beta & \alpha \\ \beta & \gamma & \alpha & \alpha & \gamma & \beta \\ \gamma & \alpha & \beta & \beta & \alpha & \gamma \end{bmatrix} \cdots \begin{bmatrix} \alpha & \beta & \gamma & \gamma & \beta & \alpha \\ \beta & \gamma & \alpha & \alpha & \gamma & \beta \\ \gamma & \alpha & \beta & \beta & \alpha & \gamma \end{bmatrix}$$

where there are r–1 repetitions of the array $\begin{bmatrix} \alpha & \beta & \gamma & \gamma & \beta & \alpha \\ \beta & \gamma & \alpha & \alpha & \gamma & \beta \\ \gamma & \alpha & \beta & \beta & \alpha & \gamma \end{bmatrix}$

This provides an array in which each of the six ordered pairs $\alpha\beta$, $\alpha\gamma$, $\beta\gamma$, $\gamma\beta$, $\gamma\alpha$, $\beta\alpha$ occurs $3+2(r-1) = 2r+1$ times and each of the ordered pairs $\alpha\alpha$, $\beta\beta$, $\gamma\gamma$ occurs $2+2(r-1) = 2r$ times. It remains to attach the suffices $1,2,\ldots,q$ appropriately. For example, the array given in Figure 1.3 provides the 9×9 latin square obtained originally by K.B.Mertz using a computer. [See A.Hedayat and K.Afsarinejed(1978).] The array given in Figure 1.4 provides the 33×33 row complete latin square obtained by J.H.Dinitz and D.R.Stinson using an argument similar to that described above and also employing a computer to attach the suffices. (Note: In order to get K.Mertz's square in its original form it is necessary (a) to permute α, β, γ to the order γ, α, β; (b) to permute the rows r_1, r_2, r_3 to the order r_2, r_3, r_1; and (c) to transpose the square so as to make it column complete instead of row complete.) In order to provide a construction applicable to all odd orders 3q, it remains only to find an algorithm for attaching the suffices in each case.

$$\begin{array}{ccccccccc}
\alpha_1 & \beta_1 & \gamma_1 & \alpha_3 & \beta_2 & \gamma_3 & \beta_3 & \gamma_2 & \alpha_2 \\
\beta_1 & \alpha_3 & \alpha_1 & \gamma_3 & \gamma_1 & \alpha_2 & \gamma_2 & \beta_3 & \beta_2 \\
\gamma_1 & \gamma_3 & \beta_2 & \beta_3 & \alpha_3 & \beta_1 & \alpha_2 & \alpha_1 & \gamma_2
\end{array}$$

Figure 1.3

$$\begin{array}{cccccccc|cccccc|cc}
\alpha_1 & \beta_2 & \gamma_2 & \alpha_2 & \beta_4 & \gamma_3 & \beta_3 & \gamma_4 & \alpha_3 & \alpha_4 & \beta_7 & \gamma_5 & \gamma_6 & \beta_5 & \alpha_5 & \alpha_7 & \beta_6 \\
\beta_1 & \alpha_2 & \alpha_6 & \gamma_2 & \gamma_5 & \alpha_3 & \gamma_3 & \beta_5 & \beta_2 & \beta_3 & \gamma_8 & \alpha_4 & \alpha_9 & \gamma_4 & \beta_9 & \beta_4 & \gamma_{10} \\
\gamma_1 & \gamma_5 & \beta_1 & \beta_4 & \alpha_1 & \beta_{10} & \alpha_5 & \alpha_4 & \gamma_6 & \gamma_{11} & \alpha_6 & \beta_3 & \beta_8 & \alpha_2 & \gamma_{10} & \gamma_7 & \alpha_8
\end{array}$$

$$\begin{array}{cccc|cccccc|ccccc}
\gamma_8 & \gamma_7 & \beta_{10} & \alpha_6 & \alpha_9 & \beta_9 & \gamma_1 & \gamma_{10} & \beta_{11} & \alpha_{10} & \alpha_8 & \beta_1 & \gamma_9 & \gamma_{11} & \beta_8 & \alpha_{11} \\
\alpha_1 & \alpha_8 & \gamma_{11} & \beta_6 & \beta_8 & \gamma_1 & \alpha_5 & \alpha_{11} & \gamma_9 & \beta_7 & \beta_{11} & \gamma_7 & \alpha_{10} & \alpha_7 & \gamma_6 & \beta_{10} \\
\beta_2 & \beta_{11} & \alpha_9 & \gamma_2 & \gamma_9 & \alpha_3 & \beta_9 & \beta_5 & \alpha_7 & \gamma_8 & \gamma_3 & \alpha_{11} & \beta_7 & \beta_6 & \alpha_{10} & \gamma_4
\end{array}$$

Figure 1.4

The above discussion leaves open the very intriguing question as to whether row complete latin squares of any prime order exist. The authors have an open mind on this question. The smallest order in doubt is 11, but resolution of the question for this order seems to be beyond the power of present day computers.

Another question raised by one of the present authors is whether orthogonal row complete latin squares exist and whether it is possible to find a row complete latin square which has a column complete orthogonal mate. K.Heinrich(1979) has answered both of these questions in the affirmative. Firstly, we have

THEOREM 1.4 If G is a sequenceable group of order n and if there are two orderings of G, say $h_0, h_1, \ldots, h_{n-1}$ and $k_0, k_1, \ldots, k_{n-1}$, such that the elements $h_0 k_0^{-1}, h_1 k_1^{-1}, \ldots, h_{n-1} k_{n-1}^{-1}$ are again the elements of G, then we can construct a pair of orthogonal row complete latin squares based on G.

<u>Proof</u> Let $a_0, a_1, \ldots, a_{n-1}$ be a sequencing of G with partial products $b_0 = a_0$, $b_1 = a_0 a_1, \ldots, b_{n-1} = a_0 a_1 \cdots a_{n-1}$. Then the latin squares $H = (h_{ij})$ and $K = (k_{ij})$, where $h_{ij} = h_i b_j$ and $k_{ij} = k_i b_j$, are both row complete and they are also orthogonal since the i-th rows of H, K respectively are given by the permutations $\sigma_i = \begin{bmatrix} b_\ell \\ h_i b_\ell \end{bmatrix}$, $\tau_i = \begin{bmatrix} b_\ell \\ k_i b_\ell \end{bmatrix}$ of $b_0, b_1, \ldots, b_{n-1}$. Consequently, H and K are orthogonal if the permutations $\tau_i^{-1} \sigma_i = \begin{bmatrix} b_\ell \\ h_i k_i^{-1} b_\ell \end{bmatrix}$, $i = 0, 1, \ldots, n-1$, are a sharply transitive set of permutations. This is true under the condition stated in the theorem. []

<u>COROLLARY</u> Under the conditions of the theorem, we can construct a pair of orthogonal latin squares one of which is complete (and so column complete) and the other is row complete.

<u>Proof</u> We re-arrange $h_0, h_1, \ldots, h_{n-1}$ into the order $b_0^{-1}, b_1^{-1}, \ldots, b_{n-1}^{-1}$, where $b_0, b_1, \ldots, b_{n-1}$ are the partial products in the sequencing of G. If we re-arrange the rows of both squares according to the permutation thus defined, H becomes complete and K remains row complete. []

Heinrich has pointed out that a group satisfying the condition of Theorem 1.4 cannot be abelian since a sequenceable abelian group has a unique element of order 2 and the product of all its elements is this element of order two; but, if G is abelian, the product $(h_0 k_0^{-1})(h_1 k_1^{-1}) \ldots (h_{n-1} k_{n-1}^{-1})$ of its elements is the identity because the k_i's are the inverses of the h_i's in a different order. However, we have the following existence theorem.

THEOREM 1.5 If the non-abelian group G of order n = pQ, where p is a prime and Q is a product of primes all of which are greater than or equal to p, is sequenceable then there exists a set of p-1 pairwise orthogonal latin squares of order n all of which are row complete.

This theorem generalizes those called Theorems 2.3 and 2.4 in K.Heinrich(1979). To prove it, we first need a well-known lemma of group theory of which the following simple proof (given by K.A.Hirsch to one of the present authors many years ago) seems to be absent from many elementary books on Group Theory.

LEMMA 1.6 If r is a positive integer which is relatively prime to the order of a finite group G, then the mapping $\sigma : g \to g^r$ for all $g \in G$ is a permutation of G.

Proof of lemma. It is only necessary to show that if g_1 and g_2 are distinct elements of G then $g_1^r \neq g_2^r$. Suppose the contrary and let $g_1^r = g_2^r$ = f say. Then f commutes with g_1 and g_2 and so with every element of the subgroup H = $\langle g_1, g_2 \rangle$ in G. Since ord H divides ord G, r is relatively prime to ord H. The cyclic group $\langle f \rangle$ comprises elements of the centre of H and so it is normal in H. The factor group H/$\langle f \rangle$ has order prime to r but contains an element $g_1 \langle f \rangle$ whose order is a divisor of r. This is a contradiction to Lagrange's theorem unless H ≡ $\langle f \rangle$ and ord(H/$\langle f \rangle$) = 1. In the latter case, $g_1 = f^s$ and $g_2 = f^t$ for some positive integers s and t. Then we have rs ≡ rt (mod m), where m = ord f = ord H. Since r is prime to ord H, s ≡ t (mod m) and so $g_1 = g_2$. []

Proof of Theorem 1.5 Let k be an integer such that $1 \leq k < p$. Since k is less than each of the prime divisors of ord G, k is relatively prime to ord G and so, by the Lemma, the mapping $\sigma_k : g \to g^k$ for all $g \in G$ is a permutation of G. Consequently, if $g_0, g_1, \ldots, g_{n-1}$ are the elements of G, each of the sequences $g_0^k, g_1^k, \ldots, g_{n-1}^k$, $k = 1, 2, \ldots, p-1$, is an ordering of the elements of G and each pair of these orderings satisfies the condition of Theorem 1.4. The result of Theorem 1.5 follows at once and, in fact, one of the p-1 squares can be made column complete as well as row complete. □

We note that Theorem 1.5 is valid, in particular, for non-abelian groups of order p^r ($r \geq 3$, since all groups of orders p and p^2 are abelian) and for non-abelian groups of order pq, $p < q$, where p and q are primes. The latter groups are sequenceable at least when 2 is a primitive element relative to p, so arbitrarily large sets of mutually orthogonal row complete latin squares exist.

We mention one more theorem which was obtained by K. Heinrich (1979).

THEOREM 1.7 If there is a set of k mutually orthogonal latin squares of order n all of which are row complete and if there is a row complete latin square of order k, then a row complete latin square of order nk can be constructed.

Proof. We consider first the case when $k = 2$. There exists a row complete square $\begin{array}{|cc|} a & b \\ b & a \end{array}$ of order 2. Let $A = (a_{ij})$ and $B = (b_{ij})$ be the row complete orthogonal latin squares of order n based on distinct n-sets S_A and S_B of symbols. When we place these in justaposition, the ordered pairs (a_{ij}, b_{ij}) are all different and each possible pair occurs just once. Let us now replace each ordered pair (a_{ij}, b_{ij}) by a row complete latin square of order 2 whose symbols are a_{ij} and b_{ij} to obtain a latin square L of size $2n \times 2n$. In the first position in each row, we insert this square with its rows in the order a,b: that is, we insert the square $\begin{array}{|cc|} a_{ij} & b_{ij} \\ b_{ij} & a_{ij} \end{array}$. In the second position, we insert the square with its rows in such order that, for each row, the entry of the first column of the square placed in second position is from the same symbol set as the entry of the last column of the square which we placed in first position. We make a similar stipulation for each other 2×2 square which we insert. In effect this means that, if j is odd, we replace the ordered pair (a_{ij}, b_{ij}) by the row complete square $\begin{vmatrix} a_{ij} & b_{ij} \\ b_{ij} & a_{ij} \end{vmatrix}$ and, if j is even, we replace the ordered pair (a_{ij}, b_{ij}) by the row complete latin square $\begin{vmatrix} b_{ij} & a_{ij} \\ a_{ij} & b_{ij} \end{vmatrix}$. Since A and B are orthogonal, all ordered pairs of the form a,b and all ordered pairs of the form b,a, where a∈S_A and b∈S_B, occur as adjacent elements in L. Also, because both A and B are row complete, all ordered pairs of the form a,a' and b,b', where a ≠ a', b ≠ b' and a,a' ∈ S_A, b,b' ∈ S_B, occur as adjacent elements in the latin square L. Thus, L is a row complete latin square of order 2n.

It is now easy to see how to generalize this construction for the case k > 2. □

We return to the question of which classes of groups are sequenceable. As we mentioned above, for finite abelian groups the question has been completely resolved by B.Gordon while for non-abelian groups of order pq (p,q distinct odd primes, p < q) the question has been resolved only for the case when 2 is a primitive root modulo p. For non-abelian groups of order 2q, which are necessarily dihedral, the question has again been resolved only partially. The first general result concerning this class of groups was obtained by B.A.Anderson(1975) who proved that, for prime numbers p, the dihedral group D_p of order 2p is sequenceable provided that p has a primitive root r such that $3r \equiv -1$ mod p and that, in addition, a certain type of design involving a subset of the integer residues modulo p can be constructed. A year later, R.Friedlander(1976) proved that the dihedral group D_p is sequenceable for all primes $p \equiv 1$ mod 4. Somewhat later still, a student of one of the present authors proved that D_p is also sequenceable if p is a prime which is congruent to 7 modulo 8 and which has a primitive root r such that $2r \equiv -1$ mod p. [See G.B.Hoghton and A.D.Keedwell(1982).] However, D_{11} can be shown to be sequenceable by

B.A. Anderson's method despite the fact that 11 is neither congruent to 1 modulo 4 nor to 7 modulo 8. The present authors conjecture that D_n, of order 2n, is sequenceable for all odd integers n ⩾ 5, this conjecture having already been verified for all odd n ⩽ 35 in G.B. Hoghton and A.D. Keedwell(1982). Recently, further evidence for this conjecture has been obtained by B.A. Anderson(1987a), who has shown that the dihedral group D_n is sequenceable for all integers n such that 5 ⩽ n ⩽ 50: that is, for all orders between 10 and 100. (It was already known that D_6 and D_8 are sequenceable, see J. Dénes and E. Török(1970).) In the same paper, Anderson has shown that all non-abelian groups having orders between 10 and 30 are sequenceable. The latter result confirms the statement in J. Dénes and E. Török(1970) that the only non-abelian group of order 10 or less which is sequenceable is the dihedral group D_5 but disproves their statement that the only group of order 12 which is sequenceable is the dihedral group D_6. It also provides supporting evidence for the conjecture of one of the present authors, given in A.D. Keedwell(1983d), that all finite non-abelian groups of order greater than eight are sequenceable. These, and other recent results of Anderson, are discussed in more detail in the last section of this Chapter.

In proving the sequenceability of the several classes of non-abelian groups which had been shown to be sequenceable prior to the recent work (1987-8) of Anderson, the same basic idea was invoked in each case: namely, that of using <u>quotient sequencings</u>.

DEFINITION. If α is any sequence $u_1, u_2, \ldots, u_n$ of elements of a group G, we shall denote by $P(\alpha)$ its sequence of partial products $v_1 = u_1$, $v_2 = u_1 u_2$, $v_3 = u_1 u_2 u_3, \ldots, v_n = u_1 u_2 \ldots u_n$. Let G be a non-abelian group of order pq, where p is a prime less than the integer q, and suppose that G has a normal subgroup H of order q. Let $G/H = gp\{x_1, x_2, \ldots, x_p\}$. A sequence α of length pq consisting of elements of G/H is called a quotient sequencing of G if each x_i, 1 ⩽ i ⩽ p, occurs q times in both α and $P(\alpha)$.

In the case of the dihedral group D_p, we may take H as the unique normal subgroup of order p and write $D_p/H = gp\{1, x: x^2 = 1\}$. Then the

3:13 Sequenceable and R-sequenceable groups

quotient sequencing used both by B.A.Anderson (1975) and by R.J. Friedlander (1976) is the sequence comprising $\frac{1}{2}(p+1)$ 1's, followed by p x's, followed by $\frac{1}{2}(p-1)$ 1's. This has as its sequence of partial products: $\frac{1}{2}(p+1)$ 1's, followed by the sequence x,1 repeated $\frac{1}{2}(p-1)$ times, followed by $\frac{1}{2}(p+1)$ x's.

We shall now give an outline of Friedlander's proof that D_p is sequenceable for all primes $p \equiv 1 \bmod 4$.

We suppose that the quotient sequencing given above arises from a sequencing $a^{\alpha_1}, a^{\alpha_2}, \ldots, a^{\alpha_{k+1}}, ba^{\beta_1}, ba^{\beta_2}, \ldots, ba^{\beta_{2k+1}}, a^{\gamma_1}, a^{\gamma_2}, \ldots, a^{\gamma_k}$, where $p = 2k+1$ (prime) and $ab = ba^{-1}$. In that case, the corresponding partial products are $a^{\alpha_1}, a^{\alpha_1+\alpha_2}, \ldots, a^{S_\alpha}, ba^{\beta_1 - S_\alpha}, a^{\beta_2 - \beta_1 + S_\alpha}, ba^{\beta_3 - \beta_2 + \beta_1 - S_\alpha}$,
$a^{\beta_4 - \beta_3 + \beta_2 - \beta_1 + S_\alpha}, \ldots, ba^{S_\beta - S_\alpha}, ba^{S_\beta - S_\alpha + \gamma_1}, ba^{S_\beta - S_\alpha + \gamma_1 + \gamma_2}, \ldots, ba^{S_\beta - S_\alpha + S_\gamma}$,
where $S_\alpha = \alpha_1 + \alpha_2 + \ldots + \alpha_{k+1}$, $S_\beta = \beta_{2k+1} - \beta_{2k} + \beta_{2k-1} - \ldots + \beta_3 - \beta_2 + \beta_1$, and $S_\gamma = \gamma_1 + \gamma_2 + \ldots + \gamma_k$.

If the above sequence is indeed to be a sequencing of D_p, then the following four conditions must be satisfied.

(i) The integers $\alpha_1, \alpha_2, \ldots, \alpha_{k+1}, \gamma_1, \gamma_2, \ldots, \gamma_k$ must be all different, modulo p;

(ii) the integers $\beta_1, \beta_2, \ldots, \beta_{2k+1}$ must be all different, modulo p;

(iii) $\alpha_1, \alpha_1 + \alpha_2, \ldots, S_\alpha, \beta_2 - \beta_1 + S_\alpha, \beta_4 - \beta_3 + \beta_2 - \beta_1 + S_\alpha, \ldots, \beta_{2k} - \beta_{2k-1} + \ldots + \beta_2 - \beta_1 + S_\alpha$ must be distinct, modulo p; and

(iv) $\beta_1 - S_\alpha, \beta_3 - \beta_2 + \beta_1 - S_\alpha, \ldots, S_\beta - S_\alpha, S_\beta - S_\alpha + \gamma_1, S_\beta - S_\alpha + \gamma_1 + \gamma_2, \ldots, S_\beta - S_\alpha + S_\gamma$ must be distinct, modulo p.

If we write $\delta_i = \beta_{2i} - \beta_{2i-1}$ and $\epsilon_i = \beta_{2i+1} - \beta_{2i}$ for $1 \leq i \leq k$ and $\epsilon_0 = \beta_1 - S_\alpha$ then conditions (iii) and (iv) can be re-written as

(iii)' the partial sums of the sequence $\alpha_1, \alpha_2, \ldots, \alpha_{k+1}, \delta_1, \delta_2, \ldots, \delta_k$ must be distinct, modulo p; and

(iv)' the partial sums of the sequence $\epsilon_0, \epsilon_1, \ldots, \epsilon_k, \gamma_1, \gamma_2, \ldots, \gamma_k$ must be distinct, modulo p.

Also, condition (ii) is equivalent to

(ii)' the partial sums of the sequence $\epsilon_0, \delta_1, \epsilon_1, \delta_2, \epsilon_2, \ldots, \delta_k, \epsilon_k$ must be distinct, modulo p.

Friedlander showed that, for $p \equiv 1 \mod 4$, the conditions (i), (ii)', (iii)', (iv)' are all satisfied when we choose the sequences $\alpha, \gamma, \delta, \epsilon$ to be as follows:

$\{\alpha_1, \alpha_2, \ldots, \alpha_{k+1}\} = \{0, 2, 4, \ldots, 2k\}$,
$\{\gamma_1, \gamma_2, \ldots, \gamma_k\} = \{1, 3, 5, \ldots, 2k-1\}$,
$\{\delta_1, \delta_2, \ldots, \delta_k\} = \{-f, -3f, -5f, \ldots, -(2k-1)f\}$,
$\{\epsilon_0, \epsilon_1, \ldots, \epsilon_k\} = \{0, 2f, 4f, \ldots, 2kf\}$,

where f is an integer which is a non-square modulo p.

The proof that D_p is sequenceable for an infinite class of primes $p \equiv 7 \mod 8$ (described above) is somewhat similar but makes use of the quotient sequencing which consists of the sequence $1, x$ repeated p times and has as its sequence of partial products $1, x$ followed by the sequence $x, 1, 1, x$ repeated $\frac{1}{2}(p-1)$ times.

For the proof that the non-abelian group of order pq is sequenceable under the conditions mentioned earlier, the concept of quotient sequencing is again used. Suppose that p is an odd prime which has 2 as a primitive root and let σ satisfy the congruence $2\sigma \equiv 1 \pmod{p}$. Then σ also is a primitive root of p since $\sigma^h \equiv 1 \pmod{p}$ implies $2^h \equiv 1 \pmod{p}$. If G is a non-abelian group of order pq, then $q = 2ph+1$ for some integer h and G has a unique normal subgroup H of order q. Let $G/H = gp\{x: x^p = 1\}$. Then the following is a quotient sequencing of G: a sequence of $2ph$ 1's, followed by x, followed by a sequence of $2ph-1$ (p-1)-tuples

$$x^{\sigma-1}, x^{\sigma^2-\sigma}, x^{\sigma^3-\sigma^2}, \ldots, x^{\sigma^{p-2}-\sigma^{p-3}}, x^{1-\sigma^{p-2}},$$

where indices are computed modulo p, followed by a sequence

$$x^{\sigma-1}, x^{\sigma^2-\sigma}, \ldots, x^{\sigma^{p-2}-\sigma^{p-3}}, 1, x^2, x^4, x^8, \ldots, x^{2^{p-2}}, x^{1-\sigma^{p-2}}.$$

We leave the reader to check that the partial products of this quotient sequencing are $2ph$ 1's, followed by the sequence

$$x, x^\sigma, x^{\sigma^2}, \ldots, x^{\sigma^{p-2}}$$

repeated $2ph-1$ times, and then the sequence

$$x, x^\sigma, x^{\sigma^2}, \ldots, x^{\sigma^{p-2}}, x^{\sigma^{p-2}}, x^{\sigma^{p-3}}, x^{\sigma^{p-4}}, \ldots, x^{\sigma^2}, x^\sigma, x, 1.$$

The next step is to try to construct a sequencing of our group G of order pq of which the above is the quotient sequencing under the mapping $\phi : G \to G/H$.

Let r be a primitive element of GF[q]. Then, we observe that, since $q = 2ph+1$ is prime, the cyclic subgroup $H = \text{gp}\{a: a^q = e\}$ of the group $G = \text{gp}\{a,b: a^q = e, b^p = e, ab = ba^s,$ where $s^p \equiv 1 \bmod q\}$ has a partial sequencing

$$e, a^{r^h - r^{h-1}}, a^{r^{h+1} - r^h}, \ldots, a^{r^{q-2} - r^{q-3}}, a^{1 - r^{q-2}}, a^{r-1}, \ldots, a^{r^{h-2} - r^{h-3}}$$

with partial products

$$e, a^{r^h - r^{h-1}}, a^{r^{h+1} - r^{h-1}}, \ldots, a^{r^{q-2} - r^{h-1}}, a^{1 - r^{h-1}}, a^{r - r^{h-1}}, \ldots, a^{r^{h-2} - r^{h-1}}.$$

This may be used to replace the first 2ph(=q-1) 1's in the quotient sequencing. The element of H which does not appear in this partial sequencing is $a^{r^{h-1} - r^{h-2}}$ and so this element must replace the last 1 in the quotient sequencing. The element of H which does not appear in the sequence of partial products is $a^{-r^{h-1}}$ and so this element must replace the last 1 in the partial products of the quotient sequencing : that is, $a^{-r^{h-1}}$ must be the product of all the elements of G when they have been ordered in the required manner.

The remainder of the proof follows the style of Friedlander's argument for the dihedral group D_p, although it is considerably more complicated. To take the case $p = 3$ as an example, it may be supposed that the above quotient sequencing arises from a sequencing of G of the form

$$e, a^{r^h - r^{h-1}}, a^{r^{h+1} - r^h}, \ldots, a^{r^{h-2} - r^{h-3}}, ba^\alpha, ba^{\alpha_1}, b^2 a^{\beta_1}, ba^{\alpha_2}, b^2 a^{\beta_2},$$
$$\ldots, ba^{\alpha_{q-2}}, b^2 a^{\beta_{q-2}}, ba^{\alpha_{q-1}}, a^{r^{h-1} - r^{h-2}}, b^2 a^{\beta_{q-1}}, b^2 a^{\beta_q}.$$

There are then four conditions to be met by the indices α_i and β_i which are somewhat similar to those arising in Friedlander's argument, together with a fifth condition to express the fact that the product of all the elements of the sequencing is $a^{-r^{h-1}}$. It has been shown in A.D.Keedwell(1981a) that the various conditions can be met, though the proof is somewhat lengthy.

Sequencings for the non-abelian groups of orders 21, 39 and 55 have been given as illustrative examples.

Lastly, as regards infinite groups, we mention that C.V.Eynden(1978) has proved that all countably infinite groups are sequenceable.

It is also worth mentioning that B.A.Anderson and P.A.Leonard(1981) have made use of sequencings of groups for the construction of Howell designs (a generalization of Room designs). See Part II of the present book for the definition.

A number of authors have discussed other kinds of group sequences. (See also the following sections of this Chapter.) We mention just two of these. R.J.Friedlander(1980) has considered arrangements of the n distinct elements $a_1, a_2, \ldots, a_n$ of a group G of order n such that as few as possible of the partial products $a_1, a_1 a_2, \ldots, a_1 a_2 \ldots a_n$ are distinct. In contrast to this, W.Holsztyński and R.F.E.Strube(1978) have investigated sequences $a_1, a_2, \ldots, a_k$ of $k (\leq n)$, not necessarily distinct, elements of G such that as few as possible of the k elements are different and yet the partial products $a_1, a_1 a_2, \ldots, a_1 a_2 \ldots a_k$ are all distinct.

(2) <u>Quasi-complete latin squares, terraces and quasi-sequenceable groups</u>.

In the preceding section, we defined an n×n latin square to be row complete if the $n(n-1)$ ordered pairs $(\ell_{ij}, \ell_{i,j+1})$ are all distinct. We shall say that such a square is <u>quasi-row-complete</u> if these $n(n-1)$ pairs of adjacent elements which occur in the rows include each unordered pair of distinct elements exactly twice. Similarly, an n×n latin square is <u>quasi-column-complete</u> if the $n(n-1)$ adjacent pairs of elements which occur in the columns include each unordered pair of distinct elements exactly twice. A latin square is <u>quasi-complete</u> if it is both quasi-row-complete and quasi-column-complete.

In (1979a) and (1979b), G.H.Freeman suggested the use of such squares in experimental design for values of n for which complete latin squares do not exist. In (1979b), he pointed out that quasi-complete latin squares, unlike complete latin squares, exist for all finite values of n. Since a complete latin square is a fortiori quasi-complete, we only need a construction for odd values of n. The construction given by Freeman makes use of E.J.Williams's construction of a 2m×2m complete latin square [see E.J.Williams(1949) or [DK], page 82] and also of the concept of σ-symmetry introduced by P.J.Owens(1976).

DEFINITION. A latin square of even order 2m based on the elements $\{1,2,\ldots,2m\}$ is said to have σ-symmetry by rows if, in each row, the sum of the elements of the j-th and $(2m+1-j)$-th columns is $2m+1$ for $j = 1,2,\ldots,m$. σ-symmetry by columns is similarly defined.

It follows from these definitions that a latin square of even order which has σ-symmetry both by rows and by columns is completely determined by its top left-hand quarter. For example, in Figure 2.1 below, the middle square has σ-symmetry and is completely determined by the 3×3 subsquare in its top left-hand corner.

$$\begin{array}{|cccccc|} 0 & 1 & 5 & 2 & 4 & 3 \\ 1 & 2 & 0 & 3 & 5 & 4 \\ 5 & 0 & 4 & 1 & 3 & 2 \\ 2 & 3 & 1 & 4 & 0 & 5 \\ 4 & 5 & 3 & 0 & 2 & 1 \\ 3 & 4 & 2 & 5 & 1 & 0 \end{array} \qquad \begin{array}{|cccccc|} 1 & 2 & 3 & 4 & 5 & 6 \\ 2 & 4 & 1 & 6 & 3 & 5 \\ 3 & 1 & 5 & 2 & 6 & 4 \\ 4 & 6 & 2 & 5 & 1 & 3 \\ 5 & 3 & 6 & 1 & 4 & 2 \\ 6 & 5 & 4 & 3 & 2 & 1 \end{array} \qquad \begin{array}{|ccc|} 1 & 2 & 3 \\ 2 & 3 & 1 \\ 3 & 1 & 2 \end{array}$$

Figure 2.1

THEOREM 2.1 For every positive integer m, there exists a quasi-complete latin square.

Proof The 2m×2m complete latin square constructed by E.J.Williams's method has as first row

0 1 2m−1 2 2m−2 3 2m−3 ...   ... m+2 m−1 m+1 m

and each particular one of its subsequent rows is obtained from the first by adding the same integer to each element of the first row in arithmetic modulo 2m. In order that the square obtained shall be column complete as well as row complete, the second, third, ..., (2m)-th rows are obtained by adding the integers 1, 2m−1, 2, 2m−2, ..., m+1, m to the first row in the order listed. The left-hand square in Figure 2.1 illustrates the case

$m = 3$. If the Williams' complete square so obtained is now relabelled so that both its first row and its first column contain the entries 1, 2, ..., 2m, the resulting square will always be σ-symmetric. To see this, note firstly that the first row, being in natural order after the relabelling, satisfies the condition of σ-symmetry. Before the relabelling, the second row has as its first and last entries the elements which appear in the second and $[(2m+1)-2]$-th columns of the first row. The second row has as its second and $[(2m+1)-2]$-th entries the elements which appear in the fourth and $[(2m+1)-4]$-th columns of the first row. It is easy to see that in fact all pairs of entries of the second row which are symmetrically placed with respect to a vertical line through the centre of the square arise from symmetrically placed pairs of entries of the first row and that the same remark is true for each other row. Consequently, after the relabelling, the Williams' complete square has σ-symmetry by rows and, because it is symmetric, also has σ-symmetry by columns. (See Figure 2.1 for an example.)

Because of the σ-symmetry, a pair of consecutive elements $\ell_{ij} = u, \ell_{i,j+1} = v$ which appear in the top left-hand quarter of the relabelled square give rise to four ordered pairs (u,v), $(\bar{v},\bar{u})$, $(\bar{u},\bar{v})$ and (v,u) in the complete square, where $\bar{u} = (2m+1)-u$ and $\bar{v} = (2m+1)-v$. Conversely, if (u,v) is any ordered pair of consecutive elements appearing in the rows of the 2m×2m relabelled square then exactly one of the pairs (u,v), $(\bar{v},\bar{u})$, $(\bar{u},\bar{v})$ and (v,u) is in the top left-hand quarter. Suppose now that we take the m×m top left-hand quarter and in it replace each entry $w > m$ by $\bar{w} = (2m+1)-w$. Then we shall obtain a quasi-row-complete latin square of order m which, being symmetric, is also quasi-column-complete. We can do this for any odd (or even) positive integer m and this is G.H.Freeman's construction of a quasicomplete latin square of odd order, as given in (1979b). []

In (1979b), G.H.Freeman has carried out an enumeration by type of complete latin squares of orders 6, 8 and 10 and in G.H.Freeman(1981) he has attempted a similar enumeration of quasi-complete latin squares for orders up to 9. He has shown that for the orders 5 and 7 there exist complete sets of pairwise orthogonal latin squares all of which are quasi-complete, whereas for the orders 6 and 8 not even a pair of quasi-complete latin squares which are orthogonal exists. We illustrate a complete set of pairwise orthogonal 5×5 quasi-complete latin squares in Figure 2.2.

1 2 3 4 5	1 2 3 4 5	1 2 3 4 5	1 2 3 4 5
2 4 5 3 1	3 5 2 1 4	5 1 4 2 3	4 3 1 5 2
3 5 2 1 4	5 1 4 2 3	4 3 1 5 2	2 4 5 3 1
4 3 1 5 2	2 4 5 3 1	3 5 2 1 4	5 1 4 2 3
5 1 4 2 3	4 3 1 5 2	2 4 5 3 1	3 5 2 1 4

Figure 2.2

By a remarkable coincidence, at almost the same time as G.H.Freeman was carrying out the investigations we have just described in England, a parallel investigation was being carried out by G.Campbell and S.Geller at the University of Purdue in the U.S.A. The latter researchers conceived exactly the same concept as we have introduced under the name of qausi-complete latin square but they called it a <u>balanced latin square</u>. Independently of Freeman, they proved in G.Campbell and S.Geller(1980) that such squares exist for every positive integer m. They, like Freeman, carried out an enumeration of such squares and discussed their value in experimental design. Moreover, they obtained all the complete sets of pairwise orthogonal balanced latin squares of order 5 (including those exhibited in Figure 2.2) and thus provided confirmation of Freeman's enumeration. They went on to show that there exist similar complete sets of pairwise orthogonal balanced latin squares for the orders 7, 11 and 13. Finally, they made a further generalization of the concepts of complete latin square and quasi-complete latin square by defining a latin square to be <u>overall balanced</u> if, for each treatment (element) i of the square, treatment j is adjacent to it four times in all, the adjacencies being either horizontal or vertical or a combination of both. We illustrate this concept by displaying in Figure 2.3 two of the five latin squares of order four which exist with the properties that they are overall balanced but not balanced.

$$\begin{array}{|cccc|} \hline 1 & 2 & 3 & 4 \\ 4 & 3 & 1 & 2 \\ 2 & 1 & 4 & 3 \\ 3 & 4 & 2 & 1 \\ \hline \end{array} \qquad \begin{array}{|cccc|} \hline 1 & 2 & 3 & 4 \\ 3 & 4 & 2 & 1 \\ 2 & 1 & 4 & 3 \\ 4 & 3 & 1 & 2 \\ \hline \end{array}$$

Figure 2.3

We showed earlier in this Chapter that the multiplication table of a finite group G can be written in the form of a row complete latin square if and only if the group G is sequenceable (Theorem 1.3). In (1984), R.A. Bailey showed that there is an analogous group property that provides a necessary and sufficient condition that the multiplication table of the group can be written in the form of a quasi-row-complete latin square. Moreover, when a group has this property its multiplication table can in fact be written in the form of a quasi-complete latin square. (Compare Theorem 1.2).

Let $b_0, b_1, b_2, \ldots, b_{n-1}$ be a sequence comprising all the elements of a group G of order n and let $a_0 = c$ (the identity), $a_1 = b_0^{-1} b_1$, $a_2 = b_1^{-1} b_2$, $\ldots$, $a_{n-1} = b_{n-2}^{-1} b_{n-1}$. Then Bailey made the following definitions.

DEFINITIONS. If the elements $a_0, a_1, a_2, \ldots, a_{n-1}$ are all distinct, the elements $b_0, b_1, b_2, \ldots, b_{n-1}$ are said to form a <u>directed terrace</u> of G. G is then <u>sequenceable</u> with sequencing $a_0, a_1, a_2, \ldots, a_{n-1}$ and partial products $b_0^{-1} b_0 = e = a_0$, $b_0^{-1} b_1 = a_0 a_1$, $b_0^{-1} b_2 = a_0 a_1 a_2, \ldots, b_0^{-1} b_{n-1} = a_0 a_1 a_2 \cdots a_{n-1}$.

If, on the other hand, the elements $a_0, a_1, a_2, \ldots, a_{n-1}$ are such that they include

(i) one occurrence of each element x of G which satifies $x^2 = e$; and

(ii) for every other element x of G, either two occurrences of x and none of x^{-1}, or one of x and one of x^{-1}, or none of x and two of x^{-1}, the elements $b_0, b_1, b_2, \ldots, b_{n-1}$ are said to form a <u>terrace</u> of G. G is then <u>quasi-sequenceable</u> with quasi-sequencing $a_0, a_1, a_2, \ldots, a_{n-1}$.

We note that, if $(G, \cdot)$ is abelian, then $b_0^{-1}, b_1^{-1}, \ldots, b_{n-1}^{-1}$ is a terrace if and only if $b_0, b_1, \ldots, b_{n-1}$ is a terrace since then $a_i' = (b_{i-1}^{-1})^{-1} b_i^{-1} = b_i^{-1} b_{i-1} = = (b_{i-1}^{-1} b_i)^{-1} = a_i^{-1}$ and so the sequence $a_0', a_1', \ldots, a_{n-1}'$ has the same property as the sequence $a_0, a_1, \ldots, a_n$.

Bailey's main theorem is the following:

THEOREM 2.2 Let $L(\underline{b},\underline{d})$ be a latin square based on a group G such that $\ell_{ij} = b_i d_j$, where $b_0, b_1, \ldots, b_{n-1}$ and $d_0, d_1, \ldots, d_{n-1}$ are orderings of the elements of G. Then L is a quasi-row-complete latin square if $d_0, d_1, \ldots, d_{n-1}$ is a terrace for G and is a quasi-complete latin square if and only if $b_0^{-1}, b_1^{-1}, \ldots, b_{n-1}^{-1}$ and $d_0, d_1, \ldots, d_{n-1}$ are both terraces for G.

Proof The ordered pair of elements occurring in the (j-1)-th and j-th columns of any row of L are of the form (p, pu_j), where $u_j = d_{j-1}^{-1} d_j$. Consider the unordered pair (r,s), where $s = ru$. Since r occurs in each column of L, this unordered pair occurs in as many rows as there are occurrences of the element u or u^{-1} within the sequence $u_0, u_1, \ldots, u_{n-1}$. That is, it occurs in two rows since $d_0, d_1, \ldots, d_{n-1}$ is a terrace for G. This proves the first statement of the theorem.

The ordered pair of elements occurring in the (i-1)-th and i-th rows of any column of L are of the form $(b_{i-1} d_j, b_i d_j)$ or $(v_i q, q)$, where $v_i = b_{i-1} b_i^{-1} = (b_{i-1}^{-1})^{-1} b_i^{-1}$. Consider the unordered pair (r,s), where $r = vs$. Since s occurs in each row of L, this unordered pair occurs in as many columns as there are occurrences of the element v or v^{-1} within the sequence $v_0, v_1, \ldots, v_{n-1}$. That is, it occurs in two columns since $b_0^{-1}, b_1^{-1}, \ldots, b_{n-1}^{-1}$ is a terrace for G. []

COROLLARY The latin square $L(\underline{d}^{-1}, \underline{d})$ based on a group G is quasi-complete if and only if $d_0, d_1, \ldots, d_{n-1}$ is a terrace for G.

In her paper (1984), Bailey has also discussed the existence of terraces.

THEOREM 2.3 The sequence $0, n-1, 1, n-2, 2, n-3, \ldots$ forms a terrace for the cyclic group C_n of order n written additively.

Proof We exhibit the terrace together with the elements $a_i = -b_{i-1} + b_i$ of the associated quasi-sequencing, taking the cases when n is even and odd separately.

If n = 2m, we have

0 2_{m-1} $2m-1$ 2 1 2_{m-3} $2m-2$ 4 2 2_{m-5} $2m-3$ 6 3 ...
... $2r-2$ $r-1$ $2m-2r+1$ $2m-r$ ... $m-2$ 3 $m+1$ $2m-2$ $m-1$ 1 m.
This is a directed terrrace.

If n = 2m+1, we have

0 2_m $2m$ 2 1 2_{m-2} $2m-1$ 4 2 2_{m-4} $2m-2$ 6 3 ...
... $2r$ r $2m-2r$ $2m-r$ ... ... $m+2$ $2m-2$ $m-1$ 2 $m+1$ $2m$ m.
This is not a directed terrace. []

COROLLARY For every positive integer n, there exists a quasi-complete latin square. (Compare Theorem 2.1.)

THEOREM 2.4 An elementary abelian 2-group (other than the cyclic group C_2) is not terraced.

Proof If G is an elementary abelian 2-group, then $x = x^{-1}$ for every $x \in G$. Consequently, G is terraced if and only if it is sequenceable. But G does not have a unique element of order 2 so, by B. Gordon's theorem (Theorem 2.3.3 of [DK]), G cannot be sequenceable. []

The following two theorems of Bailey are useful for constructing terraces.

THEOREM 2.5 If B is a terraced group and C is a cyclic group of odd order, then the direct product B×C is terraced.

Proof Let $\mathfrak{b} = (b_1, b_2, \ldots, b_n)$ be a terrace for B and let $\mathfrak{c} = (c_1, c_2, \ldots, c_p)$ be the terrace for C given in Theorem 2.3, so that $c_{2r+1} = c_{2r}^{-1}$ for $r \geq 1$. Then the following sequence is a terrace for B×C.

Case when ord B = 2m

$(b_1, c_1), (b_2, c_1^{-1}), (b_3, c_1), \ldots, (b_{2m-1}, c_1), (b_{2m}, c_1^{-1})$,
$(b_{2m}, c_2^{-1}), (b_{2m-1}, c_2), (b_{2m-2}, c_2^{-1}), \ldots, (b_2, c_2^{-1}), (b_1, c_2)$,
$(b_1, c_3), (b_2, c_3^{-1}), (b_3, c_3), \ldots, (b_{2m-1}, c_3), (b_{2m}, c_3^{-1})$,
$(b_{2m}, c_4^{-1}), (b_{2m-1}, c_4), (b_{2m-2}, c_4^{-1}), \ldots, (b_2, c_4^{-1}), (b_2, c_4)$,
... ...
$(b_1, c_p), (b_2, c_p^{-1}), (b_3, c_p), \ldots, (b_{2m-1}, c_p), (b_{2m}, c_p^{-1})$.

Case when ord B = 2m+1

$(b_1,c_1), (b_2,c_1^{-1}), (b_3,c_1), \ldots, (b_{2m},c_1^{-1}), (b_{2m+1},c_1),$
$(b_{2m+1},c_2), (b_{2m},c_2^{-1}), (b_{2m-1},c_2), \ldots, (b_2,c_2^{-1}), (b_1,c_2),$
$(b_1,c_3), (b_2,c_3^{-1}), (b_3,c_3), \ldots, (b_{2m},c_3^{-1}), (b_{2m+1},c_3),$
$(b_{2m+1},c_4), (b_{2m},c_4^{-1}), (b_{2m-1},c_4), \ldots, (b_2,c_4^{-1}), (b_1,c_4),$
$\ldots \quad \ldots \quad \ldots \quad \ldots \quad \ldots \quad \ldots \quad \ldots \quad \ldots \quad \ldots$
$(b_1,c_p), (b_2,c_p^{-1}), (b_3,c_p), \ldots, (b_{2m},c_p^{-1}), (b_{2m+1},c_p).$

We will show that the first of the above two sequences (case when ord B = 2m) is a terrace. The proof for the case when ord B is odd is similar.

Since the group B×C has no elements of order two, we wish to show that the set S formed by the elements of the sequence $(b_1,c_1)^{-1}(b_2,c_1^{-1})$, $(b_2,c_1^{-1})^{-1}(b_3,c_1), \ldots, (b_{2m-1},c_p)^{-1}(b_{2m},c_p^{-1})$ has the property that, for each element h of B×C, either h occurs twice, or h^{-1} occurs twice, or there is one occurrence each of h and h^{-1} in the set S. S will then be a quasi-sequencing of B×C.

The elements of S can be listed as follows:
$(b_1^{-1}b_2, c_1^{-2}), (b_2^{-1}b_3, c_1^2), (b_3^{-1}b_4, c_1^{-2}), \ldots, (b_{2m-1}^{-1}b_{2m}, c_1^{-2}), (e, c_1c_2^{-1}),$
$(b_{2m}^{-1}b_{2m-1}, c_2^2), (b_{2m-1}^{-1}b_{2m-2}, c_2^{-2}), \ldots, (b_3^{-1}b_2, c_2^{-2}), (b_2^{-1}b_1, c_2^2), (e, c_2^{-1}c_3),$
$(b_1^{-1}b_2, c_3^{-2}), (b_2^{-1}b_3, c_3^2), (b_3^{-1}b_4, c_3^{-2}), \ldots, (b_{2m-1}^{-1}b_{2m}, c_3^{-2}), (e, c_3c_4^{-1}),$
$\ldots \quad \ldots \quad \ldots \quad \ldots \quad \ldots \quad \ldots \quad \ldots \quad \ldots \quad \ldots \quad \ldots$
$(b_1^{-1}b_2, c_p^{-2}), (b_2^{-1}b_3, c_p^2), (b_3^{-1}b_4, c_p^{-2}), \ldots, (b_{2m-1}^{-1}b_{2m}, c_p^{-2}).$

Let us write $d_i = c_i^2$. Then $d_1 = e$, $d_{2r+1} = d_{2r}^{-1}$ for $r \geq 1$, and $d_1, d_2, \ldots, d_p$ are the elements of C reordered (by Lemma 1.6, since C has odd order).

The elements $b_1^{-1}b_2, b_2^{-1}b_3, \ldots, b_{2m-1}^{-1}b_{2m}$ contain each non-identity element $x \in B$, not of order 2, either once and its inverse once, or twice, or not at all but its inverse twice. The elements $b_{2m}^{-1}b_{2m-1}, b_{2m-1}^{-1}b_{2m-2}, \ldots, b_2^{-1}b_1$ have the same property. If $(b_h^{-1}b_{h-1}, d)$ occurs in the 2r-th row of the list of elements of S, then $(b_{h-1}^{-1}b_h, d)$ occurs in the (2r+1)-th row of this list. So, if (x,d) and (x',d') occur in the 2r-th row of the list, where x,x' $\in \{y, y^{-1}\}$ and d' = d or d^{-1}, then (x^{-1},d) and (x'^{-1},d') occur in the (2r+1)-th row. We conclude that,

either (y,d), (y,d) and (y^{-1},d), (y^{-1},d) occur,
or (y,d), (y,d^{-1}) and (y^{-1},d), (y^{-1},d^{-1}) occur,
or (y,d), (y^{-1},d) and (y^{-1},d), (y,d) occur,
or (y,d), (y^{-1},d^{-1}) and (y^{-1},d), (y,d^{-1}) occur.

Hence, the differences involve two occurrences of a particular element but not its inverse or one occurrence each of the element and its inverse. These are the requirements for a terrrace. []

THEOREM 2.6 Let $\underline{b} = (b_1, b_2, \ldots, b_n)$ be a terrace for a group G. Then

(i) if $g \in G$, $g\underline{b} = (gb_1, gb_2, \ldots, gb_n)$ is a terrace for G;

(ii) if σ is an automorphism for G, then $\sigma(\underline{b}) = (b_1^\sigma, b_2^\sigma, \ldots, b_n^\sigma)$ is a terrace for G;

(iii) $\underline{b}^R = (b_n, b_{n-1}, \ldots, b_2, b_1)$, obtained by reversing $\underline{b}$, is a terrace for G;

(iv) each cycle of $\underline{b}$ of form $(b_i, b_{i+1}, \ldots, b_{i-1})$ with the property that $b_{i-1}^{-1} b_i = (b_n^{-1} b_1)^{\pm 1}$ is a terrace for G.

Method (iv) gives one terrace additional to $\underline{b}$ if $b_n^{-1} b_1$ has order 2 and otherwise gives two terraces additional to $\underline{b}$. []

EXAMPLE $0_4 4_5 1_5 6_1 7_6 5_6 3_7 2$ is a terrace of $(Z_8, +)$.

It yields the further terraces

$$5_6 3_7 2_6 0_4 4_5 1_5 6_1 7 \text{ and } 3_7 2_6 0_4 4_5 1_5 6_1 7_6 5$$

Using the mappings $x \to x+3$ and $x \to x+5$, which are the additive analogues of $g\underline{b}$, these are essentially the same as

$$0_6 6_7 5_6 3_4 7_5 4_5 1_1 2 \text{ and } 0_7 7_6 5_4 1_5 6_5 3_1 4_6 2$$

By reversing these three terraces, we get three further terraces.

Bailey has conjectured that all finite groups not excluded by Theorem 2.4 are terraced (that is, are quasi-sequenceable). She has provided a listing and enumeration of terraces for all terraced groups of orders up to and including nine. She has also considered the problem of randomization of the corresponding designs for use in field experiments.

Recently, B.A.Anderson(1987b) has re-invented the concept of a terrace and has called it a 2-sequencing. If $b_0, b_1, b_2, \ldots, b_{n-1}$ is a terrace: that is, a sequence comprising all the elements of a group G of order n such that the sequence $a_0 = e$, $a_1 = b_0^{-1} b_1$, $a_2 = b_1^{-1} b_2, \ldots, a_{n-1} = b_{n-2}^{-1} b_{n-1}$ includes one occurrence of each element of order 2 in G and, for every other element x of G, includes one occurrence of each of x and x^{-1}, or two occurrences of x, or two occurrences of x^{-1}, then the sequence $a_0, a_1, \ldots, a_{n-1}$ is a 2-sequencing of G.

In B.A.Anderson (1987a, 1987b, 1987c, 198α, 198β) and in B.A.Anderson and P.A.Leonard (1988), 2-sequencings have been used to prove that certain classes of finite groups (in particular, the dicyclic groups) are sequenceable. In (198γ), Anderson has obtained a new product theorem for terraced groups. These results are discussed in more detail in the last section of this Chapter.

(3) R-sequenceable and R_h-sequenceable groups.

A property of finite groups which appears at first sight to be very similar to that of sequenceability is that of R-sequenceability, already mentioned in Section 6 of Chapter 2.

DEFINITION. A finite group $(G, \cdot)$ is called R-sequenceable if its elements $a_0 = e, a_1, a_2, \ldots, a_{n-1}$ can be ordered in such a way that the partial products $b_0 = a_0$, $b_1 = a_0 a_1$, $b_2 = a_0 a_1 a_2, \ldots, b_{n-2} = a_0 a_1 \cdots a_{n-2}$ are all different and so that $b_{n-1} = a_0 a_1 \cdots a_{n-1} = b_0 = e$.

In fact, as we shall see, the properties of sequenceability and R-sequenceability are mutually exclusive for a wide class of groups and, moreover, the interpretations of the two properties in terms of latin squares are quite different. When a finite group is sequenceable this guarantees that the latin square formed by its Cayley table is a complete latin square (see Section 1 of this Chapter). By contrast, when a finite group is R-sequenceable, this guarantees that the group has a complete mapping and so that the latin square formed by its Cayley table has an orthogonal mate (by virtue of properties (1) and (2) of Section 1 in Chapter 2).

Suppose that $(G, \cdot)$ is R-sequenceable then, using the notation of the definition above, the mapping $x \to \varphi(x)$, where $\varphi = (c)(b_0 \; b_1 \; b_2 \; \ldots \; b_{n-2})$ and c is the element of G which does not appear among the partial products $b_0, b_1, \ldots, b_{n-2}$, is a bijection of G which is a complete mapping. For, let θ denote the bijection defined by

$$\theta = \begin{bmatrix} b_0 & b_1 & \ldots & b_{n-2} & c \\ a_1 & a_2 & \ldots & a_{n-1} & e \end{bmatrix}.$$

Then $\varphi(c) = c.\theta(c)$ and $\varphi(b_i) = b_i \theta(b_i)$ for $i = 0, 1, \ldots, n-2$, as is required for a complete mapping.

R-sequenceability is quite an old concept which has appeared in the literature in several guises. L.J.Paige(1951) observed long ago that it was a sufficient condition for a group to have a complete mapping (cf. Theorem 1.4.6 of [DK]). Later, G.Ringel came across the same concept in the process of solving the Heawood Map-Colouring problem. Unfortunately, Ringel himself has been unwilling to reply to our enquiries so, for such little information as exists in print to show the connection we refer the reader to G.Ringel(1974a); G.Ringel(1974b), pages 25 and 26; and to G.Ringel and J.W.T.Youngs(1969).

As regards the question of which finite groups have the property of R-sequenceability, let us look first at abelian groups. For such groups, the product $b_{n-1} = a_0 a_1 \ldots a_{n-1}$ of all their elements is equal to the identity unless the group has a unique element of order two [G.A.Miller(1903), see also Section 1 of Chapter 2]. In the latter case, the product is equal to the unique element of order two and the group is then sequenceable [B.Gordon(1961)] but clearly cannot be R-sequenceable. This leaves us with the question whether all abelian groups which do not have a unique element of order two are R-sequenceable. L.J.Paige(1947) and L.Carlitz(1953) showed that such groups certainly have a complete mapping. R.J.Friedlander, B.Gordon and M.D.Miller(1978) tried to resolve the more general question. They succeeded in showing that the following types of abelian group are R-sequenceable:

(i) Cyclic groups of odd order;

(ii) Abelian groups of odd order whose Sylow 3-subgroup is cyclic;

(iii) Abelian groups of orders which are relatively prime to 6;

(iv) Elementary abelian p-groups, except the cyclic group C_2 of order 2;

(v) Abelian groups of type $G = C_2 \times C_{4k}$, $k \geq 1$;

(vi) Abelian groups whose Sylow 2-subgroup S is of one of the following kinds:

(a) $S = (C_2)^m$, $m > 1$ but $m \neq 3$,

(b) $S = C_2 \times C_h$, where $h = 2^k$ and either k is odd or else $k \geq 2$ is even and G/S has a direct cyclic factor of order congruent to 2 modulo 3.

Also, G.Ringel(1974a) had earlier claimed that abelian groups of type $C_2 \times C_{6k+2}$ are R-sequenceable.

For non-abelian groups, not too much is known. G.Ringel(1974a) has claimed that groups of the form $A_4 \times (C_2)^m$ are R-sequenceable, where A_4 denotes the alternating group on four symbols and C_2 the cyclic group of order 2. However, his proof has not been published and no-one else, so far as the present authors are aware, has succeeded in verifying the assertion. Consequently, the truth of the claim remains in doubt.

One of the present authors has shown that a finite dihedral group is R-sequenceable if and only if its order is a multiple of four and that the same class of non-abelian groups of order pq as are sequenceable (Section 1 of this Chapter) are also R-sequenceable. [See A.D.Keedwell(1983b).] The latter result contrasts with the fact, shown above, that for abelian groups the properties of sequenceability and R-sequenceability are mutually exclusive.

The first of these facts is very easy to see, as follows. In the first place, if the dihedral group $D_n = \langle \alpha, \beta : \alpha^n = \beta^2 = e, \alpha\beta = \beta\alpha^{-1} \rangle$ has order not divisible by four then n is odd and so an odd number of its 2n elements are of the form $\beta\alpha^i$. It follows that if we repeatedly make use of the generating relation $\alpha\beta = \beta\alpha^{-1}$, the product of all the elements of the group will simplify to the form $\beta^n \alpha^s$, where $\beta^2 = e$ and n is odd. Consequently, this product cannot be equal to the identity element e and so the group is not R-sequenceable. On the other hand, if the order of the group is divisible by four so that n is even, we are able to display an R-sequencing of the group in Figure 3.1 and thus show constructively that it is R-sequenceable.

As regards non-abelian groups of order pq such that p has 2 as a primitive root, where p and q are distinct primes with $p < q$, the proof that these groups are R-sequenceable follows the same lines as the proof

R-sequencing of D_{4h}	Partial products	R-sequencing of D_{4h-2}	Partial products
e		e	
α^{2h-1}	e	α^{2h-2}	e
$\alpha^{-(2h-2)}$	α^{2h-1}	$\alpha^{-(2h-3)}$	α^{2h-2}
α^{2h-3}	α	α^{2h-4}	α
$\alpha^{-(2h-4)}$	α^{2h-2}	$\alpha^{-(2h-5)}$	α^{2h-3}
.	α^2	.	α^2
.	.	.	.
.	.	.	.
$\alpha^{2\ell-1}$	$\alpha^{h-\ell}$	$\alpha^{2\ell-2}$	$\alpha^{h-\ell}$
$\alpha^{-(2\ell-2)}$	$\alpha^{h+\ell-1}$	$\alpha^{-(2\ell-3)}$	$\alpha^{h+\ell-2}$
.	$\alpha^{h-\ell+1}$	.	$\alpha^{h-\ell+1}$
.	.	.	.
.	α^{h+1}	.	α^{h-2}
α^{-2}	α^{h-1}	α^2	α^h
α	α^h	α^{-1}	α^{h-1}
β	$\beta\alpha^{3h}$	$\beta\alpha^{4h-3}$	$\beta\alpha^{3h-2}$
α^{-1}	$\beta\alpha^{3h-1}$	α	$\beta\alpha^{3h-1}$
α^2	$\beta\alpha^{3h+1}$	α^{-2}	$\beta\alpha^{3h-3}$
α^{-3}	$\beta\alpha^{3h-2}$	α^3	$\beta\alpha^{3h}$
.	.	.	.
.	$\beta\alpha^{3h-\ell+1}$	.	$\beta\alpha^{3h+\ell-3}$
$\alpha^{2\ell-2}$	$\beta\alpha^{3h+\ell-1}$	$\alpha^{-(2\ell-2)}$	$\beta\alpha^{3h-\ell-1}$
$\alpha^{-(2\ell-1)}$	$\beta\alpha^{3h-\ell}$	$\alpha^{2\ell-1}$	$\beta\alpha^{3h+\ell-2}$
.	.	.	.
.	$\beta\alpha^{-(2h-1)}$	.	$\beta\alpha^{2h}$

Figure 3.1 (continued over)

R-sequencing of D_{4h}	Partial products	R-sequencing of D_{4h-2}	Partial products
α^{2h-2}		α^{2h-3}	
	$\beta\alpha^{-1}$		$\beta\alpha^{4h-3}$
$\alpha^{-(2h-1)}$		$\alpha^{-(2h-2)}$	
	$\beta\alpha^{-2h}$		$\beta\alpha^{2h-1}$
α^{2h}		α^{2h-1}	
	β		β
$\beta\alpha^{2h}$		$\beta\alpha^{2h-1}$	
	α^{2h}		α^{2h-1}
$\beta\alpha^{2h+1}$		$\beta\alpha^{2h}$	
	$\beta\alpha$		$\beta\alpha$
$\beta\alpha^{2h+2}$		$\beta\alpha^{2h+1}$	
	α^{2h+1}		α^{2h}
.	.	.	.
.	.	.	.
.	$\beta\alpha^{h-1}$	.	$\beta\alpha^{h-2}$
$\beta\alpha^{4h-2}$		$\beta\alpha^{4h-5}$	
	α^{3h-1}		α^{3h-3}
$\beta\alpha^{4h-1}$		$\beta\alpha^{4h-4}$	
	$\beta\alpha^{h}$		$\beta\alpha^{h-1}$
$\beta\alpha$		β	
	α^{3h+1}		α^{3h-1}
$\beta\alpha^{2}$		$\beta\alpha$	
	$\beta\alpha^{h+1}$		$\beta\alpha^{h}$
$\beta\alpha^{3}$		$\beta\alpha^{2}$	
	α^{3h+2}		α^{3h}
.	.	.	.
.	.	.	.
.	α^{4h-1}	.	α^{4h-3}
$\beta\alpha^{2h-2}$		$\beta\alpha^{2h-3}$	
	$\beta\alpha^{2h-1}$		$\beta\alpha^{2h-2}$
$\beta\alpha^{2h-1}$		$\beta\alpha^{2h-2}$	
	e		e

Figure 3.1 (continued)

that they are sequenceable and makes use of exactly the same quotient sequencing (given in Section 1 of this chapter). Because of the length of the proof we refer the reader to A.D.Keedwell(1983b) for the details.

For the purpose of constructing orthogonal latin squares based on a given group $(G, \cdot)$, we really require a property stronger than R-sequenceability which we have called R_h-sequenceability and which we shall now explain.

When a group is R-sequenceable it has a complete mapping and consequently an orthogonal mate as explained above. One way of constructing this orthogonal mate is by the <u>column method</u> originally introduced by one of the present authors [see A.D.Keedwell(1966)] and subsequently re-expounded in Section 7.4 of [DK]. The column method makes use of the complete mapping $\phi = (c)(b_0\ b_1\ b_2\ \ldots\ b_{n-2})$ defined by the R-sequencing, where $b_i = a_0 a_1 \ldots a_i$ is the i-th partial product of the elements $a_0, a_1, \ldots, a_{n-1}$ of the R-sequencing and c is the element of G which does not occur among the partial products.

Let us suppose that L_1 is the latin square formed by the Cayley multiplication table of G and that the rows of L_1 are $I, S_1, S_2, \ldots, S_{n-1}$, when regarded as permutations of its first row. Because L_1 is latin, each column of L_1 contains each symbol of G exactly once and so the permutations $I, S_1, S_2, \ldots, S_{n-1}$ form a sharply transitive set T of permutations. The permutations $I, \phi S_1 \phi^{-1}, \phi S_2 \phi^{-1}, \ldots, \phi S_{n-1} \phi^{-1}$ also form a sharply transitive set because they are conjugate to the members of the set T. Thus, they define the rows of a second latin square L_2. Now, if $U_0, U_1, U_2, \ldots, U_{n-1}$ are permutations representing the rows of one latin square L_i as permutations of n symbols $0, 1, 2, \ldots, n-1$ and $V_0, V_1, V_2, \ldots, V_{n-1}$ are permutations representing the rows of another latin square L_j, it is easy to see that the squares L_i and L_j are orthogonal if and only if the n permutations $U_0^{-1} V_0, U_1^{-1} V_1, \ldots, U_{n-1}^{-1} V_{n-1}$ form a sharply transitive set. This was first pointed out by H.B.Mann(1942). It follows that our squares L_1 and L_2 defined above will be orthogonal provided that the n permutations $I, S_1^{-1} \phi S_1 \phi^{-1}, S_2^{-1} \phi S_2 \phi^{-1}, \ldots, S_{n-1}^{-1} \phi S_{n-1} \phi^{-1}$ form a sharply transitive set or, equivalently, that the permutations $\phi, S_1^{-1} \phi S_1, S_2^{-1} \phi S_2, \ldots, S_{n-1}^{-1} \phi S_{n-1}$ are a sharply transitive set. But this is ensured when ϕ is a complete mapping of G because, if S_i maps g to $g_i g$ for each $g \in G$ so that L_1 is the latin square shown in Figure 3.2, then it follows that

$$\begin{aligned}
\phi &= (c)(b_0\ \ b_1\ \ \ldots\ \ b_{n-2}) \\
S_1^{-1} \phi S_1 &= (g_1 c)(g_1 b_0\ \ g_1 b_1\ \ \ldots\ \ g_1 b_{n-2}) \\
S_2^{-1} \phi S_2 &= (g_2 c)(g_2 b_0\ \ g_2 b_1\ \ \ldots\ \ g_2 b_{n-2}) \\
&\ \ \vdots \\
S_{n-1}^{-1} \phi S_{n-1} &= (g_{n-1} c)(g_{n-1} b_0\ \ g_{n-1} b_1\ \ \ldots\ \ g_{n-1} b_{n-2})
\end{aligned}$$

$$L_1 = \begin{bmatrix} c & b_0 & b_1 & \cdots & b_{n-2} \\ g_1c & g_1b_0 & g_1b_1 & \cdots & g_1b_{n-2} \\ g_2c & g_2b_0 & g_2b_1 & \cdots & g_2b_{n-2} \\ \vdots & \vdots & \vdots & & \vdots \\ g_{n-1}c & g_{n-1}b_0 & g_{n-1}b_1 & \cdots & g_{n-1}b_{n-2} \end{bmatrix}$$

Figure 3.2

Since $g_ub_h = g_vb_k \Rightarrow g_ub_{h+1} \neq g_vb_{k+1}$ because $g_ub_{h+1} = g_ub_ha_{h+1} = g_vb_ka_{h+1}$ whereas $g_vb_{k+1} = g_vb_ka_{k+1}$, the above set of permutations is a sharply transitive set. Each element g∈G occurs in each one of the n permutations and has a different image (successor element) in each.

The fact that ϕ is a complete mapping means that the elements $b_0^{-1}b_1, b_1^{-1}b_2, \ldots, b_{n-2}^{-1}b_0$ are all distinct. (They are the elements $a_1, a_2, \ldots, a_{n-1}$ of the associated R-sequencing.) Suppose now that each of the permutations $\phi, \phi^2, \ldots, \phi^h$ is a complete mapping of G. Then we shall say that G is R_h-sequenceable. Precisely, we have

DEFINITION. A group $(G, \cdot)$ of order n is R_h-sequenceable if n-1 of its elements can be arranged in a sequence $b_0, b_1, \ldots, b_{n-2}$ in such a way that the set of elements $b_i^{-1}b_{i+1}$ for i = 0, 1, ..., n-2, are all distinct (where arithmetic is modulo n-1), and likewise the sets of elements $b_i^{-1}b_{i+2}, b_i^{-1}b_{i+3}, \ldots, b_i^{-1}b_{i+h}$. (In particular, a group which is R_1-sequenceable is R-sequenceable.)

If a group $(G, \cdot)$ is R_h-sequenceable, it is easy to see that the column method permits the construction of at least h+1 mutually orthogonal latin squares based on the Cayley table of G. The first square L_1 is the Cayley multiplication table of G as before and the (r+1)-th square L_{r+1} of the set has row permutations $I, \phi^r S_1 \phi^{-r}, \phi^r S_2 \phi^{-r}, \ldots, \phi^r S_{n-1} \phi^{-r}$ where r takes one of the values 1, 2, ..., h. The squares L_{r+1} and L_{s+1} are orthogonal because the permutations $I, S_1^{-1} \phi^{s-r} S_1, S_2^{-1} \phi^{s-r} S_2, \ldots, S_{n-1}^{-1} \phi^{s-r} S_{n-1}$ are a sharply transitive set, for each pair r,s taken from the set {0, 1, ..., h}.

Very little seems to be known about R_h-sequenceability of groups, h > 1. Indeed, the only work done on this subject to date seems to be that of one of the present authors [in A.D.Keedwell(1983b)] who has investigated cyclic groups. He has shown that C_9 is not R_2-sequenceable, that C_{15} is R_2-sequenceable (with 32 isomorphically distinct R_2-sequencings) but is not R_3-sequenceable, and that C_{21} is at least R_2-sequenceable. He has also remarked that the R_h-sequenceability of cyclic groups of prime order is covered by the following theorem.

THEOREM 3.1 An elementary abelian group of order p^n is R_h-sequenceable for $h = 1, 2, \ldots, p^n - 2$.

Proof If ω is a primitive root of the Galois field $GF[q]$, where $q = p^n$, the following sequence of elements of the additive group of $GF[q]$ (which is elementary abelian) has the required properties: $1, \omega, \omega^2, \ldots, \omega^{q-2}$. []

The following corollary is, of course, very well-known.

COROLLARY From the elementary abelian group of order p^n a set of p^n-1 mutually orthogonal latin squares all based on that group can be constructed.

Finally, in this section, let us mention a generalization of R-sequenceability quite different from that which we have just been discussing. R.J.Friedlander, B.Gordon and P.Tannenbaum(1981) have called a complete mapping ϕ of a group $(G, \cdot)$ k-regular if it can be expressed in the form

$$\phi = (c)(b_{11} b_{12} \ldots b_{1k})(b_{21} b_{22} \ldots b_{2k}) \ldots (b_{s1} b_{s2} \ldots b_{sk})$$

where $sk = n-1$. The fact that ϕ is a complete mapping requires that the elements $b_{ij}^{-1} b_{i,j+1}$, for $i = 1, 2, \ldots, s$ and $j = 1, 2, \ldots, k$, with the second suffix being added modulo k, should be all distinct and should cover the non-identity elements of G. The case when $k = n-1$ is the case when G is R-sequenceable.

D.F. Hsu and one of the present authors have made use of the same concept, under the name of $(k,1)$-complete mapping, to construct left neofields of pseudo-characteristic k (using the method of Theorem 6.1 of Chapter 2.) They have also generalized the concept further by defining a (k,λ)-complete mapping of a group $(G,\cdot)$ as an arrangement of the non-identity elements (or, indeed, any n-1 of the elements) of G, each used λ times, into s cyclic sequences of length k, say

$$(g_{11}\, g_{12}\, \cdots\, g_{1k})\, (g_{21}\, g_{22}\, \cdots\, g_{2k})\, \cdots\, (g_{s1}\, g_{s2}\, \cdots\, g_{sk})$$

such that the elements $g_{ij}^{-1} g_{i,j+1}$ (where $i = 1, 2, \ldots, s$; and the second suffix is added modulo k) comprise the non-identity elements of G each counted λ times. They have utilized the latter concept in the construction of block designs of Mendelsohn type with block size k. [For more details, see D.F. Hsu and A.D. Keedwell (1984, 1985).]

(4) Super P-groups

At the beginning of Chapter 2, we mentioned the concept of a P-group. If $(G,\cdot)$ is a finite group and G' its commutator subgroup (derived group) then every product of the n elements of G is in the same coset of G'. A P-group has the converse property that every element of the appropriate coset of G' is expressible as the product of all the elements of G in some order. As we mentioned before, it is now known that every finite group is a P-group. We consider here an extension of this concept.

DEFINITION. A finite group G is a super P-group if every element of one particular coset hG' of the derived group can be expressed as the product of the n elements of G in such a way that the orderings of the elements in these products are sequencings of G with the exception that, in the case that h = e, the element e of G' must be expressed as a product of the n elements of G which forms an R-sequencing of G.

One reason for interest in this concept is the following:

THEOREM 4.1 For abelian groups, the concepts of being a super P-group and of being sequenceable or R-sequenceable are the same.

Proof The derived group of an abelian group consists of the identity element alone. By a theorem due to G.A.Miller(1903), the product of all the elements of a finite abelian group (in any order) is equal to the identity unless the group has a unique element of order two. In the latter case, the product is equal to the unique element of order two. Evidently, therefore, in the former case, the group is a super P-group if and only if it is R-sequenceable and, in the latter case, it is a super P-group if and only if it is sequenceable. []

The reader will easily deduce from the results of Sections 1 and 3 of this Chapter that the classes of abelian group which are known to be super P-groups are (i) all finite cyclic groups (including C_2), (ii) abelian groups whose Sylow 2-subgroup is cyclic (since these are sequenceable), and (iii) abelian groups of all the kinds listed as being R-sequenceable in Section 3 above.

As a convenient source for reference, we list appropriate sequences for the elements of the cyclic groups.

For C_{2m}, a sequencing is

$0, 1, 2m-2, 3, 2m-4, 5, 2m-6, 7, \ldots, \ldots, 6, 2m-5, 4, 2m-3, 2, 2m-1$.

For C_{4m+1}, an R-sequencing is

$0, 4m, 2, 4m-2, 4, 4m-4, 6, 4m-6, \ldots, \ldots, 2m-2, 4m-(2m-2), 2m, 2m-1, 2m+3, 2m-3, 2m+5, \ldots, \ldots, 2m-(2m-3), 2m+(2m-1), 2m-(2m-1), 2m+1$.

For C_{4m+3}, an R-sequencing is

$0, 4m+2, 2, 4m, 4, 4m-2, 6, 4m-4, \ldots, \ldots, 2m-2, 4m-(2m-4), 2m, 2m+1, 2m+3, 2m-1, 2m+5, 2m-3, \ldots, \ldots, 2m-(2m-3), 2m+(2m+1), 2m-(2m-1), 2m+2$.

We showed earlier in this Chapter that non-abelian groups of order pq, where p and q are distinct odd primes with p < q and such that 2 is a primitive element modulo p, are both sequenceable and R-sequenceable. Because these groups are R-sequenceable, the product of their elements in any order lies in the derived group and it is not too surprising that it turns out that these groups are super P-groups. Thus, we have:

THEOREM 4.2 A non-abelian group of order pq, where p and q are distinct odd primes with p < q and p is a prime which has 2 as a primitive root, is a super P-group.

The proof is in A.D.Keedwell(1983c) but the present authors believe that it should be possible to provide an alternative simpler proof which makes use of automorphic mappings of the sequencings in the same kind of way as we describe below for dihedral groups.

For dihedral groups of singly-even order we have the following theorem.

THEOREM 4.3 If m is odd and the dihedral group D_m = $gp\{\alpha,\beta: \alpha^m=\beta^2=e, \alpha\beta=\beta\alpha^{-1}\}$ of order 2m is sequenceable, it is a super P-group.

Proof. If $b_{2m-1} = a_0 a_1 a_2 \ldots a_{2m-1}$ is a sequencing of D_m, it contains m elements of the type $\beta\alpha^x$, $x = 0, 1, \ldots, m-1$. Consequently, when m is odd, $b_{2m-1} \in D_m \backslash gp\{\alpha\}$.

Each inner automorphism of D_m maps a sequencing onto a sequencing. If $b_{2m-1} = \beta\alpha^s$, then the inner automorphism $\tau(\alpha^t): g \to \alpha^{-t} g \alpha^t$ maps the distinct partial products $b_0, b_1, \ldots, b_{2m-1} = \beta\alpha^s$ onto distinct partial products $b_0^*, b_1^*, \ldots, b_{2m-1}^* = \beta\alpha^{s+2t}$. Since 2 is prime to m, s+2t takes all values modulo m as t varies, so there exist sequencings equal to every element of the coset $D_m \backslash gp\{\alpha\}$ of the derived group $D_m' = gp\{\alpha\}$. []

From the results of Section 1 of this Chapter, we can deduce:

COROLLARY For all odd values of m, $5 \leq m \leq 35$, D_m is a super P-group. Also, for all primes $p \equiv 1 \pmod 4$ and for all primes $p \equiv 7 \pmod 8$ for which 2 has the exponent $\frac{1}{2}(p-1)$, D_p is a super P-group.

Because dihedral groups of doubly even order are R-sequenceable (as we proved in Section 3 of this chapter), the product of all their elements in any order is an element of the derived group $D_m' = gp\{\alpha^2: \alpha^m = e\}$. Each element of this derived group is mapped either onto itself or else onto its inverse by every inner automorphism of D_m. Consequently, from any one given element $\alpha^h \in D_m$ it is not possible to obtain all the others as images under inner automorphisms, and so the method of Theorem 4.3 fails to help us. However, in the case when m is twice an odd prime, there exists a set of outer automorphsims which we can make use of. Hence, we get:

THEOREM 4.4 If m is even, m = 2p, where p is an odd prime and D_m is sequenceable, then it is a super P-group.

Proof When m is even, the derived group of D_m = gp$\{\alpha,\beta: \alpha^m = \beta^2 = e, \alpha\beta = \beta\alpha^{-1}\}$ is D'_m = gp$\{\alpha^2\}$ and has order $\frac{1}{2}m$. Since every such group is R-sequenceable, all sequencing products b_{2m-1} belong to D'_m.

When m = 2p, D'_m has prime order p and all its non-identity elements are of order p. Also, in that case the mapping $\sigma_s: \alpha \to \alpha^{p+2s}$, $\beta \to \beta$ of the generators α,β of D_{2p}, defines an outer automorphism of the group for $1 \le s < p$ (except when p = 3, in which case it is the inner automorphism τ_β) since p+2s is relatively prime to 2p. The outer automorphisms σ_s, $1 \le s < p$, are transitive on the set of non-identity elements of D'_{2p} because $\alpha^{2t} \to \alpha^{4st}$ and so σ_s maps α^{2t} to α^{2u} if s satisfies 2ts ≡ u(mod p), $1 \le s < p$.

Since D_{2p} is sequenceable, there exists a sequencing product which is equal to some non-identity element of D'_{2p}. This can be mapped by an appropriate one of the outer automorphisms σ_s onto a sequencing product which is equal to any other required non-identity element of D'_{2p}. Since also D_{2p} is R-sequenceable, this is sufficient to prove that it is a super P-group. []

The dihedral group D_6 of order 12 is sequenceable (see Section 1) and this is the only group for which the requirements of Theorem 4.4 were known to be satisfied until the recent work of B.A.Anderson(1987a). We exhibit the fact that this group is a super P-group in Figure 4.1. (The column headed "S" is the sequencing.)

The authors conjecture that in fact all dihedral groups of order four times an odd prime are super P-groups. Moreover, they think it very likely that all dihedral groups of order 10 or more are sequenceable. A more general conjecture is that all metabelian groups are sequenceable.

Sequenceable and R-sequenceable groups

	(S)		(S_{T_β})	
e		e	e	
α^5	e	α	α^5	e
β	α^5	α^2	α^4	α^5
α^3	$\beta\alpha$	α^3		α^3
$\beta\alpha^2$	$\beta\alpha^4$	β	β	$\beta\alpha^3$
$\beta\alpha^4$	α^4	α^3	$\beta\alpha^3$	β
$\beta\alpha^3$	β	$\beta\alpha^2$	$\beta\alpha^4$	α^4
$\beta\alpha^5$	α^3	α^2	$\beta\alpha^5$	$\beta\alpha$
α	$\beta\alpha^2$	$\beta\alpha$	$\beta\alpha^2$	α
α^2	$\beta\alpha^3$	$\beta\alpha^4$	$\beta\alpha^3$	$\beta\alpha^2$
$\beta\alpha$	$\beta\alpha^5$	$\beta\alpha^3$	α^2	$\beta\alpha^4$
α^4	α^2	α^4	α	$\beta\alpha^5$
	e	α^5	$\beta\alpha$	α^2
		$\beta\alpha$		
		$\beta\alpha^5$		
		α^4		

Figure 4.1. D_6 is a super P-group

(5) <u>Tuscan squares and a graph decomposition problem</u>.

We begin this Section by introducing some picturesque notation which has been devised by S.W.Golomb and H.Taylor (1985).

A square n×n matrix all of whose rows are permutations of the same n symbols, say $1, 2, \ldots, n$, was called a <u>row latin square</u> in [DK], page 103, after D.A.Norton (1952) who first gave it this name. If such a square is also <u>row complete</u> : that is, if each ordered pair (i,j) occurs just once among the rows, then it provides a solution to a problem posed by E.G.Straus as to when it is possible to decompose the complete directed graph on n vertices into n directed paths with each path consisting of n-1 edges and traversing all the vertices (that is : a set of n disjoint Hamiltonian paths). Straus originally raised this question in the presence of one of the present authors (see below) and it was subsequently reported in N.S.Mendelsohn (1968).

Golomb and Taylor have re-christened squares of the first type Italian squares and those of the second kind Tuscan squares. They have generalized the second concept by defining a Tuscan-k square to be an Italian square which has the property that for any two symbols i,j and for each m from 1 to k, there is at most one row in which j is the m-th symbol to the right of i (that is : there is at most one row in which the signed distance from i to j is m). Clearly, a Tuscan-1 square is a Tuscan square.

As regards the existence of Tuscan-1 squares, there are none of orders 1,3 or 5 and the only ones of orders 2 and 4 are row complete latin squares (called Roman squares by Golomb and Taylor). T.W.Tillson(1980) proved that the problem of Straus has an affirmative answer for all odd $n \geq 7$ and so there are Tuscan-1 squares for every odd order $n \geq 7$ (as well as for every even order, see Section 1 of this Chapter). By computer search Golomb and Taylor found 466,144 standard form Tuscan-1 squares of order 7. [NOTE. For the historical record, it is worth mentioning that in T.W.Tillson's 1980 paper Straus's problem was attributed to J.C.Bermond, apparently at the insistence of the referee as it was correctly attributed in a preprint held by the present authors.]

In their 1985 paper, Golomb and Taylor have made the following further definitions.

DEFINITIONS. A Tuscan-(n-1) square is called a Florentine square. A Vatican square is a Roman square (row complete latin square) which is also Florentine. We may alternatively say that it is a Florentine square which is also a latin square.

Golomb and Taylor have observed that, for primes $p > 2$, the multiplication table of the finite field GF[p] defines a Vatican square. They have also shown that Tuscan-2 squares other than these do not exist for $n < 8$ but that there are exactly six Tuscan-2 squares in standard form of order 8.

For further enumerative results concerning these various kinds of square, the reader is referred to Golomb and Taylor's 1985 paper.

More recently, in S.W.Golomb, T.Etzion and H.Taylor(1990), the latter authors have introduced the following additional concept:

DEFINITION. A <u>circular Tuscan-k array</u> is an n×(n+1) array A in which each of the symbols $0, 1, \ldots, n$ appears exactly once in each row and in which the Tuscan-k property holds when the rows are read cyclically. In matrix notation, the rows are indexed from 1 to n and the columns are indexed from 0 to n. For each integer m from 1 to k, we have that the ordered pair $A(i,j), A(i,j+m)$ is different from the ordered pair $A(u,v)$, $A(u,v+m)$ unless $i = u$ and $j = v$, where arithmetic of indices is modulo n+1.

A circular Tuscan-n array is called a <u>circular Florentine array</u>.

Golomb, Etzion and Taylor have proved the following connection with latin squares:

THEOREM 5.1 A circular n×(n+1) Tuscan-k array A exists only when a set $\{A_1, A_2, \ldots, A_k\}$ of k mutually orthogonal latin squares of order (n+1) exists.

<u>Proof</u>. Let us label the rows of A with the integers $1, 2, \ldots, n$ and label the columns $0, 1, \ldots, n$. Each row of A contains all of the symbols $0, 1, \ldots, n$ and the pair of symbols u,v occur a distance m apart (m=1,2,...,k) in exactly one row of A when the rows are read cyclically. We define the entries of the latin squares as follows:

$A_1(i,j) = 0$ if $i = j$

= r if the symbol r occurs one step to the right of the symbol i in the r-th row of A. (Thus, the latin square A_1 is unipotent.)

For $t = 2, 3, \ldots, k$,

$A_t(i,j) = j$ if $i = j$

= h if the symbol h is t steps to the right of the symbol i in the row of A in which the symbol j is one step to the right of i.

It is easy to check that $\{A_1, A_2, \ldots, A_k\}$ do form a set of mutually orthogonal latin squares. Moreover, each of the squares $A_2, A_3, \ldots, A_k$ is idempotent and A_1, in addition to being unipotent, has the property that no intercalate involving elements of the main left-to-right diagonal can occur in it. For suppose, on the contrary, that the entry r were to occur in cells (i,j) and (j,i). This would require that both of the ordered pairs (i,j) and (j,i) occurred in row r of A, an impossibility since A is row-latin. □

Golomb, Etzion and Taylor have shown that a circular Tuscan-n array A of size n×(n+1) can be constructed whenever n+1 is a prime by inserting h in the s-th column of the r-th row of A if s is the unique solution of the congruence s≡rh mod(n+1).

For example, when n=4, the 4×5 array A shown in Figure 5.1 is obtained.

REMARK. Theorem 5.1 above may be contrasted with a theorem due to one of the present authors (see Theorem 7.4.1 of [DK]) which shows that the existence of an (n+1)×n latin rectangle which has the cyclic Tuscan-k property is a sufficient condition for the existence of k+1 mutually orthogonal latin squares of order (n+1).

An (n+1)×n latin rectangle with the cyclic Tuscan-(n-1) property can be constructed when n+1 is prime by taking as first row of the array the sequence $1, t, t^2, \ldots, t^{n-1}$, where t is a primitive element for the prime in question, and obtaining each of the remaining rows by adding 1 to each element of the row which precedes it, modulo (n+1).

For example, when n=4, a primitive root of 5 is 2 and we get the 5×4 array shown in Figure 5.2.

0	1	2	3	4
0	3	1	4	2
0	2	4	1	3
0	4	3	2	1

Figure 5.1

1	2	4	3
2	3	0	4
3	4	1	0
4	0	2	1
0	1	3	2

Figure 5.2

The construction of Golomb, Etzion and Taylor just described gives an n×(n+1) circular Florentine array whenever n+1 is a prime. These authors have shown that, by contrast, if n is an odd integer greater than 1, no n×(n+1) circular Florentine array exists.

In their (1990) paper, they have also investigated existence questions for all the different types of array which we have defined in this Section. Most of their actual constructions use the so-called polygonal path construction in which the symbols of the array are represented as the vertices of a graph. We call attention to one of their results which provides an interesting link with the subject matter of Chapter 9 of this book on "Latin Squares and Codes", namely:

THEOREM 5.2. An n×n polygonal path Vatican square exists if and only if an n×∞ singly-periodic Costas array exists.

For the concept of a singly periodic Costas array, we refer the reader to this later Chapter.

$$\begin{bmatrix} 0 & 1 & 5 & 2 & 4 & 3 \\ 1 & 2 & 0 & 3 & 5 & 4 \\ 2 & 3 & 1 & 4 & 0 & 5 \\ 3 & 4 & 2 & 5 & 1 & 0 \\ 4 & 5 & 3 & 0 & 2 & 1 \\ 5 & 0 & 4 & 1 & 3 & 2 \end{bmatrix} \rightarrow \begin{matrix} \underline{6} & 0 & 1 & 5 & 2 & 4 & 3 \\ 6 & 1 & 2 & 0 & 3 & \underline{5} & \underline{4} \\ 6 & 2 & 3 & 1 & 4 & \underline{0} & \underline{5} \\ 6 & 3 & \underline{4} & \underline{2} & 5 & 1 & 0 \\ 6 & 5 & 4 & 3 & 0 & \underline{2} & \underline{1} \\ 6 & 5 & 0 & 4 & 1 & \underline{3} & 2 \end{matrix} \rightarrow \begin{bmatrix} 0 & 1 & 5 & 2 & 4 & 3 & 6 \\ 4 & 6 & 1 & 2 & 0 & 3 & 5 \\ 5 & 6 & 2 & 3 & 1 & 4 & 0 \\ 2 & 5 & 1 & 0 & 6 & 3 & 4 \\ 1 & 6 & 4 & 5 & 3 & 0 & 2 \\ 3 & 2 & 6 & 5 & 0 & 4 & 1 \\ 6 & 0 & 5 & 4 & 2 & 1 & 3 \end{bmatrix}$$

Tuscan square   augmented Tuscan square   new Tuscan square

Figure 5.3

In T. Etzion (1986), a very detailed investigation of Tuscan-1 squares has been made and of the solutions to the Straus problem which they provide. A considerable number of new constructions have been devised many of which exploit the method used by T.W. Tillson (1980), which Etzion has christened "the cutting method". It may be described as follows: Given a Tuscan square T of order n defined on the symbols $0, 1, \ldots, n-1$, adjoin an additional left-most column all of whose entries are the symbol n. Select a permutation $a_0 = n, a_1, a_2, \ldots, a_{n-1}$ of the symbols $0, 1, \ldots, n$ which has the property that each row of the augmented Tuscan square T contains exactly one of the pairs $(a_i, a_{i+1}), i = 0, 1, \ldots, n-2$, as adjacent elements. Shift each row cyclically until the second element of its pair becomes the left-most symbol of that row and the first element of its pair becomes the right-most symbol. Finally adjoin the sequence $a_0, a_1, \ldots, a_{n-1}$ as an additional row to obtain a Tuscan square of order $n+1$.

Tillson proved in effect that this construction is always possible when T is appropriately chosen if n is any even integer greater than four. See Figure 5.3 for an illustration of the process.

(6) More results on the sequencing and 2-sequencing of groups.

Since the earlier parts of this Chapter were written, a considerable number of further results on the sequencing and 2-sequencing of groups have been obtained. Many of these results are dependent on the concept of a symmetric sequencing which we now define:

DEFINITION. Let $(G, .)$ be a finite group of even order $2n$ with identity e and with a unique element z of order 2. A sequencing $e, a_1, a_2, \ldots, a_{2n-1}$ of G is called a symmetric sequencing if and only if $a_n = z$ and, for $1 \leq i \leq n-1$, $a_{n+i} = a_{n-i}^{-1}$.

We note that, because z is the unique element of order 2 in G, it necessarily lies in the centre of G. (Because $g^{-1}zg$ is an element of order 2 for every $g \in G$, we have $g^{-1}zg = z$ and so z is an element of the centre.) It follows that a symmetric sequencing

$$e, a_1, a_2, \ldots, a_{n-1}, z, a_{n-1}^{-1}, a_{n-2}^{-1}, \ldots, a_1^{-1}$$

has a product sequencing of the form

$$e, b_1, b_2, \ldots, b_{n-1}, b_{n-1}z, b_{n-2}z, \ldots, b_1z, z.$$

The concept of a symmetric sequencing was originally introduced in B.A.Anderson(1976), where it was used to prove:

THEOREM 6.1. If G is a sequenceable group of odd order $2m+1$ then $G \times C_2$, where C_2 denotes the cyclic group of order 2, has a symmetric sequencing.

Proof. By hypothesis, G possesses a sequencing $e, a_1, a_2, \ldots, a_{2m}$

with partial products e, $b_1 = ea_1$, $b_2 = ea_1a_2,\ldots,b_{2m} = ea_1a_2,\ldots,a_{2m}$. Let T be an arbitrarily chosen transversal of the set $J = \{(x,x^{-1}) : x \in G\setminus\{e\}\}$. We form a sequence $c_0 = (e,0)$, $c_1 = (a_1,i_1)$, $c_2 = (a_2,i_2),\ldots,c_{2m} = (a_{2m},i_{2m})$ of $2m+1$ of the elements of $G \times C_2$ as follows: If $a_h \in T$, then $c_h = (a_h,1)$; if $a_h \notin T$, then $c_h = (a_h,0)$. Thus each $i_h \in \{0,1\}$. Next define $c_{2m+1} = (e,1)$ and $c_{2m+1+h} = c_{2m+1-h}^{-1} = (a_{2m+1-h}^{-1}, i_{2m+1-h})$. Then the elements $c_0, c_1, \ldots, c_{4m+1}$ form the group $G \times C_2$ under the operation $c_h + c_k = (a_h a_k, i_h + i_k \bmod 2)$. Moreover, the sequence $c_0, c_1, c_2, \ldots, c_{4m+1}$ is easily seen to be a symmetric sequencing of this group with c_{2m+1} as the unique element of order 2. □

In the same paper, Anderson gave a proof of the following:

THEOREM 6.2. *If the group G has a symmetric sequencing and B is an abelian group such that the orders of G and B are relatively prime, then the direct product $G \times B$ has a symmetric sequencing.*

However, the proof then given contained an error. A corrected version has been given in B.A.Anderson(1987c) which, unfortunately, is considerably more complicated than the original version as several different cases have to be considered. The idea of the proof is to generalize B.Gordon's construction of a (symmetric) sequencing for any abelian group with a unique element of order 2. (See B.Gordon(1961) or Theorem 2.3.3 of [DK].) Because of the length of the argument, we refer the reader to Anderson's 1987 paper for the details.

Anderson's 1976 paper contained another important concept: that of an <u>even starter</u>.

DEFINITION. Let $(G,.)$ be a finite group of even order $2n$ with identity e and with a unique element z of order 2. Then the set
$$E = \{\{x_1,y_1\}, \{x_2,y_2\}, \ldots, \{x_{n-1}, y_{n-1}\}\}$$
of $n-1$ unordered pairs of elements of G is called a <u>left even starter</u> of G if

(i)   every non-identity element of G except one, which we denote by m, occurs as an element of some pair of E, and

(ii) every non-identity element of G except z occurs in the set $\{x_i^{-1}y_i, y_i^{-1}x_i : 1 \leq i \leq n-1\}$. (A <u>right even starter</u> is similarly defined with the set $\{x_iy_i^{-1}, y_ix_i^{-1} : 1 \leq i \leq n-1\}$ replacing that of (ii) above. However, we shall not require the latter concept.)

Suppose that E is a left even starter of the group $(G, .)$ with z as unique element of order 2, that $E^* = E \cup \{\{e, m\}\}$ and that $Y = \{\{x, xz\} : x \in G\}$. Then we notice that, provided that $m \neq z$, E^* and Y are disjoint one-factors of the complete graph K_{2n} whose vertices are labelled by the elements of G.

EXAMPLE. The sequence $0, 4, 8, 1, 3, 5, 7, 9, 2, 6$ is a symmetric sequencing of the cyclic group C_{10} when written additively. The sequence of partial sums is $0, 4, 2, 3, 6, 1, 8, 7, 9, 5$. A left even starter for G is
$$E = \{\{6,3\}, \{8,7\}, \{2,4\}, \{9,5\}\},$$
with $m = 1$. Then,
$$E^* = \{\{0,1\}, \{6,3\}, \{8,7\}, \{2,4\}, \{9,5\}\}$$
and
$$Y = \{\{1,6\}, \{3,8\}, \{7,2\}, \{4,9\}, \{5,0\}\}.$$

We note that, in the above example, $E^* \cup Y$ is a Hamiltonian circuit of the complete graph K_{10}. More generally, we have the following theorem from B.A. Anderson (1976):

<u>THEOREM 6.3</u>. The group $(G, .)$ of even order 2n with identity element e and with a unique element z of order 2 has a symmetric sequencing if and only if it has a left even starter such that $E^* \cup Y$ is a Hamiltonian circuit of the complete graph K_{2n} whose vertices are labelled by the elements of G.

<u>Proof</u>. Suppose first that G has a symmetric sequencing with partial product sequence $e, b_1, b_2, \ldots, b_{n-1}, b_{n-1}z, \ldots, b_1z, z$.

Let $E = \{\{z, b_1z\}, \{b_1, b_2\}, \{b_2z, b_3z\}, \{b_3, b_4\}, \ldots, \{b_{n-3}z, b_{n-2}z\}, \{b_{n-2}, b_{n-1}\}\}$ if n is odd, and
$E = \{\{z, b_1z\}, \{b_1, b_2\}, \{b_2z, b_3z\}, \{b_3, b_4\}, \ldots, \{b_{n-3}, b_{n-2}\}, \{b_{n-2}z, b_{n-1}z\}\}$ if n is even.

Thus, $m = b_{n-1}z$ when n is odd and $m = b_{n-1}$ when n is even.

Since $(b_{h-1}z)^{-1}(b_h z) = a_h$ and $b_h^{-1} b_{h+1} = a_{h+1}$, it is easy to see that E is a left even starter. Also, since $Y = \{\{e,z\}, \{b_1 z, b_1\}, \{b_2, b_2 z\}, \ldots, \{b_{n-1}, b_{n-1} z\}\}$ when n is odd and $Y = \{\{e,z\}, \{b_1 z, b_1\}, \{b_2, b_2 z\}, \ldots, \{b_{n-1} z, b_{n-1}\}\}$ when n is even, we see that $E^* \cup Y$ defines the edges of a Hamiltonian circuit of K_{2n}.

Conversely, suppose that G has a left even starter E such that $E^* \cup Y$ is a Hamiltonian circuit of K_{2n}. We shall write

$$E = \{\{mz, h_2\}, \{h_2 z, h_3\}, \{h_3 z, h_4\}, \ldots, \{h_{n-1} z, z\}\},$$

since $Y = \{\{h_2, h_2 z\}, \{h_3, h_3 z\}, \ldots, \{h_{n-1}, h_{n-1} z\}, \{z, e\}\}$. (Compare the example above.) The Hamiltonian circuit allows us to assign an ordering to the elements of G by picking a starting vertex and a direction round the circuit. Since $E^* = E \cup \{\{e, m\}\}$, e and m are adjacent vertices of the Hamiltonian circuit. We take e as starting vertex and e→m as starting direction. Then the elements of G form a sequence H as follows:

$$e, m, mz, h_2, h_2 z, \ldots, h_{n-1}, h_{n-1} z, z.$$

We use this sequence H to define the sequence P of partial products $e, b_1, b_2, \ldots, b_{n-1}, b_{n-1} z, b_{n-2} z, \ldots, b_1 z, z$ for a symmetric sequencing of G.

If n is odd, then m is the (n+1)-th element of P, mz is the n-th element, h_2 is the (n−1)-th element, $h_2 z$ is the (n+2)-th element, h_3 is the (n+3)-th element, $h_3 z$ is the (n−2)-th element, h_4 is the (n−3)-th element and the remainder of the sequence is constructed outwards from the middle elements h_2, mz, m by changing sides each time the next element of the sequence H is z times its predecessor. The resulting sequence P is as follows:

$$e, h_{n-1}, \ldots, h_5 z, h_4, h_3 z, h_2, mz, m, h_2 z, h_3, h_4 z, h_5, \ldots, h_{n-1} z, z.$$

To see that this is the sequence of partial products for a symmetric sequencing, we construct the elements of the associated sequencing. These are $a_0 = e$, $a_1 = h_{n-1}$, $a_2 = b_1^{-1} b_2 = h_{n-1}^{-1} h_{n-2} z$, $a_3 = b_2^{-1} b_3 = (h_{n-2} z)^{-1} h_{n-3} = h_{n-2}^{-1} h_{n-3} z, \ldots, a_{n-i} = h_{i+1}^{-1} h_i z, \ldots, a_n = (mz)^{-1} m = z,$ $\ldots, a_{n+i} = (h_i z)^{-1} h_{i+1} = a_{n-i}^{-1}, \ldots, a_{2n-1} = h_{n-1}^{-1}$. The elements $a_{n-1}, a_{n-2}, \ldots, a_2, a_1$ are all different because E is a left even starter and none of them is equal to $z(=a_n)$ because E and Y have no pair in common.

If n is even, we construct the sequence P from the sequence H by taking m as the n-th element of P, mz as the (n+1)-th element, h_2 as the (n+2)-th element, $h_2 z$ as the (n−1)-th element and completing the construction outwards from the middle elements m, mz, h_2 as before by changing sides each time the next element of the sequence H is z times its predecessor. The resulting sequence P is as follows:

$$e, h_{n-1}, \ldots, h_5, h_4 z, h_3, h_2 z, m, mz, h_2, h_3 z, h_4, h_5 z, \ldots, h_{n-1} z, z.$$

Again it is easy to construct the elements of the associated symmetric sequencing. []

Next we prove:

THEOREM 6.4. If $(G, .)$ is a group of order $2n$ with a unique element z of order 2 and E is a left even starter for G, then E induces a one-factorization $F(E)$ on K_{2n+2}.

Proof. Let e denote the identity element of G and m the unique non-identity element of G which does not appear in E (as before) and let ∞_1 and ∞_2 be two additional symbols. For each element g of G, we define $g.\infty_i = \infty_i = \infty_i.g$, $i = 1$ or 2. Let $E^+ = E \cup \{\{e, \infty_1\}, \{m, \infty_2\}\}$ and let $Y^+ = Y \cup \{\{\infty_1, \infty_2\}\}$. Also, if $E = \{\{x_1, y_1\}, \{x_2, y_2\}, \ldots, \{x_{n-1}, y_{n-1}\}\}$, let gE denote the set of pairs $\{\{gx_1, gy_1\}, \{gx_2, gy_2\}, \ldots, \{gx_{n-1}, gy_{n-1}\}\}$ and let $gE^+ = gE \cup \{\{g, \infty_1\}, \{gm, \infty_2\}\}$.

Let $F(E) = \{gE^+, g \in G\} \cup Y^+$. Then it is obvious that each element of $F(E)$ is a one-factor of K_{2n+2}. Since $F(E)$ comprises $2n+1$ one-factors, it is only necessary to show that each edge of K_{2n+2} occurs in one of these one-factors. Edges which involve a vertex ∞_1 or ∞_2 clearly belong to $F(E)$. Let $\{g_1, g_2\}$ be an edge for which g_1, g_2 are distinct elements of G. If $g_2 = g_1 z$, then $\{g_1, g_2\} \in Y^+$. If $g_2 \neq g_1 z$, then, by definition of E, there exists a pair $\{x_i, y_i\} \in E$ such that $g_1^{-1} g_2 = x_i^{-1} y_i$ (or $y_i^{-1} x_i$). If $h \in G$ satisfies $hx_i = g_1$ then $hy_i = g_2$ and so $\{g_1, g_2\} \in hE$. This completes the proof. []

Suppose now that the complete graph K_{2n+2} has a perfect one-factorization: that is, a one-factorization such that the union of every two one-factors is a Hamiltonian circuit of the graph. If it happens that this one-factorization is induced by a group $(G, .)$ of order $2n$ with a unique element z of order 2 as in the above theorem, then, in particular, $E^+ \cup Y^+$ will be a Hamiltonian circuit of K_{2n+2}. It follows that, if the edges $\{m, \infty_2\}, \{\infty_2, \infty_1\}, \{\infty_1, e\}$ are replaced by the edge $\{m, e\}$ and the vertices ∞_1 and ∞_2 are deleted, then the consequence is that $E^* \cup Y$ (where E^* and Y are defined as for Theorem 6.3) is a Hamiltonian circuit of K_{2n} and so the group $(G,.)$ must possess a symmetric sequencing.

B.A.Anderson(1987b) used this relationship to deduce from some graph-theoretical results given in E.Seah and D.R.Stinson(1987) that the dicyclic group $Q_6 = \langle a,b: a^6 = e, b^2 = a^3, ba = a^{-1}b \rangle$ of order 12 must possess a symmetric sequencing. He was subsequently able to prove that every dicyclic group Q_{2n} (of order 4n) has a symmetric sequencing. As we remarked at the end of Section 2 of this Chapter, for that purpose he made use of the concept of a terraced group which we discussed in that Section.

We recall that a <u>terrace</u> of a group $(G,.)$ of order 2n is a listing $b_0 = e, b_1, b_2, \ldots, b_{2n-1}$ of all the elements of G such that the sequence $a_0 = e, a_1 = b_0^{-1}b_1, a_2 = b_1^{-1}b_2, \ldots, a_{2n-1} = b_{2n-2}^{-1}b_{2n-1}$ includes one occurrence of each element of order 2 in G and, for every other element x of G, includes one occurrence of each of x and x^{-1}, or two occurrences of x, or two occurrences of x^{-1}. When a terrace exists, Anderson has called the sequence $a_0, a_1, a_2, \ldots, a_{2n-1}$ associated with it a <u>2-sequencing</u>.

In the case when $(G,.)$ has a unique element z of order 2, we will denote by Z the normal subgroup $\langle e,z \rangle$ of order 2 generated by z and we will denote by σ the canonical homomorphism $g \to gZ$ from G to G/Z.

Suppose now that the group G has a symmetric sequencing
$$e, a_1, a_2, \ldots, a_{n-1}, z, a_{n-1}^{-1}, a_{n-2}^{-1}, \ldots, a_1^{-1}$$
which we will denote by S and that the associated sequence P of partial products is
$$e, b_1, b_2, \ldots, b_{n-1}, b_{n-1}z, b_{n-2}z, \ldots, b_1z, z.$$
We will denote the sequence $e, a_1, a_2, \ldots, a_{n-1}$ by A and we will denote the images under σ of the elements of A by $\alpha_0, \alpha_1, \alpha_2, \ldots, \alpha_{n-1}$ respectively and the images under σ of the elements $e, b_1, b_2, \ldots, b_{n-1}$ by $\beta_0, \beta_1, \beta_2, \ldots, \beta_{n-1}$. We will denote these two sequences of elements of G/Z by Σ and Π respectively.

Then we have the following three important theorems from B.A.Anderson(1987b):

THEOREM 6.5. Let $(G,.)$ be a finite group of order $2n$ which has a unique element z of order 2 and which possesses a symmetric sequencing S. Then the sequence Σ defined as above is a 2-sequencing of G/Z.

Proof. Since σ is a homomorphism, $b_i\sigma \cdot a_{i+1}\sigma = b_{i+1}\sigma$ for $0 \leq i \leq n-2$. That is, $\beta_i \alpha_{i+1} = \beta_{i+1}$. So Π is the sequence of partial products of the sequence Σ. Also, from the sequence P we see that $b_j \neq b_i z$ for any integers i,j and so all elements of the sequence Π are distinct. Thus, Π contains all the elements of G/Z.

Let t be the number of pairs γ, γ^{-1} such that $\gamma = \{g, gz\} \in \Sigma$ and $\gamma^{-1} \neq \gamma$. [Note that $\gamma^{-1} = \gamma \Rightarrow \{g^2, g^2 z\} = \{e, z\}$ and so $g^2 = z$.] Now, at least one of the elements g or g^{-1} and at least one of the elements gz or $g^{-1}z$ must appear in the sequence A since if neither g nor g^{-1} appeared in A then neither of g^{-1} nor g would appear in the sequence $z, a_{n-1}^{-1}, a_{n-2}^{-1}, \ldots, a_1^{-1}$ and S would not be a sequencing of G. Similarly, if neither gz nor $g^{-1}z$ appeared in A, S would not be a sequencing of G. If g or gz appears in A, then γ appears in Σ. If g^{-1} or $g^{-1}z$ appears in A, then γ^{-1} appears in Σ. It follows that either γ or γ^{-1} both appear in Σ or γ appears twice or γ^{-1} appears twice.

Next, consider the $n-2t$ elements $\gamma \in \Sigma$ such that $\gamma = \gamma^{-1}$. For each such element $\gamma = \{g, gz\}$ except the identity element $\{e, ez\}$, we have $g^2 = z$. It follows that $gz = g^{-1}$ and so $\gamma = \gamma^{-1}$ implies that $\gamma = \{g, g^{-1}\}$. Since exactly one of the elements g or g^{-1} appears in the sequence A because S is a symmetric sequencing, it follows that γ occurs exactly once in the sequence Σ. This completes the proof that Σ is a 2-sequencing of G/Z.
☐

Theorem 6.5 suggests that the following definition will be useful.

DEFINITION. Let $(G,.)$ be a finite group of order $2n$ with a unique element z of order 2 and let σ be the canonical homomorphism from G to G/Z, where $Z = \langle e, z \rangle$ as before. If G/Z possesses a 2-sequencing Σ: $\alpha_0, \alpha_1, \ldots, \alpha_{n-1}$ and if $A: a_0 = e, a_1, a_2, \ldots, a_{n-1}$ is a sequence of elements of G such that $a_i \sigma = \alpha_i$ for $i = 0, 1, \ldots, n-1$, then A is called a lifting of Σ.

THEOREM 6.6. Suppose that A is a lifting of Σ, then A can be extended to a symmetric sequencing of the group G if and only if the elements $a_1, a_2, \ldots, a_{n-1}$ form a transversal of J, where $J = \{(x, x^{-1}) : x \in G \setminus \{e, z\}\}$.

Proof. Suppose first that the elements $a_1, a_2, \ldots, a_{n-1}$ of G do form a transversal of J. We can extend Σ to form the sequence
$$S: e, a_1, a_2, \ldots, a_{n-1}, z, a_{n-1}^{-1}, a_{n-2}^{-1}, \ldots, a_1^{-1}.$$
Then S is an ordering of G with an associated partial product sequence
$$P: e, b_1, b_2, \ldots, b_{n-1}, b_{n-1}z, b_{n-2}z, \ldots, b_1 z, z.$$
To show that P contains all the elements of G, it is necessary and sufficient to show that the elements $e, b_1, b_2, \ldots, b_{n-1}$ form a transversal of $Y = \{\{x, xz\} : x \in G\}$. Now, $b_i = ea_1 a_2 \cdots a_i$ and σ is a homomorphism such that $b_i \sigma = \alpha_0 \alpha_1 \cdots \alpha_i = \beta_i$ say, where $\beta_i = \{b_i, b_i z\}$. Since Σ is a 2-sequencing of G/Z, the sequence $\beta_1, \beta_2, \ldots, \beta_{n-1}$ of partial products of Σ contains each element $\{g, gz\}$ of G/Z exactly once. Consequently, the elements $e, b_1, b_2, \ldots, b_{n-1}$ do indeed form a transversal of Y.

Conversely, if A can be extended to a symmetric sequencing S of the group G, then it is evident from the form of S that the elements $a_1, a_2, \ldots, a_{n-1}$ form a transversal of J. []

THEOREM 6.7. Let $(G, .)$ be a finite group of order 2n with a unique element z of order 2 and let $\sigma : g \to gZ$, where $Z = \langle e, z \rangle$. Suppose that $\Sigma: \alpha_0, \alpha_1, \ldots, \alpha_{n-1}$ is a 2-sequencing of G/Z and that the number of elements of order 2 in G/Z is k. Then there are exactly $2^{(n+k-1)/2}$ ways in which the 2-sequencing Σ can be lifted so as to yield a symmetric sequencing of S.

Proof. Suppose first that $\gamma = \{g, gz\} \in \Sigma$ and that $\gamma^{-1} \neq \gamma$. Then either γ or γ^{-1} appears twice in Σ or else each of γ and γ^{-1} appears once. If $\gamma = \{g, gz\}$ appears twice, then these two occurrences can be lifted to g and gz in the sequence $A: e, a_1, a_2, \ldots, a_{n-1}$ and g and gz will form a transversal of the set $\{\{g, g^{-1}\}, \{gz, g^{-1}z\}\}$ as they are required to do. There are two ways to do this, since the first occurrence of γ in Σ may be lifted either to g or gz. If γ and γ^{-1} each appear once in Σ and if $\gamma = \{g, gz\}$ is lifted to g, then $\gamma^{-1} = \{g^{-1}, g^{-1}z\}$ must be lifted to $g^{-1}z$ since the lifted elements g and $g^{-1}z$ are required to form a transversal of $\{\{g, g^{-1}\}, \{gz, g^{-1}z\}\}$. Similarly, if $\gamma = \{g, gz\}$ is lifted to gz, then $\gamma^{-1} =

$\{g^{-1}, g^{-1}z\}$ must be lifted to g^{-1}. Thus, there are again two ways in which the lifting may be carried out in such a way that the lifted elements form a transversal of J, as required by Theorem 6.6.

Suppose now that $\gamma = \gamma^{-1}$, but that $\gamma \neq \{e, ez\}$. Then γ appears just once in the 2-sequencing Σ. Also, in this case $\{g, gz\} = \{g^{-1}, g^{-1}z\}$ and so we have $\gamma = \{g, g^{-1}\}$. Consequently, γ can be lifted either to g or to g^{-1} since either lifting will provide a lifted element which is a transversal of $\{g, g^{-1}\}$. If there are k such elements γ, there are 2^k ways in which all of them can be lifted.

Also, when $\gamma \neq \gamma^{-1}$, the sequence Σ contains $(n-k-1)/2$ pairs γ, γ or γ, γ^{-1} or γ^{-1}, γ^{-1} and, as we showed above, each of these pairs can be lifted in 2 ways to elements of the sequence A. Finally, the element $\{e, ez\}$ of Σ must be lifted to the element e of A and so the total number of liftings of Σ to A is $2^k \cdot 2^{(n-k-1)/2} = 2^{(n+k-1)/2}$. □

We are now in a position to consider the sequenceability of the general dicyclic group $Q_{2n} = \langle a, b : a^{2n} = e, b^2 = a^n, ab = ba^{-1}\rangle$ of order 4n.

We note that Q_{2n} has a unique element $z = a^n$ of order 2, that the centre of Q_{2n} is $Z = \langle e, z\rangle$ and that the quotient group Q_{2n}/Z is isomorphic to the dihedral group D_n of order 2n.

It follows from Theorems 6.5 and 6.7 that Q_{2n} has a symmetric sequencing if and only if the dihedral group D_{2n} has a 2-sequencing. Moreover, since D_{2n} has n+1 elements of order 2 when n is even and n elements of order 2 when n is odd, there are $2^{3n/2}$ ways of carrying out the lifting in the first case and $2^{(3n-1)/2}$ ways in the second.

In B.A.Anderson(1987b), that author provided a somewhat involved proof of the fact that every dihedral group D_n, with n odd, n⩾3, is 2-sequenceable and hence that Q_{2n} has a symmetric sequencing when n is an odd integer greater than one. However, in B.A.Anderson(198α), a much more elegant proof was given. We describe this latter proof in the next theorem.

<u>THEOREM 6.8</u>. Every dihedral group of order twice an odd number has a 2-sequencing.

Sequenceable and R-sequenceable groups

Proof. Let $D_{2m+1} = \langle \alpha, \beta : \alpha^{2m+1} = \beta^2 = e, \alpha\beta = \beta\alpha^{-1} \rangle$ denote a dihedral group of order $2(2m+1)$ and let $e, \alpha^{t_1}, \alpha^{t_2}, \ldots, \alpha^{t_{2m}}$ be a terrace of the cyclic group $C_{2m+1} = \langle \alpha : \alpha^{2m+1} = e \rangle$. (We note that C_{2m+1} is not sequenceable but that it is terraced. See Theorem 2.3 of this Chapter.) We form a corresponding terrace of D_{2m+1} by inserting the entries $\beta\alpha^{t_{2h}}$ and $\beta\alpha^{t_{2h+1}}$ between each entry $\alpha^{t_{2h}}$ in an odd position and the following entry $\alpha^{t_{2h+1}}$ in an even position of the terrace of C_{2m+1} to obtain the sequence

$$e, \beta, \beta\alpha^{t_1}, \alpha^{t_1}, \alpha^{t_2}, \beta\alpha^{t_2}, \beta\alpha^{t_3}, \alpha^{t_3}, \ldots, \alpha^{t_{2m-1}}, \alpha^{t_{2m}}, \beta\alpha^{t_{2m}}.$$

Let $0, s_1, s_2, \ldots, s_{2m}$ be the 2-sequencing of C_{2m+1} which corresponds to the terrace $0, t_1, t_2, \ldots, t_{2m}$ when C_{2m+1} is written additively. Then, since the equations

$$\alpha^{t_{2h}} w = \beta\alpha^{t_{2h}}, \quad \beta\alpha^{t_{2h}} x = \beta\alpha^{t_{2h+1}},$$

$$\beta\alpha^{t_{2h+1}} y = \alpha^{t_{2h+1}}, \quad \alpha^{t_{2h+1}} z = \alpha^{t_{2h+2}},$$

have solutions $w = \beta\alpha^{2t_{2h}}$, $x = \alpha^{t_{2h+1} - t_{2h}} = \alpha^{s_{2h+1}}$,

$y = \beta\alpha^{2t_{2h+1}}$, $z = \alpha^{t_{2h+2} - t_{2h+1}} = \alpha^{s_{2h+2}}$ respectively,

the 2-sequencing $e, \beta, \alpha^{s_1}, \beta\alpha^{2t_1}, \alpha^{s_2}, \beta\alpha^{2t_2}, \alpha^{s_3}, \beta\alpha^{2t_3}, \ldots, \beta\alpha^{2t_{2m-1}}, \alpha^{s_{2m}},$ $\beta\alpha^{2t_{2m}}$ has the above sequence as its sequence of partial sums and so it is a 2-sequencing. □

EXAMPLE. Since $0, 6, 1, 5, 2, 4, 3$ is a terrace of the cyclic group C_7 written additively, with $0, 6, 2, 4, 4, 2, 6$ as corresponding 2-sequencing, it follows that

$$e, \beta, \beta\alpha^6, \alpha^6, \alpha, \beta\alpha, \beta\alpha^5, \alpha^5, \alpha^2, \beta\alpha^2, \beta\alpha^4, \alpha^4, \alpha^3, \beta\alpha^3$$

is a terrace of the dihedral group D_7 with corresponding 2-sequence as follows:

$$e, \beta, \alpha^6, \beta\alpha^5, \alpha^2, \beta\alpha^2, \alpha^4, \beta\alpha^3, \alpha^4, \beta\alpha^4, \alpha^2, \beta\alpha, \alpha^6, \beta\alpha^6.$$

From Theorem 6.8, it follows at once that the dicyclic group Q_{2n} of order 4n has a symmetric sequencing for every odd integer n≥3. However, in his paper (198a), Anderson has shown that a simple direct proof of this result is possible, as follows:

THEOREM 6.9. The dicyclic group $Q_{2n} = \langle a,b: a^{2n} = e, b^2 = a^n, ab = ba^{-1}\rangle$ of order 4n has a symmetric sequencing for every odd integer n≥3.

<u>Direct proof</u>. Let $0, s_1, s_2, \ldots, s_{n-1}, n, -s_{n-1}, -s_{n-2}, \ldots, -s_1$ be a symmetric sequencing of the cyclic group C_{2n} when written additively and let $0, t_1, t_2, \ldots, t_{n-1}, t_{n-1}+n, t_{n-2}+n, \ldots, t_1+n, n$ be the associated sequence of partial sums. (Note: The sequencings which were constructed by B. Gordon (1961) for any abelian group with a unique element of order 2 are symmetric, as was first observed by B.A. Anderson (1976). In the case of the cyclic group C_{2n}, Gordon's construction gives the symmetric sequencing which is defined by $t_{2j} = -j, t_{2j+1} = j+1$, for $j = 0, 1, \ldots, n-1$, where $t_{n+r} = t_{n-1-r}+n$.) From the symmetric directed terrace $e, a^{t_1}, a^{t_2}, \ldots, a^{t_{n-1}}, a^{t_{n-1}+n}, \ldots, a^{t_1+n}, a^n$ of the multiplicative cyclic group $C_{2n} = \langle a: a^{2n} = e\rangle$, we construct a corresponding terrace of Q_{2n} by inserting the entries $ba^{t_{2h}}$ and $ba^{t_{2h+1}}$ between each entry $a^{t_{2h}}$ in an odd position and the following entry $a^{t_{2h+1}}$ in an even position of the terrace to obtain the sequence $e, b, ba^{t_1}, a^{t_1}, a^{t_2}, ba^{t_2}, ba^{t_3}, a^{t_3}, \ldots, a^{t_{n-1}}, ba^{t_{n-1}}, ba^{t_{n-1}+n}, \ldots, a^{t_1+n}, ba^{t_1+n}, ba^n, a^n$.

Since the equations
$$a^{t_{2h}}w = ba^{t_{2h}}, \quad ba^{t_{2h}}x = ba^{t_{2h+1}},$$
$$ba^{t_{2h+1}}y = a^{t_{2h+1}}, \quad a^{t_{2h+1}}z = a^{t_{2h+2}},$$

have solutions $w = ba^{2t_{2h}}$, $x = a^{s_{2h+1}}$, $y = ba^{2t_{2h+1}+n}$, $z = a^{s_{2h+2}}$, the above sequence is the sequence of partial sums of the symmetric sequence $e, b, a^{s_1}, ba^{2t_1+n}, a^{s_2}, ba^{2t_2}, a^{s_3}, ba^{2t_3+n}, a^{s_4}, \ldots, a^{s_{n-1}}, ba^{2t_{n-1}}, a^n, ba^{2t_n+n}, a^{s_n}, ba^{2t_n+1}, \ldots, a^{s_{2n-1}}, ba^{2t_{2n-1}+n}$. To see that this sequencing is symmetric, it is only necessary to note, for example, that $ba^{2t_n+n} = ba^{2t_{n-1}+n} = (ba^{2t_{n-1}})^{-1}$ since $t_n = t_{n-1}+n$, that $ba^{2t_{2n-1}+n} = ba^n = b^{-1}$ and, generally, that $ba^{2t_{n+r}+n} = (ba^{2t_{n-1-r}})^{-1}$, since $t_{n+r} = t_{n-1-r}+n$, $t_{2n-1} = n$ and $b^2 = a^n$. □

EXAMPLE. The cyclic group C_6 has $0,1,4,3,2,5$ as a symmetric sequencing when written additively with $0,1,5,2,4,3$ as sequence of partial sums. It follows that the following sequence is the sequence of partial products corresponding to a symmetric sequencing of the dicyclic group Q_6:
$$e, b, ba, a, a^5, ba^5, ba^2, a^2, a^4, ba^4, ba^3, a^3.$$
The associated symmetric sequencing is
$$e, b, a, ba^5, a^4, ba^4, a^3, ba, a^2, ba^2, a^3, ba^3.$$

It still remains to show that the dicyclic group Q_{2n} has a symmetric sequencing for every even integer $n>2$. Anderson achieved this result in two steps. First, by generalizing his original (and more complicated) proof that D_n, with n odd, is 2-sequenceable, he proved in B.A. Anderson (198α) that D_n is 2-sequenceable when n is a multiple of 4. The main result of that paper is

THEOREM 6.10. If $n \geqslant 3$ and the cyclic group C_{2n} has a special symmetric sequencing, then the dihedral group D_{2n} (of order 4n) has a 2-sequencing.

A symmetric sequencing or a 2-sequencing of the cyclic group C_n is called special if, when C_n is written additively, the first three terms of the sequencing take the form $0, 2c, -c$ for some $c \in C_{2n}$. (The first three terms of the sequence of partial sums are then $0, 2c, c$.) Anderson has shown that, for $n \geqslant 3$, C_{2n} has a special symmetric sequencing if and only if C_n has a special 2-sequencing. Also, he has shown that, if $n \geqslant 5$ is odd, then C_{2n} has a special 2-sequencing. Since a special symmetric sequencing is a special 2-sequencing, it follows that, for all $n \geqslant 3$, C_{2n} has a special symmetric sequencing. By this somewhat involved and lengthy argument, Theorem 6.10 is proved.

Secondly, in B.A.Anderson(1983), the author has generalized the concepts of special symmetric sequencing and special 2-sequencing as follows:

DEFINITIONS. (1) Let $e, a_1, a_2, \ldots, a_{n-1}$ be a 2-sequencing of the cyclic group C_n written additively and let $e, b_1, b_2, \ldots, b_{n-1}$ be the sequence of partial sums. Then the 2-sequencing is called a 2_d-sequencing if, for some integer i, $1 \leq i \leq n-2$, we have $a_i = c$ and $a_{i+1} = 2c$ or, alternatively, $a_i = 2c$ and $a_{i+1} = c$, where $c \in C_n$ and $c \neq 2c \neq 0$.

(2) Let $e, a_1, a_2, \ldots, a_{2n-1}$ be a symmetric sequencing of the cyclic group C_{2n} written additively and let $e, b_1, b_2, \ldots, b_{2n-1}$ be the sequence of partial sums. Then the symmetric sequencing is called a symmetric d-sequencing if, for some integer i, $2 \leq i \leq 2n-2$, we have $a_i = c$ and $a_{i+1} = 2c$ or, alternatively, $a_i = 2c$ and $a_{i+1} = c$, where $c \in C_{2n}$ and $c \neq 2c \neq 0$.

Using these definitions, Anderson has proved:

THEOREM 6.11. If the cyclic group C_{2n} has a symmetric-d-sequencing then the dihedral group D_{2n} has a 2-sequencing.

THEOREM 6.12. For all $n \geq 2$, C_{2n} has a symmetric d-sequencing.

The proofs of Theorems 6.11 and 6.12 involve the analysis of many different cases. When we combine them, we reach the desired conclusion that the dihedral group D_n has a 2-sequencing for every even integer n and hence, combining this with Theorem 6.8 (or using Theorem 6.9 directly), we conclude that every dicyclic group Q_{2n} except the quaternion group Q_4 and Klein's four-group has a symmetric sequencing.

One of the present authors conjectures that a very much shorter proof of the above result should be possible using a modification of the construction of R-sequencings of the dihedral group D_n, n even, given in A.D.Keedwell(1983b) to obtain 2-sequencings of that group.

From Theorem 6.2, it is immediate to deduce that, if B is a finite abelian group or odd order and $n \geq 3$, then the direct product $Q_{2n} \times B$ has a symmetric sequencing. In B.A.Anderson and P.A.Leonard(1988), the authors have shown that, despite the fact that Q_4 itself is not sequenceable, it is nonetheless true that $Q_4 \times B$ has a symmetric sequencing where B is any abelian group of odd order: that is, every finite Hamiltonian group with a unique element of order 2 has a symmetric sequencing.

In parallel with the above theoretical work, Anderson has devised a computer algorithm for testing whether a given non-abelian group of fairly small order is or is not sequenceable. With the aid of this algorithm, he has shown that all of the 86 non-abelian groups whose orders lie between 10 and 32 are sequenceable and that those nine of these groups which have a unique element of order 2 possess symmetric sequencings. In addition, he has shown that the dihedral groups whose orders lie between 34 and 100 inclusive are sequenceable. Also, the symmetric and alternating groups on five symbols are sequenceable. These results seem to confirm the conjecture of one of the present authors that most non-abelian groups are sequenceable. For details of the computer algorithm and the results, see B.A.Anderson(1987a) and (1987c).

In Section 2 of this Chapter, it was proved (Theorem 2.5) that, if B is a group which is 2-sequenceable and C is a cyclic group of odd order, then the direct product $B \times C$ is 2-sequenceable. This result has recently been generalized in B.A.Anderson(198γ) where it has been shown that, if B is 2-sequenceable and H is of odd order and has a starter-translate 2-sequencing, then the direct product $B \times H$ is 2-sequenceable.

DEFINITION. Let H be a group of odd order n which is 2-sequenceable. Let $e, a_1, a_2, \ldots, a_{n-1}$ be a 2-sequencing of H whose sequence of partial products is $e, b_1, b_2, \ldots, b_{n-1}$. If both of the sequences $S = \{a_2, a_4, \ldots, a_{n-1}\}$ and $T = \{a_1, a_3, \ldots, a_{n-2}\}$ are transversals of the set of pairs $J = \{(x, x^{-1}) : x \in H \setminus \{e\}\}$, then the 2-sequencing is called a <u>starter-translate 2-sequencing</u>. (Note that, when this is the case, the set $\{\{b_1, b_2\}, \{b_3, b_4\}, \ldots, \{b_{n-2}, b_{n-1}\}\}$ is a left-starter for H and also the set

$\{\{e,b_1\},\{b_2,b_3\},\ldots,\{b_{n-3},b_{n-2}\}\}$ is a left-translate by b_{n-1} of the left-starter $\{\{b_{n-1}^{-1},b_{n-1}^{-1}b_1\},\{b_{n-1}^{-1}b_2,b_{n-1}^{-1}b_3\},\ldots,\{b_{n-1}^{-1}b_{n-3},b_{n-1}^{-1}b_{n-2}\}\}$.)

The construction of the 2-sequencing of B×H makes strong use of the fact that $\{\{b_1,b_2\},\{b_3,b_4\},\ldots,\{b_{n-3},b_{n-2}\}\}$ is a left-starter for H.

As Anderson has shown in his paper, some of the groups of odd order pq (where p and q are distinct odd primes) which were shown to be sequenceable in A.D.Keedwell(1981a) and which we discussed in the first Section of this Chapter, have starter-translate sequencings and so can be used in conjunction with the above direct product theorem to enlarge the known spectrum of terraced groups.

We end this section with a brief mention of several further recent results. F.K.Hwang(1983) has given a construction for a totally symmetric complete latin square of any even order n based on the cyclic group of that order. R.A.Bailey(1986) has given an alternative construction which may be applied to give such a square based on any sequenceable abelian group. This construction may be summarized as: in the cell of the g_i^θ-th row and g_j^θ-column of the required totally symmetric latin square L, put the entry $(g_i^{-1}g_j^{-1})^\theta$. Here, it is assumed that the rows and columns of L are labelled by the elements of the given sequenceable abelian group $G = \langle g_1,g_2,\ldots,g_n\rangle$ taken in order and the mapping θ is defined as the inverse of the mapping $\phi:g\to g^\phi$ which maps the sequence $g_1,g_2,\ldots,g_n$ onto the sequence of partial sums of a sequencing (that is: maps it onto a directed terrace).

D.J.Street(1986) has given a construction which combines several complete latin squares to give larger two-dimensional designs which are balanced with respect to nearest neighbours.

In C.K.Nilrat and C.E.Praeger(1988), the authors have shown how 2^m essentially different directed terraces for the group $C_{2m+1}\times C_2$ (m⩾1) may be constructed. These directed terraces (sequencings) in turn yield a set of 2^{2m} complete latin squares of order 4m+2 no two of which are isoplanar. (The authors call two latin squares _isoplanar_ if one can be obtained from the other by a permutation of the symbols.) The question of how to obtain a valid randomization set of squares for use in experimental design has also been addressed.

In R.A.Bailey and C.E.Praeger(1988), the above construction of directed terraces has been explored further and applied to groups of the form H×C_2 where H is any terraced (2-sequenceable) group of odd order. This work has some overlap with that of B.A.Anderson described earlier in this Section though it was carried out independently.

ADDED IN PROOF:

(1) The problems stated below Theorem 1.1 in this Chapter appeared as Problems 2.1 and 2.5 in [DK].

ADDITIONAL REMARKS

With reference to the remarks following Theorem 1.1 of this Chapter and also in A.D.Keedwell(1976b), D.Cohen and T.Etzion(1989) have recently provided an explicit construction for a row-complete latin square which cannot be made column-complete by any permutation of its rows, for every even order greater than 8.

In addition to the papers mentioned in the foregoing Chapter about designs which are balanced with respect to nearest neighbours in various ways, there are many others. We list a few of those which seem most relevant: K.Afsarinjed and A.Hedayat(1975,1978), B.S.Alimenta(1962), M.S.Bartlett(1978), J.V.Bradley(1958), S.R.Eckert, T.W.Hancock, A.Mayo and G.N.Wilkinson(1983), G.H.Freeman(1988), T.R.Houston (1966), C.Huang and A.Rosa(1975), F.K.Hwang and S.Lin(1977), A.D.Keedwell(1984b), J.C.Kiefer and H.P.Wynn(1981), J.P.Morgan (1988a,b), D.J.Street and A.P.Street(1985).

In Section 3.3 of the Chapter, we mentioned that the concept of R-sequenceability has appeared in the literature in several different guises. One which we omitted to mention is its use by D.F.Hsu(1980) to construct proper XMP-neofields. (Neofields were defined in Section 2.6 of the present book. An XMP-neofield is one which does not possess the inverse, commutativity or exchange-inverse properties.) In connection with this construction, Hsu has obtained R-sequencings of the cyclic groups C_{4m+1} and C_{4m+3} which are different from those exhibited in Section 3.4 of the Chapter. [For these R-sequencings and for more details of their use in constructing XMP-neofields, see pages 128-141 of D.F.Hsu(1980).]

A survey paper, now somewhat out-of-date, on some of the material of this Chapter is A.D.Keedwell(1981b).

CHAPTER 4

LATIN SQUARES WITH AND WITHOUT SUBSQUARES OF PRESCRIBED TYPE

(K. Heinrich)

This chapter surveys the construction of latin squares which are required to satisfy prescribed subsquare conditions. In the first section, the basic constructions which are to be used throughout the rest of the chapter are summarized and notation is introduced for the various kinds of latin squares which are to be studied.

In Section 2, latin squares which have no intercalates (subsquares of order 2) are constructed and progress in solving the more difficult problem of constructing latin squares without subsquares of any size (except 1×1) is described. In particular, it is pointed out that if there exists a perfect 1-factorization of the complete graph K_{2n} on 2n vertices then there exists a latin square of order 2n-1 which has no proper subsquares. (A perfect 1-factorization of a graph G is a 1-factorization with the property that the union of any two of the 1-factors is a Hamiltonian cycle.)

In Section 3, the existence of so-called "subsquare complete" latin squares is discussed and latin squares having the maximum possible number of intercalates are constructed. Next, h-homogeneous latin squares (that is, latin squares in which each cell lies in h intercalates) are studied. Finally, in this section, two problems posed by L. Fuchs concerning the existence of quasigroups with subquasigroups of prescribed types and some related questions are discussed. (The two problems of Fuchs appeared as Problems 1-8 and 1-9 in [DK].) K. Heinrich herself (the author of the present chapter) has solved the first problem completely and has made a very substantial contribution to the solution of the second.

In the final section of the chapter, pairs and larger sets of orthogonal latin squares which are incomplete but which can be completed by the adjunction of suitable latin subsquares (which may themselves be orthogonal) are constructed. Many of the constructions described in this section are the work of the author of this chapter.

(1) <u>Introduction</u>

We denote by LS(n) a latin square of order n. If A is an LS(n) then (i,j) denotes the cell in the ith row and jth column, and $e_A(i,j)$ the entry in that cell. We say that A is based on the elements of the set N if $e_A(i,j) \in N$ and write e(A) = N. Unless otherwise specified N = {1,2,...,n}; particularly when $e_A(i,j)$ is defined modulo n.

A <u>transversal</u> T of A is a set of n cells, one from each row and column, which contain distinct entries. If A consists of n cell-disjoint transversals $T_1,\ldots,T_n$ then the LS(n) B in which $e_B(i,j) = r$ if $(i,j) \in T_r$ is said to be <u>orthogonal</u> to A. (Clearly e(A)×e(B) = N×N.) We say that A and B are a pair of orthogonal latin squares of order n; or POLS(n). We state without proof the well known result of R.C. Bose, S.S. Shrikhande and E.T. Parker (1959). (See also Section 4 of this chapter.)

THEOREM 1.1* There exists a pair of orthogonal latin squares of order n (POLS(n)) for all n, n ≠ 2, 6.

If in a latin square A of order n the k^2 cells defined by k rows and k columns form a latin square of order k it is called a latin <u>subsquare</u> of A. Note that the cells of a subsquare need not be contiguous, as is exhibited by the order 8 latin square of Figure 1.1 in which subsquares of orders 2 and 3 are indicated. If B is a

---

*Editors' footnote: A very short proof of this result has recently been given by L. Zhu (1982). However, this paper contains two unfortunate errata as follows. Page 49, last line should read "Note that we have proved that there exist only two distinct kth power residues other than unity when p = 3k+1. In the case when p = 3k-1, we show that the cubes of ...". Page 51, line 4 should be "... then $\alpha^{3x} = \alpha^t$(mod p) or ...".

4:3 Latin squares with and without subsquares

subsquare of A we denote by $e_A(B)$ the entries of B.

$$\begin{vmatrix} 1 & 2 & 7 & 3 & 5 & 4 & 6 & 8 \\ 4 & 1 & ② & 6 & 3 & ⑤ & 8 & 7 \\ 3 & 6 & 1 & 4 & 2 & 8 & 7 & 5 \\ 5 & 8 & 3 & 2 & 1 & 7 & 4 & 6 \\ 2 & 3 & 8 & 1 & 7 & 6 & 5 & 4 \\ 8 & 7 & 6 & 5 & 4 & 3 & 2 & 1 \\ 6 & 4 & ⑤ & 7 & 8 & ② & 1 & 3 \\ 7 & 5 & 4 & 8 & 6 & 1 & 3 & 2 \end{vmatrix}$$

Figure 1.1

We are interested in the existence of latin squares with particular subsquare properties. At this point we would like to remark that the reader will find all the early results on subsquares in [DK] as well as considerable reference on other aspects of latin squares mentioned in this chapter.

We denote by LS(n;k) a latin square of order n with a subsquare of order k and ask for what n and k does an LS(n;k) exist. Clearly an LS(n;k) always exists if k = 1 or n. (When k ≠ 1 or n we call the subsquare <u>proper</u>.) The complete answer is given in Theorem 1.2 and was first proven by T. Evans (1960). But first we need to describe the construction methods referred to as <u>prolongation</u> and <u>direct product</u>.

Prolongation enables us to produce from an LS(n) with k cell-disjoint transversals an LS(n+k;k). Let A be a latin square of order n based on the symbols 1,2,...,n and with disjoint transversals $T_1, T_2, \ldots, T_k$. Add k new rows and columns to A and in the new order n+k array if (i,j) ∈ T_r put $e_B(i,j) = n+r$, $e_B(i,n+r) = e_B(n+r,j) = e_A(i,j)$ and if (i,j) ∉ T_r put $e_B(i,j) = e_A(i,j)$. Finally, in the remaining order k subarray insert an LS(k) based on the set {n+1,n+2,...,n+k}. We say that the transversal T_r has been "projected onto" row and column n+r. (Note that we could have projected T_r onto row n+r and column n+s.) To illustrate, let A be the LS(7) defined by $e_A(i,j) = 2j-i \pmod 7$ with $T_r = \{(i,i+r): 1 \le i \le 7\}$ for $1 \le r \le 3$. The latin square of order 10 constructed by projecting T_r onto row and column 7+r is shown in Figure 1.2.

1	8	9	10	2	4	6	3	5	7
7	2	8	9	10	3	5	4	6	1
6	1	3	8	9	10	4	5	7	2
5	7	2	4	8	9	10	6	1	3
10	6	1	3	5	8	9	7	2	4
9	10	7	2	4	6	8	1	3	5
8	9	10	1	3	5	7	2	4	6
2	3	4	5	6	7	1	8	9	10
3	4	5	6	7	1	2	10	8	9
4	5	6	7	1	2	3	9	10	8

Figure 1.2

The direct product of two latin squares A and B of orders m and n, respectively, is denoted A×B and has order mn. It is defined by $e_{A \times B}((r-1)m+i, (s-1)m+j) = (e_B(r,s)-1)m + e_A(i,j)$. One sees that each entry in B has simply been replaced by a copy of A and so A×B has subsquares of orders m and n. Note that our replacement of entries in B by copies of A need not be uniform and that we could in fact use different latin squares of order m. We call this a <u>non-uniform</u> product. An example of uniform replacement is shown in Figure 1.3. We also remark that if A_1 and A_2 are POLS(m), and B_1 and B_2 are POLS(n), then $A_1 \times B_1$ and $A_2 \times B_2$ are POLS(mn).

$$A = \begin{array}{|ccc|} \hline 1 & 2 & 3 \\ 3 & 1 & 2 \\ 2 & 3 & 1 \\ \hline \end{array} \qquad B = \begin{array}{|cc|} \hline 1 & 2 \\ 2 & 1 \\ \hline \end{array} \qquad A \times B = \begin{array}{|cccccc|} \hline 1 & 2 & 3 & 4 & 5 & 6 \\ 3 & 1 & 2 & 6 & 4 & 5 \\ 2 & 3 & 1 & 5 & 6 & 4 \\ 4 & 5 & 6 & 1 & 2 & 3 \\ 6 & 4 & 5 & 3 & 1 & 2 \\ 5 & 6 & 4 & 2 & 3 & 1 \\ \hline \end{array}$$

Figure 1.3

THEOREM 1.2. A latin square of order n with a proper subsquare of order k, an LS(n;k), exists if and only if $k \leq \lfloor \frac{n}{2} \rfloor$.

Proof. Given an LS(n;k) we can assume the subsquare is in the lower right corner and is based on the elements {1,2,...,k}. In row one each of these elements must lie in the first n-k cells and so $k \leq \lfloor \frac{n}{2} \rfloor$.

For the converse Theorem 1.1 gives us for all m ≠ 2,6 an LS(m) with m cell-disjoint transversals. Prolongation then yields an LS(n;k) for all $k \leq \lfloor \frac{n}{2} \rfloor$ except for LS(4;2), LS(8;2), LS(9;3), LS(10;4), LS(11;5) and LS(12;6). The direct product of an LS(2) with itself and with an LS(6) yields, respectively, an LS(4;2) and an LS(12;6). In Figure 1.4 two squares of order 6 are given. The entries, 1,2,3 and 4 in each define four cell-disjoint transversals in the other. From this we can, using prolongation, construct an LS(8;2), an LS(9;3) and an LS(10;4). An LS(11;5) is given in Figure 1.5. □

Although a latin square is equivalent to a quasigroup a subsquare is not equivalent to a subquasigroup. For example, if the last three rows and columns of Figure 1.1 are labelled 6,7 and 8, then the order 3 subsquare is not a subquasigroup. So when ([DK, p.486]) it was asked if for all sufficiently large n there exist quasigroups with no proper subquasigroups it is clear that this was not the intended question. (To construct such a quasigroup is trivial as pointed out by several authors including J. Dénes (1983).) The problem of constructing latin squares with no proper subsquares is more difficult and still not completely solved. This problem will be considered in Section 2, as will the existence of latin squares with no order 2 subsquares.

```
5 6 3 4 1 2        6  1  2  3  4  5  : 7  8  9  10 11
2 1 6 5 3 4        5  7  1  2  3  4  : 8  9  10 11 6
6 5 1 2 4 3        4  5  8  1  2  3  : 9  10 11 6  7
4 3 5 6 2 1        3  4  5  9  1  2  : 10 11 6  7  8
1 4 2 3 5 6        2  3  4  5  10 1  : 11 6  7  8  9
3 2 4 1 6 5        1  2  3  4  5  11 : 6  7  8  9  10
                   ............................
                   7  8  9  10 11 6  : 1  2  3  4  5
1 2 5 6 3 4        8  9  10 11 6  7  : 5  1  2  3  4
6 5 1 2 4 3        9  10 11 6  7  8  : 4  5  1  2  3
4 3 6 5 1 2        10 11 6  7  8  9  : 3  4  5  1  2
5 6 4 3 2 1        11 6  7  8  9  10 : 2  3  4  5  1
2 4 3 1 5 6
3 1 2 4 6 5                    Figure 1.5

   Figure 1.4
```

L. Fuchs (see for example J. Dénes (1983)) has asked two interesting questions concerning the size and position of subsquares in latin squares. These and similar problems will be discussed in Section 3. Finally, in Section 4 we look at the existence of orthogonal latin squares with common subsquares and some generalizations of this.

However, before we go on we must look at two more constructions for latin squares. These are a generalization of direct product and the use of pairwise balanced designs.

The generalized direct product has now become a very powerful tool in the study of latin squares, and in particular of orthogonal latin squares. The first step in the generalization was made by A. Sade (1960) (and in D.J. Crampin and Hilton (1975a) it is shown that this gives a proof of Theorem 1.1 for all but 27 values of n). This was then extended by R.M. Wilson (1974) and it is his extension which led several authors to modify and generalize the construction even further. A very general form of the direct product was finally given by D.R. Stinson (1981) (encompassing an earlier generalization of A.E. Brouwer and G.H.J. van Rees (1982)). We will describe only a part of this construction. Variations will be explained as the need arises.

We denote by t-POLS(n) a set of t pairwise orthogonal latin squares of order n. When t = 2 these are POLS(n), and when t = 1 are LS(n). Suppose we have a set of t latin squares of order n, each based on $N = S_1 \cup S_2 \cup \ldots \cup S_m$, where $S_i \cap S_j = \phi$ and $s_i = |S_i|$, and each with a subsquare of order s_i based on S_i in cells $S_i \times S_i$, $1 \le i \le m$, so that for any two squares A and B of the set if these m subsquares are deleted (leaving "holes"), then

$$\{(e_A(i,j), e_B(i,j)\colon (i,j) \in M\} = M, \text{ where } M = N \times N \setminus \left(\bigcup_{i=1}^{m} S_i \times S_i\right).$$

We call the resulting squares a set of t <u>incomplete</u> pairwise orthogonal latin squares denoted $t\text{-IPOLS}(n;,s_1,s_2,\ldots,s_m)$. If for each i, $1 \le i \le m$, there exist $t\text{-POLS}(s_i)$ we can "fill in" the holes and the result is a set of t-POLS(n) which we denote by $t\text{-POLS}(n;s_1,s_2,\ldots,s_m)$ to indicate the common subsquares. If t = 2 we refer to these respectively as $\text{IPOLS}(n;s_1,s_2,\ldots,s_m)$ and $\text{POLS}(n;s_1,s_2,\ldots,s_m)$; and if t = 1 they are $\text{ILS}(n;s_1,s_2,\ldots,s_m)$ and $\text{LS}(n;s_1,s_2,\ldots,s_m)$. Finally, if each of the LS(n) in a set of t-POLS(n) has the transversal $T = \{(i,i)\colon 1 \le i \le n\}$, we denote the squares by $t\text{-POLS}^*(n)$, $\text{POLS}^*(n)$ and $\text{LS}^*(n)$. If a $\text{POLS}^*(n)$ consists of a latin square of order n and its transpose it is called a self-orthogonal latin square of order n; an SOLS(n). The existence of SOLS(n) was resolved by R.K. Brayton, D. Coppersmith and A.J. Hoffman (1976).

<u>THEOREM 1.3</u>. For all n ≠ 2,3 or 6 there is a self-orthogonal latin square of order n, an SOLS(n).

In the following results we need to note that an IPOLS(m;0) is simply a POLS(m) and an IPOLS(m+a;a,0) is an IPOLS(m+a;a).

THEOREM 1.4. Given 3-POLS(p) and an IPOLS($q+a_i;a_i$), $i=1,2,\ldots,p$, then there is an IPOLS(pq+a;a), where $a = \sum_{i=1}^{p} a_i$.

Proof. We will give the construction but omit the verification. Let the 3-POLS(p) be A,B and C. The entries in C define common transversals $T_1, T_2, \ldots, T_p$ in A and B. To each of A and B add $a = \sum_{i=1}^{p} a_i$ new rows and columns. In each IPOLS($q+a_i;a_i$) let the entries be $1, 2, \ldots, q+a_i$ and let $q+1, q+2, \ldots, q+a_i$ be the entries of the deleted subsquares which are in the last a_i rows and columns. Replace the cell (u,v) of T_i in A and in B by the first q rows and columns of IPOLS($q+a_i;a_i$) in which the entry j, $1 \leq j \leq q$, is replaced by $(e_A(u,v)-1)q + j$ (respectively $(e_B(u,v)-1)q + j$) and the entry j, $q+1 \leq j \leq q+a_i$ is replaced by the entry $(p-1)q + \sum_{k=1}^{i-1} a_k + j$. Rows and columns $q+1, q+2, \ldots, q+a_i$ of IPOLS($q+a_i;a_i$) become rows and columns $pq + \sum_{k=1}^{i-1} a_k + 1$, $pq + \sum_{k=1}^{i-1} a_k + 2, \ldots, pq + \sum_{k=1}^{i} a_k$; with the same replacement of entries. □

An illustrative example is given in Figure 1.6 in which $p = q = 4$, $a_1 = a_2 = 2$, $a_3 = a_4 = 0$ (so $a = 4$). There exist three POLS(4) as shown. We also use two pairs of IPOLS (4+2;2) (which are given in Figure 1.4 on deletion of the entries in cells (5,5),(5,6),(6,5) and (6.6)) and two pairs of IPOLS(4+0;0) = POLS(4). Hence we can construct the IPOLS (20;4) as shown.

COROLLARY 1.5. Given 3-POLS(p), IPOLS ($q+a_i;a_i$), $i=1,2,\ldots,p$, and a POLS(a) where $a = \sum_{i=1}^{p} a_i$, there is an IPOLS(pq+a;p) if $q > 2a_i$ for all i, and an IPOLS(pq+a;q) if $a_i = 0$ for some i.

COROLLARY 1.6. Given 3-POLS(p), IPOLS($q+a_i;a_i$), $i=1,2,\ldots,p$ and an IPOLS($a;a_i$) for some i, where $a = \sum_{i=1}^{p} a_i$, then we can construct an IPOLS($pq+a;q+a_i$) and an IPOLS($pq+a;a_i$).

COROLLARY 1.7. Given 3-POLS(p), IPOLS($q+a_i;c,a_i$), $i=1,2,\ldots,p$, and a POLS(a), where $a = \sum_{i=1}^{p} a_i$, then there is an IPOLS($pq+a;pc$).

(We remark that the construction of A. Sade (1960) is Theorem 1.4 with only $a_1 \neq 0$.)

THEOREM 1.8. Given 4-POLS(p) and IPOLS($q+a_i+b_j;a_i,b_j$), $i=1,2,\ldots,p$ and $j=1,2,\ldots,p$, we can construct an IPOLS($pq+a+b;a,b$), where $a = \sum_{i=1}^{p} a_i$ and $b = \sum_{i=1}^{p} b_i$, and an IPOLS($pq+a+b;a$) if $b \notin \{2,3,6\}$.

Proof. Again only the construction will be given. Let the 4-POLS(p) be A,B,C and D. The entries in C and D define common transversals in A and B denoted $T_1,T_2,\ldots,T_p$ and $R_1,R_2,\ldots,R_p$. Note that T_i and R_j have exactly one cell in common. Add a+b new rows and columns to A and B and replace cell (u,v) by a copy of IPOLS($q+a_i+b_j;a_i,b_j$) if cell $(u,v) \in T_i \cap R_j$. This replacement is made as before. (Assume each IPOLS($q+a_i+b_j;a_i,b_j$) has entries $1,2,\ldots,q,q+1,\ldots,q+a_i,q+a_i+1,\ldots,q+a_i+b_j$ so that $q+1,q+2,\ldots,q+a_i$ were the entries of the deleted order a_i subsquares and $q+a_i+1,q+a_i+2,\ldots,q+a_i+b_j$ were the entries of the deleted order b_j subsquares where LS(a_i) was in rows and columns $q+1,q+2,\ldots,q+a_i$, and LS(b_j) was in the last b_j rows and columns.) Then the first q rows and columns of the IPOLS($q+a_i+b_j;a_i,b_j$) go into cell (u,v) of A

$$\begin{array}{|cccc|} 1 & 4 & 2 & 3 \\ 3 & 2 & 4 & 1 \\ 4 & 1 & 3 & 2 \\ 2 & 3 & 1 & 4 \end{array} \quad \begin{array}{|cccc|} 1 & 3 & 4 & 2 \\ 4 & 2 & 1 & 3 \\ 2 & 4 & 3 & 1 \\ 3 & 1 & 2 & 4 \end{array} \quad \begin{array}{|cccc|} 1 & 2 & 4 & 3 \\ 2 & 1 & 3 & 4 \\ 4 & 3 & 1 & 2 \\ 3 & 4 & 2 & 1 \end{array}$$

Figure 1.6

```
17 18  3  4·19 20 15 16· 5  8  6  7· 9 12 10 11· 1  2·13 14
 2  1 18 17·14 13 20 19· 7  6  8  5·11 10 12  9· 3  4·15 16
18 17  1  2·20 19 13 14· 8  5  7  6·12  9 11 10· 4  3·16 15
 4  3 17 18·16 15 19 20· 6  7  5  8·10 11  9 12· 2  1·14 13
..........................................................
19 20 11 12·17 18  7  8·13 16 14 15· 1  4  2  3· 5  6· 9 10
10  9 20 19· 6  5 18 17·15 14 16 13· 3  2  4  1· 7  8·11 12
20 19  9 10·18 17  5  6·16 13 15 14· 4  1  3  2· 8  7·12 11
12 11 19 20· 8  7 17 18·14 15 13 16· 2  3  1  4· 6  5·10  9
..........................................................
13 16 14 15· 1  4  2  3·17 18 11 12·19 20  7  8· 9 10· 5  6
15 14 16 13· 3  2  4  1·10  9 18 17· 6  5 20 19·11 12· 7  8
16 13 15 14· 4  1  3  2·18 17  9 10·20 19  5  6·12 11· 8  7
14 15 13 16· 2  3  1  4·12 11 17 18· 8  7 19 20·10  9· 6  5
..........................................................
 5  8  6  7· 9 12 10 11·19 20  3  4·17 18 15 16·14 13· 1  2
 7  6  8  5·11 10 12  9· 2  1 20 19·14 13 18 17·15 16· 3  4
 8  5  7  6·12  9 11 10·20 19  1  2·18 17 13 14·16 15· 4  3
 6  7  5  8·10 11  9 12· 4  3 19 20·16 15 17 18·14 13· 2  1
..........................................................
 1  4  2  3· 5  8  6  7· 9 12 10 11·13 16 14 15·17 20 18 19
 3  2  4  1· 7  6  8  5·11 10 12  9·15 14 16 13·19 18 20 17
..........................................................
 9 12 10 11·13 16 14 15· 1  4  2  3· 5  8  6  7·20 17 19 18
11 10 12  9·15 14 16 13· 3  2  4  1· 7  6  8  5·18 19 17 20
```

```
 1  2 17 18· 9 10 19 20·13 15 16 14· 5  7  8  6· 3  4·11 12
18 17  1  2·20 19  9 10·16 14 13 15· 8  6  5  7· 4  3·12 11
 4  3 18 17·12 11 20 19·14 16 15 13· 6  8  7  5· 1  2· 9 10
17 18  4  3·19 20 12 11·15 13 14 16· 7  5  6  8· 2  1·10  9
..........................................................
13 14 19 20· 5  6 17 18· 1  3  4  2· 9 11 12 10· 7  8·15 16
20 19 13 14·18 17  5  6· 4  2  1  3·12 10  9 11· 8  7·16 15
16 15 20 19· 8  7 18 17· 2  4  3  1·10 12 11  9· 5  6·13 14
19 20 16 15·17 18  8  7· 3  1  2  4·11  9 10 12· 6  5·14 13
..........................................................
 5  7  8  6·13 15 16 14· 9 10 17 18· 1  2 19 20·11 12· 3  4
 8  6  5  7·16 14 13 15·18 17  9 10·20 19  1  2·12 11· 4  3
 6  8  7  5·14 16 15 13·12 11 18 17· 4  3 20 19· 9 10· 1  2
 7  5  6  8·15 13 14 16·17 18 12 11·19 20  4  3·10  9· 2  1
..........................................................
 9 11 12 10· 1  3  4  2· 5  6 19 20·13 14 17 18·15 16· 7  8
12 10  9 11· 4  2  1  3·20 19  5  6·18 17 13 14·16 15· 8  7
10 12 11  9· 2  4  3  1· 8  7 20 19·16 15 18 17·13 14· 5  6
11  9 10 12· 3  1  2  4·19 20  8  7·17 18 16 15·14 13· 6  5
..........................................................
 2  4  3  1· 6  8  7  5·10 12 11  9·14 16 15 13·17 19 20 18
 3  1  2  4· 7  5  6  8·11  9 10 12·15 13 14 16·20 18 17 19
..........................................................
14 16 15 13·10 12 11  9· 6  8  7  5· 2  4  3  1·18 20 19 17
15 13 14 16·11  9 10 12· 7  5  6  8· 3  1  2  4·19 17 18 20
```

Figure 1.6(cont'd)

(respectively B); entry t, $1 \le t \le q$, is replaced by $(e_A(u,v)-1)q + t$ (respectively $(e_B(u,v)-1)q + t)$; entry t, $q+1 \le t \le q+a_i$, is replaced by $pq + \sum_{k=1}^{i-1} a_k + (t-q)$; and entry t, $q+a_i+1 \le t \le q+a_i+b_j$ is replaced by $pq+a+\sum_{k=1}^{j-1} b_k + t - (q+a_i)$. Rows and columns $q+1, q+2, \ldots, q+a_i$ of the IPOLS$(q+a_i+b_j; a_i, b_j)$ become rows and columns $pq+\sum_{k=1}^{i-1} a_k+1, \ldots, pq+\sum_{k=1}^{i} a_k$, and rows and columns $q+a_i+1, q+a_i+2, \ldots, q+a_i+b_j$ become rows and columns $pq+a+\sum_{k=1}^{j-1} b_k+1, \ldots, pq+a+\sum_{k=1}^{j} b_k$ with the same entry replacement. □

Again, a large number of corollaries are possible. We give only two of them.

COROLLARY 1.9. Given 4-POLS(p), IPOLS$(q+a_i+b_j; a_i, b_j)$, $i=1,2,\ldots,p$ and $j=1,2,\ldots,p$, an IPOLS$(a;a_k)$ and an IPOLS$(b;b_m)$ for some k and m where $a = \sum_{i=1}^{p} a_i$ and $b = \sum_{j=1}^{p} b_j$, then there exists an IPOLS$(pq+a+b; q+a_k+b_m)$.

COROLLARY 1.10. Given 4-POLS(p), IPOLS$(q+a_i+b_j; c, a_i, b_j)$, $i=1,2,\ldots,p$ and $j=1,2,\ldots,p$, a POLS(a) and a POLS(b) where $a = \sum_{i=1}^{p} a_i$ and $b = \sum_{j=1}^{p} b_j$, there is an IPOLS$(pq+a+b; pc)$.

To apply these constructions we need to know of the existence of 3-POLS(n) and 4-POLS(n). The first of the next theorems is due to S.M.P. Wang (1978) and S.M.P. Wang and R.M. Wilson (1978) but a more accessible proof can be found in W.D. Wallis (1984). In addition 3-POLS(14) were recently found by D.T. Todorov (1985). The second theorem is a result of A.E. Brouwer (1979).

THEOREM 1.11. For all $n \notin \{2,3,6,10\}$, there exist 3-POLS(n).

THEOREM 1.12. For all n ∉ {2,3,4,6,10,14,18,20,22,24,26,28,30,33,34, 38,42,44,52}, there exist 4-POLS(n).

In the application of Theorems 1.4 and 1.8 when the value of any a_i or b_j is not given it is assumed to be zero.

The next construction is closely related to these and was first described to the author by L. Zhu (personal communication).

THEOREM 1.13. Given an IPOLS$(n;s_1,s_2,\ldots,s_m)$ where $\sum_{i=1}^{m} s_i = n$ and an IPOLS$(s_i+a;a)$ for each s_i, $i=1,2,\ldots,m$, then there exists an IPOLS$(n+a;a)$ and an IPOLS$(n+a;s_i+a)$.

Proof. The proof is a simple generalization of the earlier ideas. "Project" from the m subsquares on the main diagonal. □

Various constructions related to this theorem have also been given by J.H. Dinitz and D.R. Stinson (1983).

The last construction we wish to describe is the use of pairwise balanced block designs and since we will rarely employ this method only a brief description will be given.

A pairwise balanced block design of index 1, PBD(v;K) is a collection of blocks, consisting of elements from a v-set, whose sizes are in K and with the additional property that every distinct pair of elements lies in exactly one block. For example the blocks of a PBD(12;{3,4}) are shown in Figure 1.7. (It is also a PBD(12;{3,4,7}).)

{1, 2, 3}   {1, 4, 7, 10}   {3, 5, 8, 10}   {2, 6, 9, 10}
{4, 5, 6}   {2, 5, 7, 11}   {1, 6, 8, 11}   {3, 4, 9, 11}
{7, 8, 9}   {3, 6, 7, 12}   {2, 4, 8, 12}   {1, 5, 9, 12}
{10, 11, 12}

Figure 1.7

THEOREM 1.14. Given a PBD(v;K) and a POLS*(k) whenever there is a block of size k, then there exists an IPOLS(v;k) for each k which occurs as a block size in the design.

Proof. Construct the squares as follows. If $\{b_1, b_2, \ldots, b_k\}$ is a block of the design then in rows and columns $b_1, b_2, \ldots, b_k$ insert a copy of POLS*(k) based on these entries in which $e(b_i, b_i) = b_i$. The result is immediate. □

Modifications can be made to this construction (see for example W.D. Wallis and L. Zhu (1983)) and, of course, it can be generalized to t-POLS(n).

(2) <u>Without subsquares</u>

One should notice that the latin square constructions described in Section 1 all produce subsquares. We now wish to discuss the problem of constructing latin squares without certain subsquares. Until now only two problems of this type have been considered.

A. Kotzig, C.C. Lindner and A. Rosa (1975) asked if for all n, $n \neq 2, 4$, there exists an LS(n) with no order 2 subsquares. Such squares were termed N_2-latin squares and interest in them arose as they could be used to construct sets of disjoint Steiner triple systems. In fact, to construct a set of (v+1)/2 pairwise disjoint Steiner triple systems of order v, $v \equiv 3$ or 7 (mod 12) and $v > 7$, they needed an LS((v+1)/2) with at least one column no cell of which was contained in a subsquare of order 2. Clearly N_2-latin squares have this property.

A second problem, posed by A.J.W. Hilton, asks if for all sufficiently large n there exists an LS(n) with no proper subsquares. Following the above notation we will call this an N_∞-latin square.

We will exhibit a complete solution to the first problem and then describe constructions for N_∞-latin squares giving a partial solution to the second. In both problems we use mainly a product construction.

M. McLeish (1975) has remarked that if A and B are N_2-latin squares then so too is their product. As pointed out by R.H.F. Denniston (1978) this now provides a straightforward recursive construction for N_2-latin squares of all orders n, $n \neq 2, 4$, once N_2-latin squares of order 8,16,32,2k+1,2(2k+1), and 4(2k+1), where

$k \geq 1$, have been constructed. Historically, however, progress proceeded along the following lines.

A. Kotzig, C.C. Lindner and A. Rosa (1975) began by constructing N_2-latin squares for all $n \neq 2^a$ and then M. McLeish (1975) gave constructions based on direct and generalized direct product which included the cases $n = 2^a$ for $a \geq 6$ and $a \neq 7,8$ or 13. Using prolongation (later exploited more fully in M. McLeish (1980)) she was able to produce N_2-latin squares of orders 2^7 (and hence by direct product 2^{13}) and 2^8. One easily sees that every LS(2) and LS(4) has an order 2 subsquare and so only the cases $n = 8,16$ and 32 remained. An N_2-latin square of order 8 proved to be the most difficult to find. R.H.F. Denniston (1978) has shown (by computer search) that there are exactly three non-isomorphic N_2-latin squares of order 8. However, the first N_2-latin square of order 8 (as shown in Figure 2.1) was constructed by E. Regener (and appears in A. Kotzig and J. Turgeon (1976)). In that same paper A. Kotzig and J. Turgeon gave a general construction for N_2-latin squares of order n provided $n \not\equiv 0 \pmod{3}$ and $n \not\equiv 3 \pmod{5}$, by a prolongation involving the projection of three transversals from the Cayley table of the cyclic group of order n - 3. This included the cases $n = 16$ and 32, and so the problem was solved. Since then M. McLeish (1980) has given a direct construction for N_2-latin squares of all orders $n \geq 12$, $n \neq 14$, 20 or 30. The construction consists of a prolongation involving the projection of s transversals from the Cayley table of the cyclic group of order n-s for suitably chosen n and s.

And now the proof.

LEMMA 2.1. There exist N_2-latin squares of orders 8,16 and 32.

Proof. An N_2-latin square of order 8 is shown in Figure 2.1

We will describe the Kotzig-Turgeon construction for N_2-latin squares of order n+3 when n is odd, $n \not\equiv 0 \pmod{3}$ and $n \not\equiv 0 \pmod{5}$. Let A be an LS(n) in which $e_A(i,j) = i+j \pmod{n}$. In A there are transversals $T_i = \{(2j-3i+6,j): 1 \leq j \leq n\}$, $1 \leq i \leq 3$. Apply prolongation using these by projecting T_i onto row and column n+i and inserting the LS(3), B, with $e_B(i,j) = 2-(i+j) \pmod{3}$ and e(B) =

{n+1,n+2,n+3}. Putting n = 13 and n = 29 yields N_2-latin squares of orders 16 and 32. □

1	2	3	4	5	6	7	8
2	3	1	5	6	7	8	4
3	1	4	6	7	8	2	5
4	6	8	2	1	3	5	7
5	8	2	7	3	4	6	1
6	5	7	1	8	2	4	3
7	4	5	8	2	1	3	6
8	7	6	3	4	5	1	2

Figure 2.1

LEMMA 2.2. For every odd n there is an N_2-latin square.

Proof. The Cayley table of the cyclic group of odd order is an N_2-latin square. In fact the Cayley table of the cyclic group of prime order has no proper subsquare. □

LEMMA 2.3. There are N_2-latin squares of orders $2(2k+1)$ and $4(2k+1)$, $k \geq 1$.

Proof. The construction (due to A. Kotzig, C.C. Lindner and A. Rosa (1975)) is in each case simply a non-uniform product.

Let A, B and C be LS(2k+1) defined by $e_A(i,j) \equiv i-j+1 \pmod{2k+1}$ $e_B(i,j) \equiv i+j-1 \pmod{2k+1}$ and $e_C(i,j) \equiv i+j-2 \pmod{2k+1}$. These three squares are isotopic to the Cayley table of the cyclic group of order 2k+1 and so are N_2-latin squares. Define a latin square X as the non-uniform product of A, B and C with the latin square E of order 2 in which $e_E(i,j) \equiv i+j-1 \pmod 2$ so that cells (1,1) and (2,2) are replaced by A, (1,2) is replaced by B and (2,1) by C. Then X is an N_2-latin square of order $2(2k+1)$.

We similarly construct Y, an N_2-latin square of order $4(2k+1)$, by the non-uniform product of A,B and C with F, the LS(4) shown with Y in Figure 2.2. In Y, D_i denotes a copy of D on the appropriate element set. Notice that any order 2 subsquare in Y must lie in one of the four order 2(2k+1) subsquares and so Y has no order 2 subsquares. □

$$F = \begin{vmatrix} 1 & 2 & 3 & 4 \\ 2 & 1 & 4 & 3 \\ 4 & 3 & 1 & 2 \\ 3 & 4 & 2 & 1 \end{vmatrix} \qquad Y = \begin{vmatrix} A_1 & B_2 & A_3 & B_4 \\ C_2 & A_1 & C_4 & A_3 \\ B_4 & A_3 & A_1 & B_2 \\ A_3 & C_4 & C_2 & A_1 \end{vmatrix}$$

Figure 2.2

THEOREM 2.4. There exists an N_2-latin square of order n unless n = 2 or 4 and in these cases such a square does not exist.

Proof. The proof follows immediately from Lemmas 2.1, 2.2, 2.3, the remark that the direct product of two N_2-latin squares produces another, and the fact that N_2-latin squares of orders 2 and 4 are impossible. □

We now turn to the existence of latin squares with no proper subsquares, N_∞-latin squares. To date the best general result is that for all $n \neq 2^a 3^b$ there exists an N_∞-latin square.

One easily discovers (for example from the list of LS(4) and LS(6) given in [DK, p.129-137]) that every latin square of order 4 or 6 has a proper subsquare, and that all three of the N_2-latin squares of order 8 are in fact N_∞-latin squares.

A. Rosa (and others) have noted that if there exists a perfect 1-factorization of the complete graph on 2n vertices, K_{2n}, then there exists an N_∞-latin square of order 2n-1. (A perfect 1-factorization of a graph G is a 1-factorization of G with the property that the union of any two of the 1-factors is a hamiltonian cycle.) Given a perfect 1-factorization of K_{2n}, with vertices $1,2,\ldots,2n$, consisting of the 1-factors $F_1, F_2, \ldots, F_{2n-1}$, where edge $\{i,2n\}$ is in F_i, define the N_∞-latin square A by $e_A(i,j) = k$ if $i \neq j$ and $\{i,j\} \in F_k$, and $e_A(i,i) = i$. Observe that if A has a proper subsquare B containing no cell of the main diagonal of A, then any two elements in B can yield a cycle of length at most twice the order of B, contradicting the fact that they correspond to a hamilton cycle. Since A is symmetric, if B has an entry on the main diagonal, then B is symmetric about the main diagonal and has odd order. It is easy to see that any two elements yield a cycle of length at most one more than the order of B, again a contradiction.

Perfect 1-factorizations of K_{2n} are known to exist when $2n \in \{p+1, 2p: p \text{ prime}\} \cup \{16, 28, 36, 50, 244, 344, 1332, 6860\}$. These are described in A. Kotzig (1964), B.A. Anderson (1973), (1974), E. Seah and D.R. Stinson (1987), E.C. Ihrig, E. Seah and D.R. Stinson (1987), B.A. Anderson and D. Morse (1974), and a recent personal communication from Z. Kiyasu and M. Kobayashi respectively. Note that while perfect 1-factorizations yield N_∞-latin squares the converse is not true. The smallest unresolved case for perfect 1-factorizations is K_{40}, whereas the smallest order for N_∞-latin squares is 16.

The first results on the existence of N_∞-latin squares were obtained by K. Heinrich (1980) who showed that for distinct primes p and q, there is an N_∞-latin square of order n = pq, n ≠ 6. L.D. Andersen and E. Mendelsohn (1982) generalized this construction to obtain N_∞-latin squares of all orders $n \neq 2^a 3^b$.

We shall give the construction of L.D. Andersen and E. Mendelsohn (1982); the case n = pq being contained in it. Because of its length and detail the proof will not be given. We begin by defining three latin squares of order t each isotopic to the Cayley table of the cyclic group of order t. Recall that when t is a prime these squares have no proper subsquares.

Let A(t), B(t) and C(t) be latin squares defined by $e_{A(t)}(i,j) = i+j-1 \pmod{t}$, $e_{B(t)}(i,j) = 4-(i+j) \pmod{t}$ and

$$e_{C(t)}(i,j) = \begin{cases} t & \text{if } i+j \equiv 1 \pmod{t} \\ 2 & \text{if } i+j \equiv 3 \pmod{t} \\ 3-(i+j) \pmod{t} & \text{otherwise.} \end{cases}$$

Notice that in each square cells (1,1) and (1,2) contain the entries 1 and 2. Using these we construct N_∞-latin squares of orders n = pm where p is prime and p ≥ 5.

We perform a non-uniform product of squares of order p and A(m) to produce a latin square P(p,m) of order pm. We then exchange some of the entries of P(p,m) so that the result is an N_∞-latin square of order pm. Let q be the smallest prime divisor of m. Replace cell (r,s) in A(m) by a copy of C(p) if $1 \leq r \leq \frac{m}{q}$ and $m - \frac{m}{q} + 1 \leq s \leq m$; by a copy of B(p) if $\left\lfloor \frac{m+1}{2} \right\rfloor + 1 \leq r \leq \left\lfloor \frac{m+1}{2} \right\rfloor + \frac{m}{q}$ and $m - \frac{m}{q} + 1 \leq m$; and by a copy of A(p) in all other cases. (See Figure 2.3.)

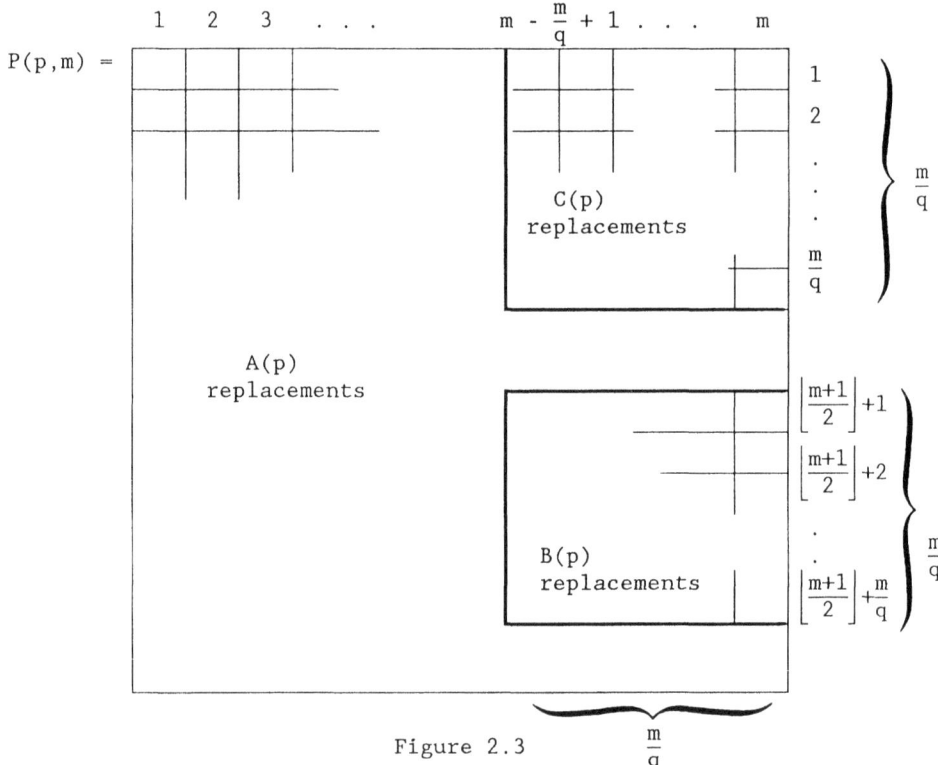

Figure 2.3

Because Cayley tables were used all proper subsquares of P(p,m) can be located. Having done this we look at the m × 2m subarray S(p,m) consisting of the first two cells in the first row of each A(p), B(p) and C(p). Clearly S(p,m) is a latin rectangle in which each row contains the entries 1,2,p+1,p+2,...,(m-1)p+1,(m-2)p+2. From P(p,m) we now construct the N_∞-latin square of order pm, D(p,m), by permuting the rows of S(p,m) according to the permutation ρ = (1 2 3 ... m). Clearly D(p,m) is a latin square and it is not difficult to show that all subsquares in P(p,m) have been destroyed. Unfortunately, considerable work is involved in showing that this permutation of S(p,m) does not introduce subsquares.

It still remains to construct N_∞-latin squares of order $n = 2^a 3^b$ where $n \geq 16$. An N_∞-latin square of order 12 was recently found by P. Gibbons and E. Mendelsohn (personal communication) after a computer search. The square is shown in Figure 2.4.

1	2	3	4	5	6	7	8	9	10	11	12
2	3	4	5	6	1	8	9	10	11	12	7
3	1	5	2	7	8	4	10	6	12	9	11
4	5	6	7	1	9	11	12	8	3	2	10
5	6	2	8	10	7	9	11	12	4	1	3
6	12	8	1	3	10	2	7	11	9	4	5
7	8	1	10	12	11	5	4	2	6	3	9
8	9	11	3	4	12	10	6	5	1	7	2
9	11	7	12	2	5	1	3	4	8	10	6
10	7	12	11	9	4	6	1	3	2	5	8
11	4	10	9	8	3	12	2	7	5	6	1
12	10	9	6	11	2	3	5	1	7	8	4

Figure 2.4

(3) <u>With subsquares</u>

We now turn our attention to latin squares having particular subsquare properties.

F.P. Hiner and R.B. Killgrove (unpublished manuscript) defined a <u>subsquare complete</u>, SC, latin square A to be one in which whenever $e_A(i,j) = e_A(k,\ell)$, $(i,j) \neq (k,\ell)$, there is a proper subsquare of A containing the cells (i,j) and (k,ℓ). In an α-β- ... -γ-SC latin square all proper subsquares have orders $\alpha, \beta, \ldots$, or γ. The study of such latin squares arose from a problem concerning projective planes as outlined below.

Associated with a projective plane of order n is a set of n-1 latin squares of order n, $D_1, D_2, \ldots, D_{n-1}$, with the property that for any two columns r and s, the n(n-1) ordered pairs $(e_{D_j}(i,r), e_{D_j}(i,s))$ $1 \leq i \leq n$, $1 \leq j \leq n-1$, are all distinct. These n-1 squares are called digraph-complete latin squares and the details of their construction are given in [DK, p.286]. (We note, however, that they arise as conjugates by row and symbol exchange from the complete set of n-1 pairwise orthogonal latin squares of order n.)

A projective plane is said to be singly generated if it has a quadrangle whose extension is the whole plane (see R.B. Killgrove (1964) for a more explicit definition) and non-singly generated if there is no such quadrangle. R.B. Killgrove (1964) has conjectured that every finite non-Desarguesian plane is singly generated, and in support of this has verified the result for planes of order less than 12. Making use of Bruck's theorem ([P. Dembowski (1968), p.145]) he showed that for planes of order at most 25 the problem was reduced to the study of 2-3-SC latin squares because, in these cases, if the plane is singly generated each of the squares in the resulting digraph complete set must be a 2-3-SC latin square. R.B. Killgrove and E. Milne (1974) showed that (up to isotopism) 2-3-SC latin squares of orders 4,6,7,8,9 and 10 were unique, and that there are exactly three of order 12. In Figure 3.1 a 2-3-SC latin squares of order 12 is given; it has a 2-3-SC latin subsquare of order 6. Since there is no 2-3-SC latin square of order 5 the smallest unresolved case is that of order 11.

We will now look at the more general results of F.P. Hiner and R.B. Killgrove.

THEOREM 3.1. A 2-SC latin square is the Cayley table of the elementary abelian group of order 2^a.

A proof was given by Hiner and Killgrove and that proof is

reproduced in K. Heinrich and W.D. Wallis (1981).

THEOREM 3.2. There are 2-3-SC latin squares of orders 2^a-1 and 3^b+1 where $a \geq 2$ and $b \geq 1$.

Proof. Let C be the Cayley table of the elementary abelian group of order $n = 2^a$. Delete the last row and column from C and replace $e_C(i,i)$ by $e_C(i,n)$ for $1 \leq i \leq n-1$. This results in a 2-3-SC latin square of order 2^a-1. (Note that this process is simply "anti-prolongation". It has been called "contraction" in [DK] following earlier authors.)

1	2	3	4	5	6	7	8	9	10	11	12
2	3	1	5	6	4	8	9	7	11	12	10
3	1	2	6	4	5	9	7	8	12	10	11
4	6	5	1	3	2	12	11	10	9	8	7
5	4	6	2	1	3	11	10	12	8	7	9
6	5	4	3	2	1	10	12	11	7	9	8
7	9	8	12	11	10	1	3	2	6	5	4
8	7	9	11	10	12	2	1	3	5	4	6
9	8	7	10	12	11	3	2	1	4	6	5
10	12	11	9	8	7	6	5	4	1	3	2
11	10	12	8	7	9	5	4	6	2	1	3
12	11	10	7	9	8	4	6	5	3	2	1

Figure 3.1

To construct a 2-3-SC latin square of order 3^b+1 we apply prolongation to D, the Cayley table of the elementary abelian group of order $n = 3^b$, projecting the transversal $T = \{(i,i): 1 \leq i \leq n\}$ onto row and column n+1. (Note that D is in fact a 3-SC latin square.) □

Using a non-uniform product Hiner and Killgrove were also able to construct 2-3-SC latin squares of orders 12 (as shown in Figure 3.1) and 18.

Apart from these results there are essentially no others on the existence of α-β- ... - γ-SC latin squares, except those arising immediately from the Cayley table of a finite gorup in which $\alpha,\beta,\ldots,\gamma$ are the orders of the group elements.

H.W. Norton (1939) referred to a latin subsquare of order 2 as an intercalate, and we will use this term. K. Heinrich and W.D. Wallis (1981) have studied the problem of determining the maximum number, $I(n)$, of intercalates possible in an LS(n).

THEOREM 3.3. If n is even $I(n) \leq n^2(n-1)/4$, and if n is odd $I(n) \leq n(n-1)(n-3)/4$.

Proof. Choose two rows of an LS(n). If n is even these two rows contain at most $n/2$ intercalates, and if n is odd at most $(n-3)/2$. □

THEOREM 3.4. For even n, $I(n) = n^2(n-1)/4$ if and only if $n = 2^a$.

Proof. The proof follows from Theorem 3.1 as to attain the bound the square must be 2-SC. □

THEOREM 3.5. For odd n, $I(n) = n(n-1)(n-3)/4$ if and only if $n = 2^a-1$.

This bound is attained only if in the latin square every pair of rows accounts for $(n-3)/2$ intercalates. Hiner and Killgrove in their unpublished manuscript showed that such a latin square is possible only when $n = 2^a-1$ and the square is constructed as in Theorem 3.2. (The proof is also given by K. Heinrich and W.D. Wallis (1981).)

A. Kotzig and J. Zaks (1983) have shown that when $n \equiv 1$ or 2 (mod 4) there are better upper bounds on $I(n)$. Their method makes use of the connection between latin squares and 1-factors of graphs and relies on earlier work of A. Kotzig (1958) and (1970). In particular, in the first of those papers Kotzig proved the following result.

THEOREM 3.6. Let G be a finite regular graph of degree $2m+1$ with $2n$ vertices and with a 1-factorization consisting of the 1-factors $F_1, F_2, \ldots, F_{2m+1}$. Let x_{ij} be the number of cycles in $F_i \cup F_j$. Then

$$\sum_{1\le i<j\le 2m+1} x_{ij} \equiv mn \pmod 2.$$

From this Kotzig derived a result concerning latin squares. Let R be a t×n, t odd, latin rectangle on the symbols $1,2,\ldots,n$. Construct a regular graph of degree t on 2n vertices $a_1, a_2, \ldots, a_n$, $b_1, b_2, \ldots, b_n$ as follows. With the columns of R labelled $1,2,\ldots,n$ row i defines a permutation ρ_i so that $\rho_i(j) = e_R(i,j)$, $1 \le j \le n$, and this defines the n edges $\{a_j, b_{e_R(i,j)}\}$, $1 \le j \le n$, which form the 1-factor F_i, $1 \le i \le t$. Clearly a cycle in $F_i \cup F_j$ corresponds to two rows of a possible latin subsquare in R; and in particular, a cycle of length 4 corresponds to an intercalate. We also note that x_{ij} is the number of cycles in the permutation σ_{ij} where $\sigma_{ij}(e_R(i,k)) = e_R(j,k)$ and that a 2-cycle in this permutation corresponds to an intercalate in the latin rectangle.

COROLLARY 3.7. In any latin square of order n, for distinct i,j and k, $x_{ij} + x_{jk} + x_{ik} \equiv n \pmod 2$.

This corollary is an immediate consequence of the previous remarks and Theorem 3.6 and it enabled A. Kotzig and J. Zaks (1983) to improve the known bounds on $I(n)$. Their result is Theorem 3.8.

THEOREM 3.8. If $n \equiv 1 \pmod 4$, then $I(n) \le (n-1)(n^2-3)/4$ and if $n \equiv 2 \pmod 4$, $I(n) \le n^2(n^2-4n+6)/4$.

Proof. Let $n = 4k+1$ and consider three rows, i,j and k, of an LS(n), A. A permutation defined by two of these rows has at most 2k cycles (or 2k-1 intercalates). But, as $x_{ij} + x_{jk} + x_{ki} \equiv 1 \pmod 2$, then amongst the three rows there are at most $2(2k-1)+(2k-2) = (2k-1)+(4k-3)$ intercalates. Now choose two rows of A, yielding 2k-1 intercalates, and construct triples of rows by adding to these from the remaining 4k-1 rows. This gives at most $(2k-1)+(4k-3)(4k-1)$ intercalates. Delete these two rows and repeat the argument. Then
$$I(n) \le 2k(2k-1) + (4k-3)\sum_{i=1}^{2k}(2i-1) = 2k(8k^2-4k-1).$$
A similar argument can be applied when $n \equiv 2 \pmod 4$. □

In K. Heinrich and W.D. Wallis (1981) it was shown that, for all n, I(n) is bounded below by cn^3 where c is a constant. This result is obtained via the product construction of Theorem 3.9 and several special constructions.

THEOREM 3.9. If $I(n) \geq k$ and $I(m) \geq \ell$, then $I(mn) \geq m^2k + n^2\ell + 4k\ell$.

Proof. Take the direct product of squares of orders m and n, and then count intercalates. □

Recently, A. Hobbs, A. Kotzig and J. Zaks (1982) have studied what they call h-homogeneous latin squares. An LS(n) is h-homogeneous if every cell lies in exactly h intercalates where, of course, $0 \leq h \leq n-1$. From the earlier results one sees immediately that an (n-1)-homogeneous LS(n) exists if and only if $n = 2^a$; an (n-2)-homogeneous LS(n) cannot exist; a 1-homogeneous LS(n) exists if and only if n is even; and a 0-homogeneous LS(n) exists if and only if $n \neq 2$ or 4. They proved the following result.

THEOREM 3.10. An (n-3)-homogeneous LS(n) exists if and only if n = 3,4,6,8,12 or 16.

The proof involves first the introduction of a regular graph for each of the symbols in the square. From this sufficient information about the structure of the square is obtained to allow the authors then, in a series of claims, to obtain the result.

The 2-3-SC latin square of order 12 shown in Figure 3.1 is also 9-homogeneous. Moreover, it is the unique 9-homogeneous latin square of order 12. We remark that from the bounds on I(n) we know that an (n-3)-homogeneous LS(n) cannot exist except perhaps when n = 3,6 or 4k. By similar counting (as claimed in A. Hobbs, A. Kotzig and J. Zaks (1982)) if n is odd, an h-homogeneous LS(n) can exist only when $h \equiv 0 \pmod 4$ as $hn^2/4$ must be integral.

Also constructed in their paper (using a non-uniform product) are an (n-k)-homogeneous LS(n) for all odd k and n = k,2k and 4k, and an (n-k+1)-homogeneous LS(n) for all even k and n = k,2k and 4k. In fact (as pointed out by Kotzig) their 3m-homogeneous LS(4m) implies I(4m) ≥ $12m^3$ which improves the lower bound of $8m^3$ given in K. Heinrich and W.D. Wallis (1981).

Finally, the next theorem is stated by Hobbs, Kotzig and Zaks and may prove quite useful in further work on this problem.

THEOREM 3.11. The direct product of a p-homogeneous latin square and a q-homogeneous latin square is one which is (pq+p+q)-homogeneous.

The proof is straightforward.

More recently M. d'Angelo and J.M. Turgeon (1982) have studied the relationship between finite groups and homogeneous latin squares. The study of the relationship between groups and homogeneous latin squares has been continued by A.M. Hobbs and A. Kotzig (1983). In that paper they construct examples of both 6- and 12-homogeneous latin squares as well as give several infinite classes of new homogeneous latin squares.

In earlier work by R.B. Killgrove, E.T. Parker and D.I. Kiel (1978) 1-homogeneous LS(8)s had been considered. In this work, a study of projective planes, they required 1-homogeneous LS(8)s which could be represented as order 4 arrays each cell of which was an order 2 subsquare; that is, an intercalate. In all six such squares exist. In Figure 3.2 two 1-homogeneous LS(8)s are shown and it is the one on the left which has the additional property. R.B. Killgrove now plans to list all 1-homogeneous LS(8)s.

The Cayley table C of the elementary abelian group of order n = 2^a has arisen several times in this discussion, and we would like to mention another of its properties. R.A. Bailey (1982) has shown that an LS(n) when considered as a set of n^2 cells with three block systems has an automorphism group of exactly four orbits only when it is C or the table of the cyclic group of order 3. (An automorphism is a permutation of the cells which preserves each block system and if n > 2 the automorphism gorup necessarily has at least four orbits on unordered pairs of cells.)

$$\begin{array}{|cccccccc|} 1 & 2 & 3 & 4 & 5 & 6 & 7 & 8 \\ 2 & 1 & 4 & 3 & 6 & 5 & 8 & 7 \\ 3 & 4 & 5 & 6 & 7 & 8 & 2 & 1 \\ 4 & 3 & 6 & 5 & 8 & 7 & 1 & 2 \\ 5 & 6 & 7 & 8 & 2 & 1 & 4 & 3 \\ 6 & 5 & 8 & 7 & 1 & 2 & 3 & 4 \\ 7 & 8 & 2 & 1 & 4 & 3 & 6 & 5 \\ 8 & 7 & 1 & 2 & 3 & 4 & 5 & 6 \end{array} \qquad \begin{array}{|cccccccc|} 1 & 2 & 3 & 4 & 5 & 6 & 7 & 8 \\ 2 & 1 & 4 & 3 & 6 & 5 & 8 & 7 \\ 3 & 4 & 5 & 7 & 1 & 8 & 6 & 2 \\ 4 & 3 & 6 & 8 & 2 & 7 & 5 & 1 \\ 5 & 6 & 8 & 2 & 7 & 3 & 1 & 4 \\ 6 & 5 & 7 & 1 & 8 & 4 & 2 & 3 \\ 7 & 8 & 1 & 6 & 4 & 2 & 3 & 5 \\ 8 & 7 & 2 & 5 & 3 & 1 & 4 & 6 \end{array}$$

Figure 3.2

We next look at progress made on two problems of L. Fuchs (cited in [DK, p.486]).

The first of these is "Given n does there always exist a quasigroup of order n containing subquasigroups of every order k, $k \leq \lfloor \frac{n}{2} \rfloor$?" The upper bound on k is imposed by Theorem 1.2. A solution to this problem is given by K. Heinrich (1977d) who proved the following theorem.

THEOREM 3.12. An LS(n) with subsquares of every order k, $k \leq \lfloor \frac{n}{2} \rfloor$, is possible if and only if $1 \leq n \leq 7$ or n = 9,11 or 13.

Such an LS(13) is shown in Figure 3.3. The subsquares are indicated and the rows and columns have been labelled to show that in fact it is a quasigroup with subquasigroups of each order k, $1 \leq k \leq 6$. In fact only when n = 11 is there no quasigroup with subquasigroups as described.

CORROLARY 3.13. A quasigroup of order n with subquasigroups of every order k, $k \leq \lfloor \frac{n}{2} \rfloor$, exists if and only if $1 \leq n \leq 7$ or n = 9 or 13.

(Some earlier related work on subquasigroups can be found in C.

Hobby, H. Rumsey and P.M. Weichsel (1960). Also, D.W. Wall (1957) has studied subquasigroups and their intersection.)

The proof of the theorem relies on the observation that if two subsquares have common cells, then they must intersect in a subsquare. Subsquares are inserted into an order n array beginning with one of order $\lfloor \frac{n}{2} \rfloor$. Three cases are considered: n = 2m, 4m+1, or 4m+3. The first is eliminated for n ≥ 8 on attempting to insert a subsquare of order $\lfloor \frac{n}{2} \rfloor$-1; and the second and third are elminated on the attempted insertion of subsquares of orders $\lfloor \frac{n}{2} \rfloor$-1 and $\lfloor \frac{n}{2} \rfloor$-2 when n ≥ 17. (Unfortunately the reason given in the paper for eliminating the case n = 15 is incorrect; but it is easily eliminated.)

	13	12	11	10	7	8	9	2	1	3	4	5	6
13	1	10	2	5	4	3	6	13	11	12	9	7	8
12	13	4	1	2	5	6	3	10	12	9	8	11	7
11	2	1	3	11	6	5	4	12	13	8	7	9	10
10	6	2	12	1	3	4	5	11	10	7	13	8	9
7	4	5	6	3	1	2	8	9	7	10	12	13	11
8	3	6	5	4	9	1	2	7	8	13	11	10	12
9	5	3	4	6	2	7	1	8	9	11	10	12	13
2	10	13	11	12	8	9	7	1	2	5	3	6	4
1	11	12	13	10	7	8	9	2	①	3	4	5	6
3	12	9	8	7	10	13	11	4	3	1	6	2	5
4	9	8	7	13	11	12	10	3	5	6	1	4	2
5	7	11	9	8	13	10	12	6	4	2	5	1	3
6	8	7	10	9	12	11	13	5	6	4	2	3	1

Figure 3.3

The other more difficult problem of Fuchs is the following. "Let n = $n_1 + n_2 + \ldots + n_k$ be a partition of n. When is it possible to construct a quasigroup of order n with disjoint subquasigroups of orders $n_1, n_2, \ldots, n_k$?" In terms of latin squares this asks for an LS(n) with disjoint (by virtue of row, column and symbol) subsquares of orders $n_1, n_2, \ldots, n_k$; that is, an LS($n; n_1, n_2, \ldots, n_k$) where $\sum_{i=1}^{k} n_i = n$. When such a square exists we will say that it realizes the partition. For example, an LS(7;1,1,1,2,2) is given in Figure 3.4.

$$\begin{vmatrix} 1 & 4 & 5 & 6 & 7 & 2 & 3 \\ 3 & 2 & 6 & 7 & 1 & 5 & 4 \\ 2 & 7 & 3 & 1 & 6 & 4 & 5 \\ 7 & 6 & 2 & 4 & 5 & 3 & 1 \\ 6 & 3 & 7 & 5 & 4 & 1 & 2 \\ 4 & 5 & 1 & 2 & 3 & 6 & 7 \\ 5 & 1 & 4 & 3 & 2 & 7 & 6 \end{vmatrix}$$

Figure 3.4

The first results on this problem were obtained by J. Dénes and E.K. Pásztor (1963) who showed that the squares can be constructed (by direct product) for n = a + a + ... + a = ka, k ≠ 2; and that there is no square realizing the partition n = 2m = m + 1 + 1 + ... + 1.

We know of only one general necessary condition.

LEMMA 3.14. Given $n = n_1 + n_2 + \ldots + n_k$ where $k \geq 3$ and $n_1 \leq n_2 \leq \ldots \leq n_k$, if there is an LS($n; n_1, n_2, \ldots, n_k$), then $n_k \leq n_1 + n_2 + \ldots + n_{k-2}$.

Proof. The result arises on trying to complete a row of the subarray A in Figure 3.5. □

Unfortunately this condition is not sufficient as is seen from the fact that, for example, there is no LS(2a+2b+3c;a,b,c,c,a+b+c) when a ≤ b < c. This is simply one of many examples and indicates that it will be quite difficult to determine the necessary conditions.

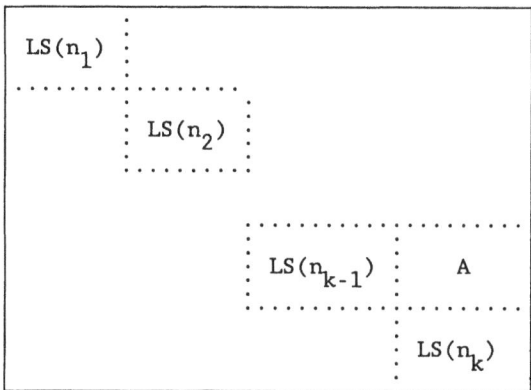

Figure 3.5

However, many results have been obtained; the case $k \leq 4$ having been completely resolved. We begin with a construction lemma.

LEMMA 3.15. Given a POLS(k), there is an LS(mk+a;a), $1 \leq a \leq mk$, and an LS(mk+a;m,m,...,m,a), $1 \leq a \leq m(k-1)$.

The proof is simply an adaptation of Theorem 1.4. Note that for the second part of the lemma one of the a_i is zero.

THEOREM 3.16. There is an LS(n) realizing the partition $n = n_1 + n_2 + n_3$ if and only if $n_1 = n_2 = n_3$.
    Proof. The proof follows from Lemma 3.14 and a direct product construction. □

Partitions of size 4 were resolved in K. Heinrich (1977c) as follows.

THEOREM 3.17. There is an LS(n) realizing the partition $n = n_1 + n_2 + n_3 + n_4$ if and only if $n_1 = n_2 = n_3$ and $1 \leq n_4 \leq 2n_1$.
    Proof. When $n = n_1 + n_2 + n_3 + n_4$ and $n_1 = n_2 = n_3$, $1 \leq n_4 \leq 2n_1$ construct the LS($n;n_1,n_2,n_3,n_4$) from Lemma 3.15 with $k = 3$, $m = n_1$ and $a = n_4$.

On the other hand suppose there is an LS(n) realizing $n = n_1 + n_2 + n_3 + n_4$. Look at the square in two ways as shown in Figure 3.6 where L_i is the subsquare of order n_i.

In every column of A_1 each element of $e(L_4)$ occurs, and in every column of A_2 each element of $e(L_3)$ occurs. So between them A_1 and A_2 contain $(n_1 + n_2)(n_3 + n_4) - 2n_3n_4$ entries from $e(L_1) \cup e(L_2)$. Now arguing on A_3 and A_4 we get $(n_1 + n_2)(n_3 + n_4) - 2n_3n_4 = 2n_1n_2$ or $(n_1 + n_2)(n_3 + n_4) = 2(n_1n_2 + n_3n_4)$.

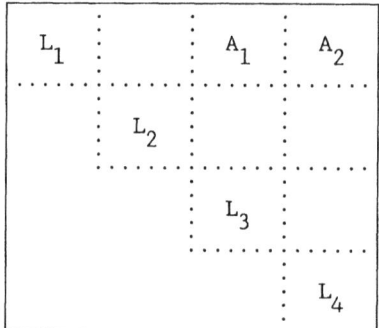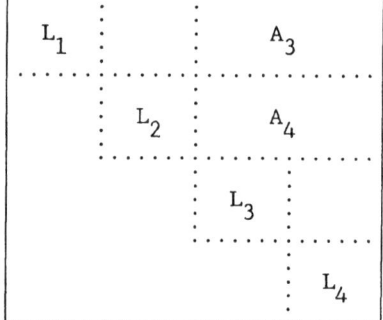

Figure 3.6

Permuting the n_i we get $(n_1 + n_3)(n_2 + n_4) = 2(n_1n_3 + n_2n_4)$. Subtracting these equations yields $(n_1-n_4)(n_2-n_3) = 0$ and similarly $(n_1 - n_2)(n_3 - n_4) = 0$. From these it follows without loss of generality that $n_1 = n_2 = n_3$. Now by Lemma 3.14, $1 \le n_4 \le 2n_1$. □

Further results have arisen from latin square embedding (see Chapter 8 for a survey of embedding problems). D.E. Daykin and R. Häggkvist (1984) have studied the problem of completing partial latin rectangles and have shown that when $n = 16m$ a partial latin square in which each row, column and symbol has been used at most $\frac{\sqrt{m}}{32}$ times can be completed. From this we deduce the next result.

THEOREM 3.18. If $n = 16m = n_1 + n_2 + \ldots + n_k$, where $n_1 \le n_2 \le \ldots \le n_k \le \frac{\sqrt{m}}{32}$, then there is an $LS(n; n_1, n_2, \ldots, n_k)$.

Also C.A. Rodger (1984) has shown that if R is a partial LS(m), $m \geq 10$, with $e(R) \subseteq \{1,2,\ldots,t\}$, and f is a non-negative integral valued function on the symbols $\{1,2,\ldots,t\}$ such that $\sum_{j=1}^{t} f(j) = t - m$, then R can be embedded in an LS(t) on these symbols in which symbol j occurs f(j) times in the diagonal of the LS(t) outside R (cells (i,i), $m + 1 \leq i \leq t$) for all $t \geq 2m+1$ if and only if:

(1) $N_R(j) \geq 2n - t + f(j); \quad 1 \leq j \leq t;$
(2) $N_R(j) = m$ implies $f(j) \neq t - m - 1;$ and
(3) When R is an LS(n), $e(R) = \{1,2,\ldots,n\}$ and $t = 2m+1$, then $\sum_{j=1}^{n} f(j) \neq 1.$

(Here $N_R(j)$ is the number of times symbol j occurs in R.) This result yields the following theorem.

THEOREM 3.19. There is an LS(n) realizing the partition $n = 1 + 1 + \ldots + 1 + n_1 + n_2 + \ldots + n_r$ where $\sum_{i=1}^{r} n_i \leq \frac{n-1}{2}$.

Another approach, as taken in K. Heinrich (1982), uses the ideas of Theorem 1.4 and Lemma 3.15 to study partitions in which the n_i can take only two values. We will now state the main results of that paper. Let $P(a,b;s,t)$, $a < b$, denote the partition $n = sa + tb = a + a + \ldots + a + b + b + \ldots + b$.

THEOREM 3.20. The partition $P(a,b;1,t)$ can be realized if and only if $t \geq 3$, and the partition $P(a,b;s,1)$ can be realized if and only if $s \geq 3$ and $(s-1)a \geq b$.

THEOREM 3.21. All partitions $P(a,b;s,t)$ with $s,t \geq 3$ can be realized.

THEOREM 3.22. The partitions $P(a,b;s,2)$ and $P(a,b;2,t)$ can be realized when s and t are sufficiently large.

Whilst the first of these theorems is easily verified the second requires considerable work. To indicate the technique used we will show how to realize the partitions P(a,ka;s,t), when s,t ≥ 3. This will also indicate the reasons why this method fails in the realization of P(a,b;s,t) when either s = 2 or t = 2.

THEOREM 3.23 All partitions P(a,ka;s,t), s,t ≥ 3, can be realized.

Proof. We look only at P(1,k;s,t), s,t ≥ 3, as if these can be realized so (by use of a direct product) can P(a,ka;s,t), s,t ≥ 3.

If k = 2 let A be an SOLS(s+t), s+t ≠ 6, and choose t transversals in A each intersecting the main diagonal in exactly one cell. Project once from each. In the empty array insert an LS*(t). The result is a realization of P(1,2;s,t). Should s+t = 6 (implying s = t = 3), apply the same procedure but use the LS*(6) of Figure 3.7. (Three transversals intersecting the main diagonal are indicated.)

$$\begin{vmatrix} \underline{\underline{1}} & 6 & 5 & 2 & \underline{4} & \overline{3} \\ 5 & \overline{2} & 6 & 1 & \underline{\underline{3}} & 4 \\ \overline{4} & 1 & 3 & \underline{\underline{6}} & 2 & 5 \\ 3 & \underline{5} & 2 & 4 & \overline{6} & 1 \\ 6 & 4 & \overline{1} & 3 & 5 & \underline{\underline{2}} \\ 2 & 3 & \underline{4} & \overline{5} & 1 & \underline{6} \end{vmatrix}$$

Figure 3.7

Suppose now that k ≥ 3 and s ≤ k(t-1). If t ≠ 6, from the direct product of an LS(k) with A, one of a POLS(t), project a total of s times from t-1 disjoint transversals in A and insert an LS*(s). If t = 6 and s ≥ k, s ≠ k + 2, product an LS*(k) with A, one of a POLS(7). Project s-k times from up to six transversals in A and insert an LS*(s-k). If t = 6 and s < k or s = k + 2 use the square of Figure 3.7 and the two transversals indicated in Figure 3.8 and repeat the above procedure.

Now, if s ≥ kt + 1 form the direct product of an LS(k) with A, one of a POLS(s), and project kt times inserting a realization of P(1,k;0,t) in the empty array.

$$\begin{vmatrix} 1 & 6 & \underline{5} & 2 & \overline{4} & 3 \\ 5 & 2 & 6 & \overline{1} & \underline{3} & 4 \\ 4 & 1 & 3 & \underline{6} & 2 & \overline{5} \\ 3 & 5 & \overline{2} & 4 & 6 & \underline{1} \\ \overline{6} & \underline{4} & 1 & 3 & 5 & 2 \\ \underline{2} & \overline{3} & 4 & 5 & 1 & 6 \end{vmatrix}$$

Figure 3.8

This leaves only the case $s = k(t-1) + r$, $1 \le r \le k$. Product an $LS^*(k)$ with one of a POLS($2t-1$) and project r times from one of the transversals, inserting an $LS^*(r)$, provided $r \ne 2$. If $r = 2$ product an $LS^*(k)$ with one of a POLS($2t-2$) (use the square of Figure 3.7 if $2t-2=6$), project $k+2$ times from two transversals and insert an $LS^*(k+2)$ in the empty array. □

The author is aware of only one other study related to these problems of Fuchs and that is by T.H. Straley (1972). She was interested in Stein quasigroups (that is, quasigroups satisfying $x(xy) = yx$ which also arise from Steiner triple systems) with a specified number of subquasigroups of a given order. Her main result was that, given a Steiner quasigroup V of order v and v Steiner quasigroups of order q, q > v, which pairwise intersect in the same quasigroup of order p, then if q > vp and q-p is not divisible by the order of any proper subquasigroup of V, there is a Steiner quasigroup of order v(q-p)+p containing a copy of each of the v quasigroups of order q and no other order q subquasigroups.

(4) **With subsquares and orthogonal**

L. Euler (1782) constructed the IPOLS(6;2) shown in Figure 1.4. He was at the time looking for a POLS(6). Being unable to find one and knowing a POLS(2) did not exist, he conjectured that there were no POLS(n), n ≡ 2(mod 4). His conjecture was further supported when G. Tarry (1900) showed the non-existence of a POLS(6). The first counter example was published in March 1959 by R.C. Bose and S.S. Shrikhande who, using pairwise balanced designs, constructed a POLS(22). The smallest case n = 10 (as given in Figure 1.2) was published by E.T. Parker in April of that year. Finally, in a joint paper in 1960, the three authors showed the existence of a POLS(n) for all n ≠ 2,6.

E.T. Parker (1959) had in fact constructed an IPOLS ((3q-1)/2, (q-1)/2) for prime power q and q ≡ 3(mod 4); the case q = 7 yielding the POLS(10). Later [E.T. Parker (1962)] he studied necessary conditions for the existence of t-POLS(n;k) obtaining the following result.

THEOREM 4.1. If there exist t-POLS(n;k), then n ≥ (t+1)k.

Proof. Suppose that in each of the t squares the order k subsquare is in the lower right-hand corner and is based on the elements of K. In the first row of each square the elements of K must lie in the first n-k columns and over all squares the columns occupied by these elements must be distinct. Therefore tk ≤ n-k and the result follows. □

COROLLARY 4.2. If there exist t-IPOLS(n;k) then n ≥ (t+1)k.

COROLLARY 4.3. If there exists an SOLS(n;k) (or an ISOLS(n;k)) then n ≥ 3k + 1.

Proof. The proof is as in Theorem 4.1 except that only n-k-1 cells in the first row of both the square and its transpose are available. □

However, specific study of the existence of t-IPOLS(n;k) did not take place for several years although K. Yamamoto (1954) had constructed an IPOLS(4m+2;2) when 4m+1 is a prime power. The problem was taken up by J.D. Horton (1971) and (1974) who showed that an IPOLS(n;k) exists for fixed k and all sufficiently large n. At about the same time D.J. Crampin and A.J.W. Hilton (1975b) obtained essentially the same result by showing that for $k \neq 2, 3$ or 6, an SOLS(n;k) exists whenever $n > 36373 \times 5^2 2^{16} (p_1^{e_1} p_2^{e_2} \ldots)^2 k^2$ where $e_i = 0$ if $p_i | k$ or $p_i > \max\{p_j: p_j | k\}$, and $p_i^{e_i} < k < p_i^{e_i+1}$ if $p_i \nmid k$ and $p_i < \max\{p_j: p_j | k\}$. Prolongation, direct product and Sade's generalized direct product were used in the proofs.

A great improvement on these bounds was made by D.A. Drake and H. Lenz (1980) who showed, using pairwise balanced designs, that for all $n \geq 4k + 3$ and $k \geq 304$ an SOLS(n;k) exists. L. Zhu (1984b) then improved their result to show the existence of an IPOLS(n;k) for $n > 3k + 6$ and $k \geq 304$. As for the small cases, J.D. Horton had shown that an IPOLS(n;2), $n \geq 6$, exists for all but perhaps ten values of n and solutions for these remaining values of n were later constructed by K. Heinrich (1977b). W.D. Wallis and L. Zhu (1983) then showed the existence of an IPOLS(n;k) for all $n \geq 3k$ with $k = 3, 4$ or 5. The work was completed in two more papers. L. Zhu (1984c) showed that if n or k is odd and $k \neq 1, 2$, or 6 then a POLS(n;k) exists (and hence an IPOLS(n;k)) whenever $n \geq 3k$, and K. Heinrich and L. Zhu (1986) showed that if n and k are even and $k \neq 2$, then an IPOLS(n;k), $n \geq 3k$, exists. More recently, G.B. Belyavskaya and A.V. Nazarok (1987) have constructed pairs of orthogonal quasigroups of order 3k+i, $i = 0, 1, 2$, $k \neq 2, 6$, with orthogonal subquasigroups of order k.

We will now prove that an IPOLS(n;k) exists whenever $n \geq 3k$ except when $n = 6$ and $k = 1$. We begin with a more general form of E.T. Parker's result, as given in K. Heinrich (1977a) but also published by other authors; in particular K. Yamamoto (1961) constructed POLS(3k+1,k) for $k \equiv 3 \pmod 4$.

THEOREM 4.4. There exist IPOLS(3k+1;k) for all k.

Proof. Let A be an LS(2k+1) with $e_A(i,j) \equiv 2j-i \pmod{2k+1}$ and transversals $T_i = \{(j, j+i): 1 \leq j \leq 2k+1\}$, $1 \leq i \leq k$. Use these transversals to prolongate, projecting T_i onto row and column 2k+1+i. The result is an ISOLS(3k+1;k). □

(Note that if $k \notin \{2, 3, 6\}$ the result of Theorem 4.4 also follows from Corollary 1.5.)

At this point we should remark that prolongation had already been used successfully by F. Ruiz and E. Seiden (1974), A. Hedayat and E. Seiden (1971) and (1974), A. Hedayat (1978) and L. Zhu (1982) to produce new types of POLS(n), particularly when n ≡ 2(mod 4). Also J.D. Horton (1974) used this method to construct IPOLS(n;k). The next result is a restatement of Theorem 1.1

THEOREM 4.5. For all n ≠ 2,6, there exists an IPOLS(n;1).

LEMMA 4.6. An IPOLS(n;2) exists for 6 ≤ n ≤ 16.
  Proof. An IPOLS(6;2) was given in Figure 1.4. For each of the others we give, in Figure 4.1, the entries in cells
{(1,i): 1 ≤ i ≤ n} ∪ {(j,1): n-1 ≤ j ≤ n} in the form of a vector of length n+2. The vector is
  (e(1,1), e(1,2),...,e(1,n-2); e(1,n-1), e(1,n); e(n-1,1), e(n,1)).
The entries in cells (i,j) when 1 ≤ i,j ≤ n-2 are determined by e(i,j) = e(i-1,j-1)+1(mod n-2) if e(i,j) is different from n-1 and n, and e(i,j) = e(i-1,j-1) if e(i,j) = n-1 or n (where i-1 and j-1 are also calculated modulo n-2 on the symbols 1,2,...,n-2), and in the remaining cells by e(i,j) ≡ e(i-1,j)+1(mod n-2) and e(j-1) ≡ e(j,i-1)+1(mod n-2) where j = n-1 or n and 1 ≤ i ≤ n-2. When only one vector is given for the IPOLS(n;2), then it is in fact an ISOLS(n;2). □

Clearly the idea of Lemma 4.6 can be generalized to construct IPOLS(n;k) by giving the entries in cells (i,j), 1 ≤ j ≤ n, and (i,1), n-k+1 ≤ i ≤ n, for each square (unless it is symmetric) and employing the same rules but modulo n-k.

THEOREM 4.7. There exist IPOLS(n;2) for all n ≥ 6.
  Proof. When n = 7m, m ≠ 1,2,6, take the direct product of a POLS(7) and a POLS(m), replacing one copy of the POLS(7) by an IPOLS(7;2). Cases m = 1 and 2 are given in Lemma 4.6 and when m = 6 we have n = 42 = 14·3 and a similar direct product of a POLS(14) by a POLS(3) can be applied.

ISOLS(7;2)   :   (1,6,7,2,4;3,5;2,3)

IPOLS(8;2)   :   (1,7,4,6,3,8;2,5;4,6)

(1,5,7,3,8,2;4,6;2,5)

ISOLS(9;2)   :   (1,8,9,5,7,2,6;3,4;6,5)

ISOLS(10;2) :   (1,3,9,6,4,10,2,5;7,8;5,7)

ISOLS(11;2) :   (1,3,5,7,9,2,6,10,11;4,8;8,7)

ISOLS(12;2) :   (1,4,11,8,12,5,3,6,10,7;2,9;4,6)

ISOLS(13;2) :   (1,4,8,3,6,10,2,11,12,13,9;5,7;9,8)

ISOLS(14;2) :   (1,3,8,10,13,14,6,11,5,12,9,4;7,2;8,10)

ISOLS(15;2) :   (1,14,15,5,8,11,3,2,13,7,6,10,12;4,9;3,7)

ISOLS(16;2) :   (1,14,10,5,13,3,12,7,11,15,16,4,2,9;8,6;11,5)

Figure 4.1

If $n = 7m + i$, $1 \leq i \leq 3$, and $m \neq 1,2,3$ or 6 apply Theorem 1.4 (here an SOLS(m) is sufficient) with $p = m$, $q = 7$ and $a_1 = i$. If $i = 2$ we are done. Otherwise replace one copy of the POLS(7) by an IPOLS(7;2). The cases $m = 1$ and 2 except for $n = 17$ are given in Lemma 4.6. So we still have the cases $n \in \{17,22,23,24,43,44,45\}$. By direct product and Theorem 4.4 there exist IPOLS(n;k) where $(n;k) \in \{(22;7), (24;8), (43;14), (44;11), (45;9)\}$ and inserting an IPOLS(k;2) yields the result. An IPOLS(17;2) and an IPOLS(23;2) are constructed from Theorem 1.2 with $p = 5$ or 7, $q = 3$ and $a_1 = a_2 = 1$.

Finally, when $n = 7m + i$, $4 \leq i \leq 6$ and $m \neq 1,2,3,6,10$ apply Theorem 1.4 with $p = m$, $q = 7$ and $a_1 = a_2 = 2$, $a_3 = i - 4$. If $i = 6$ insert an IPOLS(6;2) and if $i \neq 6$ insert a POLS(i) and replace one of the copies of POLS(7) by an IPOLS(7,2). The case $m = 1$ is given in Lemma 4.6. The remaining values of n are $n \in \{18,19,20,25,26,27,46,47,48,74,75,76\}$. Direct product and replacement yield IPOLS(n;2) for $n \in \{25,27,48,75\}$. The existence of IPOLS(19;6) and IPOLS(46;15) yield IPOLS(n;2) when $n = 19$ and 46, and direct product then gives $n = 76 = 4 \cdot 19$. Cases $n = 26, 47$ and 74 are dealt with using Theorem

1.4 with p = 8, 15 or 24, q = 3 and $a_1 = a_2 = 1$. This theorem also yields an IPOLS(18;2) when p = q = 4 and $a_1 = a_2 = 1$. An IPOLS(20;2) comes from Theorem 1.8 with p = 5, q = 3, $a_1 = a_2 = 1$ and $b_1 = b_2 = b_3 = 1$. □

To construct IPOLS(n;k), k > 2, we need a series of lemmas. The idea of this proof comes from the early work of L. Zhu (1984a) and (1984b).

Let S_3 denote the set {1,2,3,6,10}, and let $S_4 = S_3 \cup$ {4,14,18,20,22,24,26,28,30,33,34,38,42,44,52} recalling that if $p \notin S_3$ then there exist 3-POLS(p), and if $p \notin S_4$ then there exist 4-POLS(p).

LEMMA 4.8. If $n \geq 7k$ and $k \notin S_3$, then an IPOLS(n;k) exists.

Proof. Let $n = k\ell+i$, $\ell \geq 7$ and $0 \leq i < k$. If k+i = 6, then either k = 4 and i = 2, or k = 5 and i = 1. In these cases apply Corollary 1.5 with p = k, q = ℓ and $a_1 \in \{1,2\}$. Should $a_1 = 2$ replace one of the IPOLS($\ell+2$;2) with a POLS($\ell+2$) so that the order k subsquare is preserved.

In all other cases put $n = k(\ell-1)+(k+i)$ and apply Corollary 1.5 with p = k, q = $\ell-1$, $a_1 = a_2 = \ldots = a_i = 2$ and $a_{i+1} = a_{i+2} = \ldots = a_k = 1$. □

LEMMA 4.9. If $4k \leq n < 5k$ and $k \notin S_3$, then there exists a IPOLS(n;k).

Proof. Write n = 4k+i, $0 \leq i \leq k-1$. If $i \neq 2$ or 6 use Corollary 1.5 with p = k, q = 4 and $a_1 = a_2 = \ldots = a_i = 1$. When i = 2 and k $\neq$ 4 begin with Corollary 1.5 setting p = 4, q = k and $a_1 = 2$, and then insert a POLS(k+2). To construct an IPOLS(18;4), use Corollary 1.6 with p = 5, q = 3 and $a_1 = a_2 = a_3 = 1$.

When i = 6 (implying k ≥ 7) use Corollary 1.6 with p = 4, q = k and $a_1 = a_2 = a_3 = 2$; insert first a POLS(k+2) and then an IPOLS(6;2). □

LEMMA 4.10. If $k \notin S_3$ and $3k \leq n < 4k$, then there exists an IPOLS(n;k).

Proof. Put n = 3k+i, $0 \leq i \leq k-1$. For $i \neq 2$ or 6 the result follows easily from Theorem 1.4 with p = k, q = 3 and $a_1 = a_2 = \ldots =$

$a_i = 1$.

Now for the cases $i = 2$ and $i = 6$. First suppose $k \notin S_4$ and apply Theorem 1.8 with $p = k$, $q = 3$, $a_1 = 1$ and $b_1 = b_2 = \ldots = b_{i-1} = 1$. Fifteen values of k in $S_4 \setminus S_3$ remain. If $k \in \{14, 18, 22, 24, 26, 30, 34, 38, 42\}$ we use Corollary 1.10 with $p = k/2$, $q = 6$, $c = 2$, $a_1 = 1$ and $b_1 = b_2 = \ldots = b_{i-1} = 1$. (Note that this also works for $k = 10$.)

In Figure 4.2 is the IPOLS(14;4) constructed by L. Zhu (1984a). For $k \in \{20, 28, 44, 52\}$ again use Corollary 1.10 with $p = k/4$, $q = 12$, $c = 4$, $a_1 = 1$ and $b_1 = b_2 = \ldots = b_{i-1} = 1$. This leaves only $k = 33$ which also follows from Corollary 1.10 but with $p = 11$, $q = 9$, $c = 3$, $a_1 = 1$ and $b_1 = b_2 = \ldots = b_{i-1} = 1$. (Note that an ISOLS(11;3) can be constructed from the vector $(1,9,2,10,8,3,11,4;6,5,7;7,4,2)$.)

LEMMA 4.11. If $k \notin S_3$ and $5k \leq n < 7k$, then there exists an IPOLS(n;k).

Proof. Put $n = 5k+i$, $0 \leq i \leq 2k-1$. If i is even, say $i = 2r$, $0 \leq r \leq k-1$, apply Corollary 1.5 with $p = k$, $q = 5$ and $a_1 = a_2 = \ldots = a_r = 2$ unless $2r = 2$ or 6. If $2r = 2$ or 6, replace one of the IPOLS(7;2) with a POLS(7), and insert an IPOLS(6;2) if $2r = 6$.

1	13	10	14	11	3	4	2	12	5	6	7	9	8
4	2	13	1	14	5	3	12	6	11	7	8	10	9
6	5	3	13	2	4	12	7	11	14	8	9	1	10
5	7	6	4	13	12	8	11	14	3	9	10	2	1
12	6	8	7	5	9	11	14	4	13	10	1	3	2
13	9	14	11	4	10	2	3	1	12	5	6	8	7
8	14	11	3	12	13	9	1	2	10	4	5	7	6
14	11	2	12	9	7	13	8	10	1	3	4	6	5
11	1	12	8	10	14	6	13	7	9	2	3	5	4
10	12	7	9	8	11	14	5	13	6	1	2	4	3
9	10	1	2	3	8	7	6	5	4				
2	3	4	5	6	1	10	9	8	7				
3	4	5	6	7	2	1	10	9	8				
7	8	9	10	1	6	5	4	3	2				

Figure 4.2

1	4	11	5	8	14	12	13	6	10	2	9	3	7
14	2	5	11	6	12	13	7	1	9	3	10	4	8
12	14	3	6	11	13	8	2	10	7	4	1	5	9
13	12	14	4	7	9	3	1	8	11	5	2	6	10
10	13	12	14	5	4	2	9	11	8	6	3	7	1
3	11	4	7	9	10	14	12	13	5	1	8	2	6
11	3	6	8	4	2	9	14	12	13	10	7	1	5
2	5	7	3	13	11	1	8	14	12	9	6	10	4
4	6	2	13	12	1	11	10	7	14	8	5	9	3
5	1	13	12	14	3	10	11	9	6	7	4	8	2
7	8	9	10	1	6	5	4	3	2				
9	10	1	2	3	8	7	6	5	4				
6	7	8	9	10	5	4	3	2	1				
8	9	10	1	2	7	6	5	4	3				

Figure 4.2 (cont'd)

Suppose now that i is odd.

When k is even, say k = 2t, we apply Corollary 1.5 with p = 5 and q = k = 2t. All earlier results will be used and we note in particular that there is always an IPOLS(m;r) for even m, m ≥ 3r and even r, r ≠ 6,10; and that for t ≠ 7 we have constructed an IPOLS(3t-1;t-1) (for t = 10 see Lemma 4.10). In each of the following cases two options are given so that we can avoid "holes" of orders 6 and 10.

If 1 ≤ i ≤ t put a_1 = i-1, a_2 = 1 or a_1 = i-3, a_2 = 2, a_3 = 1. If t+1 ≤ i ≤ 2t and t is odd put a_1 = i-t, a_2 = t or a_1 = i-t-2, a_2 = t, a_3 = 2, and if t is even put a_1 = i-t-1, a_2 = t, a_3 = 1 or a_1 = i-t-3, a_2 = t, a_3 = 2, a_4 = 1. If 2t+1 ≤ i ≤ 3t put a_1 = i-2t-1, a_2 = a_3 = t, a_4 = 1 or a_1 = i-2t+1, a_2 = a_3 = t-1, a_4 = 1. Here we have a problem when i = 3t and i-2t-1 = 6 or 10 as neither case applies. We need an IPOLS(13t;2t) when t = 7 or 11 so take the product of an IPOLS(13;2) and a POLS(t). If 3t+1 ≤ i ≤ 2k-1 and t is odd, put a_1 = i-3t, a_2 = a_3 = a_4 = t or a_1 = i-3t+2, a_2 = t, a_3 = a_4 = t-1. Again we cannot have both i = 4t-1 and i-3t = 6 or 10. The IPOLS(14t-1;2t) for $t = 7, 11$ are constructed from Corollary 1.9 with $p = t$, $q = 2t-1$, $a_1 = a_2 = a_3 = 1$ and $b_1 = b_2 = b_3 = 1$. Finally, if $3t+1 \le i \le 2k-1$ and t is even put $a_1 = i-3t+1$, $a_2 = a_3 = t$, $a_4 = t-1$ or $a_1 = i-3t+3$, $a_2 = a_3 = a_4 = t-1$. This latter construction fails if $i = 4t-1$ and $i-3t+1 = 6$ or 10. Instead apply Theorem 1.8 to construct IPOLS(14t-1;2t) when $t = 6$ by putting $p = 7$, $q = 10$, $a_1 = a_2 = \ldots = a_6 = 2$, $b_1 = 1$ and when $t = 10$ by putting $p = 11$, $q = 10$, $a_1 = a_2 = \ldots = a_{10} = 2$, $b_1 = b_2 = \ldots = b_9 = 1$.

The case k odd, $k = 2t+1$, follows from Corollary 1.5, essentially as above, but with $p = 5$ and $q = 2t+1$. Note that we now also have the existence of IPOLS(m;r) for odd m, $m \ge 3r$ and even r, $r \ne 6$ or 10. We shall also need IPOLS(3t;t-1) and we have their existence except when $t = 7$ or 11.

The cases $1 \le i \le 2t$ follow as before. For the remainder we make the same choice of a_i but find that thirteen cases are not covered. These are given in Table 4.1 and all use either Theorem 1.4 or 1.8. □

LEMMA 4.12. If $k \in \{3, 6, 10\}$, then there exists an IPOLS(n;k).

Proof. First note that by Lemmas 4.8, 4.9, 4.10 and 4.11 there exists an IPOLS(n;3k) for $n \ge 9k$, $k \in \{3, 6, 10\}$, and hence an IPOLS(n;k) for $n \ge 9k$ and $k \in \{3, 6, 10\}$ (an IPOLS(18;6) is easily constructed using an ISOLS (6;2)). Thus, in each case only finitely many values of n need be resolved.

When $k = 3$ we need to construct an IPOLS(n;3), $9 \le n \le 26$. A direct product construction yields all n divisible by 3. An ISOLS(10;3) was given in Theorem 4.4 and an ISOLS(11;3) was given in Lemma 4.10. The cases $n \in \{13, 16, 18, 19, 22, 25, 26\}$ follow by Corollary 1.5 as each can be expressed in the form $n = 3t+i$, $i \ne 2$ or 6, $1 \le i \le t-1$, so we set $p = t$, $q = 3$ and $a_1 = a_2 = \ldots = a_i = 1$. The cases $n = 17$ and 23 follow from Corollary 1.9 with $p = 5$ and $p = 7$, respectively, $q = 3$ and $a_1 = b_1 = 1$.

An IPOLS(20;3) also comes from Corollary 1.9 but with $p = 5$, $q = 3$, $a_1 = 1$ and $b_1 = \ldots = b_4 = 1$. Finally an IPOLS(14;3) (as is given in A. Hedayat and E. Seiden (1974)) is shown in Figure 4.3.

n	k	p	q	a_i's and b_j's
62	9	5	10	$a_1 = 1$, $a_2 = \ldots = a_5 = 2$, $b_1 = b_2 = b_3 = 1$
76	11	5	13	$a_1 = 6$, $a_2 = a_3 = 2$, $a_4 = 1$
88	13	25	3	$a_1 = a_2 = \ldots = a_{13} = 1$
96	15	27	3	$a_1 = a_2 = \ldots = a_{15} = 1$
102	15	29	3	$a_1 = a_2 = \ldots = a_{15} = 1$
106	15	29	3	$a_1 = a_2 = \ldots = a_{15} = 1$, $b_1 = \ldots = b_4 = 1$
118	17	8	12	$a_1 = 1$, $a_2 = \ldots = a_5 = 4$, $b_1 = \ldots = b_5 = 1$
132	19	11	10	$a_1 = 1$, $a_2 = \ldots = a_{10} = 2$, $b_1 = b_2 = b_3 = 1$
144	21	41	3	$a_1 = a_2 = \ldots = a_{21} = 1$
144	23	11	11	$a_1 = 5$, $a_2 = \ldots = a_{10} = 2$
148	23	5	25	$a_1 = 12$, $a_2 = a_3 = 5$, $a_4 = 1$
154	23	13	10	$a_1 = \ldots = a_4 = 5$, $a_5 = a_6 = a_7 = 1$, $b_1 = 1$
158	23	45	3	$a_1 = a_2 = \ldots = a_{23} = 1$

Table 4.1

Clearly, for $k = 6$ the remaining values of n are $19 \leq n \leq 53$ and n not divisible by 3. (If $3 | n$ use Corollary 1.7 (as all a_i are zero we only need POLS(3)) with $p = 3$, $q = n/3$ and $c = 2$.) When $n = 25$ or 31 use Corollary 1.7 with $p = 4$ or 5, $q = 6$, $c = 2$ and $a_1 = 1$. Then insert either an IPOLS(8;2) or an IPOLS(10;2). For $n = 32$ apply Corollary 1.10 as above with $p = 5$, $q = 6$, $c = 2$ and $a_1 = b_1 = 1$. Theorem 1.4 with $p = 4$, 5 or 7, $q = 4$ and $a_1 = a_2 = a_3 = 2$ yields an IPOLS(n;6) for $n = 22$, 26 and 34, while Corollary 1.6 with $p = q = 4$, $a_1 = a_2 = a_3 = 2$ and $a_4 = 1$ yields an IPOLS(23;6). The remaining IPOLS(n;6) are also constructed from Corollary 1.6; in each case with $q = 4$ and $a_i \in \{0,1,2\}$. If $n = 28$ or 29 put $p = 5$, if $35 \leq n \leq 41$ put $p = 7$, and if $43 \leq n \leq 53$ put $p = 9$. Finally, an ISOLS(19;6) was given in Theorem 4.4 and an IPOLS(20;6) was constructed by L. Zhu (1984b) and comes from the vectors (1,3,15,16,17,18,2,14,5,19,10,9,11,

4:3 *Latin squares with and without subsquares*

20;7,12,6,13,8,4;8,4,3,9,6,5) and (1,15,13,12,14,9,16,7,17,11,18,19,
20,2;8,3,4,9,5,6;7,8,6,4,12,13) with calculations modulo 14.

11	1	2	12	13	5	6	14	8	9	10	4	7	3
4	5	12	13	8	9	14	11	1	2	3	7	10	6
8	12	13	11	1	14	3	4	5	6	7	10	2	9
12	13	3	4	14	6	7	8	9	10	11	2	5	1
13	6	7	14	9	10	11	1	2	3	12	5	8	4
9	10	14	1	2	3	4	5	6	12	13	8	11	7
2	14	4	5	6	7	8	9	12	13	1	11	3	10
14	7	8	9	10	11	1	12	13	4	5	3	6	2
10	11	1	2	3	4	12	13	7	8	14	6	9	5
3	4	5	6	7	12	13	10	11	14	2	9	1	8
7	8	9	10	12	13	2	3	14	5	6	1	4	11
5	2	10	7	4	1	9	6	3	11	8			
6	3	11	8	5	2	10	7	4	1	9			
1	9	6	3	11	8	5	2	10	7	4			

12	13	2	3	4	5	6	7	14	9	10	11	1	8
13	4	5	6	7	8	9	14	11	1	12	2	3	10
6	7	8	9	10	11	14	2	3	12	13	4	5	1
9	10	11	1	2	14	4	5	12	13	8	6	7	3
1	2	3	4	14	6	7	12	13	10	11	8	9	5
4	5	6	14	8	9	12	13	1	2	3	10	11	7
7	8	14	10	11	12	13	3	4	5	6	1	2	9
10	14	1	2	12	13	5	6	7	8	9	3	4	11
14	3	4	12	13	7	8	9	10	11	1	5	6	2
5	6	12	13	9	10	11	1	2	3	14	7	8	4
8	12	13	11	1	2	3	4	5	14	7	9	10	6
11	9	7	5	3	1	10	8	6	4	2			
3	1	10	8	6	4	2	11	9	7	5			
2	11	9	7	5	3	1	10	8	6	4			

Figure 4.3

We next need to construct IPOLS(n;10) for $31 \leq n \leq 89$ and n not divisible by 5, as if 5 divides n we can apply Corollary 1.7 with $p = 5$, $q = n/5$ and $c = 2$. When $n = 34$, or $38 \leq n \leq 46$ apply Corollary 1.4 with $p = 4$, $q = 8$, $a_1 = 2$ and $a_i \in \{0,1,2,3,4\}$, $2 \leq i \leq p$, as required. Repeat this for $47 \leq n \leq 58$ but with $p = 5$; for $62 \leq n \leq 82$ with $p = 7$; and for $83 \leq n \leq 89$ with $p = 9$. This leaves seven cases. Corollary 1.5 with $p = 10$, $q = 3$ and $a_1 = 1$ gives $n = 31$ and $n = 32$, 36 were given in the proof of Lemma 4.10. For $n = 33$ use Corollary 1.7 with $p = 5$, $q = 6$, $c = 2$ and $a_1 = a_2 = a_3 = 1$. Corollary 1.6 with $p = 4$, $q = 9$ and $a_1 = 1$ yields an IPOLS(37;10). For an IPOLS(59;10) apply Corollary 1.5 with $p = 5$, $q = 10$, $a_1 = a_2 = a_3 = 3$ and for an IPOLS(61;10) repeat but with $a_4 = 2$. □

Thus the existence of an IPOLS(n;k) is now completely solved. Very closely related to this problem is the question of the existence of an ISOLS(n;k) when $n \geq 3k+1$. In fact it was this question that was studied both by D.J.Crampin and A.J.W. Hilton (1975b), and by D.A. Drake and H. Lenz (1980). The question of their existence has not yet been completely solved. K. Heinrich and L. Zhu (1987) have shown that an ISOLS(n;k) exists for all $n \geq 3k+1$, (n;k) ≠ (6;1), (8;2), except perhaps when (n;k) ∈ {(6m+2;2m): $m \geq 2$} ∪ {(6m+6;2m): $m \geq 2$}. Very recently K. Heinrich, L. Wu and L. Zhu (personal communication) have constructed all ISOLS(6m+6;2m). However, no ISOLS(6m+2;2m) is known and in fact computer search shows that there is no ISOLS(8;2). With this in mind it is interesting to note that D.R. Stinson (1986) has recently constructed 3-IPOLS(8;2).

An ISOLS(n;k) is an ILS(n;k) which is orthogonal to its transpose; the second square is obtained from the first by exchanging rows and columns. F.E. Bennett and L. Zhu (1987) have studied the problem of constructing COILS(n;k), which is an ILS(n;k) with an orthogonal mate obtained from it by exchanging rows and entries, and have shown that for $n \geq 3k+1$ except for finitely many exceptions a COILS(n;k) exists.

At this point we should mention that apart from the work done directly on constructing IPOLS(n;k) and ISOLS(n;k) their existence has arisen in many other contexts. F.E. Bennett and N.S. Mendelsohn

(1980) in their study of the existence of Stein quasigroups showed that for all $n \geq 191$ (and for many smaller n) there exists a Stein quasigroup of order n with Stein subquasigroups of orders 4 and 5. Since a Stein quasigroup yields a self-orthogonal latin square of order n this result implies the existence of ISOLS(n;4) and ISOLS(n;5) for $n \geq 191$. In an unpublished manuscript F.E. Bennett extended this to construct an ISOLS(n;4) for all $n \geq 13$ except perhaps for $n \in \{14, 18, 19, 23, 26, 27, 30, 35\}$, and an ISOLS(n;5) for all $n \geq 16$ except perhaps when $n \in \{19, 22, 23, 26, 27, 28, 30, 34, 38\}$. He obtained similar results for $k = 7, 8$ and 9. (We also remark that in C.C.Lindner, E. Mendelsohn, N.S. Mendelsohn and B. Wolk (1979) it was pointed out that the graph of the conjugates of such a Stein quasigroup realizes a 6-cycle.)

About the same time D.A. Drake and J.A. Larson (1983), while studying the existence of PBD(n;L) where $L = Z^+ \setminus \{2, 3, 6\}$ (see Section 1 for the definition), showed that for all $n \in \{13, 16, 17, 20, 21, 22, 24, 25, 28, 29\} \cup \{m : m \geq 31\}$ there exists a PBD(n;L) with a block of size 4 and that for all other n except perhaps $n = 30$, the design is impossible. They also showed that for all $n \in \{17, 20, 21, 24, 25, 29, 31, 32, 33\} \cup \{m : m \geq 35\}$ there exists a PBD(n;L) with a block of size 5 and such a design for all other n except perhaps $n = 28, 30$ and 34 is impossible. Of course, these results immediately yield ISOLS(n;4) and ISOLS(n;5).

Also related to IPOLS(n;k) are what are called r-orthogonal quasigroups. Two LS(n) are r-orthogonal if when they are placed in juxtaposition, exactly r distinct ordered pairs are obtained, $n \leq r \leq n^2$. So, for example, an IPOLS(n;k) yields an $(n^2 - k^2 + r)$ - orthogonal LS(n) if two r-orthogonal LS(k) exist. These r-orthogonal latin squares have been studied extensively by G.B. Belyavskaya (1976), (1977) and (1979) (see also Chapter 6). In particular she studied the existence of an IPOLS(n;2) so as to construct a pair of $(n^2 - 2)$ - orthogonal LS(n). Note that pairs of $(n^2 - 1)$-orthogonal LS(n) do not exist. She showed [G.B. Belyavskaya (1976)], using an algebraic method, that an IPOLS(n;2) exists when $n \equiv 0, 2 \pmod{6}$.

Very little work has been done on constructing t - IPOLS(n; $s_1, s_2, \ldots, s_m$) for $t > 2$ or $m > 1$. Of course, when $t = 2$ we have

IPOLS($n;s_1,\ldots,s_m$) where $s_i = 1$, $1 \leq i \leq m$ and $1 \leq m \leq n$, and $n \neq 2,3$ or 6. J.H. Dinitz and D.R. Stinson (1983) have shown that there exists an IPOLS($2m;s_1,\ldots,s_m$) with $s_i = 2$, $1 \leq i \leq m$, if and only if $m \geq 4$ and that there exist 6-IPOLS($2m;s_1,\ldots,s_m$) with $s_i = 2$, $1 \leq i \leq m$, for several primes m. (Note that in P.J. Schellenberg, D.R. Stinson, S.A. Vanstone and J.W. Yates (1981) the main recursive construction for Howell designs depends on the existence of an IPOLS($n;s_1,s_2,\ldots,s_m$) where $n = \sum_{i=1}^{m} s_i$.)

Sets of t-IPOLS($2m;s_1,\ldots,s_m$), $s_i = 2$, $1 \leq i \leq m$, have also been studied by R.C. Mullin and D.R. Stinson (1984) who have constructed 3-IPOLS($2m;s_1,\ldots,s_m$), where m is odd $s_i = 2$, $1 \leq i \leq m$, and have shown that with only a few possible exceptions they always exist. When m is even L. Zhu (1984d) has constructed all but finitely many 3-IPOLS($2m;s_1,s_2,\ldots,s_m$) where $s_i = 2$, $1 \leq i \leq m$. In fact the three squares constructed always consist of a square and its transpose, while the third is symmetric. These arrays have several applications and, for example, were used by C.C. Lindner and D.R. Stinson (1984) to construct Steiner pentagon systems. A related design, an SOLSSOM($n;k$), which consists of an ISOLS($n;k$) and a symmetric LS($n;k$) which together form 3-IPOLS($n;k$), has been used in C.C. Lindner, R.C. Mullin and D.R. Stinson (1983) to construct resolvable orthogonal arrays which are invariant under the Klein group.

Finally, we comment on the existence of t-IPOLS($n;k$). Clearly any set of orthogonal latin squares constructed by prolongation or some kind of product will have subsquares. Probably the most surprising result is that of A.E. Brouwer (unpublished manuscript) who has constructed 4-IPOLS(10;2) which is shown in D.R.Stinson (1986). L. Zhu (1983) has shown that t-IPOLS($n;k$) exist for $n \geq (t+1)k$ with only a finite number of exceptions when $3 \leq t \leq 5$ and $7 \leq k \leq 9$ (other partial results on 3-IPOLS($n;k$) have been obtained by E. Seiden and C.-J. Wu (1976)), while D.A. Drake and H. Lenz (1980) had much earlier shown that for $n \geq 7k+7$ and $k \geq 837$, there exist 5-IPOLS($n;k$).

(5) Acknowledgement.

The research for this chapter was financially supported by the Natural Sciences and Engineering Research Council of Canada under Grant A-7829. The author is grateful for the hospitality of the University of Newcastle, Australia, where this work was begun.

ADDITIONAL REMARKS BY THE EDITORS.

With reference to Theorem 1.12 and also the statement which precedes Lemma 4.8, we draw the attention of the reader to the fact that, since this Chapter was written, T. D. Todorov(1989) has constructed four POLS of order 20.

With reference to the remark in Section 2 about an N_∞-latin square of order 12, see P. Gibbons and E. Mendelsohn(1987). Also, with reference to the remark in Section 4 about the construction by K. Heinrich et al of all ISOLS(6m+6; 2m), see K. Heinrich, L. Wu and L. Zhu(199α) and with reference to the remark made near the end of Section 4 about the construction of 4-IPOLS(10; 2), see A. E. Brouwer(1984).

CHAPTER 5

RECURSIVE CONSTRUCTIONS OF MUTUALLY ORTHOGONAL LATIN SQUARES (A.E.Brouwer)

In the historical development of the theory of latin squares, one of the most intriguing questions and strongest motivations for research has been the famous conjecture of Euler postulating the non-existence of pairs of orthogonal latin squares of order 4k+2. The resolution of this question in the late 1950's led to the study of the values of the function $N(v)$, which denotes the maximum number of latin squares of order v in a pairwise orthogonal set, and more recently to the study also of the related function n_r, which denotes the least positive integer such that, for all $v > n_r$, sets of r mutually orthogonal latin squares of order v exist. Thus, for example, the resolution of the Euler conjecture in the negative amounted to proving that $N(v) \geq 2$ for all $v \neq 2$ or 6 or, equivalently, that $n_2 = 6$.

One of the most interesting questions of the present time and one which has been a stimulus for many current researchers is the question whether $N(10)$ is or is not greater than two. This question has been made even more tantalizing by the construction in 1978 of a set of four latin squares of order 10 which are almost pairwise orthogonal by A.E.Brouwer, the author of the present chapter. [See A.E.Brouwer(1984) for details.] Other interesting results obtained since the publication of [DK] have been the construction of a set of four mutually orthogonal latin squares of order 15 by P.J.Schellenburg, G.H.J.Van-Rees and S.A.Vanstone(1978) and the construction of a set of three mutually orthogonal latin squares of order 14 by D.T.Todorov(1985).

It is probably true to say that A.E.Brouwer has done more work on improving the known bounds for $N(v)$ and n_r than almost anyone else and we were therefore delighted when he agreed to provide this contribution to the present book.

(1) **Introductory definitions**.

Let S be a fixed set (of "symbols") of cardinality n. We say that two n×n matrices $A = (a_{ij})$ and $B = (b_{ij})$ with entries in S are <u>orthogonal</u> if for all $(s,t) \in S \times S$ there is a unique position (i,j) such that $a_{ij} = s$ and $b_{ij} = t$. In this Chapter we shall be interested in constructing large sets of pairwise orthogonal matrices.

The connection with Latin squares is as follows: given a set $A^{(k)}$ ($k = 1, \ldots, r$) of pairwise orthogonal matrices we may (after permuting the n^2 positions) assume that $A^{(1)}$ has constant rows and $A^{(2)}$ has constant columns. Now each $A^{(k)}$ ($k = 3, \ldots, r$) is a Latin square, and we have found $r-2$ mutually orthogonal Latin squares.

Conversely, given $r-2$ mutually orthogonal Latin squares we can add two orthogonal matrices, one with constant rows and one with constant columns and get a set of r pairwise orthogonal matrices.

EDITORS' REMARK. To accord with the definitions of Chapter 6 two matrices related as above should be called n^2-<u>orthogonal</u>. In Chapter 11 of [DK], where the connections between orthogonal latin squares, matrices and arrays described here were also given, they were called <u>n-orthogonal</u>.

Clearly, for the concept of orthogonality the matrix structure does not play a rôle: that is, we might as well talk about orthogonal vectors of length n^2. If we define an <u>orthogonal array</u> $OA(n,r)$ of order n and depth r to be an $r \times n^2$ matrix over S such that any two rows are orthogonal, then an $OA(n,r)$ is equivalent to a set of r pairwise orthogonal matrices of order n. (Similarly one might consider orthogonal arrays <u>of strength t</u>, that is $r \times n^t$ matrices A over S such that for any t rows $i_1, \ldots, i_t$ and any t symbols $s_1, \ldots, s_t$ there is a unique column j such that $a_{i_k j} = s_k$ ($1 \leq k \leq t$). See also page 190 of [DK]. Unless the contrary is explicitly mentioned, all orthogonal arrays in this Chapter will have strength 2.)

A more geometric picture is obtained by regarding the columns of an orthogonal array as the lines of a geometry. If R is the set of rows of the orthogonal array A, then take as point set the set $R \times S$, and let the line L_j corresponding to the j-th column be the set $L_j = \{(i, a_{ij}) : i \in R\}$. What we get is called a transversal design. (This concept is defined in the next Section.)

(2) Pairwise balanced designs - definitions.

A pairwise balanced design is a set X (of points) together with a set B of subsets of X (called blocks) such that for some integer λ each 2-subset $\{x,y\}$ of X is contained in precisely λ blocks. The number λ is called the index of the design, and unless specified otherwise we shall always assume that $\lambda = 1$. (In particular we need not worry about the possibility of repeated blocks.) When $\lambda = 1$ two blocks have at most one point in common, and the blocks are sometimes called lines and the design a linear space (not to be confused with the linear spaces from linear algebra).

More generally, a partial linear space or near-linear space [see L.Batten(1986)] is a set X (of points) together with a set L of subsets of X (called lines) such that two points are joined by at most one line.

A set of blocks C of a pairwise balanced design is called a clear set if the blocks of C are pairwise disjoint. It is called a parallel class if C is a partition of X.

If ∞ is an element not in X then there is a natural 1-1 correspondence between pairwise balanced designs (of index unity) on $X \cup \{\infty\}$ and pairwise balanced designs with designated parallel class C on X: if the latter has block set B the the former has block set $(B \backslash C) \cup C^*$ where $C^* = \{C \cup \{\infty\} : C \in C\}$.

If a pairwise balanced design (X,B) has several pairwise disjoint parallel classes $C_j (1 \leq j \leq k)$ then one obtains a new pairwise balanced design by "adding points at infinity": find k new points $\infty_1, \ldots, \infty_k$ and put

$$\bar{X} = X \cup \{\infty_1, \ldots, \infty_k\}, \quad \bar{B} = (B \backslash \bigcup_{j=1}^{k} C_j) \cup \bigcup_{j=1}^{k} C_j^* \cup \{\{\infty_1, \ldots, \infty_k\}\},$$ where

$C_j^* = \{C \cup \{\infty_j\} : C \in C_j\}$. (An example is the construction of a projective plane from an affine plane by adding "a line at infinity").

A pairwise balanced design (X,B) is called resolvable if B can be partitioned into parallel classes.

A group divisible design is a set X (of points), a partiton G of X (the subsets of which are called groups) and a collection B of subsets of X (the blocks) such that $(X, B \cup G)$ is a pairwise balanced design of index unity. (In other words, we have a pairwise balanced design with designated parallel class, and decide to call the subsets of this parallel class groups instead of blocks. There should be no confusion with the algebraic concept of group). (Several more general versions of the concept of group divisible design are in common use. A fairly standard definition says that a group divisible design with indices λ_1 and λ_2 is a set X, a partition G of X and a collection B of blocks such that if $x, y \in X$ are two points in the same group $G \in G$ then $\{x, y\}$ is in λ_1 blocks, otherwise $\{x, y\}$ is in λ_2 blocks. H.Hanani(1975b) takes $\lambda_1 = 0$, $\lambda_2 = \lambda$. We take $\lambda_1 = 0$, $\lambda_2 = 1$. For the general concept and, more generally, for partially balanced incomplete block designs see D.Raghavarao(1971)).

A pairwise balanced design with blocks of size k on $|X| = v$ points is denoted by $B(k;v)$. A group divisible design with blocks of size k and groups of size m on v points is denoted by $GD(k,m;v)$. A transversal design $TD(r;n)$ is a group divisible design $GD(r,n;rn)$.

If several block sizes may occur, we write $B(K;v)$ when each occurring block size is a member of the set K, and similarly $GD(K,M;v)$. [For a study of transversal designs with index $\lambda > 1$ see H.Hanani(1975a)].

(3) Simple constructions for transversal designs

As we have seen, the three concepts of set of (r-2) mutually orthogonal Latin squares and orthogonal array (of depth r) and transversal design (with r groups) are equivalent. We shall mostly use the language of transversal designs. We shall denote the set of all n such that a $TD(r;n)$ exists by $TD(r)$.

THEOREM 3.1 The integers $0, 1 \in TD(r)$ for all r. If $n \in TD(r)$, $n > 1$, then $n \geq r-1$, and $r-1 \in TD(r)$ if and only if there exists a projective plane of order r-1.

Proof. A $TD(r;0)$ is a design with no points and no blocks. A $TD(r;1)$ is a design with r points, all in the unique block. If $n > 1$ and $r \geq 2$

then let B be a fixed block, G a fixed group and x a point not in B∪G. The r-1 blocks on x meting B meet G in distinct points, so n = $|G| \geq r-1$. If n = r-1 then any two blocks meet, and adding a point at infinity to the parallel class formed by the groups produces a projective plane with lines of size r.

Conversely, given a projective plane with lines of size r, removal of a point yields a TD(r;r-1). []

EDITORS' REMARK. An alternative geometrical picture is obtained by observing that the concepts of <u>geometric net</u> of order n with r parallel classes (defined in Section 1 of Chapter 11) and transversal design TD(r;n) are dual concepts. The rn lines of the net are the rn points of the transversal design, the pencils of r lines each through the n^2 finite points of the net form the n^2 blocks of the design and the r parallel classes of lines of the net (each of which contains n lines) correspond to the r groups of the design each of which contains n points. We have n≥r-1 always since, if G is a fixed parallel class of the net, B a fixed point of the net (necessarily on one of the lines g of G) and x a line not incident with B and not belonging to the parallel class G, then the r-1 lines through B distinct from g (one from each parallel class other than G) meet x in distinct points.

$$\begin{array}{ccccccccc}
\alpha_1 & \alpha_1 & \alpha_1 & \beta_1 & \beta_1 & \beta_1 & \gamma_1 & \gamma_1 & \gamma_1 \\
\alpha_2 & \beta_2 & \gamma_2 & \alpha_2 & \beta_2 & \gamma_2 & \alpha_2 & \beta_2 & \gamma_2 \\
\alpha_3 & \beta_3 & \gamma_3 & \beta_3 & \gamma_3 & \alpha_3 & \gamma_3 & \alpha_3 & \beta_3 \\
\alpha_4 & \beta_4 & \gamma_4 & \gamma_4 & \alpha_4 & \beta_4 & \beta_4 & \gamma_4 & \alpha_4
\end{array}$$

Figure 3.1

EXAMPLE (illustrating Theorem 3.1). The columns of the orthogonal array OA(4;3^2) given in Figure 3.1 are the (nine) blocks of a group divisible design GD(4,3;12) whose four groups are the following: $\{\alpha_1,\beta_1,\gamma_1\}, \{\alpha_2,\beta_2,\gamma_2\}, \{\alpha_3,\beta_3,\gamma_3\}, \{\alpha_4,\beta_4,\gamma_4\}$. Thus, 3∈TD(4) and this corresponds to the fact that a projective plane of order 3 (or 4-net of order 3) exists. The two orthogonal latin squares or four orthogonal 3×3 matrices which correspond to this plane are shown in Figure 3.2.

$$\begin{array}{|ccc|}\hline \alpha_1 & \alpha_1 & \alpha_1 \\ \beta_1 & \beta_1 & \beta_1 \\ \gamma_1 & \gamma_1 & \gamma_1 \\ \hline \end{array} \quad \begin{array}{|ccc|}\hline \alpha_2 & \beta_2 & \gamma_2 \\ \alpha_2 & \beta_2 & \gamma_2 \\ \alpha_2 & \beta_2 & \gamma_2 \\ \hline \end{array} \quad \begin{array}{|ccc|}\hline \alpha_3 & \beta_3 & \gamma_3 \\ \beta_3 & \gamma_3 & \alpha_3 \\ \gamma_3 & \alpha_3 & \beta_3 \\ \hline \end{array} \quad \begin{array}{|ccc|}\hline \alpha_4 & \beta_4 & \gamma_4 \\ \gamma_4 & \alpha_4 & \beta_4 \\ \beta_4 & \gamma_4 & \alpha_4 \\ \hline \end{array}$$

Figure 3.2

THEOREM 3.2 [K.A.Bush(1952), H.F.MacNeish(1922)]. If $m, n \in TD(r)$ then $mn \in TD(r)$.

Proof. Given r pairwise orthogonal matrices $A^{(i)}$ over a symbol set S and r pairwise orthogonal matrices $B^{(i)}$ over a symbol set T ($1 \leq i \leq r$), the r matrices $A^{(i)} \times B^{(i)}$ over the symbol set $S \times T$ will be pairwise orthogonal. []

COROLLARY. Let $n = \prod_i p_i^{e_i}$ be the factorization of n into prime powers. Then $n \in TD(r)$ for $r = \min_i (p_i^{e_i}+1)$.

Proof. If q is a prime power then there exists a projective plane of order q and hence $q \in TD(r)$ for $r \leq q+1$. []

THEOREM 3.3. [E.T.Parker(1959), R.C.Bose and S.S.Shrikhande(1960)]. Let (X, B) be a pairwise balanced design such that for each $B \in B$ we have $|B| \in TD(r+1)$. Then $|X| \in TD(r)$.

Proof. If R is an r-set then construct a transversal design with point set $R \times X$ and groups $\{y\} \times X$ for $y \in R$ with subdesigns $TD(r; |B|)$ on $R \times B$ for each $B \in B$, taking care that each of these subdesigns contains the blocks $R \times \{b\}$ for $b \in B$. This will yield the required design. For each $B \in B$, the $TD(r; |B|)$ with parallel class that we need is obtained from the given $TD(r+1; |B|)$ by throwing away one group and taking as parallel class the blocks that used to contain a fixed point of this thrown-away group. []

This construction can be strengthened in many ways. First of all one can weaken the hypothesis "$|B| \in TD(r+1)$" to "there exists a $TD(r; |B|)$ with a parallel class", and strengthen the conclusion to "there exists a $TD(r; |X|)$ with a parallel class".

5:7   *Recursive constructions*   155

Let us call a transversal design $TD(r;n)$ with e pairwise disjoint parallel classes a $TD_e(r;n)$. Adding points at infinity shows that a $TD_n(r;n)$ exists if and only if a $TD(r+1);n)$ exists.

A second variation on Theorem 3.3 requires $|B| \in TD_1(r)$ for all $B \in B \backslash C$ where C is a clear set of blocks, and $|B| \in TD(r)$ for $B \in C$. (Now the conclusion is $|X| \in TD(r)$). It even suffices to ask that C be an almost clear set: that is, that for each $C \in C$ there is at most one $x \in C$ such that x is member of more than one block of C.

A third version is the following.

THEOREM 3.4. [E.T.Parker(1959), R.C.Bose and S.S.Shrikhande(1960)]. Let (X,B) be a pairwise balanced design such that B has a partition $\{B_j\}_j$, where each family B_j has blocks of constant size k_j and is either a partition of X or a symmetric 1-design on X. (Such a design is called separable). Assume that $|B| \in TD(r)$ for each $B \in B$. Then $|X| \in TD(r)$.

Proof. By the previous argument, we have $|X| \in TD(r-1)$. We shall show that a $TD(r-1;n)$ (where $n = |X|$) can be constructed so as to possess n pairwise disjoint parallel classes; then adding points at infinity will show that $n \in TD(r)$. Indeed, if B_j is a partition of X then for each $B \in B_j$ construct a transversal design $TD_k(r-1;k)$ (where $k = k_j$) with pointset $R \times B$ and groups $\{y\} \times B$, $y \in R$, where R is a fixed $(r-1)$-set. If we number the parallel classes of each of these designs from 1 to k (and make sure that the "verticals" $R \times \{b\}$ belong to parallel class 1 for all b) then the union of the parallel classes which have a given number is a parallel class on $R \times X$.

On the other hand, if B_j is a symmetric 1-design on X then we cannot construct the transversal designs on $R \times B$ ($B \in B_j$) independently. Instead, let N be the point-block incidence matrix of (X,B_j) and write N as the sum of $k = k_j$ permutation matrices N_t ($1 \leq t \leq k$). We may regard N_t as a 1-1 correspondence $\phi_t: B_j \to X$.

Let B_0 be a fixed block in B_j and construct a $TD_1(r-1,k)$ on $R \times B_0$ containing the verticals. For each non-vertical block T of this design construct a parallel class $\{T_B: B \in B_j\}$ with $T_{B_0} = T$ and such that for each $r \in R$ the transversal T_B contains the point $(r, \phi_t B)$, where t is determined by $(r, \phi_t B_0) \in T$.

In this way we "transport" the transversal design on $R \times B_0$ and construct isomorphic copies on $R \times B$ for all $B \in B_j$, but in such a way that the entire collection of blocks is resolvable into parallel classes.

Taking all the blocks found in this way, and the groups $\{y\} \times X$, $y \in R$ yields the required design. □

As a modification to the previous idea of transporting a transversal design around a symmetric 1-design, suppose (X,B) is as in Theorem 3.4, and that B_j is a symmetric 1-design. This time construct a $TD_k(r-1, k+1)$ on $R \times (B_0 \cup \infty)$ where ∞ is a new element. Repeating the previous construction, we find, for each point (r, ∞), k almost parallel classes of transversals (disjoint apart from the common point (r, ∞)); for the i-th of these almost parallel classes relabel the point (r, ∞) as (r, ∞_i) ($1 \leq i \leq k$) so that different almost parallel classes have different points (r, ∞_i) in common. This yields:

THEOREM 3.5. Let (X,B) be a pairwise balanced design such that B contains pairwise disjoint families B_j such that (X, B_j) is a symmetric 1-design with block size k_j. Suppose that $|B| \in TD(r+1)$ for $B \in B \setminus \bigcup_j B_j$ and that $|B+1| \in TD(r+1)$ for $B \in \bigcup_j B_j$. Finally suppose that $\sum_j k_j \in TD(r)$. Then $|X| + \sum_j k_j \in TD(r)$.

We shall call this construction "adding points at infinity to a symmetric 1-design". Note that this terminology is misleading: we do not construct a pairwise balanced design on n+k points, but only a transversal design with groups of that size.

(3)* Examples

Let $N(\nu)$ be the maximum number of mutually orthogonal Latin squares of order ν. We have $N(0) = N(1) = \infty$, $N(q) = q-1$ for prime powers q and $N(\nu) \leq \nu-1$ for arbitrary ν. The statements $\nu \in TD(r)$ and $N(\nu) \geq r-2$ are equivalent.

(i) We may apply Theorem 3.3 with a projective plane as design

(X,B). This yields for prime powers q that $N(q^2+q+1) \geq N(q+1)$. Usually this bound is bad, but when q+1 is also a prime power we get $N(q^2+q+1) \geq q$.

EXAMPLES: $N(21) \geq 4$, $N(57) \geq 7$, $N(273) \geq 16$, $N(993) \geq 31$.

(ii) If q is a prime power then there exists a $2 \text{-} (q^3+1, q+1, 1)$ design (a "unital", the isotropic points and hyperbolic lines in the projective plane $PG(2,q^2)$ with a unitary polarity). This design is resolvable with q^2 parallel classes, and adding q^2 points at infinity yields a pairwise balanced design $B(\{q+2, q^2\}; q^3+q^2+1)$. In the case when q+2 also is a prime power, this yields $N(q^3+q^2+1) \geq q$.

EXAMPLE: $N(393) \geq 7$.

(iii) If q is an even prime power then there exists a resolvable $2\text{-}(\frac{1}{2}q(q-1), \frac{1}{2}q, 1)$ design (where points and blocks are the exterior lines and points of a hyperoval in $PG(2,q)$) with q+1 parallel classes. Thus we find a $B(\{\frac{1}{2}q, \frac{1}{2}q+1, x\}; \frac{1}{2}q(q-1)+x)$ by adding x points at infinity $(0 \leq x \leq q+1)$, where blocksize $\frac{1}{2}q+1$ does not occur for $x = 0$ and blocksize $\frac{1}{2}q$ does not occur for $x = q+1$.

EXAMPLES:
$N(120) \geq 7$ (q = 16, design resolvable),
$N(136) \geq 7$ (q = 16, x = 16, one parallel class of blocks of size 8),
$N(504) \geq 7$ (q = 32, x = 8)
$N(528) \geq 15$ (q = 32, x = 32, one parallel class of blocks of size 16),
$N(2016) \geq 31$ (q = 64, design resolvable).

(iv) Useful pairwise balanced designs can often be constructed from a projective or affine plane by throwing away a suitably chosen set of points.

Throwing away one point from $PG(2,q)$, we obtain a $B(\{q, q+1\}; q^2+q)$ where the blocks of size q form a parallel class. If q+1 is a prime power, then it follows that $N(q^2+q) \geq q-1$.

EXAMPLES: $N(20) \geq 3$, $N(72) \geq 7$, $N(272) \geq 15$, $N(992) \geq 30$.

Starting with $AG(2,q)$ instead we find (if $q-1$ and q are prime powers) that $N(q^2-1) \geq q-2$.

EXAMPLES: $N(24) \geq 3$, $N(63) \geq 6$, $N(80) \geq 7$, $N(288) \geq 15$, $N(1023) \geq 30$.

Throwing away x points from one line of $PG(2,q)$ or $AG(2,q)$, we find a $B(\{q+1-x,q,q+1\};q^2+q+1-x)$ or $B(\{q-x,q-1,q\};q^2-x)$. In this way one gets
$N(54) \geq 4$ (q = 7, x = 3), $N(280) \geq 7$ (q = 17, x = 9),
$N(264) \geq 7$, $N(265) \geq 8$, $N(267) \geq 10$ (q = 16, x = 9,8,6),
$N(285) \geq 12$ (q = 17, x = 4)
$N(993-x) \geq 31-x$ (q = 31, x = 3,5,7,13,15,19,21,23),
$N(1024-x) \geq 31-x$ (q = 32, x = 7,9,13,16,24).

If $q \equiv 0$ or 1 (mod 3) then $PG(2,q)$ contains a subconfiguration isomorphic to $AG(2,3)$, and removing that yields a $B(\{q-2,q,q+1\};\nu-9)$. For q = 31 this shows that $N(984) \geq 27$.

Throwing away x points from a (hyper)oval in $PG(2,q)$ or $AG(2,q)$ yields a $B(\{q-1,q,q+1\};q^2+q+1-x)$ or a $B(\{q-2,q-1,q\};q^2-x)$ for $x \leq q+1$ (or $x \leq q+2$ if q is even). Since 7,8,9 are three consecutive prime powers, we we find with q = 8: $N(66) \geq 5$, $N(68) \geq 5$, $N(69) \geq 6$, $N(70) \geq 6$, and with q = 9: $N(74) \geq 5$, $N(75) \geq 5$, $N(76) \geq 5$, $N(78) \geq 6$.

Note that the blocks of size 7 form a clear set in $B(\{7,8,9\};70)$ and in $B(\{7,8,9\};78)$ and an almost clear set in $B(\{7,8,9\};69)$. (After writing this I found that L.Zhu(1984e) had made the same observation.)

(v) Continuing in this vein, we note that $PG(2,q^2)$ has a partition into Baer subplanes, and taking t of those produces a $B(\{t,q+t\};t(q^2+q+1))$ where the collection of blocks of size q+t forms a symmetric 1-design and the collection of blocks of size t is resolvable into $q^2-q+1-t$ parallel classes. This yields many useable pairwise balanced designs.

EXAMPLES: $N(189) \geq 8$ (q = 4, t = 9),
             $N(253) \geq 12$ (q = 4, t = 12, add one point at infinity to
                                  get an almost clear set of blocks of size 13),

$N(357) \geq 9$ ($q = 5$, $t = 11$, add 16 points at infinity to the symmetric 1-design).

[For more details, see A.E.Brouwer(1980a)].

(vi) Adding $q+1$ points at infinity to the symmetric 2-design $PG(2,q)$, we find if both $q+1$ and $q+2$ are prime powers: $N((q+1)^2+1) \geq q$.

EXAMPLES: $N(10) \geq 2$, $N(65) \geq 7$.

(vii) From a Singer difference set we find a separable subdesign $B(\{9,13,16\};469)$ in $PG(2,37)$. It follows that $N(469) \geq 8$. [Again, see A.E.Brouwer(1980a)].

In these examples I have listed virtually all the instances I know of cases in which the pairwise balanced design construction yields the best known bound on $N(\nu)$, and in which the pairwise balanced design is not itself a (truncated) transversal design. (In fact, in case (iv), it is a truncated transversal design). In the next Section we shall see that one can do better with a transversal design as ingredient than with a general pairwise balanced design at starting point.

(4) <u>Wilson's Construction</u>.

Applying the construction of Theorem 3.5 to a transversal design $TD(m+1;t)$ of which one group has been truncated to size h (so that we have a $GD(\{m,m+1\},\{t,h\};mt+h)$) we obtain the following results:
If $N(t) \geq m-1$ then, for $0 < h < t$, $N(mt+h) \geq \min\{N(t), N(h), N(m)-1, N(m+1)-1\}$
If $N(t) \geq m-2$, then $N(mt) \geq \min\{N(m)-1, N(t)\}$.

Clearly the second bound is worse than MacNeish's bound. The first one is always worse (or at least not better) than Wilson's bound [see Theorem 2.3 of R.M.Wilson(1974a)] which is as follows:

THEOREM 4.1. If $0 < h < t$, then $N(mt+h) \geq \min\{N(m), N(m+1), N(t)-1, N(h)\}$. (For, if $m-1 \leq N(t)$ then $N(m)-1 < N(t)$, so in the first bound the minimum cannot be $N(t)$).

This bound and Wojtas's bound [see M.Wojtas(1977)] given in Theorem 4.2 below, are two of the best lower bounds for $N(\nu)$ known.

THEOREM 4.2.   If $0 < h < t$, then $N(mt+h) \geq \min\{N(m), N(m+1), N(m+h), N(t)-h\}$.

Both the Wilson bound and the Wojtas bound can be obtained from special cases of Wilson's construction, which we shall now describe.

Ingredients for the construction.

(1) A transversal design $TD(k+\ell; t)$ of which ℓ groups have been truncated, so that k groups have size t and the remaining groups have size $h_i (1 \leq i \leq \ell)$ where clearly $0 \leq h_i \leq t$. We denote the union of the ℓ truncated groups by H so that $|H| = \sum_{i=1}^{\ell} h_i = h$ say

(2) Transversal designs $TD(k; h_i)$ for $1 \leq i \leq \ell$.

(3) Transversal designs $TD(k; m+|B \cap H|)$ for each block B from the $TD(k+\ell; t)$ with $|B \cap H|$ pairwise disjoint blocks.

We construct a $TD(k; mt+h)$ in the obvious way (given the ingredients and the result to be obtained) as follows:

Let the $TD(k+\ell; t)$ have groups $G_1, G_2, \ldots, G_k, H_1, H_2, \ldots, H_\ell$. Then the constructed designs will have groups $(G_j \times M) \cup (H \times \{j\})$ for $j = 1, 2, \ldots, k$, all of size $mt+h$, where M is an arbitrary set of cardinality M. Put ingredients (2) on the sets $H_i \times K$, $1 \leq i \leq \ell$, where $K = \{1, 2, \ldots, k\}$. For each block B from ingredient (1), the set $(B \setminus H) \times M \cup (B \cap H) \times K$ has cardinality $k(m+|B \cap H|)$: put ingredient (3) on this set in such a way that the groups of this design are subsets of the design to be constructed and so that, for each $b \in B \cap H$, the set $\{b\} \times K$ is a block. It is straightforward to check that this construction works.

Theorem 4.1 is obtained by taking $\ell = 1$ and $h_1 = h$. Theorem 4.2 is obtained by taking $\ell = h$ and $h_i = 1$, $1 \leq i \leq \ell$. If we take $\ell = 2$, $h_1 = u$ and $h_2 = v$, we get the following theorem [see Theorem 2.4 of R.M.Wilson(1974a)].

THEOREM 4.3. If $0<u,v<t$, then
$$N(mt+u+v) \geq \min\{N(m), N(m+1), N(m+2), N(u), N(v), N(t)-2\}.$$

(4)* Examples

First we give the following specific examples which use the constructions described in Section 4 above.

(i) $N(95) \geq 6$ follows from Theorem 3.5 using a truncated $TD(9;11)$ since $95 = 8.11+7$.

(ii) $N(33) \geq 3$ follows from Theorem 4.1 since $33 = 4.8+1$.

$N(84) \geq 6$ follows from Theorem 4.1 since $84 = 7.11+7$.

(iii) $N(91) \geq 7$ follows from Theorem 4.2 since $91 = 8.11+3$.

(iv) $N(94) \geq 6$ follows from Theorem 4.3 since $94 = 7.11+8+9$.

(v) $N(90) \geq 6$ [M.Wojtas(1980b)] follows since $90 = 6.11+8+8+8$ and we can truncate a $TD(9;11)$ in such a way that each block meets H in at least one point. (In general, taking $\ell = 3$, one can obtain the condition that $B \cap H = \phi$ for all B. Certainly this is the case when $h_1 \leq h_2$ and $(t-h_1)(t-h_2) < h_3$.) Another example is $N(796) \geq 7$ since $796 = 70.11+8+8+8$.

(vi) $N(135) \geq 7$ [A.E.Brouwer(1978)] follows since $135 = 8.16+7$ and we can truncate a $TD(15;16)$ in such a way that each block meets the set H in $0,1$ or 3 points, and $h_i = 1$ ($1 \leq i \leq 7$). In fact we may take H to be a Fano subplane of $PG(2,16)$.

(vii) $N(164) \geq 6$ follows since $164 = 7.23+3$ and we can take $h = 3$, $h_1 = h_2 = h_3 = 1$, $|B \cap U| \leq 2$. In fact, for $h \leq t$ and t a prime power, we can take H to be part of an oval in $PG(2,t)$ and obtain $N(mt+h) \geq \min\{N(m), N(m+1), N(m+2), N(t)-h\}$. [See R.M.Wilson(1974a) Theorem 2.5 and A.E.Brouwer(1979)].

(viii) We obtain a variant of Theorem 4.3 due to G.H.J.Van Rees(1978) by taking v points, no three on a line, all on different groups and with $t > \binom{v}{2}$. Then we can add w points all in one group and get
$$N(mt+w+v) \geq \min\{N(m), N(m+1), N(m+2), N(w), N(t)-v-1\} \text{ for } t \geq w+\binom{v}{2}.$$
A table of examples of this construction is given in Figure 4.1. I know of no example with $v \neq 2$ where this construction yields the best known lower bound.

n	m	t	w	v	lower bound for $N(n)$
1554	81	19	13	2	8
1884	81	23	19	2	8
2046	81	25	19	2	8
2298	99	23	19	2	8
2694	99	27	19	2	8
4622	271	17	13	2	12
4776	207	23	13	2	8

Figure 4.1

(5) <u>Weighting and Holes</u>

As was noted by M.Wojtas(1980a) and D.R.Stinson(1979b) in certain special cases and by A.E.Brouwer and G.H.J.Van Rees(1982) in general, one may generalize Wilson's construction by giving weights to the points of H.

In this way one can construct a transversal design $TD(k; mt+ \sum_{h \in H} m_h)$, where m_h is the weight of h (h∈H). Ingredient(1) is unchanged, and ingredients(2) and (3) are replaced by:

(2') Transversal designs $TD(k; \sum_{h \in H_i} m_h)$ for $1 \leq i \leq \ell$.

(3') For each block B from the $TD(k+\ell;t)$, a transversal design $TD(k; m+ \sum_{h \in B \cap H} m_h)$ with pairwise disjoint subdesigns $TD(k; m_h)$ (h∈B∩H).

(The construction is entirely analogous to that in Section 4.)

But one may go further: the only reason one needs the subdesigns in ingredient(3'), is to throw them out in order not to cover certain pairs twice; in other words, what actually is needed is a transversal designs with holes: that is, one needs

(3") $TD(k; m+ \sum_{h \in B \cap H} m_h) - \sum_{h \in B \cap H} TD(k; m_h)$ and ingredient(3") may well exist while ingredient(3') does not.

Let us formally define the concept of "transversal design with holes". The above discussion shows that what we have in mind looks like a transversal design from which a collection of pairwise disjoint subdesigns has been removed.

DEFINITION. A transversal design with holes $TD(k;v) - \sum_{i=1}^{r} TD(k;u_i)$ consists of a set X of cardinality kv (the set of points), a partition G of X into k groups of v elements each, pairwise disjoint subsets Y_i of X ($1 \leq i \leq r$) of cardinality ku_i (the holes) such that $|Y_i \cap G| = u_i$ for each $G \in G$ and each i, $1 \leq i \leq r$, and a collection B of subsets of X of cardinality k (the blocks) such that no block meets a group or a hole in more than one point, and any two points not in the same group or hole are in a unique block.

It follows that $|B| = v^2 - \sum_{i=1}^{r} u_i^2$. For $r = 0$ the concept "transversal design with zero holes" coincides with the usual transversal design. In the case $u_i = 1$ for all i, $1 \leq i \leq r$, then $TD(k;v) - rTD(k;1)$ (in an obvious extension of the notation) exists if and only if a $TD(k;v)$ with r pairwise disjoint blocks exists, showing that (3") generalises (3). If $TD(k;u_i)$ exists we may put it on set Y_i and thus "plug" the hole Y_i, obtaining a transversal design with r-1 holes. Conversely, if a transversal design (with holes) has a subdesign (disjoint from all the holes) we can unplug it and obtain a transversal design with r+1 holes. Not all holes can be filled: J.D. Horton (1974), who introduced the concept "transversal design with one hole" under the name "incomplete array" constructed a $TD(4;6)-TD(4;2)$, but neither a $TD(4;6)$ nor a $TD(4;2)$ exists. (Also A.E. Brouwer (1978) constructed a $TD(6;10)-TD(6;2)$, while not even a $TD(5;10)$ is known.)

As a very important special case (with $\ell = 1$), the following theorem was obtained by A.E. Brouwer (1979):

THEOREM 5.1. If $t = \sum_{j=1}^{p} h_j$ and if the designs $TD(k+1;t)$, $TD(k; \sum_{j=1}^{p} m_j h_j)$ and (for $j = 1, \ldots, p$) $TD(k;m+m_j)-TD(k;m_j)$ all exist, then also a $TD(k;mt+ \sum_{j=1}^{p} m_j h_j)$ exists.

Instead of making holes $TD(k;m_h)$ in the ingredients (3") corresponding to all blocks B on the point $h \in H$ we may leave one such ingredient alone and make a hole in the ingredient (2') corresponding to the group containing h. For the general formulation of this construction and for more details about the construction of transversal designs with holes, see A.E.Brouwer and G.H.J.Van Rees(1982), especially Theorem 1.2. The most important special case is the following theorem:

THEOREM 5.2. [A.E.Brouwer(1980b)]. If $w = \sum_{i=1}^{\ell} w_i$ and TD$(k+\ell;t)$, TD$(k;m)$, TD$(k;m+w)$ and (for $j = 1,\ldots,\ell$) TD$(k;m+w_j)$-TD$(k;w_j)$ all exist, then also TD$(k;mt+w)$ exists.

(5)* Examples.

We give three different illustrations of our constructions.

(i)   We show that $N(5467) \geq 15$. The construction uses a distribution of holes as described following Theorem 5.1. Noting that $5467 = 19.271+289+29$ we apply the construction with $k = 17$, $t = 19$, $m = 271$, $\ell = 2$, $h_1 = 17$, $h_2 = 13$; $289 = 17.17$: the points in H_1 all get weight 17; $29 = 1.17+12.1$: one point x_0 in H_2 gets weight 17, the twelve others weight 1. We need the following ingredients:

(1)   TD$(19;19)$ exists since 19 is prime.

(2)   TD$(17;289)$-17TD$(17;17)$ exists, e.g. by the construction of Theorem 3.5 on the affine plane AG$(2,17)$.

TD$(17;29)$ exists since 29 is prime.

(3)   TD$(17;271)$ exists since 271 is prime.

TD$(17;272)$-TD$(17;1)$ exists by MacNeish: $272 = 16.17$.

TD$(17;288)$-TD$(17;17)$ exists since Wilson's construction for TD$(17;288)$ using $288 = 16.7+16$ yields a design with subdesign TD$(17;17)$.

TD$(17;289)$-TD$(17;17)$-TD$(17;1)$ exists, and is found from AG$(2,17)$.

TD$(17;305)$-TD$(17;17)$ exists since Wilson's construction for TD$(17;305)$ using $305 = 16.19+1$ yields a design with subdesign TD$(17;17)$.

For the standard distribution of holes we would have needed TD$(17;305-2$TD$(17;17)$, but it is not obvious how to obtain this ingredient. Therefore we cover the pairs in the km_h-subsets corresponding to points $h \in H_1$ in the designs corresponding to the (unique) block B containing h and x_0. This yields the required TD(17;5467).

(ii) We show that $N(4738) \geq 8$. (Previously, this was the largest unknown value for 8 squares; it follows that $n_8 \leq 4242$). We note that $4738 = 271.17+(125=7.17+6)+6 \times 1$. Apply the construction with $k = 10$, $t = 17$, $m = 271$, $\ell = 7$, $h_1 = 13$, $h_2 = h_3 = h_4 = h_5 = h_6 = h_7 = 1$; in H_1 give seven points weight 17 and six points weight 1. Give all other points in H weight 1. Choose the six points on $H \setminus H_1$ on a single block B where $B \cap H_1 = \phi$.

(iii) We show that $N(10618) \geq 15$ and that $N(10632) \geq 15$. (These were the largest unknown values for 15 squares; it follows that $n_{15} < 10000$).

$10618 = 435.23+(293=2.16+9.29)+(320=20.16)$

$10632 = 435.23+(128=8.16)+(499=4.16+15.29)$

Ingredients:	23,128,293,499 are prime powers.
$320 = 16.19+16$	shows $N(320) \geq 15$.
$435 = 16.27+3 \times 1$	shows $N(435) \geq 15$.
$451 = 16.27+19$	shows the existence of TD(17;451)-TD(17;16).
$464 = 16.29$	shows the existence of TD(17;464)-TD(17;29).
$467 = 16.29+3 \times 1$	shows the existence of TD(17;467)-2TD(17;16).

(6) <u>Asymptotic Results</u>.

S. Chowla, P. Erdos and E. G. Straus (1960) showed that $\lim_{v \to \infty} N(v) = \infty$. Consequently, we may define n_r to be $\max\{v, N(v) < r\}$ for all $r \geq 2$ as in the Introduction. In fact, S. Chowla et al showed that $n_r < (3r)^{91}$, a result that was improved by K. Rogers (1964) to $n_r < r^{42}$, by Y. Wang (1966) to $n_r < r^{26}$, by R. M. Wilson (1974a) to $n_r < r^{17}$ and by T. Beth (1983) to $n_r < r^{14.8}$, all for sufficiently large r.

For small values of r, explicit upper bounds for n_r have been obtained. The current state of affairs (in 1985) is as follows:

$n_2 = 6$ [R.C. Bose, S.S. Shrikhande and E.T. Parker (1960)]

$n_3 \leq 14$ [S.M.P. Wang and R.M. Wilson (1978)]

$n_4 \leq 52$ [R. Guérin (1966)]

$n_5 \leqslant 62$ [H.Hanani(1970)]
$n_6 \leqslant 76$ [M.Wojtas(1980b)]
$n_7 \leqslant 780$, $n_9 \leqslant 5842$, $n_{10} \leqslant 7222$ [A.E.Brouwer and G.H.J.Van Rees(1982)]
$n_8 \leqslant 4216$, $n_{11} \leqslant 7222$, $n_{12} \leqslant 7286$, $n_{13} \leqslant 7288$, $n_{14} \leqslant 7874$, $n_{15} \leqslant 8360$, $n_{30} \leqslant 52502$ [A.E.Brouwer, unpublished.]

The proofs are by the constructions given above (together with some explicit constructions for small v) coupled with some number theory (trivial for fixed r, sieve methods for large r) required to show that sufficiently large numbers can be written in a suitable form.

EDITORS' REMARK. The construction by D.T.Todorov(1985) of three mutually orthogonal latin squares of order 14 may have affected some of these values. In particular, we have $n_3 \leqslant 10$. The table of values of N(v) given in the next Section has been revised to take account of Todorov's result.

(7) <u>Table of values of N(v) up to v = 200</u>.

The value of v is obtained by adding the column number to the row number in Figure 7.1. The corresponding minimum value of N(v) is in the body of the table. When v>100 is a prime power, the entry in the table is "C", denoting a complete set of mutually orthogonal latin squares, to save space. To complete the values, we remark that $N(200) \geqslant 7$.

ADDITIONAL REMARKS BY THE EDITORS

Subsequent to his construction of three mutually orthogonal latin squares (m.o.l.s.) of order 14 in 1985, D.T.Todorov(1989) has obtained four m.o.l.s. of order 20. Also, M.Peters and R.Roth(1987) have obtained four m.o.l.s. of order 24. As regards asymptotic results for N(v), M.G.Lu(1985) has obtained the bound $N(v) \geqslant v^{10/143} - 2$ which is a slight improvement on the bounds of R.M.Wilson(1974a) and T.Beth(1983).

In the original draft of this Chapter, A.E.Brouwer listed several papers in his bibliography which were not cited in his text. These are J.H.Dinitz and D.R.Stinson(1983), A.L.Dulmage, D.M.Johnson and

Recursive constructions

	0	1	2	3	4	5	6	7	8	9
0	-	-	1	2	3	4	1	6	7	8
10	2	10	5	12	3	4	15	16	3	18
20	4	4	3	22	4	24	3	16	3	28
30	3	30	31	4	3	4	4	36	3	4
40	4	40	3	42	3	4	4	46	4	48
50	6	4	3	52	4	5	7	7	5	58
60	4	60	4	6	63	7	5	66	5	6
70	6	70	7	72	5	5	5	6	6	78
80	7	80	8	82	6	6	6	6	7	88
90	6	7	6	6	6	6	7	7	6	8
100	8	C	6	C	7	7	6	C	6	C
110	6	6	7	C	6	7	6	8	6	6
120	7	C	6	6	6	C	6	C	C	7
130	6	C	6	7	6	7	7	C	6	C
140	6	7	6	10	8	7	6	7	6	C
150	6	C	7	8	8	7	6	C	6	6
160	7	7	6	C	6	7	6	C	7	C
170	6	8	6	C	6	6	10	9	6	C
180	6	C	6	6	7	8	6	10	6	8
190	6	C	7	C	6	7	6	C	6	C

Figure 7.1

N.S.Mendelsohn(1961), J.H.van Lint(1974), W.H.Mills(1977), R.C.Mullin, P.J.Schellenburg, D.R.Stinson and S.A.Vanstone(1978,1980) D.R.Stinson(1978,1979a), K.Szajowski(1976), R.M.Wilson(1974b), M.Wojtas(1981). Two more papers closely related to this subject matter are Y.Wang(1964) and L.Zhang(1963). The latter paper gives, among other things, a construction of s−1 m.o.l.s. of order s from a set of s−2 of that order. (Compare Section 5 of Chapter 2.)

The subject of the construction of m.o.l.s. really deserves more than one Chapter because it has many aspects. In particular, many papers have been written on the subject of the construction of sets of m.o.l.s. of special types such as doubly diagonal latin squares (see [DK], pages 194-214) and Knut Vik designs. Since the latter subject became an active subject of consideration after [DK] was written, we list the relevant papers as follows: K.Afsarinajed(1986,1987); A.O.L.Atkin, L.Hay and R.G.Larson(1977); A.Hedayat(1977); A.Hedayat and W.T.Federer(1975). Also, as regards the subject of doubly diagonal latin squares, just as these Additional Remarks were being typed, news arrived that the question of existence of pairs of orthogonal doubly diagonal latin squares has been resolved in the affirmative for all $n \neq 2,3,6$. The construction makes use of the latin square of order 10 with 5504 transversals which we mentioned in Section 2 of Chapter 2. For the details, see J.W.Brown, F.Cherry, L.Most, M.Most, E.T.Parker and W.D.Wallis(199α). This result solves Problems 6.1 of [DK] in the affirmative.

Another recent topic is the construction of m.o.l.s. which form a so-called power set: that is, the squares of the set can be expressed in the form $L, L^2, \ldots, L^r$, where, if $L = (\alpha_1, \alpha_2, \ldots, \alpha_n)$ and $M = (\beta_1, \beta_2, \ldots, \beta_n)$ are two latin squares of order n, we define their product LM to be the square $(\alpha_1\beta_1, \alpha_2\beta_2, \ldots, \alpha_n\beta_n)$. Here, $\alpha_1, \alpha_2, \ldots, \alpha_n$ represent the rows of L as permutations of n symbols $1, 2, \ldots, n$ from natural order and the β's are similarly defined. We refer the reader in particular to two papers which at the time of writing are still only in preprint form: namely, J.Dénes, G.L.Mullen and S.J.Suchower(199α) and H.Y.Song(199α).

An unexpected application of the bound on the maximum number of m.o.l.s. of a given order is given in D.G.Rogers(199α).

CHAPTER 6

r-ORTHOGONAL LATIN SQUARES (G.B.Belyavskaya)

At the time when [DK] was published, the generalization of orthogonality of latin squares and quasigroups known as r-orthogonality had barely begun to be investigated. The major contributor to this subject in the past ten years has been G.B.Belyavskaya, who is the author of the present chapter.

In her survey of the subject she first considers some general properties of r-orthogonality. She shows the interrelation between r-orthogonality and partial admissibility and she deduces criteria for the existence of an r-orthogonal mate of a latin square. Next, she treats the concepts of near-orthogonality and perpendicularity as special cases of r-orthogonality. She introduces the notion of the spectrum of partial orthogonality, establishes its main properties and hence obtains the spectra of partial orthogonality for latin squares of small order. Finally, she discusses r-orthogonal sets of latin squares, the possibility of their practical utilization in coding theory and some problems raised thereby.

EDITORS' NOTE: Because this Chapter of necessity contains frequent references to Belyavskaya's own papers on r-orthogonality, we have cited these as [B(1976)] and [B(1977)] throughout the present Chapter.

(1) Some weaker modifications of the concept of orthogonality.

As is well known, two latin squares of order n defined on the same symbol set are called orthogonal if, when the squares are juxtaposed, each of the n^2 possible ordered pairs of symbols occurs in just one of the n^2 pairs of corresponding cells. Many works are devoted to the investigation of the properties of orthogonal latin squares and to the construction of maximal sets of such squares (see, for example [DK]).

In parallel with the investigation of orthogonality, various weaker versions of the concept have been proposed and studied. Thus L.Euler (1779), having failed to construct a pair of orthogonal latin squares of order 6, himself constructed an incomplete graeco-latin square (see Figure 1.1) comprising $34 = 6^2 - 2$ different ordered pairs and two vacant cells. The missing ordered pairs were (2,5) and (4,6) whereas the pairs required to complete the two separate latin squares would be (4,5) and (2,6). When these latter pairs are inserted, we obtain two superimposed latin squares which incorporate latin subsquares constructed from the four elements 2,4,5,6 and occupying the same four cells in each of the two squares : namely the cells of the second and fourth rows and the first and third columns. The condition for the two superimposed latin squares to be orthogonal is broken in just these 2×2 subsquares.

$$E = \begin{array}{|cccccc} 11 & 22 & 33 & 44 & 55 & 66 \\ 26 & 31 & 45 & 63 & 14 & 52 \\ 34 & 65 & 12 & 56 & 23 & 41 \\ & 54 & & 32 & 61 & 13 \\ 53 & 16 & 64 & 21 & 42 & 35 \\ 62 & 43 & 51 & 15 & 36 & 24 \end{array}$$

Figure 1.1

A graeco-latin square of this kind was later called an incomplete Euler square of Euler type by K.Yamamoto(1954). It is easy to see that, by suitably renumbering the symbols in one of the squares, the squares can be transformed so that the 2×2 subsquare whose cell entries break the condition of orthogonality is the same for each of the two squares. K.Heinrich(1977b) called two latin squares with this property near-orthogonal. (See also Chapter 4 of the present book).

In other words, two latin squares of order n, consisting of the same set Q of elements, are called near-orthogonal if they contain a common latin subsquare of order two, constructed from the subset S = {a,b} of elements say, and if when these two latin squares are superimposed but with the common subsquare omitted, all the ordered pairs of the set $(Q \times Q) \setminus (S \times S)$ appear in the superimposed squares.

J.D.Horton(1974) has discussed a weakened form of near-orthogonality via the concept of incomplete orthogonal arrays which we shall consider in detail in Section 6 of this Chapter. Also, he has considered an analogue of the near-orthogonal situation: namely the case when two latin squares have a common latin subsquare of size $k \times k$ ($k \geq 2$) whose elements are taken from a set S of cardinal k and it is required that, when the two squares are superimposed with the common subsquare omitted, all the ordered pairs of the set $(Q \times Q) \setminus (S \times S)$ shall appear. We remark that two latin squares are said to have a common subsquare if they have latin subsquares which consist of the same elements and occupy the same positions in the two latin squares although the subsquares themselves may be different and what is more, may be orthogonal.

Another well-known weakened version of the condition of orthogonality is the concept of perpendicularity (see [DK], pages 436 and 464) introduced for symmetric latin squares.

Two symmetric latin squares of order n are called perpendicular if, when they are superimposed, exactly $\frac{1}{2}n(n+1)$ different ordered pairs of symbols occur among the n^2 ordered pairs of corresponding cells.

This is the largest number of different ordered pairs which can occur when two symmetric latin squares of order n are superimposed. The concept of orthogonality and the several alternative weaker versions of the concept mentioned above have natural generalizations which we introduce in the next section of this Chapter.

(2) r-Orthogonal latin squares and quasigroups

DEFINITION. Two latin squares of order n defined on the same set Q are called r-orthogonal if, when they are superimposed, exactly r different ordered pairs of the set $Q \times Q$ occur among the n^2 ordered pairs of cells.

Evidently, we must have $n \leq r \leq n^2$. When $r = n^2$ the squares are orthogonal, while near-orthogonality and perpendicularity are special cases of r-orthogonality for which $r = n^2-2$ and $r = \frac{1}{2}n(n+1)$ respectively.

The concept of r-orthogonality was introduced in [B(1976)] and the form of incomplete orthogonality considered by J.D.Horton(1974) (which we mentioned above) is a special case of (n^2-k+r)-orthogonality of n×n latin squres for which we suppose that, in place of a common k×k subsquare, the two squares contain r-orthogonal k×k latin subsquares in corresponding positions, where $k \leq r \leq k^2$. We necessarily have $k \leq \frac{1}{2}n$ because k is the order of a latin subsquare of a latin square of order n.

In Figure 1.2, we display a pair of 12-orthogonal latin squares of order four defined on the elements 1,2,3,4.

$$\begin{vmatrix} 1 & 2 & 3 & 4 \\ 3 & 4 & 1 & 2 \\ 2 & 1 & 4 & 3 \\ 4 & 3 & 2 & 1 \end{vmatrix} \qquad \begin{vmatrix} 3 & 4 & 2 & 1 \\ 1 & 3 & 4 & 2 \\ 4 & 2 & 1 & 3 \\ 2 & 1 & 3 & 4 \end{vmatrix}$$

Figure 1.2

As is well known ([DK], page 174) the concepts of orthogonal latin squares and of orthogonal quasigroups are closely connected. There is an analogous connection between r-orthogonal latin squares and r-orthogonal quasigroups.

DEFINITION. Two quasigroups (Q,.) and (Q,*) are called <u>r-orthogonal</u> if the set $\{(x.y, x*y) : x, y \in Q\}$ contains exactly r different ordered pairs of the set Q×Q; in other words, if the system of equations x.y = a, x*y = b has a (not necessarily unique) solution for exactly r different ordered pairs $(a,b) \in Q \times Q$.

It is evident that any two quasigroups (Q,.) and (Q,*) of order n must be r-orthogonal for some particular r, $n \leq r \leq n^2$.

In [B(1976)], the present author established the following algebraic criterion for r-orthogonality of two quasigroups, generalizing the analogous criterion for orthogonality given in Lemma 2 of V.D.Belousov(1968). First we need a definition.

DEFINITION. Let $Q(A)$ and $Q(B)$ be two quasigroups. Their right product $C = AB$ is the binary operation on Q given by $C(x,y) = A(x,B(x,y))$. In fact, the groupoid $Q(C)$ is always a right quasigroup: that is, the equation $C(a,y) = b$ can be solved uniquely for y for all $a, b \in Q$. We shall say that the right quasigroup $Q(C)$ is of type r if the equation $C(x,a) = b$ is soluble (not necessarily uniquely) for exactly r different ordered pairs $a, b \in Q$.

THEOREM 2.1. The quasigroups $Q(A)$ and $Q(B)$ are r-orthogonal if and only if the right quasigroup AB^{-1} (or BA^{-1}) is of type r, where the binary operation B^{-1} is defined by the statement that $B^{-1}(x,z) = y$ if and only if $B(x,y) = z$.

Proof. When we solve the pair of equations $A(x,y) = a$, $B(x,y) = b$ we get $y = B^{-1}(x,b)$ from the second equation. Substituting this into the first equation, we obtain $A(x, B^{-1}(x,b)) = a$ or, equivalently, $AB^{-1}(x,b) = a$. Consequently, the existence of a solution of the above pair of equations for the ordered pair $(a,b) \in Q \times Q$ is equivalent to existence of a solution of the equation $C(x,b) = a$, where $C = AB^{-1}$. []

It is easy to show that there do not exist right quasigroups of order n of types $n+1$ or n^2-1 and so

COROLLARY. Pairs of $(n+1)$- or (n^2-1)-orthogonal quasigroups of order n do not exist.

As regards the special case of n-orthogonality, we have proved the following result in [B(1976)].

THEOREM 2.2. The pair of quasigroups $(Q, .)$ and $(Q, *)$ of order n are n-orthogonal if and only if there exists a permutation α of the set Q such that $x*y = \alpha(x.y)$.

Proof. Let L_u, L_v be the latin squares which represent the multiplication tables of the n-orthogonal quasigroups $(Q, .)$ and $(Q, *)$. When L_u and L_v are juxtaposed, only n ordered pairs of symbols from Q occur among the n^2 cells which we may suppose without loss of generality to be the pairs (u_i, v_i), where $i = 1, 2, \ldots, h$ and $u_i, v_i \in Q$. If α is the permutation of Q defined by $u_i \to v_i$, then $x*y = \alpha(x.y)$ for all $x, y \in Q$.

The converse is trivial. []

In the investigation of r-orthogonality, the following notion of (k,s)-invertibility of a quasigroup (from [B(1976)]), is useful.

Let $Q(.)$ be a quasigroup of order n, and suppose that k,s are positive integers such that $1 \leq k \leq s \leq n$. Let A, B, C be three subsets of the set Q such that $|A| = k$, $|B| = |C| = s$.

DEFINITION. A quasigroup $Q(.)$ is called (k,s)-invertible with respect to A,B,C if the equations $x.c = a$, $b.y = a$ have solutions $x \in B$, $y \in C$ for all $a \in A$, $b \in B$, $c \in C$.

We remark that the triplet A, B, C of subsets is ordered in this definition.

The (k,s)-invertibility of a quasigroup with respect to A, B, C means that the associated latin square has the property that within it we can choose s rows $b_1, b_2, \ldots, b_s$ (this is the set B) and s columns $c_1, c_2, \ldots, c_s$ (this is the set C) such that every element of the set $A = \{a_1, a_2, \ldots, a_k\}$ is contained in every row $b_i \in B$ and in every column $c_i \in C$ ($i = 1, 2, \ldots, s$) of the subtable B×C of size s×s of the latin square.

In Figure 2.2, we give an example of a quasigroup of order 8 which is $(3,4)$-invertible with respect to the subsets $A = \{1,2,3\}$, $B = \{2,3,6,7\}$,

	1	2	3	4	5	6	7	8
1	4	6	2	1	5	8	3	7
2	(2)	(3)	6	5	8	(4)	7	(1)
3	(6)	(1)	8	7	4	(3)	5	(2)
4	5	4	1	2	3	7	6	8
5	7	8	3	4	1	5	2	6
6	(3)	(2)	4	6	7	(1)	8	(5)
7	(1)	(7)	5	8	6	(2)	4	(3)
8	8	5	7	1	3	6	2	4

Figure 2.2

$C = \{1,2,6,8\}$. The elements of the subtable B×C are denoted by bracketing. It is easy to check that every element of $A = \{1,2,3\}$ is contained in every row and in every column of this subtable.

We mention some properties of (k,s)-invertibility.

(1) Any quasigroup of order n is (k,n)-invertible for all k, with $1 \leq k \leq n$, and $(1,s)$-invertible for all s: $1 \leq s \leq n$.

(2) If a quasigroup of order n is (k,s)-invertible, where $s < n$, then it is (k_1,s)-invertible for all k_1 satisfying $1 \leq k_1 \leq k$.

(3) A (k,s)-invertible (with respect to A, B, C) quasigroup of order n, where $s < n$, is $(k,n-s)$-invertible with respect to A, Q\B, Q\C, whence $k \leq n/2$.

(4) A quasigroup with a proper subquasigroup or with a proper latin subsquare of order k is (k,k)-invertible. In the first case $A = B = C$ but in the second case we only have $|A| = |B| = |C|$.

For the investigation of r-orthogonality of groups it will be useful to have the following lemma, proved in [B(1976)].

LEMMA 2.1. If a group of order n contains an element of order s_0, $s_0 \neq 1$, then it is $(2,s_0)$-invertible. Conversely, if a group of order n is $(2,s)$-invertible for some $s < n$, then it contains at least one element distinct from the identity of order less than or equal to s.

It is easy to see that in a (k,s)-invertible (with respect to A, B, C) quasigroup any two elements of the set A may be exchanged within the subtable B×C without affecting the quasigroup property. This fact is used in [B(1976)] to prove the following theorem.

THEOREM 2.3. If a quasigroup of order n is (k,s)-invertible, where $k \geq 2$, $s < n$, then it has an $(n+k_1)$-orthogonal mate for each k_1 satisfying $2 \leq k_1 \leq k$.

Proof. If the quasigroup $(Q,.)$ is (k,s)-invertible, then the s×s subsquare B×C of the multiplication table L of $(Q,.)$ contains every symbol of the set A in each row and in each column. Consequently, if any k_1 of these symbols are permuted cyclically within this subtable, the latin square property is unaffected and we get a new quasigroup $(Q,*)$ which is $(n+k_1)$-orthogonal to $(Q,.)$. □

Since for every $n \geq 4$ there exists a latin square of order n with a latin subsquare of any prescribed order $k \leq n/2$ (see Theorem 1.2 of Chapter 4), then property (4) of (k,s)-invertibility and Theorem 2.3 imply the following theorem, proved in [B(1976)].

THEOREM 2.4. For every $n \geq 4$ there exists an $(n+k)$-orthogonal pair of quasigroups of order n, if $2 \leq k \leq n/2$.

In her paper [B(1976)], the author established the following criterion for the existence of an $(n+2)$-orthogonal mate of a quasigroup of order n.

THEOREM 2.5. A quasigroup of order n has an $(n+2)$-orthogonal mate if and only if it is $(2,s)$-invertible for some particular value of s, $s < n$.

Proof. The sufficiency of this condition follows from Theorem 2.3. We prove the necessity.

Let $(Q,.)$ be a quasigroup whose multiplication table is given by a latin square L_u whose distinct entries are denoted by $u_1, u_2, \ldots, u_n$ and whose row and column borders have entries $r_1, r_2, \ldots, r_n$ and $c_1, c_2, \ldots, c_n$ respectively. Suppose that L_u has an $(n+2)$-orthogonal mate L_v whose n distinct entries are denoted by $v_1, v_2, \ldots, v_n$ and that, when the two squares are placed in juxtaposition, each of the ordered pairs (u_i, v_i), $i = 1, 2, \ldots, n$, occurs among the cells of the juxtaposed pair (L_u, L_v). Since the squares L_u and L_v are $(n+2)$-orthogonal, there must be two further ordered pairs which occur in the cells of (L_u, L_v). Suppose that (u_h, v_j) is one of these pairs and that it occurs in each of the s rows $r_1, r_2, \ldots, r_s$. Since each u_i occurs just once in each of these rows and since only two pairs distinct from the pairs (u_i, v_i) occur anywhere in (L_u, L_v), it follows that the symbol pair (u_j, v_h) also occurs in each of these s rows. Since there are s occurrences of the ordered pair (u_h, v_j) in (L_u, L_v), this pair must also occur in each of s columns $c_1, c_2, \ldots, c_s$ and so consequently must the pair (u_j, v_h) also. It now follows that, if $A = \{u_h, u_j\}$, $B =$

$\{r_1, r_2, \ldots, r_s\}$ and $C = \{c_1, c_2, \ldots, c_s\}$, then the equations $x.c = a$, $b.y = a$ are soluble in $(Q,.)$ for all $a \in A$, $b \in B$, $c \in C$. []

Applying Lemma 2.1, we get the following corollary of theorem 2.5.

COROLLARY. A group of prime order p has no (p+2)-orthogonal mate.

We remark that in section 3 an alternative criterion will be given for a quasigroup of order n to have an (n+2)-orthogonal mate (equivalent to the condition of theorem 2.5).

(3) Partial admissibility of quasigroups, its connection with r-orthogonality

As is well-known, orthogonality of latin squares has very close connections with the concept of transversal of a latin square.

We recall that a transversal of a latin square of order n is a set of n cells, one in each row, one in each column and such that no two of the cells contain the same symbol (see [DK]).

Also, to every transversal of a latin square there corresponds a complete mapping of the associated quasigroup, and conversely.

The permutation θ of the set Q is called complete for the quasigroup $Q(.)$, if the mapping θ' : $\theta'(x) = x.\theta(x)$ of the set Q into itself is also a permutation.

A quasigroup is said to be admissible, if it has at least one complete mapping [cf. V.D.Belousov(1967b)].

The concept of transversal has the following generalization, considered in G.B.Belyavskaya and A.F.Russu(1976) [and also in S.K.Stein(1975), see Section 6 of Chapter 2].

DEFINITION. A chain of a latin square of order n is a set of n cells, taken one from each row and one from each column, but such that these cells do not necessarily contain distinct elements. The elements which lie in the n cells are called the elements of the given chain.

The number of different elements which occur in the chain is called the rank of the chain.

Evidently the rank t of an arbitrary chain of a latin square of order n satisfies the condition $1 \leq t \leq n$. As examples, the bracketed elements in the latin square given in Figure 3.1 form a chain of rank 3, while the cells in positions (1,2), (2,1), (3,4) and (4,3) form a chain of rank 1 comprising the element 2 alone.

$$\begin{array}{|cccc} (1) & 2 & 3 & 4 \\ 2 & 3 & (4) & 1 \\ 3 & 4 & 1 & (2) \\ 4 & (1) & 2 & 3 \end{array}$$

Figure 3.1

To every chain of a quasigroup $Q(.)$ there corresponds a unique permutation θ of the set Q, as in the case of a transversal. However, the mapping $θ'(x) = x.θ(x)$, $x \in Q$, is in general a mapping of the set Q into itself. We shall say in this case that the chain is defined by the mapping θ, that the elements $θ'(x)$, $x \in Q$, are the elements of the chain defined by the mapping θ, and that the element $θ'(x)$ is in the position $(x, θ(x))$ of the Cayley table of the quasigroup $Q(.)$. It is evident that every permutation θ of the set Q defines a chain in any quasigroup given on the set Q. Also, if a mapping θ defines a chain of rank t in the quasigroup $Q(.)$ then $|θ'(Q)| = t$, where $θ'(Q) = \{θ'(x) : x \in Q\}$ and $θ'(x) = x.θ(x)$. A transversal of a quasigroup (latin square) of order n is a chain of rank n.

DEFINITION. A quasigroup is called <u>t-admissible</u>, if it has at least one chain of rank t.

We remark that the concept of a chain of rank t has a close connection with the concept of a partial transversal of length t (cf. [DK]): a partial transversal of length t in a latin square is certainly a part of some chain (and may be part of more than one chain) of rank $t' \geq t$. Thus the existence of a partial transversal of length t in a latin square implies the existence of a chain of rank $t' \geq t$. On the other hand, the existence of a chain of rank t' implies the existence of a partial transversal of each length t for which t ≤ t'.

In G.B.Belyavskaya(1982a) and in G.B.Belyavskaya and A.F.Russu(1976) some properties of partial admissibility (t-admissibility) of quasigroups were established and the concept of the spectrum of partial admissibility of a quasigroup was introduced.

DEFINITION. The spectrum of partial admissibility of a quasigroup of order n is the set of values t from $\{1,2,\ldots,n\}$ for which this quasigroup is t-admissible.

The spectrum of partial admissibility of a quasigroup $Q(.)$ will be denoted by S_Q.

We list the spectra of partial admissibility of cyclic groups G of order n ≥ 3, which were determined in G.B.Belyavskaya and A.F.Russu(1976).

$$S_G = \begin{cases} \{1,2,\ldots,n-2,n\}, & \text{if n is an odd composite number.} \\ \{1,3,4,\ldots,n-2,n\}, & \text{if n is an odd prime number.} \\ \{1,2,\ldots,n-2,n-1\}, & \text{if n is an even integer.} \end{cases}$$

In G.B.Belyavskaya(1982a) the spectra of partial admissibility of groups G of odd order n ≥ 3 were determined: namely,

$S_G = \{1,3,4,\ldots,n-2,n\}$ if n is a prime number,
$S_G \supseteq \{1,2,\ldots,n-2,n\}$ if n is a composite number.

As a consequence of the investigations made in that paper and in G.B.Belyavskaya and A.F.Russu(1976) it is also known that $S_G = \{1,2,\ldots,n-2,n\}$ if G is an abelian group of odd composite order n. However, the question whether any non-abelian group of odd composite order n is (n-1)-admissible remains open. (The smallest such group has order 21.)

In what follows we shall see that the concepts of t-admissibility and r-orthogonality are interconnected. Thus, knowledge of the spectrum of partial admissibility of a certain quasigroup makes it easier to answer the question: "For what values of r has this quasigroup an r-orthogonal mate?" We proceed to describe the connection between partial admissibility and r-orthogonality.

By Theorem 5.1.1 in [DK], a latin square of order n has an orthogonal mate if and only if it has n disjoint transversals. The analogous criterion for the case of r-orthogonality is the following theorem from [B(1977)]. First of all, we remark that two chains of a quasigroup $Q(.)$ defined by the mappings θ_1 and θ_2 will be called disjoint if $\theta_1(x) \neq \theta_2(x)$ for every x in Q; that is, if the chains have no common cells. Also, the <u>rank-sum</u> of two chains is the sum of their separate ranks.

THEOREM 3.1. A quasigroup $Q(.)$ of order n has an r-orthogonal mate if and only if it has n disjoint chains whose rank-sum is equal to r.

Proof. Suppose that the quasigroups $Q(.)$ and $Q(\circ)$ of order n are r-orthogonal, and suppose that a chain of rank 1 consisting of the (single) element a_i (i = 1, 2, ..., n) in the quasigroup $Q(\circ)$ is defined by the mapping θ_{a_i} of the set Q: that is,

$$x \circ \theta_{a_i}(x) = a_i, \quad \text{for all } x \in Q.$$

Then the chains defined by the elements a_i and a_j ($a_i \neq a_j$) are disjoint since $\theta_{a_i}(x) \neq \theta_{a_j}(x)$ for every $x \in Q$. Because the quasigroups $Q(.)$ and $Q(\circ)$ are r-orthogonal the mappings θ_{a_i} (where i runs over all the values 1, 2, ..., n) define n disjoint chains whose rank-sum is equal to r.

Conversely, suppose that a quasigroup $Q(.)$ of order n has n disjoint chains whose rank-sum is r, and that these chains are defined by the mappings $\theta_1, \theta_2, ..., \theta_n$. Then $\theta_i(x) \neq \theta_j(x)$ for every $x \in Q$, when $i \neq j$. We define an operation $(\circ)$ on the set Q in the following manner: $x \circ \theta_i(x) = a_i$ for i = 1, 2, ..., n. It is easy to check that $Q(\circ)$ is a quasigroup which is r-orthogonal to the quasigroup $Q(.)$. This completes the proof. []

We consider the condition of Theorem 3.1 in more detail. Let us suppose that the quasigroups $Q(.)$ and $Q(\circ)$ of order n are r-orthogonal, and that p_i ($0 \leq p_i \leq n$) denotes the number of chains of rank i ($1 \leq i \leq n$) among the n disjoint chains of the quasigroup $Q(.)$ which we know to exist in this quasigroup by Theorem 3.1. Then we have

$$\sum_{i=1}^{n} p_i = n, \quad \sum_{i=1}^{n} i p_i = r.$$

Let $n^2 - r = d$. The number d is called the <u>defect</u> of orthogonality of a quasigroup of order n (cf. [B(1977)]).

Taking into account the fact that $\sum_{i=1}^{n} p_i = n$, the equality $\sum_{i=1}^{n} i p_i = r$ can be written as

$$\sum_{i=1}^{n} (n-i) p_i = d \quad \text{or as} \quad \sum_{k=1}^{n-1} k p_{n-k} = d. \tag{A}$$

The last equality is the most convenient to use when we investigate r-orthogonality with small defect d, that is, when r is near to n^2. On the other hand, for the investigation of r-orthgonality when r is near to n (the order of the quasigroup) it is helpful to write r in the form $r = n+k$. Furthermore, we write the required equalities concerning the ranks of the chains, as follows:

$$\sum_{i=1}^{n} p_i = n, \quad \sum_{i=2}^{n} (i-1) p_i = k. \tag{B}$$

For example, applying the equality (A) for $d = 4$, we get

$$p_{n-1} + 2p_{n-2} + 3p_{n-3} + 4p_{n-4} = 4, \text{ whence } p_{n-i} = 0 \text{ if } i > d.$$

We also have $\sum_{i=1}^{n} p_i = n$ and $0 \leq p_i \leq n$.

We list all the a priori possible cases of chain rank combinations which these relations permit and then make use of certain relations among the chain ranks (of the n chains) which follow from properties of the quasigroup to discard non-realizable ones. For the case $d = 4$, we find that a priori there are the following 5 possible cases:

(1) $p_{n-1} = 4$, $p_{n-2} = p_{n-3} = p_{n-4} = 0$, $p_n = n-4$;
(2) $p_{n-1} = 2$, $p_{n-2} = 1$, $p_{n-3} = p_{n-4} = 0$, $p_n = n-3$;
(3) $p_{n-1} = 1$, $p_{n-3} = 1$, $p_{n-2} = p_{n-4} = 0$, $p_n = n-2$;
(4) $p_{n-4} = 1$, $p_{n-1} = p_{n-2} = p_{n-3} = 0$, $p_n = n-1$;
(5) $p_{n-2} = 2$, $p_{n-1} = p_{n-3} = p_{n-4} = 0$, $p_n = n-2$.

Analysing them, we establish that the cases (3) and (4) are not realizable in the quasigroup.

Analogously, if r is near to the order n of the quasigroup, we use the formula (B). For instance, in the case when $r = n+5$ we obtain

$$p_2 + 2p_3 + 3p_4 + 4p_5 + 5p_6 = 5, \quad \sum_{i=1}^{n} p_i = n,$$

and $p_i = 0$ if $i > 6$.

We give a table (see Figure 3.2) of results which were obtained in [B(1977)] by the abovementioned procedure. These results give a precise form to the condition of Theorem 3.1 for the cases $r = n^2-k$ and $r = n+k$, for $k = 1,2,3,4,5$. In the second column of this table we list the conditions required for the existence of n mutually disjoint chains. If more than one condition is listed, this means that for the existence of an r-orthogonal mate of the quasigroup $Q(.)$ of order n the existence of n chains satisfying any one of the conditions is sufficient. In the third column we state alternative criteria for existence (some of which will be proved later). (We recall that p_i denotes the number of chains of rank i in the set of n disjoint chains.)

Next we prove Theorems 3.2 and 3.3 which are mentioned in Figure 3.2 as remarks and which were proved in [B(1977)] with the aid of the conditions obtained above and listed in the second column of Figure 3.2.

THEOREM 3.2. A quasigroup of order n has an (n+2)-orthogonal mate if and only if it is 2-admissible.

Proof. Suppose that the quasigroup $(Q,.)$ of order n has an (n+2)-orthogonal mate $(Q,*)$ and that L_u, L_v are the corresponding latin squares. By relabelling the symbols in L_v, we may suppose without loss of generality that the ordered pairs in the superimposed latin squares L_u, L_v are the pairs $(1,1),(2,2),\ldots,(n,n),(1,2),(2,1)$ (cf. Theorem 2.5). The cells which contain the symbol 1 in L_v define a chain of length 2 in L_u. Likewise the cells which contain the symbol 2 in L_v define a second chain of length 2 in L_u disjoint from the first. (cf. Figure 3.2.)

The converse is evident. []

6:15   *r-Orthogonal latin squares*

Value of r	Conditions required for the existence of an n-tuple of chains.	Notes
n^2	$p_n = n$	Exists for any $n \neq 2, 6$ ([DK])
n^2-1	-	Does not exist
n^2-2	$p_{n-1} = 2$, $p_n = n-2$	Exists for any $n \geq 6$ (See Section 5)
n^2-3	$p_{n-1} = 3$, $p_n = n-3$	
n^2-4	(1) $p_{n-1} = 4$, $p_n = n-4$, or (2) $p_{n-2} = 2$, $p_n = n-2$, or (3) $p_{n-1} = 2$, $p_{n-2} = 1$, $p_n = n-3$	
n^2-5	(1) $p_{n-1} = 5$, $p_n = n-5$ or (2) $p_{n-1} = 1$, $p_{n-2} = 2$, $p_n = n-3$ or (3) $p_{n-1} = 3$, $p_{n-2} = 1$, $p_n = n-4$	
n	$p_1 = n$	Exists for any order
n+1	-	Does not exist
n+2	$p_2 = 2$, $p_1 = n-2$	Such an n-tuple of chains exists iff the quasigroup is $(2,s)$-invertible for $s \leq n$ (Theorem 2.5). This is equivalent to 2-admissibility of the quasigroup (Theorem 3.2)
n+3	$p_2 = 3$, $p_1 = n-3$	Such an n-tuple of chains exists iff the quasigroup is $(3,s)$-invertible; $s \leq n$ (Theorem 3.3)
n+4	(1) $p_3 = 1$, $p_2 = 2$, $p_1 = n-3$ or (2) $p_2 = 4$, $p_1 = n-4$	
n+5	(1) $p_3 = 2$, $p_2 = 1$, $p_1 = n-3$ or (2) $p_2 = 5$, $p_1 = n-5$ or (3) $p_3 = 1$, $p_2 = 3$, $p_1 = n-4$	

Figure 3.2

The following theorem is an analogue of Theorem 2.5 for (n+3)-orthogonality.

THEOREM 3.3. A quasigroup of order n has an (n+3)-orthogonal mate if and only if it is (3,s)-invertible for some $s<n$.

Proof. Let a quasigroup $Q(.)$ of order n be (3,s)-invertible for some $s<n$. Then, by Theorem 2.3, it has an (n+3)-orthogonal mate. Conversely, if a quasigroup $Q(.)$ of order n has an (n+3)-orthogonal mate, then by the conditions of Figure 3.2 it has an n-tuple of pairwise disjoint chains, three of which have rank 2 and the others have rank 1. Suppose that the elements $a,b,c \in Q$ belong to the chains of rank 2 (we note that these are the elements which are not contained in the n–3 chains of rank 1) and that these chains are defined by the permutations θ_1, θ_2, and θ_3 of the set Q. In addition suppose that $x.\theta_1(x) = a$ if $x \in Q_1$ and that $x.\theta_1(x) = b$ if $x \in Q \setminus Q_1$. Let $|Q_1| = s$, then the element a is contained in just one of the remaining two chains of rank 2 (defined by the mapping θ_2 or θ_3) and occurs in it (n–s) times. (It cannot be contained in both of these chains. In fact, because these chains have rank 2, they each contain the element c and one of them must also contain the element b s times). Let us suppose that the element a belongs to the chain defined by the permutation θ_2, then $x.\theta_2(x) = a$ if $x \in Q \setminus Q_1$ and $x.\theta_2(x) = c$ if $x \in Q_1$. We remark that $x.\theta_2(x) \neq b$ if $x \in Q_1$, because in the contrary case the chain defined by the mapping θ_3 would contain only the element c and thus would have rank 1. Consequently, $x.\theta_3(x) = c$ if $x \in Q \setminus Q_1$ while $x.\theta_3(x) = b$ if $x \in Q_1$. Let C be the set of columns which contain the element a in the first chain (so that $C = \theta_1(Q_1)$ where $\theta(Q_1) = \{\theta(x) : x \in Q_1\}$). Then this same set of columns contains the element c in the second chain (because the remaining columns contain a). That is, $C = \theta_2(Q_1)$. In the third chain, the elements c and b occur and so b occurs in the set C of columns which do not contain c. That is, $C = \theta_3(Q_1)$. It now follows that the quasigroup $(Q,.)$ is (3,s)-invertible with respect to $A = \{a,b,c\}$, $B = Q_1$ and $C = \theta_1(Q_1) = \theta_2(Q_1) = \theta_3(Q_1)$, which proves the theorem. []

We mention now several other results proved in [B(1977)] with the aid of conditions from Figure 3.2.

THEOREM 3.4. A group of prime order n does not have an (n+k)-orthogonal mate for k = 2,3,4,5.

Proof. From the description of the spectra of partial admissibility of cyclic groups (which were obtained at the beginning of this section), it follows that a group of prime order n ⩾ 3 fails to be 2-admissible. Also, it is easy to check that a group of order n = 2 is not 2-admissible. Since each of the conditions required for an n-tuple of disjoint chains when r = n+k, (k = 2,3,4,5) given in Figure 3.2 implies the existence of at least one chain of rank 2 in the n-tuple of chains, the stated result follows. []

We note that the assertion of Theorem 3.4 for k = 2 was also established in the Corollary of Theorem 2.5.

THEOREM 3.5. An abelian group of order n (n>2) does not have an (n^2-2)-, (n^2-3)-, or (n^2-5)-orthogonal mate. An abelian group of order n ≠ 4 has an (n^2-4)-orthogonal mate if and only if it has n disjoint chains of which two have rank n-2 and the remainder are transversals.

Proof. Utilizing conditions from Figure 3.2, we establish that an abelian group of order n ≠ 3,5 which has an (n^2-k)-orthogonal mate for k = 2,3,5, must be (n-1)-admissible and admissible (i.e. n-admissible) at the same time, while in the case of n = 3,5 the group must be (n-1)-admissible. We note that groups of order 3,5 are admissible but that, as a consequence of results due to L.J.Paige(1947), an abelian group of order n cannot be admissible and (n-1)-admissible at the same time. Therefore, the first part of the theorem is proved. For the same reason, if n ≠ 3,4, of the three conditions listed in Figure 3.2 for the case when r = n^2-4, there remains only the condition (2), which contains the second part of the theorem. For n = 3, there remain the conditions (2) and (3), which are not realizable in a group of order 3. []

REMARK. As is well-known, there are two non-isomorphic groups of order n = 4: the cyclic group and the Klein group. In [B(1977)] it was shown that both of these groups have (n^2-4) = 12-orthogonal mates.

(4) Spectra of partial orthogonality of latin squares (quasigroups).

As is well-known, there are pairs of orthogonal latin squares of order n for all $n > 2$, $n \neq 6$ (see [DK]).

In the investigation of r-orthogonality it is natural to raise the following question: "For what values of r ($n \leq r \leq n^2$) do there exist r-orthogonal latin squares (quasigroups) of order n?"

In connection with this question, the following concepts were introduced in G.B.Belyabskaya(1982b).

DEFINITION. The <u>spectrum of partial orthogonality</u> for the class of all quasigroups of order n is the set of all values r ($n \leq r \leq n^2$), for which there exists a pair of r-orthogonal quasigroups of order n.

This spectrum will be denoted by R_n.

Analogously the spectrum of partial orthogonality of a single quasigroup $Q(.)$ [$Q(A)$] may be considered. This will be denoted by $R(Q)$ [$R(A)$].

In Sections 2 and 3 we have already derived the following information concerning the spectrum of partial orthogonality:

(1) $n+1, n^2-1 \notin R_n$, $n \in R_n$ for all n [Theorem 2.1];

(2) $n+k \in R_n$, if $n \geq 4$, $2 \leq k \leq n/2$ [Theorem 2.4];

(3) if G_n is a group of prime order n then, for the values $k = 2,3,4,5$, $n+k \notin R(G_n)$ [Theorem 3.4];

(4) if $\tilde{G}_n$ is an abelian group of order $n > 2$ then n^2-2, n^2-3, $n^2-5 \notin R(\tilde{G}_n)$ [Theorem 3.5].

Also, as we remarked above, it is well-known that $n^2 \in R_n$ for any $n \neq 2,6$. Next we establish the following additional information.

THEOREM 4.1. If G is a t-admissible group of order n, then $nt \in R(G)$.

Proof. Since G is t-admissible it has a chain of rank t. Let θ be the permutation of G which corresponds to this chain so that $x.\theta(x) = \theta'(x)$, $|\theta'(G)| = t$. It then follows that $x.(\theta(x).a) = \theta'(x).a$. That is, $x.R_a\theta(x) = R_a\theta'(x)$ and so the mappings $R_a\theta$, $a \in G$, where $R_ax = x.a$, define an n-tuple of disjoint chains, each of which has rank t.

Consequently, by Theorem 3.1, the group G has an nt-orthogonal mate. (This theorem generalizes Theorem 1.4.2 in [DK].) []

From this theorem, by utilizing the description of the spectrum of partial admissibility of cyclic groups (given in Section 3), and also the fact that there exists a cyclic group of any order n, we obtain the following

COROLLARY. $R_n \supseteq \{nt: t = 1, 2, \ldots, n-2, n\}$, if n is an odd composite integer; $R_n \supseteq \{nt: t = 1, 3, 4, \ldots, n-2, n\}$ if n is an odd prime integer; $R_n \supseteq \{nt: t = 1, 2, \ldots, n-1\}$ if n is an even integer. For every integer n one has $R_n \supseteq \{nt: t = 1, 3, 4, \ldots, n-2\}$.

In the description of the connection between the spectra of partial orthogonality R_m and R_n given in G.B.Belyavskaya(1983), the construction of crossed product of quasigroups [see V.D.Belousov(1967b)] is applied, which corresponds to the sum composition method of construction of latin squares. (See [DK], page 436.)

Let P(.) be a quasigroup defined on the set $P = \{u, v, w, \ldots\}$. To every ordered pair $(u, v) \in P \times P$ we make correspond some quasigroup from the system $Q(\Sigma)$ of quasigroups defined on the set Q by the mapping

$$(u, v) \to A_{u,v} \in \Sigma.$$

We define an operation $(_o)$ on the set $M = P \times Q$ as follows

$$(u, a)_o (v, b) = [u \cdot v, A_{u,v}(a, b)], \quad u, v \in P; \ a, b \in Q.$$

$M(_o)$ will be a quasigroup which is said to be the <u>crossed product</u> of the quasigroup P(.) with the system $Q(\Sigma)$ of quasigroups.

The following theorem establishes a connection between the spectrum of partial orthogonality of the class of all quasigroups of order m and that of the class of all quasigroups of order n, and it gives a method of construction of r-orthogonal quasigroups of order mn, for some values of r satisfying $mn \leq r \leq (mn)^2$.

THEOREM 4.2. If $s \in R_m$ and $r_i \in R_n$, $i = 1, 2, \ldots, s$, then $\sum_{i=1}^{s} r_i \in R_{mn}$.

Proof. $s \in R_m$ implies the existence of an s-orthogonal pair of quasigroups $(Q, .)$, $(Q, *)$ of order m. Let I_u and I_v be the latin squares which represent the multiplication tables of these quasigroups and let (u_i, v_i), $i = 1, 2, \ldots, s$ be the s distinct ordered pairs which occur among the cells of the superimposed pair (L_u, L_v). (Note here that $m \leq s \leq m^2$ and that we are assuming that $Q = \{1, 2, \ldots, m\}$.) Let $Q_1, Q_2, \ldots, Q_m$ be a collection of m pairwise disjoint sets of symbols, each of cardinal n. On each of these sets Q_h, we construct an n×n latin square $L(Q_h)$ in such a way that the pair of latin squares $L(Q_{u_i})$ $L(Q_{v_i})$ is an r_i-orthogonal pair for each i, $i = 1, 2, \ldots, s$.

Next, if $i \cdot j = k$ in the square I_u, we replace the element k of the cell (i,j) by the latin square $L(Q_k)$ to obtain an mn×mn latin square M_u defined on the symbol set $Q_1 \cup Q_2 \cup \ldots \cup Q_m$. Likewise, if $i*j = k$ in the square I_v, we replace the element k of the cell (i,j) by the latin square $L(Q_k)$ to obtain an mn×mn latin square M_v defined on the same symbol set. Since we have chosen the m latin squares $L(Q_h)$ in such a way that the pair of squares defined on the symbol sets Q_{u_i}, Q_{v_i} is an r_i-orthogonal pair, it is clear that the mn×mn latin squares M_u and M_v must be $(\sum_{i=1}^{s} r_i)$-orthogonal. []

We note that the latin square M_u is the multiplication table of a quasigroup which is the crossed product of $(Q, .)$ with quasigroups defined on the sets $Q_1, Q_2, \ldots, Q_m$.

Because the direct product of quasigroups is a special case of the crossed product (in the case when $\Sigma = \{A\}$), Theorem 4.2 implies the following corollary.

COROLLARY (1) If the quasigroup C is a direct product of the quasigroups A and B, and if $s \in R(A)$ and $r_i \in R(B)$ for $i = 1, 2, \ldots, s$, then $\sum_{i=1}^{s} r_i \in R(C)$.

COROLLARY (2) Let H be a normal subgroup of the group G, let $s \in R(G/H)$, $r_i \in R(H)$ for $i = 1, 2, \ldots, s$. Then $\sum_{i=1}^{s} r_i \in R(G)$.

Indeed, as we proved in G.B.Belyavskaya(1983) (and the same fact follows easily from the extension theory of groups, see M.Hall(1959), Chapter 15), a group G having a normal subgroup H is isomorphic to the crossed product of the factor group G/H and of the system of quasigroups which are isotopic to the group H and which consequently each have the same spectrum of partial orthogonality as does H.

We note that these consequences are generalizations of the corresponding assertions for the case of orthogonal quasigroups (cf. [DK]).

We now list the spectra of partial orthogonality for quasigroups of small orders n = 2,3,4,5 which were determined in G.B.Belyavskaya(1982b) making use of the information contained in Figure 3.2:

$R_2 = \{2\}$, $R_3 = \{3,9\}$, $R_4 = \{4,6,8,9,12,16\}$,
$R_5 = \{5,7,10,11,12,13,14,15,16,17,18,19,21,25\}$, and
$R(G_4) = \{4,6,8,9,12\}$, $R(G_4^*) = \{4,6,8,12,16\}$,
$R(G_5) = \{5,11,13,15,17,19,25\}$,

where G_4, G_4^*, G_5 are respectively the cyclic group of order 4, the Klein group of order 4, and the group of order 5.

This list and also theorem 4.2 were utilized in G.B.Belyavskaya(1982b) to obtain the following information concerning R_8, R_9, R_{12}, and R_{15} (we use $\langle k, \ell \rangle$ to denote the set $\{k,k+1,\ldots,\ell\}$):

$R_8 \supseteq \langle 10,22 \rangle \cup \{25,28,62\} \cup \{8t: 1 \leq t \leq 8\}$ possibly excluding the integer 19.

$R_9 \supseteq \{9+6k: 0 \leq k \leq 12\} \cup \{9t: 2 \leq t \leq 8\}$,

$R_{12} \supseteq \langle 12,144 \rangle$, excluding the integers 13 and 143 and (possibly also) 35,39,131,133,135,138,139,141,

$R_{15} \supseteq \langle 15,225 \rangle$, excluding the integers 16 and 224, and (possibly also) 206,208,209,220,222.

For further information about the spectra of partial orthogonality, see Theorem 5.2, the Corollary to Theorem 6.1, and the results about near-orthogonal and perpendicular latin squares given in Section 5.

(5) **Near-orthogonal and perpendicular latin squares**.

The concept of near-orthogonality was defined in Section 1, where we were describing some of the various weak forms of orthogonality which have been studied.

We recall that two latin squares of order n are near-orthogonal if they have a common latin 2×2 subsquare and if, when they are superimposed, there are exactly n^2-2 distinct ordered pairs of elements. Near-orthogonality is a special case of (n^2-2)-orthogonality. However, there exist (n^2-2)-orthogonal latin squares of order n which contain 2×2 subsquares occupying the same cells but consisting of different elements whereas, in the case that the squares are near-orthogonal, not only do they contain a 2×2 subsquare occupying the same cells but also these squares contain the same elements. It is evident that, by relabelling the elements of one of these latin subsquares, they can be made to contain the same elements. It follows from Figure 3.2 that a quasigroup of order n has an (n^2-2)-orthogonal mate if and only if it has n pairwise disjoint chains, two of which have rank n-1 and the remainder of which are transversals. In this connection, the following, as yet still open, question has been raised: Does or does not the existence of an n-tuple of mutually disjoint chains of the above kind in a quasigroup imply the existence in its Cayley table of a 2×2 subsquare (whose cells are contained in the two chains of rank n-1), or is a more general situation possible such that the repeated elements of the two chains of rank n-1 do not form a latin subsquare?

In other words, it is an open question whether or not there exist a pair of (n^2-2)-orthogonal latin squres which fail to be near-orthogonal in whatever way their elements are relabelled.

Recently, the spectrum of near-orthogonality (that is, the listing of those orders for which there exist pairs of near-orthogonal quasigroups) has been completely determined by the efforts of J.D.Horton and K.Heinrich. In J.D.Horton(1974) it was shown that pairs of near-orthogonal latin squares of order n can be constructed for all $n \geq 33$. The same construction was used in Horton's Ph.D. Thesis [J.D.Horton(1971)] to find pairs of near-orthogonal latin squares for all orders $n \geq 6$ except n = 8, 10, 12, 14, 16, 17, 20, 23, 24, 32. Finally, K.Heinrich(1977b) gave examples of pairs of near-orthogonal latin squares of order n for all remaining orders n ⩾ 6. (For more details, see Chapter 4 of the present book.)

Consequently, near-orthogonal quasigroups (and thus (n^2-2)-orthogonal quasigroups) exist for every order n ⩾ 6. If n < 6, such pairs do not exist.

Certain other methods of construction of near-orthogonal latin squares are known. Thus, K.Yamamoto(1954) gave a method for constructing near-orthogonal latin squares of order n = 4m+2 (m⩾1), where n-1 is a prime power, by means of prolongation of the Cayley table of the additive group of the Galois field GF[n-1], (cf. [DK]). In G.B.Belyavskaya(1979) near-orthogonal latin squares of order n ⩾ 6 were constructed by prolongation of an abelian group of order n-1, where n-1 is an odd integer not divisible by 3. However, both of these methods produce near-orthogonal latin squares of even order only.

In G.B.Belyavskaya(1979), it is shown among other things that groups of odd order have no near-orthogonal mates. Also, it follows from Theorem 3.5 that abelian groups of order n > 2 do not have (n^2-2)-orthogonal mates. Hence, among the groups only non-abelian groups of even order may have a near-orthogonal mate. But the question of when such pairs exist remains open. In connection with this question, we notice that the criterion for the existence of an (n^2-2)-orthogonal pair of quasigroups of order n (given in Figure 3.2) implies that the quasigroup must be both admissible (that is, have a transversal) and (n-1)-admissible. Both of the non-abelian groups of order 8 (one of which is defined by the defining relations $a^4 = e$, $b^2 = e$, $ab = ba^3$; the other by the relations $a^4 = e$, $a^2 = b^2$, $ab = ba^3$; see, for example, M.Hall(1959)) have these properties.

Computations by a computer showed that, surprisingly, each of these groups has the same number 384 of transversals and the same number 512 of chains of rank 7. (In fact, the computer showed that, for the first group, $t_1 = 8$, $t_2 = 296$, $t_3 = 1568$, $t_4 = 8368$, $t_5 = 16640$, $t_6 = 12544$, $t_7 = 512$, $t_8 = 384$, where t_i denotes the number of chains of rank i. For the second group, $t_1 = 8$, $t_2 = 104$, $t_3 = 2208$, $t_4 = 7408$, $t_5 = 17408$, $t_6 = 12288$, $t_7 = 512$, $t_8 = 384$.)

Next we establish some results concerning perpendicular latin squares.

We recall that a pair of symmetric latin squares of order n is said to be <u>perpendicular</u> if superimposition of their Cayley tables gives $n(n+1)/2$ different ordered pairs of their elements. Perpendicularity is a special case of $n(n+1)/2$-orthogonality because this same number of ordered pairs may be obtained by superimposition of latin squares one (or both) of which fails to be symmetric. (Any pair of the latin squares exhibited in Figure 6.1 provides an example of the latter situation.)

There is a connection between the investigation of perpendicular latin squares and that of the so-called perpendicularly separable quasigroups. Every perpendicularly separable quasigroup defines some skew Room square [as is shown, for example, in D.Steedley(1976)].

The following definitions are taken from D.Steedley(1974)].

DEFINITION. By a <u>separation</u> of a quasigroup $Q(_o)$ is meant a partition of $(Q \times Q) \setminus \{(x,x) : x \in Q\}$ into disjoint sets Q_1 and Q_2 such that (x,y) is in Q_1 if and only if (y,x) is in Q_2. We will denote such a partition by $S(Q_1, Q_2)$.

DEFINITION. A quasigroup $Q(_o)$ is called <u>separable</u> if it has a separation $S(Q_1, Q_2)$ and if also the groupoids $Q(o_1)$ and $Q(o_2)$ are quasigroups, where the operations "o_1", "o_2" are defined as follows:

(1) $x o_i x = x_o x$ for all x in Q, and

(2) $x o_i y = y o_i x = x_o y$ if (x,y) is in Q_i (i = 1,2).

It may happen that the quasigroups $Q(o_1)$, $Q(o_2)$ are perpendicular. In this case the quasigroup $Q(_o)$ is called <u>perpendicularly separable</u>.

Separable quasigroups were studied in A.D.Keedwell(1978), B.T.Rumov(1982) and D.Steedley(1974) and (1976).

Some information about the orders of quasigroups which admit perpendicular mates is contained in the following

<u>THEOREM 5.1</u> If the canonical factorization $\prod_{i=1}^{\nu} p_i^{\alpha_i}$ of the number v has the propery that $p_i^{\alpha_i} \equiv 1 \pmod{k}$, $p_i > 2$ and k odd, then there exists a

quasigroup of order v which has a separation into a pair of perpendicular quasigroups.

This theorem and the following one about r-orthogonality were proved in B.T.Rumov(1982).

THEOREM 5.2. Under the hypotheses of Theorem 5.1, there exist r-orthogonal quasigroups of order v for each value of r in the set
$\{v(v+1)/2 - skv: s = 0,1,\ldots,(v-1)/2k\}$, where $k = 2s+1$.

Symmetric idempotent perpendicular latin squares (in which the elements of the main left-to-right diagonal are in natural order) deserve special attention because they are equivalent to Room squares, as is shown in Theorem 6.4.1 of [DK]. For this reason, a considerable number of authors have made a study of perpendicular Steiner quasigroups.

A Steiner quasigroup is a quasigroup which satisfies the identities $x^2 = x$, $x.xy = y$, $yx.x = y$. From the equivalence of such quasigroups with Steiner triple systems (page 74 of [DK]) it follows that, if n is the order of a Steiner quasigroup, then $n \equiv 1$ or $3 \pmod 6$. Steiner quasigroups are commutative and, in consequence of this, a pair of such quasigroups cannot be orthogonal but they may be perpendicular. Therefore, it is natural to ask : For what values of n ($n \equiv 1$ or $3 \pmod 6$) do pairs of perpendicular Steiner quasigroups exist?

A pair of perpendicular Steiner quasigroups is equivalent to a pair of orthogonal Steiner triple systems (see [DK]), so we recall briefly some of the results mentioned in [DK] about orthogonal Steiner triple systems. R.C.Mullin and E.Nemeth(1969) proved that if n is a prime power and $n \equiv 1 \pmod 6$, then orthogonal pairs do exist, and, in (1970), they showed that orthogonal pairs of order 9 do not exist. N.S.Mendelsohn(1970) proved the existence of an idempotent ($x^2 = x$), semisymmetric (x.yx = y) quasigroup which is separable into a pair of perpendicular Steiner quasigroups for every prime order $p \equiv 1 \pmod 6$. He also showed that perpendicular Steiner quasigroups exist for all orders $n \equiv 1 \pmod 6$ except when n has odd powers of primes p which are congruent to

2 modulo 3 as factors. D.Steedley(1974) has shown that a semisymmetric, idempotent quasigroup with a separating automorphism can be separated into a pair of perpendicular Steiner quasigroups. Here a cyclic automorphism α of an idempotent, semisymmetric quasigroup $Q(_o)$ is called separating if, for all $a,b \in Q$, the equality $a\alpha^k = a_o b$ implies $b\alpha^k = (a_o b)_o b$ and $b\alpha^\ell = a_o b$ implies $a\alpha^\ell = a_o(a_o b)$, where $0 \leq k, \ell \leq n-1$. In A.Rosa(1974) a pair of perpendicular Steiner quasigroups of order 27 is given, thus showing that such quasigroups exist for at least one value of $n \equiv 3 \pmod 6$. In addition to these results, we note that C.C.Lindner and N.S.Mendelsohn(1973) proved the following theorem, using the generalized singular direct product of quasigroups (see [DK]).

THEOREM 5.3. Suppose that $p_1, p_2, \ldots, p_t$ are prime numbers congruent to 1 modulo 3 and that $q_1, q_2, \ldots, q_u$ are odd prime numbers congruent to 2 modulo 3. Then, if $v = p_1^{a_1} p_2^{a_2} \ldots p_t^{a_t} q_1^{2b_1} q_2^{2b_2} \ldots q_u^{2b_u}$, where a_i, b_i are nonnegative integers, there exists a pair of perpendicular Steiner quasigroups of order v.

These results show that the solution of the question of existence of pairs of perpendicular Steiner quasigroups of orders $v \equiv 1 \pmod 6$ can be reduced to solving the problem for the case when $v = pq$, where p,q are odd prime numbers congruent to 2 modulo 3.

In K.B.Gross(1975), a generalization of the construction of R.C.Mullin and E.Nemeth(1969) was used to construct sets of t mutually orthogonal Steiner triple systems of order $q = p^k$, where p is a suitable prime. In particular, this author has proved that if $q = 1+6 \cdot 2^s \cdot t$, where t is odd, is the order of a finite field then there exist $t = \lfloor T/2^{2s+3} \rfloor$ mutually perpendicular Steiner quasigroups, where $T = (2q - 4\sqrt{q} - 4)/9$. If we put $s = 0$, we can deduce that when $q = 1+6t > 173$ is the order of a finite field, there exists a set of at least four mutually perpendicular Steiner quasigroups. The author has also listed the sizes of the sets of mutually perpendicular Steiner quasigroups which he has been able to construct for values of $q \leq 173$.

(6) r-orthogonal sets of latin squares

According to the well-known definition, a set of latin squares is said to be mutually orthogonal if every pair of distinct latin squares of the set is orthogonal.

Analogously an r-orthogonal set of latin squares can be defined as a set each pair of which is r-orthogonal.

To an r-orthogonal set of latin squares there corresponds an r-orthogonal system of quasigroups. In Figure 6.1 we give an example of a 15-orthogonal set consisting of five latin squares of order 5. The first square of this set corresponds to the cyclic group of order 5 and the others are obtained by permutation of rows of the first square.

```
1 2 3 4 5    1 2 3 4 5    1 2 3 4 5    1 2 3 4 5    1 2 3 4 5
2 3 4 5 1    4 5 1 2 3    3 4 5 1 2    2 3 4 5 1    5 1 2 3 4
3 4 5 1 2    3 4 5 1 2    5 1 2 3 4    4 5 1 2 3    2 3 4 5 1
4 5 1 2 3    2 3 4 5 1    4 5 1 2 3    5 1 2 3 4    3 4 5 1 2
5 1 2 3 4    5 1 2 3 4    2 3 4 5 1    3 4 5 1 2    4 5 1 2 3
```

Figure 6.1

From the theory of orthogonal sets of latin squares (see [DK]), it is known that the existence of s−2 mutually orthogonal latin squares of order n is equivalent to the existence of an $s \times n^2$ array $OA(n,s)$, whose entries are the numbers $1, 2, \ldots, n$ and such that each two rows of the array are orthogonal submatrices.

Analogously, an r-orthogonal set of s−2 latin squares of order n is equivalent to an $s \times n^2$ array of the following kind.

DEFINITION. An $s \times n^2$ matrix of n elements will be called an r-orthogonal array (and denoted by $OA(n,s,r)$) if it satisfies the following conditions:

(1) the first and second rows have the following forms:
   111...1222...2333...3...nnn...n and
   123...n123...n123...n...123...n respectively;

(2) when any one of the remaining s–2 rows is placed in juxtaposition to the first or the second row all the n^2 ordered pairs of elements occur;

(3) juxtaposing any two of the remaining s–2 rows gives rise to exactly r distinct ordered pairs.

It is easy to see that an $OA(n,s,r)$ defined in this way is equivalent to an r-orthogonal set of latin squares containing s–2 latin squares of order n. Conditions (1) and (2) guarantee that the entries of the ith row of the array permit the construction of a latin square of order n for i = 3,4,...,s; while condition (3) guarantees that any two of these s–2 latin squares are r-orthogonal.

J.D.Horton(1974) introduced the concept of an <u>incomplete orthogonal array</u> denoted by $IA(n,k,s)$, as an array of size $s \times (n^2-k^2)$ defined on a set S of n elements and containing a subset T of k elements such that juxtaposition of any two rows of the array gives all the distinct ordered pairs of the set $(S \times S) \setminus (T \times T)$.

In analysing what the existence of such an array means, it can be remarked that it corresponds to the existence of an r-orthogonal set of s–2 latin squares for $r \geq n^2-k^2+k$. To see this, we observe firstly that we can construct from the array a system of s–2 incomplete latin squares of order n having a k×k empty subsquare in common (cf. Section 1) and that this subsquare can be filled with elements of the set T without violating the conditions for each of the n×n squares to be latin squares. If, in each of the s–2 squares we fill the empty subsquare with the same k×k latin square based on the set T, then we obtain an (n^2-k^2+k)-orthogonal set of s–2 latin squares of order n. If, on the other hand, a u-orthogonal set of s–2 k×k latin squares exists ($k \leq u \leq k^2$), we can fill the s–2 empty subsquares with the s–2 members of this set and so obtain an (n^2-k^2+u)-orthogonal set of s–2 latin squares of order n.

Therefore, if there exists an incomplete orthogonal array $IA(n,k,s)$ and there exists a u-orthogonal set of latin squares of order k, ($k \leq u \leq k^2$) which contains s–2 latin squares, then there exists an (n^2-k^2+u)-orthogonal set of s–2 latin squares of order n.

For the special case k = 2, we have an $IA(n,2,s)$ and this corresponds to a set of s–2 near-orthogonal latin squares of order n.

In J.D.Horton(1974) the following result was proved.

THEOREM 6.1. If q is a power of a prime number and d is a proper divisor of q-1, then an IA(q+d,d,4) exists.

Since, as shown above, an incomplete orthogonal array IA(n,k,4) determines a pair of (n^2-k^2+u)-orthogonal latin squares of order n which contain u-orthogonal latin subsquares of order k ($k \leq u \leq k^2$) if these exist, Theorem 6.1 has the following corollary.

COROLLARY (1). If, under the hypotheses of Theorem 6.1, there exists a pair of u-orthogonal latin squares of order d, where $d \leq u \leq d^2$, then there exist a pair of $(q^2+2qd+u)$-orthogonal latin squares of order q+d.

In particular, we have

COROLLARY (2). Under the hypotheses of Theorem 6.1, there exist pairs of $(q^2+2qd+d)$-orthogonal latin squares of order q+d.

Let L(k,s) denote the smallest number with the property that $n \geq L(k,s)$ implies the existence of an IA(n,k,s). In J.D.Horton(1974), the author has proved that the number L(k,4) is finite for any k.

From the theory of orthogonal latin squares it is known that the maximum number of latin squares of order n in an orthogonal set is n-1. This naturally raises the question: What is the maximum number of latin squares of order n in an r-orthogonal set, where $r < n^2$? In what follows we shall describe what is known about this question.

An n-orthogonal set of latin squares of order n can evidently contain an infinite number of latin squares because any latin square of order n is n-orthogonal to itself. If we require that the latin squares of such a set are distinct, then we get n! as the maximum number of such latin squares by Theorem 2.2.

The following step towards the solution of the general problem was made in G.B.Belyavskaya(1983), where an upper bound for the number of latin squares in an (n+2)-orthogonal set of latin squares of order n was determined.

THEOREM 6.2. Let $\Sigma = \{A_1, A_2, \ldots, A_k\}$ be an (n+2)-orthogonal set of quasigroups of order n, then $k \leq 2^{\lfloor n/2 \rfloor - 1}$, where $\lfloor r \rfloor$ denotes the integer part of r.

In the proof of this theorem various criteria for (n+2)-orthogonality were used, for instance (2,s)-invertibility and its properties (see Section 2).

DEFINITION. An (n+2)-orthogonal set of latin squares of order n is said to be <u>complete</u> if it contains $2^{\lfloor n/2 \rfloor - 1}$ latin squares.

In the same paper, G.B.Belyavskaya(1983), the existence of complete (n+2)-orthogonal systems of quasigroups of order n associated with some specified group (that is, which include that group as one member of the set of quasigroups) was investigated. In particular, the following theorems were proved.

THEOREM 6.3. Let G be a group of order n which has a proper subgroup of order s, then there exists an (n+2)-orthogonal system of quasigroups associated with this group and containing $k = 2^{(n/s) - 1}$ quasigroups.

THEOREM 6.4. The maximum number of quasigroups in an (n+2)-orthogonal system which is associated with a group of order n is equal to $2^{(n/p) - 1}$, where $p(\neq 1)$ is the smallest prime divisor of n.

In the proof of the latter theorem, Lemma 2.1 is utilized.

COROLLARY. For every even n there exists a complete (n+2)-orthogonal system of quasigroups of order n but for odd n complete (n+2)-orthogonal systems associated with a group of order n do not exist.

In Figure 6.2 we give an example taken from the abovementioned paper of a complete 8-orthogonal system of quasigroups of order 6 associated with the symmetric group of degree 3. (For the sake of brevity we have omitted the heading row and column.)

```
0 1 2 3 4 5      0 1 2 3 4 5      0 1 2 3 4 5      0 1 2 3 4 5
1 0 3 2 5 4      1 0 3 2 5 4      1 0 3 2 5 4      1 0 3 2 5 4
2 4 5 1 3 0      2 4 5 1 3 0      2 4 5 0 3 1      2 4 5 0 3 1
3 5 4 0 2 1      3 5 4 0 2 1      3 5 4 1 2 0      3 5 4 1 2 0
4 2 1 5 0 3      4 2 0 5 1 3      4 2 1 5 0 3      4 2 0 5 1 3
5 3 0 4 1 2      5 3 1 4 0 2      5 3 0 4 1 2      5 3 1 4 0 2
```

Figure 6.2

A.D.Keedwell(1980) tried to construct a triplet of 10×10 pairwise orthogonal latin squares such that all three of the squares comprise permutations of the same ten columns. Although he failed to construct such a triplet, he obtained examples of (n^2-4)-, (n^2-6)-, and (n^2-n)-orthogonal 10×10 triplets with this property. Moreover, he conjectured that (n^2-4)-orthogonality is best possible if n = 10 and the latin squares consist of the same columns.

In this connection, we remark that A.Brouwer(1984) has constructed a set of four (n^2-2)-orthogonal latin squares of order n = 10.

Further information about r-orthogonal sets of latin squares will be given in the next section in connection with certain problems which arise in the application of r-orthogonal sets of latin squares to the construction of codes.

To end this section, we remark finally that the upper bound for the number of latin squares in a set of pairwise perpendicular symmetric latin squares was studied in K.B.Gross, R.C.Mullin and W.D.Wallis(1973). Let $\nu(n)$ be the maximum number of symmetric latin squares of order n in a pairwise perpendicular set, then the following theorems hold true.

<u>THEOREM 6.5</u>. Let $p^k = 1+2^n t$, where t is an odd integer and p is a prime number, then $\nu(p^k) \geq \lfloor (t-1)/2^{n-1} \rfloor +1$.

THEOREM 6.6. Let m be a positive integer, then there exists a least positive integer $v(m)$ such that for all odd $v \geq v(m)$, $\nu(v) \geq m$.

The paper of K.B. Gross et al also contains a list of known lower bounds for $\nu(n)$ where $1 \leq n < 100$ and n is an odd integer.

(7) <u>Applications of r-orthogonal latin squares and problems raised thereby</u>.

In (1977), J.Dénes discussed various ways of using latin squares for the construction of codes and made the following suggestion for applying r-orthogonal sets of latin squares to the coding of information. Let $\Sigma = \{L_1, L_2, \ldots, L_k\}$ be a set of r-orthogonal latin squares of the same order n. Then each pair L_i, L_j of these squares determines a set of r distinct pairs of elements comprising those ordered pairs which are produced when the two squares L_i, L_j are superimposed. Denote this set of pairs by H_{ij}. For instance, if L_1 and L_2 are the two 12-orthogonal latin squares which are exhibited in Figure 7.1, then

$$H_{12} = \{11, 13, 14, 21, 22, 24, 31, 32, 33, 42, 43, 44\}.$$

$$L_1 = \begin{vmatrix} 1 & 2 & 3 & 4 \\ 2 & 3 & 4 & 1 \\ 3 & 4 & 1 & 2 \\ 4 & 1 & 2 & 3 \end{vmatrix} \quad L_2 = \begin{vmatrix} 1 & 2 & 3 & 4 \\ 4 & 1 & 2 & 3 \\ 2 & 3 & 4 & 1 \\ 3 & 4 & 1 & 2 \end{vmatrix}$$

Figure 7.1

(It is important to realise that r-orthogonal sets of latin squares may exist for which H_{ij} is the same set of pairs for more than one choice of the suffices i,j.) Dénes proposed the simple encoding procedure of using ij as the coded form of H_{ij}. Then, provided that the system of r-orthogonal latin squares being employed is stored both at the transmitting and receiving ends of the transmission line, the code symbol ij can be immediately decoded at the receiving end by means of the latin squares L_i and L_j.

In this connection, J.Dénes(1977) formulated three problems:

(1) What is the maximum value of k such that at least one system of

k pairwise r-orthogonal latin squares of order n exists?

(2) For which values of r do pairs of r-orthogonal latin squares of order n exist?

(3) How can one characterize those sets H_{ij} to which correspond pairs of r-orthogonal latin squares of order n, say L_i and L_j, in the way described above?

Problem (3) was re-formulated in the following way in J.Dénes (1979):

For which values of m,n and r,s does a set $\{T_1, T_2, \ldots, T_s\}$ of r-orthogonal latin squares (each defined on the same set $H = \{a_1, a_2, \ldots, a_n\}$) exist such that for at least m (distinct) r-subsets H_k^* (k = 1,2,...,m) of H^2 there exists at least one pair of latin squares, say $T_k = ||a_{ij}||$ and $T_\ell = ||b_{ij}||$, for which the ordered pairs (a_{ij}, b_{ij}), i,j = 1,2,...,n, are the elements of H_k^*?

Recently, G.Quattrocchi(1984) has investigated this problem. He has constructed some systems of r-orthogonal latin squares of order n, for particular values of m and s which satisfy the requirements. In his construction each of the s squares of the system is obtained from the same base square $||\ell_{ij}||$ by suitable permutations of its columns. The base square $||\ell_{ij}||$ is defined by $\ell_{ij} = (i+j-2)_n$, where i,j = 1,2,...,n and $(a)_n$ denotes the remainder after division of a by n.

Let $\alpha = (a_0, a_1, \ldots, a_{n-1})$ and $\beta = (b_0, b_1, \ldots, b_{n-1})$ be two permutations of the set S_n of all permutations defined on the set $\{0, 1, \ldots, n-1\}$ and let $[\alpha-\beta]_n$ denote the set $\{(a_i - b_i)_n : i \in \{0, 1, \ldots, n-1\}\}$, so that $1 \leq |[\alpha-\beta]_n| = |[\beta-\alpha]_n| \leq n$ for all $\alpha, \beta \in S_n$. Then Quattrocchi has made the following definition.

DEFINITION. Let n, k, σ be a triplet of positive integers such that $\sigma \geq 2$ and $1 \leq k \leq n$. Then a set A of permutations is called an (n,k,σ) set of permutations if (i) $A \subseteq S_n$, (ii) $|[\alpha-\beta]_n| = k$ for all $\alpha, \beta \in A$, $\alpha \neq \beta$, and (iii) $|A| = \sigma$.

The key theorem for Quattrocchi's construction as given in the abovementioned paper is the following:

THEOREM 7.1. Let $A = \{\alpha_\tau = (a_0^{(\tau)}, a_1^{(\tau)}, \ldots, a_{n-1}^{(\tau)}) : \tau \in \{0, 1, \ldots, \sigma-1\}\}$ be an (n,k,σ) set of permutations and define $\ell_{ij}^{(\tau)} = (a_{ij}^{(\tau)} + i-1)_n$ for $i,j = 0, 1, \ldots, n-1$. Then $L_\tau = ||\ell_{ij}^{(\tau)}||$ for $\tau \in \{0, 1, \ldots, \sigma-1\}$ is a set of σ kn-orthogonal latin squares of order n.

Quattrocchi has constructed a series of (n,k,σ)-sets and by means of Theorem 7.1 he has proved that

(1) for every $n \geq 5$, there exists a 3n-orthogonal set of s = 6 latin squares of order n containing m = 4 distinct sets H_k^*;

(2) for every $n \geq 5$, there exists a 3n-orthogonal set of s = n latin squares of order n containing m = n-1 distinct sets H_k^*;

(3) for every $n \geq 3$, there exists a 3n-orthogonal set of s = n-1 latin squares of order n containing m = n-2 distinct sets H_k^*;

(4) for every $n \geq 5$, there exists a 5n-orthogonal set of s = n-1 latin squares of order n containing m = n-2 distinct sets H_k^*;

(5) if n is a positive composite integer and σ is the largest proper divisor of n, there exists a 2n-orthogonal set of s = σ latin squares of order n containing m = σ-1 distinct sets H_k^*.

Finally, we notice that the example of a 15-orthogonal system of latin squares of order 5 given in Figure 6.1 has relevance to J. Dénes's problem (3) because it contains five latin squares of order 5 which contain m = 6 distinct subsets H_k^*. Namely, we have the sets H_{12}, H_{13}, $H_{14} = H_{24} = H_{45}$, H_{15}, $H_{23} = H_{35}$, and $H_{25} = H_{34}$. This is an improvement on the bound m = n-1 = 4 given by Quattocchi's result (2).

CHAPTER 7

LATIN SQUARES AND UNIVERSAL ALGEBRA (T.Evans)

"The use of algebraic techniques in combinatorics is well-established and permutations, finite fields, matrices, etc, are essential tools in the combinatorialist's kit. However, the usefulness of algebra need not be confined to the techniques and structures of classical algebra. The methods of universal algebra and the use of quite exotic algebraic structures often seem to be the appropriate way to look at some combinatorial problems. In particular, latin squares and related structures lend themselves to this approach."

The above words were written by Trevor Evans who kindly accepted the Editors' invitation to write the present Chapter and sent a first draft some considerable time ago. Unfortunately, deteriorating health subsequently left him with little energy to respond to Editorial comments and the usual suggestions for minor changes of format, etc. Indeed, it has not been possible to obtain replies to letters since May 1986 and so, in the absence of any instructions to the contrary, the Editors have assumed permission to publish the original draft. A few corrections of misprints have been made and also some other small amendments to suit the general style of the book. A short addendum listing papers relevant to the subject matter has been added.

As the author originally stated "We give here a sample of some of the results which have obtained so far. The first section gives some of the elementary concepts of universal algebra which we will need and we follow this with many examples of <u>varieties of algebras</u> where the corresponding latin squares have interesting combinatorial properties. Then we give a very simple algebraic construction of counterexamples to Euler's famous conjecture about orthogonal latin squares. The next sections continue the study of orthogonal latin squares and show a rather surprising connection between <u>free algebras</u> and orthogonal latin squares (and other combinatorial structures). We finish with a brief survey of many other interesting universal algebraic-combinatorial connections which space does not allow us to develop in detail.

(1) <u>Universal algebra preliminaries</u>

By an <u>algebra</u> A we mean a nonempty set A, the <u>elements</u> of the algebra, and an indexed family Ω of <u>operations</u>, each of which is a mapping $A^n \to A$, for some $n \geq 0$. We write $A = (A;\Omega)$. A mapping $f : A^0 \to A$ is called a <u>nullary</u> operation, $g : A \to A$ a <u>unary</u> operation, $h : A^2 \to A$ a <u>binary</u> operation, and so on. In general, we will use function notation such as $g(a_1,a_2)$ or $h(a_1,a_2,a_3)$ for the value of an operation although we will usually use superscript notation $x^1, x^*, x^{-1}, \ldots$ for unary operations and infix notation, $x.y, x+y, x-y, \ldots$ for binary operations. A nullary operation is simply a constant or fixed element of A.

Familiar examples of algebras are groups, rings, lattices, quasigroups, fields and vector spaces over a field (with scalar multiplication by each field element being thought of as a unary operation on vectors in the latter case).

We say that two algebras A, B are <u>similar</u> if they have the same family of operations. The class of all algebras of some given operation type is called a <u>similarity class</u> and its <u>type</u> may be denoted by the indexed sequence of integers corresponding to the <u>arities</u> n of the indexed family Ω of operations $A^n \to A$. For example, a semigroup (one binary operation $x.y$) has type (2), a group defined by multiplication $x.y$, inversion x^{-1} and the element e, has type $(2,1,0)$. A lattice has type $(2,2)$ and so on.

A <u>variety</u> *V (or an <u>equationally defined class</u>) of algebras is a class of similar algebras $(A;\Omega)$ consisting of all such algebras which satisfy some given set of <u>identities</u> I, each of which has the form

$$u(x_1,x_2,\ldots) = v(x_1,x_2,\ldots), \text{ for all } x_1,x_2,x_3,\ldots$$

Here, u, v are well-formed expressions built up from the operation symbols in Ω and variables $x_1, x_2, x_3, \ldots$.

In other words, a variety is a class of similar algebras which can be characterized by axioms each of which has the form of an identity.

Examples of varieties of algebras are groups, abelian groups, rings, commutative rings, lattices, distributive lattices, semigroups, commutative semigroups, loops, moufang loops, quasigroups, near-rings: in fact, almost all of the familiar classes of algebras. Note, however, that fields do not form a variety (because inversion cannot be expressed in the form of a universal identity since the zero is excluded), nor do division rings, near-fields, cancellation semigroups, ternary fields, integral domains. In the next sections we will look at a number of examples of varieties connected with latin squares.

We say that a variety *U is a subvariety of *V if every *U-algebra is a *V-algebra. For example, abelian groups form a subvariety of the variety of groups, distributive lattices form a subvariety of the variety of lattices. A common way to form a subvariety of a variety *V is to impose more identities on the algebra : indeed, this is essentially the only way.

In any similarity class of algebras we have the usual algebraic constructions at our disposal. We can form subalgebras. Also the direct product $A \times B$ can be defined in the usual fashion. If $(A; \Omega)$, $(B; \Omega)$ are similar algebras, we introduce the operations in Ω on the cartesian product $A \times B$ by

$$f((a_1, b_1), (a_2, b_2), \ldots, (a_n, b_n)) = (f(a_1, a_2, \ldots), f(b_1, b_2, \ldots, b_n))$$

for each n-ary operation f in Ω. This may be extended to the cartesian product of any sequence of similar algebras in the obvious way.

A homomophism from one algebra $A = (A; \Omega)$ into a similar algebra $B = (B; \Omega)$ is a mapping $\alpha : A \to B$ such that $\alpha : a_i \to b_i$, $i = 1, 2, 3, \ldots$, implies

$$\alpha : f(a_1, a_2, \ldots, a_n) \to f(b_1, b_2, \ldots, b_n)$$

for each n-ary operation f in Ω. If α is one-one and onto, it is called an isomorphism from A to B. A homomorphism from A into A is called an endomorphism of A and an endomorphism which is one-one and onto is called an automorphism.

A variety *V is closed under these three constructions : taking subalgebras, cartesian products and homomorphic images. There is a celebrated theorem of G.Birkhoff(1935) which states the converse : namely, "If a class C of similar algebras is closed under these constructions, then C is a variety of algebras."

This gives us a second method for constructing varieties. Instead of being given by a set of defining identities, a variety *V may be described by giving a set K of similar algebras which generate it using the above constructions. The identities which *V satisfies will then be, of course, those satisfied by every algebra in the generating set K.

(2) Varieties of latin squares

A quasigroup on a set A is usually defined as consisting of a multiplication x.y on A such that, for any a,b in A, each of the equations

$$a.x = b, \quad y.a = b$$

has a unique solution in A. The multiplication table of a quasigroup is a latin square based on A. For our purposes, we need a more complicated (but equivalent) definition. A quasigroup is an algebra consisting of a non-empty set A of elements, and three binary operations x.y, x\y, x/y called multiplication, left-division, right-division, respectively. These operations satisfy

(2.1) $$\begin{array}{ll} x.(x\backslash y) = y, \quad (x/y).y = x \\ x\backslash(x.y) = y, \quad (x.y)/y = x \end{array} \quad \text{for all } x,y \text{ in } A,$$

The first pair of equations show that equations $au = b$, $va = b$ have solutions $u = a\backslash b$, $v = b/a$ and the second pair imply the uniqueness of the solutions.

We need this more complicated definition of a quasigroup in order to show explicitly that quasigroups do form a variety of algebras. We will denote this variety by *Q.

We actually have six latin squares associated with the operations in any *Q-algebra, namely the tables for x.y, y.x, x\y, y\x, x/y, y/x. In such a set the latin squares are said to be <u>conjugates</u> or <u>parastrophes</u> of each other. The tables for x.y and y.x are said to be <u>transposes</u> of each other.

The variety of loops : that is, quasigroups with neutral element e, may be defined by the above identities and either (i) adding a further identity x\x = x/x (which implies the existence of a multiplicative neutral element) or (ii) adding a new constant (nullary) operation and the identities ex = xe = x. Note that in the varieties of quasigroups and loops, the further identities x/(y\x) = y, (y/x)\y = x hold and that in the variety of loops x/e = e\x = x, x\x = x/x = e hold.

Many varieties of quasigroups which are of interest are given by further identities involving only the multiplication operation and sometimes these identities are sufficient to allow us to define the division operations in terms of multiplication. In this case, we still usually describe the varieties as "a variety of quasigroups" rather than "a variety of groupoids".

For example, a <u>Steiner quasigroup</u> is an algebra consisting of a set S and a binary operation of multiplication xy which satisfies the laws

$$x^2 = x$$
$$xy = yx \qquad \text{for all } x,y \text{ in } S,$$
$$x(xy) = y$$

Note that x\y = x/y = xy.

Given a Steiner triple system (STS), if we define a multiplication on the set of elements S by (i) x^2 = x, and (ii) for x≠y, xy = z, where {x,y,z} is the unique triple containing x,y, we obtain a Steiner quasigroup. Conversely, given a Steiner quasigroup (S;.), if we define a set of triples on S by {x,y,z} is a triple iff x≠y and xy = z, then we obtain an STS.

These quasigroups corresponding to STS's form a variety, and so are closed under direct products. Thus, if there are STS's of orders m,n, then there is an STS of order mn.

We can construct another algebra from an STS on a set S. Add an element $e \notin S$, to S and define a multiplication on $S \cup \{e\}$ by (i) ex = xe = x, (ii) $x^2 = e$, and (iii) if $x,y \in S$ and $x \neq y$, then xy = z, where $\{x,y,z\}$ is a triple in the STS. We obtain an algebra called a <u>Steiner loop</u> (S;e,.) which satisfies the following laws

$$\left. \begin{array}{l} ex = xe = x \\ xy = yx \\ x(xy) = y \end{array} \right\} \text{ for all } x,y \text{ in } S \cup \{e\}.$$

Conversely, given a Steiner loop on a set L, we can construct an STS on $L \setminus \{e\}$ by defining $\{x,y,z\}$ to be a triple in the STS if $x \neq y$ and xy = z in L.

Since Steiner loops also form a variety and so are closed under direct products, we see that if there are STS's of orders m,n, then there is an STS of order $(m+1)(n+1)-1 = mn+m+n$.

Thus, with very little calculation, from STS's of sizes m,n, we can construct STS's of sizes mn and mn+m+n and 2m+1 (taking one loop to be the one-element Steiner loop). In fact, from the trivial one-element loop, we can construct STS's of orders $3,7,9,15,19,21,\ldots$! Unfortunately, we cannot get STS's of all possible orders by this simple method and it leaves an interesting question: Is there or are there other general algebraic constructions which will give us all possible orders?

(3) <u>Varieties of orthogonal latin squares</u>

When do the multiplication tables (A;.), (A;o) of two quasigroups define orthogonal latin squares? The requirements are that, for any a,b in Q, we must have unique solutions x,y in Q to the equations x.y = a, xoy = b. This is easily translated into "variety" language. We have two triples $(x.y, x \backslash y, x/y)$ and $(xoy, x \backslash\!\!\backslash y, x \emptyset y)$ of quasigroup operations each satisfying the identities (2.1) and the variety $*OQ^{(2)}$ defined by these six binary operations, the eight quasigroup identities (comprising (2.1) for each triple of operations), two more binary operations x△y, x∇y, and four more identities

(3.1)
$$(x.y) \triangle (xoy) = x, \quad (x.y) \triangledown (xoy) = y$$
$$(x \triangle y).(x \triangledown y) = y, \quad (x \triangle y) \circ (x \triangledown y) = y$$

determines two orthogonal latin squares on the underlying set in each algebra. Conversely, given two orthogonal latin squares, we can construct an algebra in this variety from them.

The variety *OQ$^{(2)}$ contains an algebra of order equal to any prime power greater than two since, if F is a finite field and a,b two non-zero distinct elements in F, the binary operations $f_1(x,y) = ax+y$, $f_2(x,y) = bx+y$ are quasigroup multiplications on the elements of F.

THEOREM 3.1 *OQ$^{(2)}$ contains an algebra of order $n = 2^{i_1}3^{i_2}5^{i_3}\ldots p_t^{i_t}$, a product of prime powers, for any exponents $i_1 \geq 2$, and $i_2,\ldots,i_t \geq 0$.

Proof A variety is closed under direct products. []

We can obtain the strongest form of this result by constructing the variety *OQ$^{(k)}$ of k-sets of orthogonal latin squares. We define *OQ$^{(k)}$ as follows. As operators it has k triples of quasigroup operations and k(k-1) further binary operations: that is, k^2+2k binary operations in all. As regards identities, we have four for each triple of quasigroup operations thus defining k quasigroups (or latin squares) on each algebra in the variety. The remaining k(k-1) binary operations separate into pairs, each pair playing the role of △, ∇ in (3.1) and so forcing each pair of the k quasigroup multiplications to be orthogonal. Thus, we have 2k(k-1) further identities in addition to the 4k quasigroup identities.

THEOREM 3.2 There is a variety *OQ$^{(k)}$ whose algebras consist of all k-sets of mutually orthogonal latin squares. This variety contains algebras of all orders n which are prime power products of the form $n = 2^{i_1}3^{i_2}\ldots p_t^{i_t}$, where $k+1 \leq \min p_j^{i_j}$.

Proof   If F is a finite field of order $q = p^i > 2$, then we can construct q−1 quasigroup multiplications $f_1, \ldots, f_{q-1}$ on the elements of F, by taking $f_j(x,y) = a_j x + y$, $j = 1, 2, \ldots, q-1$, where $a_1, a_2, \ldots, a_{q-1}$ are the non-zero elements of F. In this way, we get q−1 mutually orthogonal latin squares of order q. Since a variety is closed under direct products, the theorem follows immediately. []

Theorem 3.1 shows that there are a pair of orthogonal latin squares of order n for all positive integers n not of the form 4m+2. Theorem 3.2 is often stated in the form: if $n = p_1^{i_1} p_2^{i_2} \ldots p_t^{i_t}$, then there exists a k-set of mutually orthogonal latin squares of order n where $k = \min_j (p_j^{i_j} - 1)$. In this form it is often known as McNeish's Theorem. (See [DK], page 390.)

In the following section, we shall need the following theorem.

THEOREM 3.3   There is a variety *OIQ$^{(k)}$ whose algebras consist of all k-sets of mutually orthogonal idempotent latin squares. If *OQ$^{(k+1)}$ contains an algebra of order n, then so does *OIQ$^{(k)}$.

Proof   The result follows immediately from the fact that the existence of a (k+1)-set of mutually orthogonal latin squares of order n implies the existence of a k-set of mutually orthogonal latin squares of order n all of which are idempotent. To obtain them, we simply rearrange the rows of all the squares of the given (k+1)-set of squares simultaneously so that one square of the rearranged set has the same element in every cell of its leading diagonal. The remaining squares then all have their leading diagonal as a transversal. We may permute the symbols in each one of these squares individually without affecting its orthogonality to the others in such a way as to make it idempotent. [] (For more details and an example, see [DK], pages 178 and 179.)

(4) Euler's Conjecture

We have already seen that, for any $k \geq 2$, there is a variety $*OQ^{(k)}$ whose algebras may be regarded as all k-sets of mutually orthogonal latin squares and that, for $k = 2$, the corresponding variety contains algebras of all orders $n \geq 1$ where $n \not\equiv 2 \pmod 4$. This leaves unresolved the question of whether $*OQ^{(2)}$ contains algebras of orders $2,6,10,\ldots$: that is, whether there exist pairs of mutually orthogonal latin squares of these orders. The claim that there do not is the famous Euler Conjecture. It is obvious that there is no algebra of order two in $*OQ^{(2)}$ and in 1900 Tarry showed (by the method of exhaustion) that there is no algebra of order six in $*OQ^{(2)}$. The conjecture was finally settled in 1958-9-60 in a series of papers by Parker, Bose, Shrikhande who showed that there are indeed pairs of orthogonal latin squares of all orders except $1,2,6$. (For more details, see [DK], page 158 and Chapter 11.)

To show the application of universal algebra to the problem, we construct a simple counterexample to Euler's conjecture which, however, does not give the full results of Parker, Bose, Shrikhande.

For this, we need a lemma which gives a method of constructing algebras in certain special varieties. Let us recall that a <u>pairwise balanced block design</u> $(S;B_1,B_2,\ldots,B_t)$ of index unity is a finite set S and a collection of subsets $B_1,B_2,\ldots,B_t$, called <u>blocks</u>, such that, for every pair of elements x,y in S, there is exactly one block to which both x and y belong.

<u>LEMMA</u> Let $*V$ be a non-trivial variety defined by binary operations only and such that (i) all of the defining identities of $*V$ involve at most two variables, and (ii) every operation f in $*V$ is idempotent : that is, $f(x,x) = x$ for all x. Let $D = (S;B_1,B_2,\ldots,B_t)$ be a pairwise balanced block design such that for each i, there is given a $*V$-algebra on the set B_i. Then it is possible to construct a $*V$-algebra on the set S.

<u>Proof</u> We construct the $*V$-algebra on S by defining each V-operation f on S by (i) $f(x,x) = x$, and (ii) if $x \neq y$, then $f(x,y)$ is the value of $f(x,y)$ in the $*V$-algebra defined on the block to which x,y belong. []

Before giving the counterexample to Euler's conjecture, we use the above Lemma to obtain a weaker form of E.T.Parker's counterexample to MacNeish's conjecture. (See [DK], page 397.) Recall that MacNeish proved that if $n = 2^{i_1} 3^{i_2} \ldots p_t^{i_t}$, then there is a k-set of mutually orthogonal latin squares of order n where $k \leq \min_j (p_j^{i_j}-1)$. He conjectured that $\min_j (p_j^{i_j}-1)$ was the maximal number of mutually orthogonal latin squares of order n. For n = 21, using the construction of Section 3, this yields two mutally orthogonal latin squares. Parker produced four. Now use the above Lemma applied to the projective plane of order four with the lines as blocks. Here $|S| = 21$ and $|B_i| = 5$, for every i. By Theorem 3.3, *OIQ$^{(3)}$ contains algebras of order five. Hence, by the Lemma, *OIQ$^{(3)}$ contains an algebra of order twenty-one. Thus, there exist at least three mutually orthogonal idempotent latin squares of order twenty-one.

For Euler's conjecture, we need a slightly more complicated block design. From the projective plane of order seven remove three points on a line and let the remaining lines and depleted lines be our design D = $(S;B_i)$. Here, $|S| = 54$ and $|B_i|$ is 5, 7 or 8. Since *OIQ$^{(3)}$ contains algebras of orders 5,7,8 by Theorem 3.3, it contains an algebra of order fifty-four by the Lemma. Since $54 \equiv 2 \pmod 4$, the conjecture is disproved.

Note that, since varieties are closed under products, we actually have constructed infinitely many counterexamples to Euler's conjecture.

(5) Free algebras and orthogonal latin squares

Every non-trivial variety *V contains a <u>free algebra</u> $F_n(*V)$, for each cardinal n. Such an algebra contains a set of n elements $g_1, g_2, g_3, \ldots$ with the properties : (i) $F_n(*V)$ is generated by these elements, and (ii) any mapping $g_i \to a_i$, $i = 1, 2, 3, \ldots$ of these generators onto any elements $a_1, a_2, a_3, \ldots$ in an arbitrary *V-algebra $A = (A;\Omega)$ can be extended to a homomorphism of F_n into A. We say that F_n is <u>free in *V on the</u>

generators $g_1, g_2, g_3, \ldots$ or is __freely generated__ by $g_1, g_2, g_3, \ldots$. F_n is uniquely determined (to within isomorphism) by the cardinality of one of its free generating sets. The reader is probably familiar with free groups and free commutative rings (for example, polynomial rings over the integers). Free quasigroups and loops were first described by G.E.Bates (1947) geometrically in terms of 3-nets, and then by T.Evans (1951) and (1953) algebraically patterned after the theory of free groups.

Briefly, we may say that $F_n(*V)$, in a variety with operations Ω, has as its elements "words" in the generators $g_1, g_2, \ldots$ and the operations of Ω, two words $u(g_1, g_2, \ldots)$, $v(g_1, g_2, \ldots)$ being equal (or representing the same element of F_n) if they are equal as a consequence only of the defining identities of *V so that in $F_n(*V)$, $u(g_1, g_2, \ldots) = v(g_1, g_2, \ldots)$ if and only if $u(x_1, x_2, \ldots) = v(x_1, x_2, \ldots)$ is an identity holding in all *V-algebras.

We will be interested only in varieties where the finitely generated free algebras are actually finite and we will show a rather remarkable connection between properties of such free algebras and sets of mutually orthogonal latin squares. We recall from Section 1 that a variety may be given either by a set of defining identities or by a set K of similar algebras which generate *V. In this second case, the identities of *V are precisely those common to all of the algebras in K. If K consists of only one algebra, there is a particularly simple description of $F_n(*V)$. We need only the case when n is finite and, in particular, n = 2 and we restrict ourselves to this. Let $A = (A; \Omega)$ and consider the algebra of all "polynomials" $w(x, y)$ in x, y on A, where by "polynomial" we mean an expression in x, y and the operation symbols of Ω. Such an expression determines a __polynomial function__ $A^2 \to A$ on any *V-algebra $A = (A; \Omega)$ and the operations of Ω can be defined on this set of functions in this obvious way. We obtain a *V-algebra $P_2(A)$.

__LEMMA__ $P_2(A)$ is the free algebra on two generators in the variety *V. The projection functions $p_1(x, y) = x$, $p_2(x, y) = y$ form a set of free generators for $P(A)$.

We omit the proof of this result.

EXAMPLE. Let A be the 3-element Steiner quasigroup

	a	b	c
a	a	c	b
b	c	b	a
c	b	a	c

The only polynomial functions in two variables on A are given by $p_1(x,y) = x$, $p_2(x,y) = y$, $w(x,y) = xy$.

The Table for $P_2(A)$ is

	p_1	p_2	w
p_1	p_1	w	p_2
p_2	w	p_2	p_1
w	p_2	p_1	w

This is isomorphic to A. Thus A is itself the free algebra on two generators in the variety it generates.

THEOREM 5.1 Let $F_2(*V)$ be the free algebra on two generators g_1, g_2 in the variety $*V$ and let $\{g_1, g_2, w_1(g_1, g_2), \ldots, w_k(g_1, g_2)\}$ be a set of elements in $F_2(*V)$ such that every pair of elements in this set freely generates $F_2(*V)$. Then, on every finite algebra $A = (A; \Omega)$ in $*V$, the k functions $w_1(x,y), \ldots, w_k(x,y)$ on A have a set of k mutually orthogonal latin squares as tables.

Proof Let $A = (A; \Omega)$ be a finite $*V$-algebra. The free algebra $F_2(*W)$ in the variety $*W$ generated by A has as elements all polynomial functions $w(x,y)$ on A and is generated by the projection functions $p_1(x,y) = x$, $p_2(x,y) = y$.

$F_2(*W)$ is a homomorphic image of $F_2(*V)$ under the mapping $g_1 \to p_1$, $g_2 \to p_2$ and so in $F_2(*W)$ each pair of elements in $S = \{p_1, p_2, w_1(p_1, p_2), \ldots, w_k(p_1, p_2)\}$ generates $F_2(*W)$. This means that the endomorphism α: $p_1 \to u$, $p_2 \to v$ where $u, v \in S$ is an automorphism of $F_2(*W)$ and so has an inverse β: $p_1 \to s$, $p_2 \to t$ in $F_2(*W)$. This is equivalent to

(5.1)
$$u(s(p_1,p_2), t(p_1,p_2)) = p_1, \quad v(s(p_1,p_2), t(p_1,p_2)) = p_2$$
$$s(u(p_1,p_2), v(p_1,p_2)) = p_1, \quad t(u(p_1,p_2), v(p_1,p_2)) = p_2$$

in $F_2(^*W)$.

We can rewrite these equalities as statements about functions on A, namely,

(5.2)
$$u(s(x,y), t(x,y)) = x, \quad v(s(x,y)), t(x,y)) = y$$
$$s(u(x,y), v(x,y)) = x, \quad t(u(x,y), v(x,y)) = y$$

for all x,y in A.

But (5.2) is nothing more than the equations (3.1) written in function notation and so $u(x,y)$, $v(x,y)$, regarded as binary operations on A, have orthogonal tables. Thus, each pair $w_i(x,y)$, $w_j(x,y)$, $i \neq j$ give a pair of orthogonal tables on A and since, by the same argument, the tables for each $w_i(x,y)$ and the projection tables of $p_1(x,y)$, $p_2(x,y)$ are orthogonal, the table for $w_i(x,y)$ is a latin square. This concludes the proof. []

Two well-known results are simple applications of Theorem 5.1.

(i) Let *V be the variety of abelian groups of prime exponent p. Then $F_2(^*V)$, on generators x,y, is generated by every pair of elements in $\{x, y, xy, xy^2, \ldots, xy^{p-1}\}$. Hence, on the group of order p in *V, we can construct p-1 mutually orthogonal latin squares: namely, the tables of $xy, xy^2, \ldots, xy^{p-1}$.

(ii) Let *V be the variety of two-dimensional vector spaces over the field $GF[q]$, where $q = p^n$. Let λ be a primitive root of $GF[q]$ so that $1, \lambda, \lambda^2, \ldots, \lambda^{q-2}$ are the non-zero elements of $GF[q]$. Any pair of linearly independent vectors generate *V and so *V is generated by any pair of elements in the set $\{(1,0), (0,1) (1,1) (1,\lambda), (1,\lambda^2), \ldots, (1,\lambda^{q-2})\}$. If we write x for $(1,0)$, y for $(0,1)$, then the other q-1 elements in this set can be written as $x+y, x+\lambda y, \ldots, x+\lambda^{q-2} y$ and these q-1 binary operations on $GF[q]$, regarded as the vector space of dimension one over $GF[q]$, yield q-1 mutually orthogonal latin squares as their tables.

We can also use Theorem 5.1 to obtain a simple version of the well-known construction of a finite plane of order n from a sharply doubly transitive group G on $N = \{1, 2, \ldots, n\}$. For each $i \in \{1, 2, \ldots, n\}$ define a binary operation $xo_i y$ on N by

$$x o_i x = x$$
$$x o_i y = i\alpha_{x,y}$$
for all x, y in N

where $\alpha_{x,y}$ is the permutation in G which maps $1 \to x$, $2 \to y$.

Let A be the algebra on N with operations o_i, $i = 1, 2, \ldots, n$. It is routine to verify that

(i) $1, 2$ is a generating set for A (since $1 o_i 2 = i$)

(ii) any mapping $1 \to a$, $2 \to b$ can be extended to an endomorphism of A, namely $i \to a o_i b$.

Hence, A is free in the variety *V it generates and has $\{1, 2\}$ as a free generating set.

Furthermore, any pair $i \neq j$ of elements in N generate A and hence are a free generating set. By Theorem 5.1, there are $n-2$ mutually orthogonal latin squares (actually idempotent) on N and this yields a finite plane of order n.

Hence, with every sharply doubly transitive group of degree n is associated a finite plane of order n.

[EDITORS' NOTE: T. Evans did not supply a bibliography for the above chapter but the reader may like to note that the following papers are particularly relevant to its subject matter: A. Beutelspacher (1984), T. Evans (1973, 1975, 1976, 1979, 1982a, 1982b) and N. S. Mendelsohn (1969, 1975). The list is by no means complete.]

CHAPTER 8

EMBEDDING THEOREMS FOR PARTIAL LATIN SQUARES
(C.C.Lindner)

The subject of this chapter had its origins in the work of T.Evans in the late 1950's, though the first results were obtained somewhat earlier by M.Hall and H.J.Ryser. (They appear in Theorems 2.2 and 3.1 below.) From the early 1970's onwards a number of authors have taken an interest in the topic and much progress has been made. One of the first to make major contributions was C.C.Lindner, who is the author of the present account.

In the survey which follows, he first discusses embedding a latin rectangle in a latin square and then proves Evans' theorem that a partial n×n latin square can be embedded in a t×t latin square for all $t \geq 2n$. Next, he discusses the corresponding problems when the latin rectangles or squares are firstly commutative and secondly both commutative and idempotent. He goes on to consider the much harder problem which arises when it is required to embed a partial latin rectangle or square which is idempotent but is not commutative. Lindner's second main topic is that of embedding partial quasigroups which satisfy identities different from, or additional to, the commutative or idempotent laws. He follows this by giving a table summarizing all the known results and comparing them with those which are best possible. The topic of the final part of the chapter is the so-called "Evans' conjecture" that "a partial n×n latin square with at most n-1 cells occupied can always be completed to an n×n latin square" and its proof.

Lindner has a very characteristic writing style which many readers will recognize and which the Editors thought it was their duty to preserve. However, some editing seemed desirable to keep the chapter in line with the overall format of the book and the Editors have also added a short addendum.

(1) Introduction

A <u>partial</u> latin square is an n×n array such that in each row and column each of the symbols 1,2,3,...,n occurs <u>at most once</u>. The following is an example of a 4×4 partial latin square.

$$P = \begin{bmatrix} 1 & 4 & . & . \\ . & . & 3 & 4 \\ . & 1 & 2 & . \\ . & . & . & 1 \end{bmatrix}$$

An immediate observation shows that the partial latin square P cannot be completed to a latin square; i.e., the empty cells <u>cannot</u> be filled in with symbols from the set {1,2,3,4} so that the result is a latin square. A very obvious question to ask then is whether or not it is possible to complete P if we are allowed to enlarge P and, of course, introduce additional symbols. That is, does there exist a latin square Q agreeing with the partial latin square P in its upper left-hand corner? In this case P is said to be <u>embedded</u> in Q. The latin square Q given below does the trick; i.e., P is embedded in Q.

$$Q = \begin{bmatrix} 1 & 4 & 5 & 2 & 3 \\ 2 & 5 & 3 & 4 & 1 \\ 3 & 1 & 2 & 5 & 4 \\ 5 & 3 & 4 & 1 & 2 \\ 4 & 2 & 1 & 3 & 5 \end{bmatrix}$$

The fascinating problem of whether or not an arbitrary partial latin square can always be embedded in a latin square remained unsettled until 1960, when Trevor Evans proved that such was possible. In a by-now classic paper, T.Evans(1960), the author proved that any partial n×n latin square can always be embedded in some t×t latin square for evey t ≥ 2n (the best possible result of this kind). This theorem of Evans on the finite embeddability of partial latin squares (quasigroups) was the starting point for a large collection of results on the finite embeddability of various types of latin square. The object of this chapter is to give a (hopefully) comprehensive survey of a large subset of the present day state of the art. The topic of embedding (partial) latin squares has grown to the point where it is just not possible to touch on every aspect of the subject in a single chapter. What is covered, of course, has to do with the author's interest and the author apologies for all omissions. The reader interested in pursuing this subject will find plenty of references to topics not covered here in the bibliography citations.

Finally, there is a paucity of detail in much of what follows. This is by design! The author is attempting to present the flavour of the topic and not a tedious exercise in mental gymnastics. The gory details of the proofs can be found in the original papers. And so, in most cases (but not all), the proofs are only sketched.

(2) Systems of distinct representatives

In this section we collect together the tools necessary for the proofs of Ryser's and Cruse's theorems [from H.J.Ryser(1951) and A.B.Cruse(1974a) respectively] which, as we shall see, are the basis for most of the embedding theorems in this chapter.

By a system of distinct representatives (SDR) for the sets $S_1, S_2, \ldots, S_n$ is meant an n-tuple $(x_1, x_2, \ldots, x_n)$ of n distinct elements such that $x_i \in S_i$.

EXAMPLE.

$S_1 = \{1,2\}$
$S_2 = \{1,4,6\}$ Each of the following is an SDR for the sets S_1, S_2, S_3
$S_3 = \{1,2\}$ and S_4: $(1,4,2,5)$, $(2,4,1,5)$, $(2,6,1,5)$, and $(1,6,2,5)$.
$S_4 = \{1,2,5\}$

$S_1 = \{1,2\}$
$S_2 = \{1,4,6\}$ The sets S_1, S_2, S_3 and S_4^* do not have an SDR since
$S_3 = \{1,2\}$ three distinct elements are required to represent the sets
$S_4^* = \{1\}$ S_1, S_3 and S_4^* but $|S_1 \cup S_3 \cup S_4^*| = 2$.

The above example illustrates the very obvious fact that a necessary condition for a collection of sets to have an SDR is that the union of every k of them must contain at least k elements. A famous theorem due to P.Hall(1935) guarantees that this condition is, in fact, sufficient.

THEOREM 2.1 [P.Hall(1935)]. A necessary and sufficient condition for the sets $S_1, S_2, \ldots, S_n$ to have an SDR is that the union of every k of them contains at least k elements. []

For the time being, an r×n <u>latin rectangle</u> is an r×n (r ≤ n) array such that each of the symbols $1, 2, 3, \ldots, n$ occurs <u>exactly once</u> in each row and <u>at most once</u> in each column. Since the first r rows of any n×n latin square form an r×n latin rectangle, it is natural to ask whether or not every r×n latin rectangle is the first r rows of some n×n latin square. In 1945, Marshall Hall (no relation) used P.Hall's Theorem to prove that, in fact, such is the case.

THEOREM 2.2 [M.Hall(1945)]. Every r×n latin rectangle is the first r rows of some n×n latin square.

<u>Proof</u>. Let R be an r×n latin rectangle and denote by S_i the set of all symbols which do not occur in the i-th column of R. For example, in Figure 2.1, we have $S_i = \{1, 2, \ldots, n\} \setminus \{x_1, x_2, \ldots, x_r\}$.

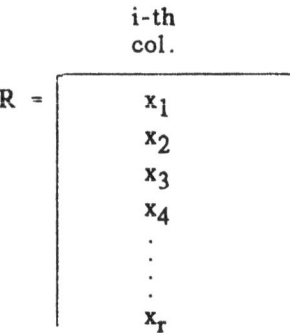

Figure 2.1

Since the union of each k of the sets S_i together contain $k(n-r)$ elements (including repetitions), at least k of these elements must be distinct otherwise one element would be repeated more than n-r times among the S_i. However, every element of $\{1,2,\ldots,n\}$ has already been used r times in the first r rows of R so the latter cannot occur. From Philip Hall's theorem it follows at once that the sets $S_1, S_2, \ldots, S_n$ have an SDR. Adding this SDR as a row of R enlarges R to an $(r+1) \times n$ latin rectangle R^*. Iteration of this procedure eventually adds n-r rows to r giving the desired extension to an n×n latin square. []

In the proof of Marshall Hall's theorem there is no restriction on the SDR for the sets $S_1, S_2, \ldots, S_n$; i.e., any SDR will do. However, quite frequently it is not only necessary to find an SDR for a collection of sets but to find an SDR which contains certain specified elements as well. In particular, this will be necessary for the proofs of Ryser's theorem on embedding latin rectangles, Cruse's theorem on embedding commutative latin rectangles (both mentioned above), and the theorem of L.D.Andersen, A.J.W.Hilton and C.A.Rodger(1982) on embedding idempotent latin squares. The following theorems are exactly what are needed for these proofs.

THEOREM 2.3 [A.J.Hoffman and H.W.Kuhn(1956)]. A necessary and sufficient condition for the sets $S_1, S_2, \ldots, S_n$ to have an SDR which includes all the elements of a set M is (in addition to these sets having an SDR) that, for every subset M' of M, at least $|M'|$ of the sets $S_1, S_2, \ldots, S_n$ have non-empty intersection with M'. []

THEOREM 2.4 [L.R.Ford and D.R.Fulkerson (1962)]. Let m≤n and let $T_1, T_2, \ldots, T_m$ and $S_1, S_2, \ldots, S_n$ be any two collections of sets having SDRs. Then there exist SDRs $(t_1, t_2, \ldots, t_m)$ of $T_1, T_2, \ldots, T_m$ and $(s_1, s_2, \ldots, s_n)$ of $S_1, S_2, \ldots, S_n$ such that $\{t_1, t_2, \ldots, t_m\} \subseteq \{s_1, s_2, \ldots, s_n\}$ if and only if the union of every i of the sets $T_1, T_2, \ldots, T_m$ <u>intersects</u> the union of every j of the sets $S_1, S_2, \ldots, S_n$ in at least i+j-n elements. []

(3) The Theorems of Ryser and Evans (on latin rectangles and squares).

Let K be an n×n latin square and denote by R the r×s upper left-hand corner of K. Suppose that K is partitioned as shown in Figure 3.1. If we

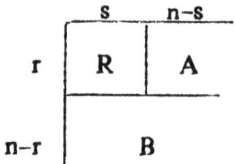

Figure 3.1

denote by $R(i)$, $A(i)$ and $B(i)$ respectively the number of occurrences in R, A and B of the symbol $i \in \{1,2,3,\ldots,n\}$, then $A(i) \leq n-s$ and $B(i) = n-r$, so $n = R(i)+A(i)+B(i) \leq R(i)+(n-s)+(n-r)$. This gives $n \leq R(i)+2n-(r+s)$, and therefore $R(i) \geq r+s-n$.

We now generalize the definition of a latin rectangle given in Section 2.

DEFINITION. An r×s latin rectangle based on $1,2,3\ldots,n$ is an r×s array such that each cell is occupied by one of the symbols $1,2,3,\ldots,n$ and such that each of these symbols occurs at most once in each row and column.

The above remarks show that a necessary condition for an r×s latin rectangle R based on $1,2,\ldots,n$ to be the upper left-hand corner of some n×n latin square is that $R(i) \geq r+s-n$ for every $i \in \{1,2,\ldots,n\}$. A remarkable theorem due to H.J.Ryser shows that this obvious necessary condition for a latin rectangle to be the upper left-hand corner of a latin square is also sufficient.

THEOREM 3.1 [H.J.Ryser(1951)]. Let R be an r×s latin rectangle based on $1,2,\ldots,n$ and denote by $R(i)$ the number of times that the symbol i occurs in R. Then R is the upper left-hand corner of an n×n latin square if and only if $R(i) \geq r+s-n$ for all $i \in \{1,2,3,\ldots,n\}$.

Proof. The proof consists of showing that an r×s (s<n) latin

rectangle R based on $1, 2, 3, \ldots, n$ and satisfying $R(i) \geq r+s-n$ for all i can be enlarged to an $r \times (s+1)$ latin rectangle R^* still based on $1, 2, 3, \ldots, n$ and such that $R^*(i) \geq r+(s+1)-n$ for all i. Iteration then extends R to an $r \times n$ latin rectangle Q based on $1, 2, 3, \ldots, n$ (the definition of this latin rectangle Q being as in section 2) and Marshall Hall's theorem (Theorem 2.2) can then be used to add n-r rows to Q, producing the desired $n \times n$ latin square.

To begin, let S_j denote the set of all symbols which do not occur in the j-th row of R. (In Figure 3.2, $S_j = \{1, 2, \ldots, n\} \setminus \{x_1, x_2, \ldots, x_s\}$.)

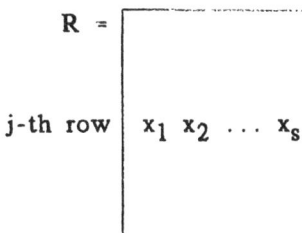

Figure 3.2

In order to enlarge R to an $r \times (s+1)$ latin rectangle R^* satisfying $R^*(i) \geq r+(s+1)-n$ for all i, we must not only find an SDR for the sets $S_1, S_2, \ldots, S_r$, but we must find an SDR containing all of the symbols which occur in R <u>exactly</u> r+s-n times. This is precisely where Hoffman and Kuhn's theorem (Theorem 2.3) enters the picture. For, if we let M be the set of all symbols which occur in R exactly r+s-n times, it is not difficult to show that (a) the sets $S_1, S_2, \ldots, S_r$ have an SDR, and (b) for every subset M' of M, at least $|M'|$ of the sets $S_1, S_2, \ldots, S_r$ have non-empty intersection with M'. (The details can be found in Ryser's paper). Hence the sets $S_1, S_2, \ldots, S_r$ have an SDR which includes all of the elements in M and Ryser's theorem follows. []

Evans' theorem is now an immediate consequence of Ryser's theorem.

THEOREM 3.2 [T.Evans(1960)]. A partial n×n latin square can be embedded in a t×t latin square for every $t \geq 2n$.

Proof. Let P be an n×n partial latin square based on $1, 2, \ldots, n$ and let $t \geq 2n$. Let N be an n×n latin square based on $n+1, n+2, \ldots, 2n$, and fill in each unoccupied cell in P with the entry in the corresponding cell of N. The result is an n×n latin rectangle P* based on $1, 2, \ldots, 2n$ and is therefore also based on $1, 2, 3, \ldots, t \geq 2n$. By Ryser's theorem, P* is the upper left-hand corner of a t×t latin square if and only if each $i = 1, 2, \ldots, t$ occurs in P* at least $n+n-t$ times. Since $t \geq 2n$, $n+n-t \leq 0$, and so this condition is trivially satisfied. Hence P* is the upper left-hand corner of a t×t latin square T and, of course, P is therefore embedded in T. []

The following example illustrates the construction used in this proof.

EXAMPLE.

$$P = \begin{bmatrix} 1 & 2 & 3 & . \\ . & 3 & 2 & 4 \\ . & 4 & 1 & . \\ . & . & . & 1 \end{bmatrix}$$

4×4 partial latin square based on 1,2,3,4. Note that P cannot be completed.

$$N = \begin{bmatrix} 5 & 6 & 7 & 8 \\ 6 & 7 & 8 & 5 \\ 7 & 8 & 5 & 6 \\ 8 & 5 & 6 & 7 \end{bmatrix}$$

4×4 latin square based on 5,6,7,8.

$$P^* = \begin{bmatrix} 1 & 2 & 3 & 8 \\ 6 & 3 & 2 & 4 \\ 7 & 4 & 1 & 6 \\ 8 & 5 & 6 & 1 \end{bmatrix}$$

4×4 latin rectangle based on 1,2,3,4,5,6,7,8 obtained from P by filling in the empty cells in P with the entries from the corresponding cells in N.

$$T = \begin{bmatrix} 1 & 2 & 3 & 8 & 4 & 5 & 6 & 7 \\ 6 & 3 & 2 & 4 & 7 & 1 & 8 & 5 \\ 7 & 4 & 1 & 6 & 2 & 8 & 5 & 3 \\ 8 & 5 & 6 & 1 & 3 & 7 & 2 & 4 \\ 4 & 7 & 8 & 5 & 1 & 2 & 3 & 6 \\ 5 & 1 & 7 & 3 & 8 & 6 & 4 & 3 \\ 3 & 8 & 5 & 2 & 6 & 4 & 7 & 1 \\ 2 & 6 & 4 & 7 & 5 & 3 & 1 & 8 \end{bmatrix}$$

8×8 latin square with P* in the upper left-hand corner whose existence is guaranteed by Ryser's Theorem, since each of 1,2,3,4,5,6,7,8 occurs in P* at least 4+4-8 = 0 times.

It is worth remarking that Evans' theorem is the best possible general embedding theorem of its kind, since it is possible to construct for every n ⩾ 4 a partial n×n latin square which cannot be embedded in a latin square less than twice its size.

(4) <u>Cruse's Theorems (on commutative latin rectangles and squares)</u>.

In C.C.Lindner(1972), the present author gave the first constructions for embedding partial commutative and partial idempotent commutative latin squares into complete squares of the same type. Not only are the proofs in that paper very tedious but they also give extremely large containing squares. Subsequently, by obtaining an analogue of Ryser's theorem for commutative latin rectangles, A.B.Cruse(1974a) was able to obtain the best possible bounds for embedding partial commutative and idempotent commutative latin squares. We, of course, present Cruse's elegant work here. So that there will be no confusion in what follows, we make the following definitions.

DEFINITIONS. A <u>partial commutative latin square</u> is a partial latin square such that, if cell (i,j) is occupied by x, then so is cell (j,i). A <u>partial idempotent latin square</u> is a partial latin square with the property that <u>all</u> of the cells on the main diagonal are occupied and furthermore cell (i,i) is occupied by i. An r×r latin rectangle based on 1,2,...,n is <u>commutative</u> provided that the cells (i,j) and (j,i) are occupied by the same symbol, and is idempotent provided that cell (i,i) is occupied by i for i = 1,2,3,...,r.

Let K be any n×n commutative latin square and denote by R the r×r upper left-hand corner of K. Also, let R(i) denote the number of occurrences in R of the symbol i. By Ryser's theorem we must have, of course, that R(i) ⩾ 2r−n for each i. However, since K is commutative, there is another not-so-obvious necessary condition which must be satisfied. Because K is commutative the totality of occurrences of each i off the main diagonal is even. Therefore, the number of times each $i \in \{1,2,3,\ldots,n\}$ occurs on the main diagonal must have the same parity as n. (To be fancy: the number of times each i occurs on the main diagonal is congruent to n (mod 2)).

It follows that if R(i) ≢ n (mod 2), then i must occur on the main diagonal of K outside R. Since there are n−r cells on the main diagonal of K outside R it follows that we can have <u>at most</u> n−r of the symbols i for which R(i) ≢ n (mod 2). In other words, we must have R(i) ≡ n (mod 2) for <u>at least</u> r of the symbols $i \in \{1,2,3,\ldots,n\}$. Therefore, two necessary conditions for an r×r commutative latin rectangle to be the upper left-hand corner of an n×n commutative latin square are (a) that R(i) ⩾ 2r−n for each $i \in \{1,2,3,\ldots,n\}$, and (b) that R(i) ≡ n (mod 2) for at least r different $i \in \{1,2,3,\ldots,n\}$. Allan Cruse has shown that these two conditions are, in fact, sufficient.

THEOREM 4.1 [A.B.Cruse(1974a)]. A necessary and sufficient condition for an r×r commutative latin rectangle based on $1,2,3,\ldots,n$ to be the upper left-hand corner of an n×n commutative latin square is that R(i) ⩾ 2r−n for each $i \in \{1,2,3,\ldots,n\}$ and that R(i) ≡ n (mod 2) for at least r different $i \in \{1,2,3,\ldots,n\}$.

<u>Proof</u>. The proof consists in showing that an r×r (r<n) commutative latin rectangle satisfying the necessary conditions to be the upper left-hand corner of any n×n commutative latin square can be enlarged to an (r+1)×(r+1) commutative latin rectangle also satisfying the necessary conditions to be the upper left-hand corner of an n×n commutative latin square.

So, let R be such an r×r commutative latin rectangle. As before, denote by S_i the set of all symbols which do not occur in the i-th row of R (which of course is equal to the set of all symbols which do not occur

8:11 Embedding theorems for partial latin squares

in the i-th column), and let M be the set of all symbols which occur in R either 2r–n or 2r–n+1 times. Further, denote by S the set of all symbols i such that $R(i) \equiv n \pmod 2$. Then the sets $S_1, S_2, \ldots, S_r$, and S have an SDR containing M or, in the case when S is empty, the sets $S_1, S_2, \ldots, S_r$ have an SDR containing M. (The details can be found in Cruse's paper.) Now construct an $(r+1) \times (r+1)$ commutative latin rectangle R^* as in Figure 4.1.

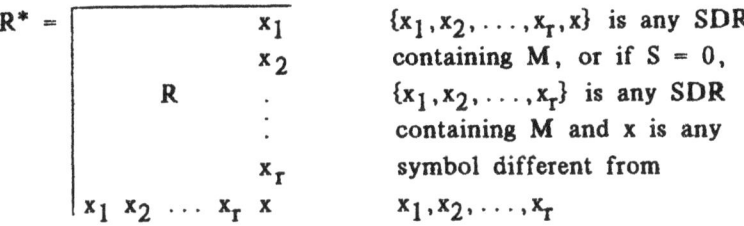

$R^* = \begin{bmatrix} & & & x_1 \\ & & & x_2 \\ & R & & \cdot \\ & & & \cdot \\ & & & x_r \\ x_1 & x_2 & \ldots & x_r & x \end{bmatrix}$

$\{x_1, x_2, \ldots, x_r, x\}$ is any SDR containing M, or if $S = 0$, $\{x_1, x_2, \ldots, x_r\}$ is any SDR containing M and x is any symbol different from $x_1, x_2, \ldots, x_r$

Figure 4.1

Trivially, $R^*(i) \geq 2(r+1)-n$ and a bit of reflection (but not too much) shows that $R^*(i) \equiv n \pmod 2$ for at least r+1 different $i \in \{1, 2, 3, \ldots, n\}$, which completes the proof. []

THEOREM 4.2 [A.B.Cruse(1974a)]. A partial n×n commutative latin square can be embedded in a t×t commutative latin square for every <u>even</u> t $\geq 2n$.

Proof. Let P be a partial n×n commutative latin square and fill in the empty cells with the entries from an n×n commutative latin square based on $n+1, n+2, \ldots, 2n$ (as in the proof of Theorem 3.2). This gives an n×n commutative latin rectangle P^* based on $1, 2, \ldots, 2n$ and therefore based on $1, 2, \ldots, t$ when t $\geq 2n$. Since $2n-t \leq 0$ it follows that $P^*(i) \geq 2n-t$. Since t is even and at least n of the symbols $1, 2, \ldots, t$ do not occur on the main diagonal of P^*, we must have $P^*(i)$ even for these symbols: that is, $P^*(i) \equiv t \pmod 2$ for at least n of $1, 2, \ldots, t$. Hence, P^* is the upper left-hand corner of a t×t commutative latin square by Theorem 4.1. []

THEOREM 4.3 [A.B.Cruse(1974a)]. An n×n partial idempotent commutative latin square can be embedded in a t×t commutative latin square for <u>every</u> t $\geq 2n$.

Proof. If t is even the proof is identical to that of Theorem 4.2. If t is odd, then, since each of $1, 2, 3, \ldots, n$ occurs on the main diagonal of P^*, each of these symbols occurs an odd number of times in P^*: that is, $P^*(i) \equiv t \pmod{2}$ for $i = 1, 2, \ldots, n$. []

THEOREM 4.4 [A.B.Cruse(1974a)]. An $n \times n$ partial idempotent commutative latin square can be embedded in a $t \times t$ idempotent commutative latin square for every odd $t \geq 2n+1$.

Proof. Let P be a partial idempotent commutative latin square of order n and suppose that P is embedded in a commutative latin square K of odd order $t \geq 2n+1$. Then, in K each symbol must occur on the main diagonal exactly once. Renaming the symbols on the diagonal of K outside P as $n+1, n+2, \ldots, t$ gives the desired embedding. []

The following definition and theorem will be needed for the embedding of partial Steiner triple systems (discussed in Section 8).

DEFINITION. A $2n \times 2n$ latin square is said to be half-idempotent if and only if cell (x, x) is occupied by x if $x \leq n$ and by $x-n$ if $x > n$.

It is a trivial matter to construct a half-idempotent commutative latin square of every order $2n$. Just rename the elements in the Cayley table of $(\mathbb{Z}_{2n}, +)$.

THEOREM 4.5. An $n \times n$ partial idempotent commutative latin square can be embedded in a $2t \times 2t$ half-idempotent commutative latin square for every $t \geq n$.

Proof. Let P be an $n \times n$ partial idempotent commutative latin square and let P^* be the $t \times t$ partial idempotent commutative latin square in Figure 4.2. Now use Theorem 4.3 to embed P^* in a $2t \times 2t$ commutative latin square M. Then each of the symbols $1, 2, \ldots, 2t$ must occur on the main diagonal of M an even number of times. Since each of $1, 2, \ldots, t$ already occurs on the main diagonal of M in P^*, each must also occur on the main diagonal of M outside P^*. A suitable permutation of the rows and columns $t+1, t+2, \ldots, 2t$ transforms M into a half-idempotent commutative latin square containing P. []

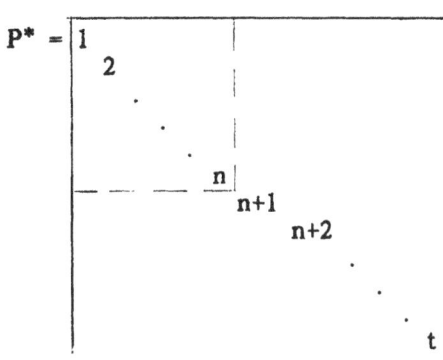

P* is a t×t partial idempotent commutative latin square and the only cells outside P which are occupied are on the main diagonal.

Figure 4.2

We end this section by remarking that all of the results in this section are the best possible with respect to the size of the containing squares.

(5) <u>Embedding idempotent latin squares</u>.

At first glance it would seem that the problem of embedding a partial idempotent latin square would be a good deal easier than the embedding of a partial idempotent commutative latin square. It turns out, however, that the dropping of commutativity makes the problem hellishly difficult (as quite a few people have come to realize over the years). One of the reasons for this difficulty is that the obvious necessary conditions for an r×r idempotent latin rectangle to be the upper left-hand corner of a t×t idempotent latin square are not sufficient. The obvious necessary conditions are that if the r×r idempotent latin rectangle R is the upper left corner of a t×t idempotent latin square K, then $R(i) \geq 2r-t$ for all $i \in \{1,2,\ldots,t\}$ and $R(j) \geq 2r+1-t$ for all $j \in \{r+1,\ldots,t\}$. (To see this, we use the argument given at the beginning of Section 3. The element j occurs at most n-r-1 times in the rectangle A since, if the square is idempotent, there is one column of A in which j cannot occur.) The example given in Figure 5.1 shows that unfortunately these necessary conditions are <u>not sufficient</u>. R <u>satisfies</u> the necessary conditions to be the upper left-hand corner of a 5×5

$$R = \begin{array}{|ccc|} \hline 1 & 5 & 4 \\ 4 & 2 & 5 \\ 5 & 4 & 3 \\ \hline \end{array}$$

Figure 5.1

idempotent latin square, but <u>obviously</u> is not!

The first construction for embedding partial idempotent latin squares was given by the present author by a very complicated construction involving Steiner triple systems and guaranteeing only that the containing square has size less than 2^{2n}. [See C.C.Lindner(1971a)]. Subsequently, A.J.W.Hilton(1973) gave a construction which was not only simpler than the one given by the author in 1971 but reduced the size of the containing square to 4n. Since the best possible general embedding is into a square of size 2n+1, people continued to struggle with this problem and finally L.D.Anderson, A.J.W.Hilton and C.A.Rodger(1982) reduced the size of the containing square to 2n+1. The proof we give here is slightly different from the original proof and is due to C.A.Rodger.

Let K be a t×t latin square which contains an r×(r+1) latin rectangle R in its top left-hand corner and other cell entries as shown in Figure 5.2.

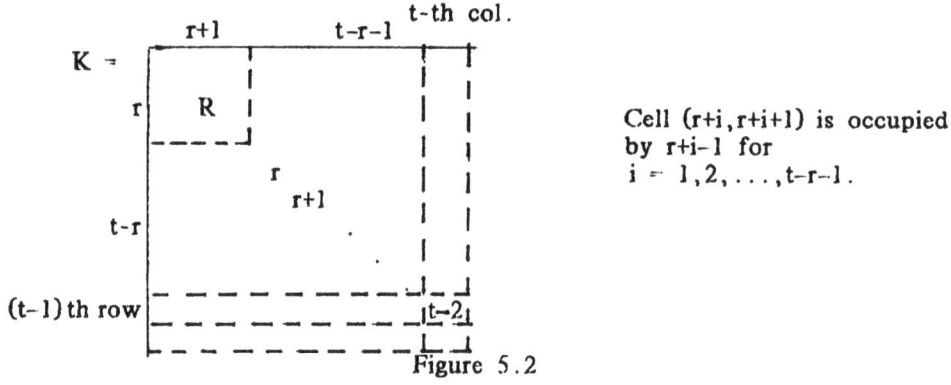

Cell (r+i, r+i+1) is occupied by r+i-1 for i = 1, 2, ..., t-r-1.

Figure 5.2

In what follows we will say that the latin rectangle R is <u>diagonally placed</u> in K. It is immediate that two necessary conditions for an r×(r+1) latin rectangle R to be diagonally placed in a t×t latin square are $R(i) \geq 2r+1-t$ for all $i \in \{1,2,\ldots,t\}$ and $R(j) \geq 2r+2-t$ for all $j \in \{r, r+1, \ldots, t-2\}$. The following result shows that these necessary conditions are sufficient.

<u>LEMMA 5.1</u> [L.D.Andersen and A.J.W.Hilton(1983)]. An r×(r+1) latin rectangle R based on 1,2,3,...,t can be diagonally placed in a t×t latin square if and only if $R(i) \geq 2r+1-t$ for all $i \in \{1,2,\ldots,t\}$ and $R(j) \geq 2r+2-t$

for all $j \in \{r, r+1, \ldots, t-2\}$.

<u>Proof</u>. The proof consists in showing that an $r \times (r+1)$ latin rectangle satisfying the necessary conditions to be diagonally placed in a $t \times t$ latin square can be enlarged to an $(r+1) \times (r+2)$ latin rectangle which also satisfies the necessary conditions to be diagonally placed in a $t \times t$ latin square. So, let R be any $r \times (r+1)$ latin rectangle satisfying the necessary conditions to be diagonally placed in a $t \times t$ latin square. Let sets A_i and B_j be defined by

$A_i = \{x : x \text{ does not appear in the i-th row and } x \neq r\}$, and
$B_j = \{y : y \text{ does not appear in the j-th column and } y \neq r\}$

Now there is no difficulty at all in showing that $A_1, A_2, \ldots, A_r$ and $B_1, B_2, \ldots, B_{r+1}$ have SDRs which enlarge R to an $(r+1) \times (r+2)$ latin rectangle R^* satisfying $R^*(i) \geq 2r+3-t$ for all $i \in \{1, 2, 3, \ldots, t\}$. However, there is considerable difficulty in finding SDRs so that the enlargement also satisfies $R^*(i) \geq 2r+4-t$ for all $i \in \{r+1, r+2, \ldots, t-2\}$. (This is where Ford and Fulkerson's Theorem 2.4 comes in handy.) To this end, let

$$f(i) = \begin{cases} 1, & \text{if } i \in \{r+1, \ldots, t-2\}, \text{ and} \\ 0, & \text{otherwise} \end{cases}$$

and define

$S = \{i : i \neq r \text{ and } R(i) = 2r+1-t+f(i)\}$, and
$T = \{j : j \neq r \text{ and } R(j) = 2r+2-t+f(j)\}$.

We need to find SDRs $(x_1, x_2, \ldots, x_r)$ of $A_1, A_2, \ldots, A_r$ and $(y_1, y_2, \ldots, y_{r+1})$ of $B_1, B_2, \ldots, B_{r+1}$ such that

$S \subseteq \{x_1, x_2, \ldots, x_r\} \cap \{y_1, y_2, \ldots, y_{r+1}\}$, and
$T \subseteq \{x_1, x_2, \ldots, x_r\} \cup \{y_1, y_2, \ldots, y_{r+1}\}$.

So set $A_i^* = A_i \times \{1\}$, $B_i^* = B_i \times \{2\}$, and $A^* = \{A_1^*, A_2^*, \ldots, A_r^*, B_1^*, B_2^*, \ldots, B_{r+1}^*\}$. Further, define a collection of sets M as follows: For each $i \in S$, place the two sets $\{(i,1)\}$ and $\{(i,2)\}$ in M, and, for each $j \in T$, place the set

$\{(j,1),(j,2)\}$ in M. It is easy to see that $|M| = m \leq 2r+1$ and that each of the collections of sets A^* and M have SDRs. Now a straightforward argument shows that the union of any x of the sets in M intersects the union of any y of the sets in A^* in at least $x+y-(2r+1)$ ordered pairs. Hence, Theorem 2.4 guarantees the existence of SDRs $((c_1,t_1),(c_2,t_2),\ldots,(c_m,t_m))$ of M, $t_i = 1$ or 2, and $((x_1,1),(x_2,1),\ldots,(x_r,1),(y_1,2),(y_2,2),\ldots,(y_{r+1},2))$ of A such that $\{(c_1,t_1),(c_2,t_2),\ldots,(c_m,t_m)\} \subseteq \{(x_1,1),(x_2,1),\ldots,(x_r,1),(y_1,2),(y_2,2),\ldots,(y_{r+1},2)\}$. It follows that $(x_1,x_2,\ldots,x_r)$ and $(y_1,y_2,\ldots,y_{r+1})$ are SDRs of $a_1,a_2,\ldots,a_r$ and $b_1,b_2,\ldots,b_{r+1}$ such that

S $\{x_1,x_2,\ldots,x_r\} \cap \{y_1,y_2,\ldots,y_{r+1}\}$, and
T $\{x_1,x_2,\ldots,x_r\} \cup \{y_1,y_2,\ldots,y_{r+1}\}$

completing the proof. []

EDITORS' NOTE: The method of proof used here is somewhat similar to that used in V.W.Bryant(1984) to prove a related embedding theorem.

THEOREM 5.2 [L.D.Andersen,A.J.W.Hilton and C.A.Rodger(1982)]. A partial n×n idempotent latin square can be embedded in a t×t idempotent latin square for every $t \geq 2n+1$.

Proof. Let P be a partial n×n idempotent latin square based on $1,2,\ldots,n$ and fill in the empty cells with the entries from the corresponding cells of an n×n latin square based on $n,n+1,\ldots,2n-1$ all of whose cells on the main diagonal are occupied by n. The result is an n×n idempotent latin rectangle P^* based on $1,2,3,\ldots,2n-1$. To begin with, it is trivial to add a first column to P^* so as to obtain an n×(n+1) latin rectangle R_1 such that t-1 occurs in cell (1,1) and $R_1(i) \geq 2n+2-t$ for all $i \in \{n+1,\ldots,t-2\}$, as shown in Figure 5.3.

8:17   *Embedding theorems for partial latin squares*   233

$$R_1 = \begin{array}{|l|} \hline t{-}1 \\ c_2 \\ c_3 \\ \vdots \\ c_n \end{array} \quad P^*$$

Note that the entries in R_1 are all from the set $\{1,2,\ldots,2n{-}1\}$ except for the entry in cell $(1,1)$ and that $t{-}1 \geq 2n$.

Figure 5.3

Now form the $(n+1) \times (n+2)$ partial latin rectangle R_2 shown in Figure 5.4.

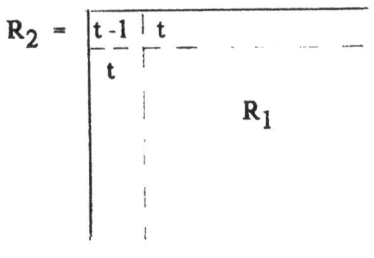

$(n+1) \times (n+2)$ partial latin rectangle based on $1,2,\ldots,t$ with all cells of the first row and column except the first two (which are occupied by $t{-}1,t$) empty.

Figure 5.4

We claim that the empty cells in R_2 can be filled in to produce an $(n+1) \times (n+2)$ latin rectangle R_3 such that $R_3(i) \geq 2n+3-t$ for all $i \in \{1,2,\ldots,t\}$ and $R_3(j) \geq 2n+4-t$ for all $j \in \{n+1,\ldots,t-2\}$. The technique of proof here is exactly the same as in the proof of Lemma 5.1. So, as before, let

$A_i = \{x : x$ does not appear in the i-th column of R_2 and $x \neq t-1, t\}$

$B_j = \{y : y$ does not appear in the j-th row of R_i and $y \neq t-1, t\}$.

Further, let

$$f(i) = \begin{cases} 1, & \text{if } i \in \{n+1,\ldots,t-2\}, \text{ and} \\ 0, & \text{otherwise} \end{cases}$$

and define

$S = \{i : i \neq t-1$ or t, and $R_1(i) = 2n+1-t+f(i)\}$, and

$T = \{j : j \neq t-1$ or t, and $R_1(j) = 2n+2-t+f(j)\}$.

If we can find SDRs $(x_3, x_4, \ldots, x_{n+2})$ of $A_3, A_4, \ldots, A_{n+2}$ and $(y_3, y_4, \ldots, y_{n+1})$ of $B_3, B_4, \ldots, B_{n+1}$ such that

$S \subseteq \{x_3, x_4, \ldots, x_{n+2}\} \cap \{y_3, y_4, \ldots, y_{n+1}\}$, and

$T \subseteq \{x_3, x_4, \ldots, x_{n+2}\} \cup \{y_3, y_4, \ldots, y_{n+1}\}$

we will have the desired completion of R_2. A straightforward application of Theorem 2.4 (as in the proof of Lemma 5.1) shows that, in fact, such SDRs always exist. Now since $R_3(i) \geq 2n+3-t$ for all $i \in \{1,2,\ldots,t\}$ and $R_3(j) \geq 2n+4-t$ for all $j \in \{n+1, n+2, \ldots, t-2\}$, R_3 can be diagonally placed in a t×t latin square D as in Figure 5.5.

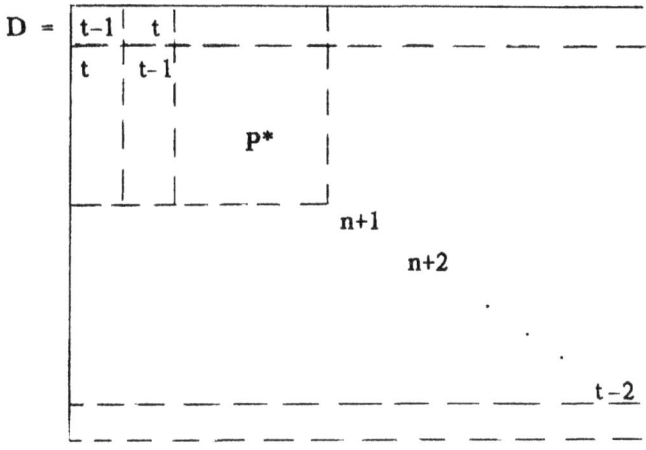

Figure 5.5

Let T denote the $(t-2) \times (t-2)$ idempotent latin rectangle based on $1, 2, \ldots, t$ which is obtained by deleting from D its first two columns and its first and last rows and let S_i denote the symbols which do not occur in the i-th column of T. Then t and t-1 will each occur exactly once among the sets $S_1, S_2, \ldots, S_{t-2}$ and furthermore they will not both occur in the same S_i. We can assume that $t \in S_1$ and that $t-1 \in S_2$. Now let t^* be a symbol which does not belong to the set $\{1,2,3\ldots,t\}$ and replace t by t^* in S_1 and t-1 by t^* in S_2. We can then use Hoffman and Kuhn's Theorem 2.3 to find two SDRs $(x_1=t^*, x_2, x_3, \ldots, x_{t-2})$ and $(y_1, y_2=t^*, y_3, \ldots, y_{t-2})$ such that $\{x_i, y_i\} = S_i$. If we replace t^* in the first SDR by t and replace t^* in the second SDR by t-1 and then adjoin these SDRs as rows to T, the resulting t×(t-2) latin rectangle K is as shown in Figure 5.6. We can now extend K to a latin square L by adjoining two additional columns with the aid of Ryser's theorem. Since neither t nor t-1 occur in the first row of T, they do not occur in the first row of K and so must occur in cells (1,t-1) and (1,t) of L. Since the last two columns of L can be interchanged without affecting K, we can assume that cell (1,t-1) is occupied by t and that

8:19 *Embedding theorems for partial latin squares* 235

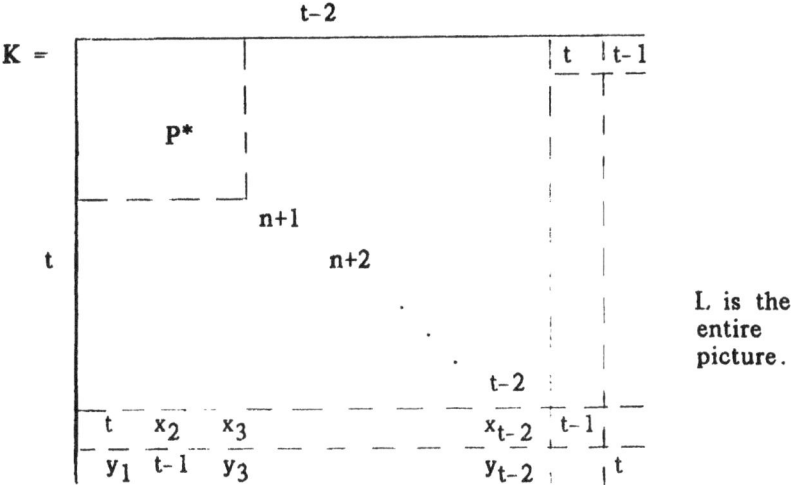

Figure 5.6

$(1,t)$ is occupied by $t-1$. This forces L to be idempotent, which completes the proof. []

As with all of the other results obtained so far, the embedding of an $n \times n$ partial idempotent latin square in a $(2n+1) \times (2n+1)$ square is the best possible embedding with respect to the size of the containing square.

EXAMPLE. The following is an example of the technique used in Theorem 5.2 to embed a partial idempotent latin square (in the case when $n = 4$ and $t = 9$).

$$P = \begin{bmatrix} 1 & 3 & . & . \\ 3 & 2 & . & 1 \\ 4 & . & 3 & . \\ . & . & 2 & 4 \end{bmatrix} \qquad P^* = \begin{bmatrix} 1 & 3 & 7 & 6 \\ 3 & 2 & 6 & 1 \\ 4 & 6 & 3 & 7 \\ 6 & 7 & 2 & 4 \end{bmatrix}$$

$$R_1 = \begin{bmatrix} 8 & 1 & 3 & 7 & 6 \\ 5 & 3 & 2 & 6 & 1 \\ 2 & 4 & 6 & 3 & 7 \\ 1 & 6 & 7 & 2 & 4 \end{bmatrix} \qquad R_2 = \begin{bmatrix} 8 & 9 & . & . & . & . \\ 9 & 8 & 1 & 3 & 7 & 6 \\ . & 5 & 3 & 2 & 6 & 1 \\ . & 2 & 4 & 6 & 3 & 7 \\ . & 1 & 6 & 7 & 2 & 4 \end{bmatrix}$$

$$R_3 = \begin{bmatrix} 8 & 9 & 7 & 1 & 4 & 5 \\ 9 & 8 & 1 & 3 & 7 & 6 \\ 4 & 5 & 3 & 2 & 6 & 1 \\ 1 & 2 & 4 & 6 & 3 & 7 \\ 5 & 1 & 6 & 7 & 2 & 4 \end{bmatrix}$$

$R_3(i) > 2$ for all i, and
$R_3(j) > 3$ for j = 5,6, and 7.

$$D = \begin{bmatrix} 8 & 9 & 7 & 1 & 4 & 5 & 2 & 3 & 6 \\ 9 & 8 & 1 & 3 & 7 & 6 & 4 & 5 & 2 \\ 4 & 5 & 3 & 2 & 6 & 1 & 8 & 7 & 9 \\ 1 & 2 & 4 & 6 & 3 & 7 & 9 & 8 & 5 \\ 5 & 1 & 6 & 7 & 2 & 4 & 3 & 9 & 8 \\ 7 & 6 & 8 & 9 & 1 & 2 & 5 & 4 & 3 \\ 2 & 7 & 5 & 8 & 9 & 3 & 1 & 6 & 4 \\ 3 & 4 & 2 & 5 & 8 & 9 & 6 & 1 & 7 \\ 6 & 3 & 9 & 4 & 5 & 8 & 7 & 2 & 1 \end{bmatrix}$$

R_3 is diagonally placed in D.

$$K = \begin{bmatrix} 1 & 3 & 7 & 6 & 4 & 5 & 2 & 9 & 8 \\ 3 & 2 & 6 & 1 & 8 & 7 & 9 & 4 & 5 \\ 4 & 6 & 3 & 7 & 9 & 8 & 5 & 1 & 2 \\ 6 & 7 & 2 & 4 & 3 & 9 & 8 & 5 & 1 \\ 8 & 9 & 1 & 2 & 5 & 4 & 3 & 7 & 6 \\ 5 & 8 & 9 & 3 & 1 & 6 & 4 & 2 & 7 \\ 2 & 5 & 8 & 9 & 6 & 1 & 7 & 3 & 4 \\ 9 & 1 & 4 & 5 & 7 & 2 & 6 & 8 & 3 \\ 7 & 4 & 5 & 8 & 2 & 3 & 1 & 6 & 9 \end{bmatrix}$$

L is the entire 9×9 latin square. P is embedded in L.

(6) Conjugate quasigroups and identities

So far as we are concerned, a quasigroup is merely a latin square with a headline and sideline. Let $(Q, \circ)$ be a quasigroup of order n and define an $n^2 \times 3$ array A by the statement that (x,y,z) is a row of A if and only if $x \circ y = z$. As a consequence of the fact that the equations $a \circ x = b$ and $y \circ a = b$ are uniquely solvable for all $a, b \in Q$, it follows that if we run our fingers down any two columns of A we get each ordered pair belonging to Q×Q <u>exactly once</u>. An array with this property is called an orthogonal array and, not too surprisingly, the construction can be reversed. That is, if A is any $n^2 \times 3$ orthogonal array (defined on a set Q) and we define a binary operation on Q by $x \circ y = z$ if and only if (x,y,z) is a row of A, then $(Q, \circ)$ is a quasigroup. If $\alpha \in S_3$ (the symmetric group on $\{1,2,3\}$) and A is any $n^2 \times 3$ orthogonal array we will denote by $A\alpha$ the orthogonal array obtained by permuting the columns of A according to α. Two orthogonal arrays A and B are said to be <u>conjugate</u> provided there is at least one $\alpha \in S_3$ such that $A\alpha = B$. Two quasigroups are said to be <u>conjugate</u> or <u>parastrophic</u> provided that their corresponding orthogonal arrays are conjugate. (See also Chapter 1 or [DK], pages 65-68.)

EXAMPLE. The quasigroups σ_1 and σ_2 shown in Figure 6.1 are conjugate because $A\sigma = B$, where $\alpha = (1\ 2\ 3)$.

(σ_1)	1	2	3
1	2	3	1
2	1	2	3
3	3	1	2

(σ_2)	1	2	3
1	3	1	2
2	1	2	3
3	2	3	1

$$A = \begin{bmatrix} 1 & 1 & 2 \\ 1 & 2 & 3 \\ 1 & 3 & 1 \\ 2 & 1 & 1 \\ 2 & 2 & 2 \\ 2 & 3 & 3 \\ 3 & 1 & 3 \\ 3 & 2 & 1 \\ 3 & 3 & 2 \end{bmatrix} \qquad B = \begin{bmatrix} 2 & 1 & 1 \\ 3 & 1 & 2 \\ 1 & 1 & 3 \\ 1 & 2 & 1 \\ 2 & 2 & 2 \\ 3 & 2 & 3 \\ 3 & 3 & 1 \\ 1 & 3 & 2 \\ 2 & 3 & 3 \end{bmatrix}$$

Figure 6.1

The <u>conjugate invariant subgroup</u> H of an orthogonal array A (and the corresponding quasigroup) is defined by $H = \{\alpha \in S_3 : A\alpha = A\}$.

In the above example, the conjugate invariant subgroup of (Q, o_1) is $\langle(13)\rangle = \{(1),(13)\}$ and the conjugate invariant subgroup of (Q, o_2) is $\langle(12)\rangle = \{(1),(12)\}$. There is an intimate connection between the conjugate invariant subgroup of a quasigroup and the identities which the quasigroup satisfies. The following tables (see Figures 6.2 and 6.3) are self explanatory (and easily verified).

(Q,o) is invariant under conjugation by	(Q,o) satisfies
(12)	$xy = yx$
(13)	$(yx)x = y$
(23)	$x(xy) = y$
(123)	$(xy)x = y$
(132)	$x(yx) = y$

Figure 6.2

The conjugate invariant subgroup of (Q,o) contains the subgroup	(Q,o) satisfies at least
$\langle(12)\rangle$	$xy = yx$ (commutative)
$\langle(13)\rangle$	$(yx)x = y$
$\langle(23)\rangle$	$x(xy) = y$
$\langle(123)\rangle$	$(xy)x = y$ and $x(yx) = y$ (semisymmetric)
S_3	$xy = yx$, $(yx)x = y$, $x(xy) = y$, $(xy)x = y$, and $x(yx) = y$ (totally symmetric)

Figure 6.3

Since the permutation corresponding to $(xy)x = y$ generates $\langle(123)\rangle$ and the permutation corresponding to $x(yx) = y$ also generates $\langle(123)\rangle$ we need only one of these identities to describe a <u>semisymmetric</u> quasigroup. We will usually use $x(yx) = y$. It is immediate that the permutations corresponding to any two of the identities $xy = yx$, $(yx)x = y$, $x(xy) = y$, and $x(yx) = y$ generate S_3 and so just two will suffice to describe a totally symmetric quasigroup. In what follows we will use the two identities $(yx)x = y$ and $x(xy) = y$.

We can now use the above observations to obtain immediately some additional embedding theorems.

THEOREM 6.1 A partial $x(xy) = y$ quasigroup of order n can be embedded in an $x(xy) = y$ quasigroup of order t for every even $t \geq 2n$.

Proof. By a partial $x(xy) = y$ quasigroup is meant a partial quasigroup $(Q, \circ)$ such that whenever $p \circ q$ is defined, then so is $p \circ (p \circ q)$ and, furthermore, $p \circ (p \circ q) = q$. So, let $(Q, \circ)$ be a partial $x(xy) = y$ quasigroup and denote by $(Q, \times)$ the (13) conjugate of $(Q, \circ)$. Then $(Q, \times)$ is a partial commutative quasigroup and so by Cruse's Theorem can be embedded in a commutative quasigroup $(V, .)$ of order t for every even $t \geq 2n$. Taking the (13) conjugate of $(V, .)$ then gives an $x(xy) = y$ quasigroup with the partial quasigroup $(Q, \circ)$ embedded in it. []

Since the identity $x^2 = x$ is invariant under conjugation (because every conjugate of a (partial) idempotent quasigroup remains a (partial) idempotent quasigroup), a bit of imagination transforms Cruse's Theorems 4.3 and 4.4 into the following theorems.

THEOREM 6.2 A partial $\{x^2 = x, x(xy) = y\}$ ($\{x^2 = x, (yx)x = y\}$) quasigroup of order n can be embedded in an $x(xy) = y$ ($(yx)x = y$) quasigroup of order t for every $t \geq 2n$. []

THEOREM 6.3 A partial $\{x^2 = x, x(xy) = y\}$ ($\{x^2 = x, (yx)x = y\}$) quasigroup of order n can be embedded in a $\{x^2 = x, x(xy) = y\}$ ($\{x^2 = x, (yx)x = y\}$) quasigroup of order t for every odd $t \geq 2n+1$. []

Since the results on commutative and idempotent commutative quasigroups are best possible, it goes without saying that the results in Theorem 6.2 are also best posible.

(7) Embedding semisymmetric and totally symmetric quasigroups.

So that there will be no confusion in what follows, we make the following definitions.

DEFINITIONS. A <u>partial semisymmetric</u> quasigroup is a partial quasigroup $(P, \circ)$ such that whenever $p \circ q$ is defined, then so are $q \circ (p \circ q)$ and $(p \circ q) \circ p$ and, furtheremore, $q \circ (p \circ q) = p$ and $(p \circ q) \circ p = q$. A <u>partial totally symmetric</u> quasigroup is a partial quasigroup $(P, \circ)$ such that whenever $p \circ q$ is defined, then so are $q \circ p$, $p \circ (p \circ q)$, $(q \circ p) \circ p$, $(p \circ q) \circ q$, and $q \circ (q \circ p)$ and, furthermore, $p \circ q = q \circ p$, $p \circ (p \circ q) = (q \circ p) \circ p = q$, and $(q \circ p) \circ q = q \circ (p \circ q) = p$.

The reader is <u>cautioned</u> that neither semisymmetric nor totally symmetric quasigroups are necessarily idempotent (or even have any idempotents for that matter.)

In everything that follows $(T, \circ_1)$ and $(T, \circ_2)$ will always be the two totally symmetric quasigroups whose multiplication tables are given in Figure 7.1.

$(\circ_1)$	1	2	3		$(\circ_2)$	1	2	3
1	2	1	3		1	1	3	2
2	1	3	2		2	3	2	1
3	3	2	1		3	2	1	3

Figure 7.1

THEOREM 7.1 [A.B.Cruse and C.C.Lindner(1975)]. A partial semisymmetric quasigroup $(P, \circ)$ of order n can be embedded in a semisymmetric quasigroup of order 3t for every t which is the order of a quasigroup containing $(P, \circ)$.

Proof Let $(P, \circ)$ be a partial semisymmetric quasigroup of order n and let $(Q, \cdot)$ be <u>any</u> quasigroup (no restrictions) of order t containing $(P, \circ)$. Let $S = Q \times \{1, 2, 3\}$ and define a binary operation $(\circ)$ on S as follows:

(a) $(x, i) \circ (y, j) = (x \circ y, i \circ_2 j)$ if $x \circ y$ is defined in $(P, \circ)$; and

(b) if x o y is not defined in (P,o):

$$(x,i)_o(y,j) \begin{cases} = (x.y, i \ o_1 \ j), \text{ if } i = j; \text{ and, in cases when } i \neq j, \\ = (a, i \ o_1 \ j), \text{ if } i \ o_1 \ j = i, \text{ and } a.x = y; \\ = (b, i \ o_1 \ j), \text{ if } i \ o_1 \ j = j, \text{ and } y.b = x. \end{cases}$$

A routine proof shows that $(S,_o)$ is a semisymmetric quasigroup. If we identify each x in P with $(x,1)$, then $(x,1)_o(y,1) = (x \ o \ y, 1)$ in $(S,_o)$ if and only if x o y is defined in (P,o). Hence (P,o) is embedded in $(S,_o)$. []

COROLLARY. A partial semisymmetric quasigroup (P,o) of order n can be embedded in a semisymmetric quasigroup of order v for every $v \geq 6n$ such that $v \equiv 0$ or 3 (mod 6).

Proof. Write $v = 6n+3k = 3(2n+k)$ and embed (P,o) in a quasigroup or order 2n+k by Evans' Theorem. []

THEOREM 7.2 [A.B.Cruse and C.C.Lindner(1975)] A partial totally symmetric quasigroup (P,o) of order n can be embedded in a totally symmetric quasigroup of order 3t for every t which is the order of a commutative quasigroup containing (P,o).

Proof Let (P,o) be a partial totally symmetric quasigroup of order n. Then (P,o) is a partial commutative quasigroup. Let $(Q,.)$ be any commutative quasigroup of order t containing (P,o) and define a binary operation $(_o)$ on $S = Q \times \{1,2,3\}$ by:

(a) $(x,i)_o(y,j) = (x \ o \ y, i \ o_2 \ j)$ if x o y is defined in (P,o); and

(b) if x o y is not defined in (P,o):

$$(x,i)_o(y,j) \begin{cases} = (x.y, i \ o_1 \ j), \text{ if } i = j; \text{ and, in cases when } i \neq j, \\ = (a, i \ o_1 \ j), \text{ if } i \ o_1 \ j = i, \text{ and } a.x = y; \\ = (b, i \ o_1 \ j), \text{ if } i \ o_1 \ j = j, \text{ and } y.b = x. \end{cases}$$

It is immediate that $(S,_o)$ is totally symmetric and, of course, (P,o) is embedded in $(S,_o)$. []

COROLLARY. A partial totally symmetric quasigroup (P,o) of order n can be embedded in a totally symmetric quasigroup of order v for every $v \geq 6n$ such that $v \equiv 0 \pmod 6$.

Proof Write $v = 6n+6k = 3(2n+2k)$ and embed (P,o) in a commutative quasigroup of order $2n+2k$ by Cruse's Theorem. []

EXAMPLE. In Figure 7.2, we give an example of the construction used in Theorem 7.2 to embed partial totally symmetric quasigroups.

An immediate observation shows that (P,o) is embedded in $(S,_o)$ and that $(S,_o)$ is totally symmetric. (In fact, three disjoint copies of (P,o) are embedded in $(S,_o)$)

COMMENTS: The results in the Corollaries to Theorems 7.1 and 7.2 are not the best possible. The best possible embedding theorem for partial semisymmetric quasigroups would be that a partial semisymmetric quasigroup of order n can always be embedded in a semisymmetric quasigroup of order t for all $t \geq 2n$. This problem remains far from settled. The best possible embedding theorem for partial totally symmetric quasigroups would be that a partial totally symmetric quasigroup of order n can be embedded in a totally symmetric quasigroup of order t for all even $t \geq n$ if n is even, and in a totally symmetric quasigroup of order t for all $t \geq 2n$ if n is odd. This problem also remains far from settled.

(o)	1	2	3
1	3	2	1
2	2	1	.
3	1	.	3

(P,o)
partial totally symmetric
quasigroup

(.)	1	2	3	4	5	6
1	3	2	1	4	5	6
2	2	1	4	5	6	3
3	1	4	3	6	2	5
4	4	5	6	2	3	1
5	5	6	2	3	1	4
6	6	3	5	1	4	2

(Q,.)

commutative
quasigroup
containing
(P,o)

∘	11 21 31 41 51 61	12 22 32 42 52 62	13 23 33 43 53 63
11	31 21 11 42 52 62	33 23 13 41 51 61	32 22 12 63 53 43
21	21 11 42 52 62 32	23 13 61 31 41 51	22 12 53 43 33 63
31	11 42 31 62 22 52	13 51 33 21 61 41	12 63 32 53 43 23
41	42 52 62 22 32 12	61 41 51 11 21 31	43 33 23 13 63 53
51	52 52 22 32 12 42	51 31 41 61 11 21	53 43 63 23 13 33
61	62 32 52 12 42 22	41 61 21 51 31 11	63 53 43 33 23 13
12	33 23 13 61 51 41	32 22 12 43 53 63	31 21 11 42 52 62
22	23 13 51 41 31 61	22 12 43 53 63 33	21 11 62 32 42 52
32	13 61 33 51 41 21	12 43 32 63 23 53	11 52 31 22 62 42
42	41 31 21 11 61 51	43 53 63 23 33 13	62 42 52 12 22 32
52	51 41 61 21 11 31	53 63 23 33 13 43	52 32 42 62 12 22
62	61 51 41 31 21 11	62 33 53 13 43 23	42 62 22 52 32 12
13	32 22 12 43 53 63	31 21 11 62 52 42	33 23 13 41 51 61
23	22 12 63 33 43 53	21 11 52 42 32 62	23 13 41 51 61 31
33	12 53 32 23 63 43	11 62 31 52 42 22	13 41 33 61 21 51
43	63 43 53 13 23 33	42 32 22 12 62 52	41 51 61 21 31 11
53	53 33 43 63 13 23	52 42 62 22 12 32	51 61 21 31 11 41
63	43 63 23 53 33 13	62 52 42 32 22 12	61 31 51 11 41 21

(S, ∘)

Figure 7.2

(8) <u>Embedding Mendelsohn and Steiner triple systems</u>.

A <u>Steiner quasigroup</u> is any quasigroup satisfying the identities

$$x^2 = x \quad \text{idempotent}$$

$$\left.\begin{array}{l} x(xy) = y \\ (yx)x = y \end{array}\right\} \text{totally symmetric}$$

and a <u>Mendelsohn qausigroup</u> is any quasigroup satisfying the identities

$$x^2 = x \quad \text{idempotent}$$

$$x(yx) = y \quad \text{semisymmetric}.$$

In other words, a Steiner quasigroup is an idempotent totally symmetric quasigroup and a Mendelsohn quasigroup is an idempotent semisymmetric quasigroup. Hence a partial Steiner quasigroup is a partial idempotent totally symmetric quasigroup and a partial Mendelsohn quasigroup is a partial idempotent semisymmetric quasigroup. The constructions given in this section are easier to describe using design vernacular. So here goes!

DEFINITIONS. A <u>Steiner triple system</u> (STS) is a pair (S,t) where S is a set (of points) and t is a collection of 3-element subsets (triples) of S such that every pair of distinct points of S belong to exactly one triple of t. If "exactly" is replaced by "at most" we have the definition of a <u>partial Steiner triple system</u>. A <u>cyclic triple</u> is a collection of three ordered pairs of the form $\{(a,b),(b,c),(c,a)\}$, where a, b and c are distinct. We will denote such a triple by (a,b,c), (b,c,a) or (c,a,b). A <u>Mendelsohn triple system</u> (MTS) is a pair (M,T) where M is a set of points and T is a collection of triples such that every ordered pair (a,b), $a \neq b$, belongs to exactly one cyclic triple of T. As with Steiner triple systems, if we replace "exactly" with "at most" we have the definition of a partial Mendelsohn triple system. The partial STS (MTS) (P,p) is embedded in the STS (MTS) (S,t) provided $P \subseteq S$ and $p \subseteq t$.

The following equivalences are quite well known.

(partial) Steiner quasigroup $(Q, \circ)$	(partial) Steiner triple system (Q, t)
Define a collection of triples t by $\{a,b,c\} \in t$ if and only if $a \circ b = b \circ a = c$, $a \circ c = c \circ a = b$, and $b \circ c = c \circ b = a$ (are defined).	Define a binary operation "$\circ$" on Q by $a \circ a = a$, for all $a \in Q$, and if $a \neq b$, $a \circ b = c$ if and only if $\{a,b,c\} \in t$.
Then (Q,t) is a (partial) triple system	Then $(Q, \circ)$ is a (partial) Steiner quasigroup

(partial) Mendelsohn quasigroup (M,o)	(partial) Mendelsohn triple system (M,T)
Define a collection of cyclic triples T by (a,b,c)∈T if and only if a o b = c, b o c = a, and c o a = b (are defined)	Define a binary operation "o" on Q by a o a = a, for all a∈Q, and if a≠b, a o b = c if and only if (a,b,c)∈T
Then (M,T) is a (partial) Mendelsohn triple system	Then (M,o) is a (partial) Mendelsohn quasigroup

It is also well-known that the spectrum for STSs is the set of all $v \equiv 1$ or 3 (mod 6) [see C.C.Lindner(1980)] and that the spectrum for MTSs is the set of all $v \equiv 0$ or 1 (mod 3) [see D.G.Hoffman and C.C.Lindner(1981)], except v = 6 (for which no such system exists). The following two constructions pack the spectrum in each case, as well as being quite useful in what follows. (More precisely, modifications of these two constructions turn out to be quite useful.)

Construction of STSs. The following two constructions are slight modifications of well-known constructions due to R.C.Bose(1939) and Th.Skolem(1958).

The Bose Construction. Let (Q,o) be an idempotent commutative quasigroup of order 2k+1 and put S = Q×{1,2,3}. Define a collection of triples t of S by:
  (a) {(x,1),(x,2),(x,3)}∈t for every x∈Q; and
  (b) if x≠y, the three triples {(x,1),(y,1),(xoy,2)},
      {(x,2),(y,2),(xoy,3)}, and {(x,3),(y,3),(xoy,1)}∈t.
Then (S,t) is an STS of order 6k+3.

The Skolem Construction. Let (Q,o) be a half-idempotent commutative quasigroup of order 2k and put S = {∞} ∪ (Q×{1,2,3}). Define a collection of triples t of S by:
  (a) {(x,1),(x,2),(x,3)}∈t for every x∈Q such that x ≤ k;

(b) for each x > k, the three triples {∞, (x,1), (x-k,2)},
{∞, (x,2), (x-k,3)}, and {∞, (x,3), (x-k,1)} ∈t; and

(c) if x≠y, the three triples {(x,1), (y,1), (x∘y,2)},
{(x,2), (y,2), (x∘y,3)}, and {(x,3), (y,3), (x∘y,1)} ∈t.

Then (S,t) is an STS of order 6k+1. Since there exists an idempotent commutative quasigroup of every order 2k+1 <u>and</u> a half-idempotent commutative quasigroup of every order 2k, the Bose and Skolem constructions produce an STS of every order v ≡ 1 or 3 (mod 6).

<u>Construction of MTSs</u>. We split the construction into two parts.

<u>v ≡ 0 (mod 3)</u>. Write v = 3k and let (Q,∘) be an idempotent quasigroup of order k. Put S = Q×{1,2,3} and define a collection of cyclic triples t by:

(a) ((x,1), (x,2), (x,3)) and ((x,2), (x,1), (x,3)) belong to t for every x∈Q; and

(b) if x≠y, the six cyclic triples ((x,1), (y,1), (x∘y,2)),
((y,1), (x,1), (y∘x,2)), ((x,2), (y,2), (x∘y,3)),
((y,2), (x,2), (y∘x,3)), ((x,3), (y,3), (x∘y,1)), and
((y,3), (x,3), (y∘x,1)) belong to t.

Then (S,t) is a MTS of order 3k.

Note: there does not exist an idempotent quasigroup of order k = 2, <u>but</u> this is not the reason for the nonexistence of an MTS of order 6.

<u>v ≡ 1 (mod 3)</u>. Write v = 3k+1, let (Q,∘) be an idempotent quasigroup of order k, and put S = {∞} ∪ (Q×{1,2,3}). Now define a collection t of cyclic triples by modifying part (a) of the v ≡ 0 (mod 3) construction by replacing the two cyclic triples ((x,1), (x,2), (x,3)) and ((x,2), (x,1), (x,3)) by the four cyclic triples (∞, (x,1), (x,2)), (∞, (x,2), (x,3)), (∞, (x,3), (x,1)), and ((x,1), (x,3), (x,2)) for each x∈Q.

Then (S,t) is an MTS of order 3k+1. (The case 3.2+1 = 7, of course, cannot be constructed in this manner. The easiest way to handle this exception is to replace each triple (x,y,z) of an STS of order 7 by the two triples (x,y,z) and (x,z,y) of an MDS of order 3.)

The first embedding theorem for partial STSs was given by

C.Treash(1971) who showed that a partial triple system of order n could be embedded in a triple system of order at most 2^{2n}. The was subsequently improved by the present author [see C.C.Lindner(1975b)], who showed that a partial STS of order n can be embedded in an STS of order v for all $v \geq 6n+1$ such that $v = 1$ or $3 \pmod 6$. Eventually, the bound was improved to all $v \geq 4n+1$ such that $v = 1$ or $3 \pmod 6$ by L.D.Andersen, A.J.W.Hilton and E.Mendelsohn(1980). The length (and complexity) of the 4n+1 embedding makes it prohibitive to reproduce here, and so we will settle for a sketch of the 6n+1 embedding followed by a statement of the 4n+1 embedding.

THEOREM 8.1 [C.C.Lindner(1975b)]. A partial Steiner triple system of order n can be embedded in a Steiner triple system of order v for every $v \geq 6n+1$ such that $v = 1$ or $3 \pmod 6$.

Proof. We split the proof into two parts.

$\underline{v = 3 \pmod 6}$. Let (P,p) be a partial STS of order n and define a partial binary operation "o" on P by:

(a) $x \circ x = x$ for all $x \in P$; and

(b) if $x \neq y$, $x \circ y$ is defined if and only if $\{x,y,z\} \in p$ and then $x \circ y = z$.

It follows that (P,o) is (at least) a partial idempotent commutative quasigroup of order n. Write $v = 6k+3$, where $k \geq n$, and let (Q,.) be an idempotent commutative quasigroup of order 2k+1 containing (P,o). Put $S = Q \times \{1,2,3\}$ and define a collection of triples t of S by:

(a) $\{(x,1),(x,2),(x,3)\} \in t$ for every $x \in Q$;

(b) for each triple $\{x,y,z\} \in p$, the three triples $\{(x,1),(y,1),(z,1)\}$, $\{(x,2),(y,2),(z,2)\}$, and $\{(x,3),(y,3),(z,3)\} \in t$;

(c) for each triple $\{x,y,z\} \in p$, the six triples listed below belong to t:

$\{(x,1),(y,2),(z,3)\}$, $\{(x,2),(y,1),(z,3)\}$, $\{(x,3),(y,1),(z,2)\}$,
$\{(x,1),(y,3),(z,2)\}$, $\{(x,2),(y,3),(z,1)\}$, $\{(x,3),(y,2),(z,1)\}$; and

(d) if $x \neq y$ and x and y do not belong to a triple of P, the three triples $\{(x,1),(y,1),(x.y,2)\}$, $\{(x,2),(y,2),(x.y,3)\}$, and $\{(x,3),(y,3),(x.y,1)\} \in t$.

It is straightforward to see that (S,t) is an STS of order $v = 6k+3$ and part (b) of this construction guarantees that three disjoint copies of (P,p) are embedded in (S,t).

$\underline{v \equiv 1 \pmod{6}}$. Write $v = 6k+1$, where $k \geq n$, but now embed $(P,\circ)$ in a half-idempotent quasigroup $(Q,.)$ of order $2k$. Put $S = \{\infty\} \cup (Q \times \{1,2,3\})$ and define a collection of triples t by:

(a) $\{(x,1),(x,2),(x,3)\} \in t$ for every $x \in Q$ such that $x \leq k$;

(b) for each $x > k$, the three triples $\{\infty,(x,1),(x-k,2)\}$, $\{\infty,(x,2),(x-k,3)\}$, and $\{\infty,(x,3),(x-k,1)\} \in t$;

(c) as part (b) of the $v \equiv 3 \pmod{6}$ case;

(d) as part (c) of the $v \equiv 3 \pmod{6}$ case; and

(e) as part (d) of the $v \equiv 3 \pmod{6}$ case.

Then (S,t) is an STS which, by part (b) of the construction, contains three disjoint copies of (P,p). Combining the above two cases completes the proof of the theorem. []

THEOREM 8.2 [L.D.Andersen, A.J.W.Hilton, and E.Mendelsohn(1980)].

A <u>partial</u> Steiner triple system of order n can be embedded in a Steiner triple system of order v for every $v \geq 4n+1$ such that $v \equiv 1$ or $3 \pmod{6}$. []

The result of Theorem 8.2 gives the best embedding so far obtained. However, the best possible result would be that a partial STS of order n can be embedded in an STS of order v for $v \geq 2n+1$. While this is surely true, no one (as yet) has produced a proof. However, if we are interested in embedding <u>complete</u> STSs, the best possible result is available. The following theorem is due to J.Doyen and R.M.Wilson. However, the proof we sketch is due to A.Stern and H.Lenz(1980).

THEOREM 8.3 [J.Doyen and R.M.Wilson(1973)]. Any Steiner triple system of order n can be embedded in a Steiner triple system of order v for every $v \geq 2n+1$ such that $v \equiv 1$ or $3 \pmod{6}$.

<u>Proof</u>. Let (S,t) be an STS of order n and let $v \geq 2n+1$ be admissible (that is: congruent to 1 or 3 $\pmod{6}$). Further, let $W = \{0,1,2,\ldots,v-n-1\}$ and $X = \{1,2,\ldots,(v-n)/2\}$. In A.Stern and

H. Lenz's paper, it is shown that it is possible to partition $X \setminus \{(v-n)/2\}$ into subsets D and D* and also produce a collection of triples T* of elements of W such that the following three statements are true:

(a) $|D| = (n-1)/2$;

(b) (W, T*) is a partial triple system; and

(c) $x \neq y$ belongs to a triple of T* if and only if $|x-y| \in D^*$.

Now if $x \neq y$ and $|x-y| \notin D^*$, then $|x-y| \in D \cup \{(v-n)/2\}$. Since $|D| = (n-1)/2$, the number of 2-element subsets $\{x,y\}$ of W such that $|x-y| \in D \cup \{(v-n)/2\}$ is precisely $n(v-n)/2$. Let G denote the set of all such 2-element subsets of W. G, of course, is precisely the set of all 2-element subsets of W which are not covered by a triple of T. In A. Stern and H. Lenz (1980), it is shown that G can always be 1-factored. Now let $F = \{F_1, F_2, F_3, \ldots, F_n\}$ be a 1-factorization of G and let α be any one-to-one mapping of $\{1, 2, \ldots, n\}$ onto S. Define a collection of triples T of $S \cup W$ as follows:

(a) the triples of t belong to T;

(b) the triples of T* belong to T; and

(c) $\{x, y, a\} \in T$ if and only if $\{x,y\} \in F_i$ and $i\alpha = a$.

It is immediate that $(S \cup W, T)$ is an STS of order v containing (S, t) as a subsystem. []

THEOREM 8.4 [A.B. Cruse and C.C. Lindner (1975)]. A partial Mendelsohn triple system can be embedded in a Mendelsohn triple system of order v for every $v \geq 6n+3$ such that $v \equiv 0$ or 1 (mod 3).

Proof. As in the proof of Theorem 8.1 for STSs, we split the proof into two parts.

$\underline{v \equiv 0 \pmod 3}$. Let (P, p) be a partial MTS of order n and define a partial binary operation "o" on P by:

(a) $x \circ x = x$ for all $x \in P$; and

(b) if $x \neq y$, $x \circ y$ is defined if and only if $(x, y, z) \in p$ and then $x \circ y = z$.

It follows that (P, o) is (at least) a partial idempotent quasigroup of order n. Write $v = 3k$, where $k \geq 2n+1$, and let (Q, .) be an idempotent quasigroup of order k containing (P, o). Put $S = Q \times \{1, 2, 3\}$ and define a collection of cyclic triples t of S by:

(a) $((x,1),(x,2),(x,3))$ and $((x,2),(x,1),(x,3))$ belong to t for every $x \in Q$;

(b) for each cyclic triple $(x,y,z) \in p$, the three cyclic triples $((x,1),(y,1),(z,1))$, $((x,2),(y,2),(z,2))$, and $((x,3),(y,3),(z,3)) \in t$;

(c) for each cyclic triple $(x,y,z) \in p$, the six cyclic triples listed below belong to t;

$((x,1),(y,2),(z,3))$, $((x,2),(y,3),(z,1))$, $((x,3),(y,1),(z,2))$, $((x,1),(y,3),(z,2))$, $((x,2),(y,1),(z,3))$, $((x,3),(y,2),(z,1))$; and

(d) If $x \neq y$ and (x,y) does not belong to a cyclic triple of p, the three cyclic triples $((x,1),(y,1),(x \cdot y,2))$, $((x,2),(y,2),(x \cdot y,3))$, and $((x,3),(y,3),(x \cdot y,1))$ belong to t.

Then (S,t) is an MTS of order $v = 3k$ and, by part (b) of the construction, three disjoint copies of (P,p) are embedded in (S,t).

<u>$v \equiv 1 \pmod{3}$</u>. Write $v = 3k+1$, where $k \geq 2n+1$, put $S = \{\infty\} \cup (Q \times \{1,2,3\})$ and define a collection of cyclic triples t precisely as in the case $v \equiv 0 \pmod{3}$ except that (a) is replaced by:

(a)* for each $x \in Q$, place the four cyclic triples $(\infty,(x,1),(x,2))$, $(\infty,(x,2),(x,3))$, $(\infty,(x,3),(x,1))$, and $((x,1),(x,3),(x,2))$ in t.

Then (S,t) is an MTS of order $3k+1$ with three disjoint copies of (P,p) embedded in it.

Combining the cases $v \equiv 0 \pmod{3}$ and $v \equiv 1 \pmod{3}$ completes the proof. []

The above theorem does not give the best possible embedding for partial Mendelsohn triple systems. The best possible result would be that a partial MTS of order n can be embedded in an MTS of order v for all admissible $v \geq 2n+1$. Just as with STSs, this is surely true. It only remains for someone to produce a proof! In the interim, the following theorem gives the best possible embedding for <u>complete</u> Mendelsohn triple systems.

THEOREM 8.5 [D.G.Hoffman and C.C.Lindner(1981)]. Any Mendelsohn triple system of order n can be embedded in a Mendelsohn triple system of order v for every v ⩾ 2n+1 such that v ≡ 0 or 1 (mod 3).

Proof. We split the proof into three parts.

n ≡ v ≡ 0 (mod 3). In this case, both n/3 and v/3 are integers and v/3 ⩾ 2(n/3)+1. Hence, by Theorem 5.2, there exists an idempotent quasigroup $(Q, \circ)$ of order v/3 containing a subquasigroup $(P, \circ)$ of order n/3. The construction of MTSs for the case v ≡ 0 (mod 3) described at the beginning of this section gives a Mendelsohn triple system of order 3(v/3) = v containing (as a consequence of the fact that the subquasigroup $(P, \circ)$ is embedded in $(Q, \circ)$) a subsystem of order 3(n/3) = n.

n ≡ v ≡ 1 (mod 3), n ≠ 7. Trivially, (n-1)/3 and (v-1)/3 are integers and (v-1)/3 ⩾ 2((n-1)/3)+1. Hence, there exists an idempotent quasigroup $(Q, \circ)$ of order (v-1)/3 containing a subquasigroup $(P, \circ)$ of order (n-1)/3. The construction of MTSs described at the beginning of this section for the case v ≡ 1 (mod 3) now gives a Mendelsohn triple system of order 3((v-1)/3)+1 = v containing a subsystem of order 3(n/3) = n. (The case n = 7 has to be handled separately, since there does not exist an idempotent quasigroup of order 2. This is a trivial matter. See G.Stern and H.Lenz(1980).)

n ≢ v (mod 3). Let W = {0,1,2,...,v-n-1}. A directed 2-factor of W is simply a set of v-n ordered pairs (a,b), a ≠ b, such that each element of W occurs exactly once as a first co-ordinate and exactly once as a second co-ordinate. In G.Stern and H.Lenz(1980), it is shown that it is possible to partition W\{0} into sets D and D* and also produce a collection of cyclic triples T* such that

(a) $|D|$ = n;

(b) (W,T*) is a partial MTS; and

(c) (x,y) belongs to a cyclic triple of T* if and only if y-x∈D*.

If G = {(a,b): b-a∈D}, then G is precisely the set of all ordered pairs of W not belonging to a cyclic triple of T*. For each $d_i \in D$, let F_{d_i} = {(a,b): b-a = d_i}. Then each F_{d_i} is a directed 2-factor of W and F = {$F_{d_1}, F_{d_2}, \ldots, F_{d_n}$} is a factorization of G into directed 2-factors. Now let (S,t) be an MTS of order n and define a collection of cyclic triples T of S∪W by:

(a) the cyclic triples of t belong to T;

(b) the cyclic triples of T* belong to T; and

(c) $(x,y,d_i) \in T$ if and only if $(x,y) \in F_{d_i}$.

Then (S∪W,T) is an MTS of order v containing (S,t) as a subsystem. Combining the above three cases completes the proof of the theorem. []

(9) Summary of embedding theorems

Partial I-quasigroup of order n with I =	Best possible embedding	Best embedding to date
(1) ϕ	all $t \geq 2n$	all $t \geq 2n$ [T. Evans (1960)]
(2) $xy = yx$		
(3) $x(xy) = y$	all even $t \geq 2n$	all even $t \geq 2n$ [A.B. Cruse (1974a)]
(4) $(yx)x = y$		
(5) $x^2 = x$, $xy = yx$		
(6) $x^2 = x$, $x(xy) = y$	all odd $t \geq 2n+1$	all odd $t \geq 2n+1$ [A.B. Cruse (1974a)]
(7) $x^2 = x$, $(yx)x = y$		
(8) $x^2 = x$	all $t \geq 2n+1$	all $t \geq 2n+1$ [L.D. Andersen, A.J.W. Hilton and C.A. Rodger (1982)]
(9) $x(yx) = y$	all $t \geq 2n$	all $t \geq 6n$ such that $t \equiv 0$ or $3 \pmod 6$ [A.B. Cruse and C.C. Lindner (1975)]
(10) $x(xy) = y$, $(yx)x = y$	all $t \geq 2n$ if n is odd, and all even $t \geq 2n$ if n is even	all $t \geq 6n$ such that $t \equiv 0 \pmod 6$ [A.B. Cruse and C.C. Lindner (1975)]
(11) $x^2 = x$, $x(yx) = y$	all $t \geq 2n+1$ such that $t \equiv 0$ or $1 \pmod 3$	all $t \geq 6n+3$ such that $t \equiv 0$ or $1 \pmod 3$ [A.B. Cruse and C.C. Lindner (1975)]
(12) $x^2 = x$, $x(xy) = y$, $(yx)x = y$	all $t \geq 2n+1$ such that $t \equiv 1$ or $3 \pmod 6$	all $t \geq 4n+1$ such that $t \equiv 1$ or $3 \pmod 6$ [L.D. Andersen, A.J.W. Hilton, and E. Mendelsohn (1980)]

Complete I-quasigroup of order n with I =	Best possible embedding	Best embedding to date
(1), (2), (3), (4), (5), (6), (7), (8), (9) and (10)	same as for partial quasigroup	same as for partial quasigroup
(11)	all $t \geq 2n+1$ such that $t \equiv 0$ or 1 (mod 3)	all $t \geq 2n+1$ such that $t \equiv 0$ or 1 (mod 3) [D.G.Hoffman and C.C.Lindner(1981)]
(12)	all $t \geq 2n+1$ such that $t \equiv 1$ or 3 (mod 6)	all $t \geq 2n+1$ such that $t \equiv 1$ or 3 (mod 6) [J.Doyen and R.M.Wilson(1973)]

A quick glance at the above tables shows that the problem of obtaining the best possible embedding for partial quasigroups of types (9), (10), (11), and (12) remains open and so too does the problem of obtaining the best possible embedding for complete quasigroups of types (9) and (10). These problems are (of course) not independent, since a solution of (9) and (10) in the partial case also settles (9) and (10) in the complete case. (A complete quasigroup is also a partial quasigroup.)

(10) <u>The Evans' conjecture (Smetaniuk's proof)</u>.

We noted in Section 1 of this Chapter that a partial latin square cannot necessarily be completed to a latin square. (That is the reason, of course, for all the interest in embedding theorems for partial latin squares.) In particular, it is a trivial matter to construct a partial n×n latin square with n cells occupied which cannot be completed to a latin square. For example, the partial 4×4 latin square P given in Figure 10.1 cannot be completed to a latin square.

$$P = \begin{array}{|cccc|} \hline 1 & 2 & 3 & . \\ . & . & . & 4 \\ . & . & . & . \\ . & . & . & . \\ \hline \end{array}$$

Figure 10.1

This observation led to the question "Can an n×n partial latin square which has n-1 or less cells occupied be completed to a latin square of order n?" In 1960, Trevor Evans (in the same paper in which he proved that a partial n×n latin square can always be embedded in a t×t latin square for every t ⩾ 2n) conjectured that a partial n×n latin square with at most n-1 cells occupied can always be completed to a latin square. This conjecture from T.Evans(1960) became widely known as the Evans' conjecture. Although experimental evidence was all in favour of the conjecture being true, a satisfactory proof eluded researchers until quite recently. (A discussion of the problem and of the steps which had been made to resolve it by the time that [DK] was published will be found in Section 3.3 of that book.)

In R.Haggkvist(1978), that author gave a proof of the Evans' conjecture for all n ⩾ 1111 (a rather dramatic number). Subsequently, two complete solutions have been given: one by L.D.Andersen and A.J.W.Hilton(1983), which is very lengthy, and the other by B.Smetaniuk(1981). It seems fitting to end this Chapter where it all began: that is, with the problem of completing latin squares. We do this by giving Smetaniuk's proof of the Evans' conjecture.

[EDITORS' NOTE: The above remarks deserve some amplification. Despite the dissimilar publication dates, the above two solutions were obtained at about the same time and quite independently. Because the proof given by Smetaniuk seemed much the simpler, publication of the Andersen and Hilton paper was delayed. However, the latter paper not only showed the truth of Evans' conjecture but also showed that in many cases a partial n×n latin square with as many as n cells occupied can be completed to a latin square, and it also provided criteria on the location of the occupied cells for this to be the case. However, somewhat later, R.M.Damerell(1983) was able to adapt Smetaniuk's construction algorithm to provide a considerably shorter proof of this additional result.

One of the present Editors later gave a simplified version of Smetaniuk's argument which is largely descriptive and which makes it conceptually easy to see how the construction works. However, according to the referee, the author considers that this descriptive argument does not cover all possible cases. We have, therefore, adopted a suggestion of the referee and added the simplified version as an appendix.]

For Smetaniuk's proof of the Evans' conjecture, we will first need a few preliminary ideas and results.

In what follows, the cells $(1,m), (2,m-1), \ldots, (m,1)$ will be called the <u>back diagonal</u> of an m×m latin square. Now, let L be an n×n latin square and denote by Y the part of L which is on and above the back diagonal. We define a partial $(n+1)\times(n+1)$ latin square P(L) as follows: Each cell on the back diagonal is occupied by n+1, the part of P(L) above the back diagonal is Y, and all cells below the back diagonal are empty. For an illustrative example, see Figure 10.2.

$$L = \begin{vmatrix} 1 & 4 & 2 & 3 & 5 \\ 5 & 3 & 1 & 4 & 2 \\ 3 & 1 & 5 & 2 & 4 \\ 2 & 5 & 4 & 1 & 3 \\ 4 & 2 & 3 & 5 & 1 \end{vmatrix} \qquad P(L) = \begin{vmatrix} 1 & 4 & 2 & 3 & 5 & 6 \\ 5 & 3 & 1 & 4 & 6 & . \\ 3 & 1 & 5 & 6 & . & . \\ 2 & 5 & 6 & . & . & . \\ 4 & 6 & . & . & . & . \\ 6 & . & . & . & . & . \end{vmatrix}$$

Figure 10.2

<u>THEOREM 10.1</u> [B.Smetaniuk(1981)] Let L be any n×n latin square. Then the $(n+1)\times(n+1)$ partial latin square P(L) can always be completed to a latin square.

<u>Proof</u>. Let $2 \le k \le n-1$ and let M be any partial latin square obtained by completing the first k columns of P(L). For each $i = n-k+2, n-k+3, \ldots, n$, denote by L_i the symbols in the first k cells of row i of L and by M_i the symbols in row i of M. If $M_i \setminus \{n+1\} \subseteq L_i$ for all $i = n-k+2, n-k+3, \ldots, n$ we will say that M is a <u>good extension</u> of P(L). The idea of the proof is to show that if M is a good extension of P(L), then column k+1 of M can be filled in so that the result is a partial latin square M* which is also a good extension of P(L). Since the second column of P(L) can trivially be completed to a good extension of P(L), iteration completes the first n columns of P(L). The (n+1)-th column can

Embedding theorems for partial latin squares

be completed by Marshall Hall's theorem, Theorem 2.2. Now let M be a good extension of P(L) and, for each symbol x ≠ n+1 in the bottom row of M, define a sequence $(r_1, r_2, \ldots, r_t)$ as follows:

(i) r_1 is the unique number such that x belongs to cell (n+1,c) of M and to cell (r_1,c) of L; and

(ii) if $r_1, r_2, \ldots, r_i$ have been defined, then r_{i+1} is defined if and only if there exists an integer d such that x belongs to cell (r_i, d) of M and to cell (r_{i+1}, d) of L.

For each $x \in \{1,2,3,\ldots,n\}$ define the <u>starting row</u> r(x) of x to be n+1 if x does not belong to the last row of M, and r_t if x does belong to the last row of M.

The (k+1)-th column of M is completed as follows: Choose the symbol x belonging to cell (n-k+1,k+1) of L and form the sequence $(x=x_1, x_2, \ldots, x_m)$ as follows:

(i) $r(x_m) = n+1$;

(ii) $r(x_m)$ is the first occurrence of n+1; and

(iii) if m ≥ 2, each x_{i+1} is the entry in cell $(r(x_i), k+1)$ of L.

Now, fill in cell $(r(x_i), k+1)$ with x_i and the remaining cells with the entries in the corresponding cells of L. It is straightforward (and not difficult) to see that the resulting partial latin square is a good extension of P(L). []

EXAMPLE.

$$L = \begin{vmatrix} 1 & 8 & 7 & 2 & 3 & 4 & 5 & 6 \\ 6 & 4 & 1 & 8 & 5 & 3 & 7 & 2 \\ 5 & 3 & 6 & 1 & 7 & 8 & 2 & 4 \\ 3 & 5 & 8 & 4 & 2 & 6 & 1 & 5 \\ 4 & 2 & 3 & 6 & 1 & 5 & 8 & 7 \\ 2 & 6 & 5 & 3 & 8 & 7 & 4 & 1 \\ 8 & 5 & 2 & 7 & 4 & 1 & 6 & 3 \\ 7 & 1 & 4 & 5 & 6 & 2 & 3 & 8 \end{vmatrix} \quad P(L) = \begin{vmatrix} 1 & 8 & 7 & 2 & 3 & 4 & 5 & 6 & 9 \\ 6 & 4 & 1 & 8 & 5 & 3 & 7 & 9 & . \\ 5 & 3 & 6 & 1 & 7 & 8 & 9 & . & . \\ 3 & 5 & 8 & 4 & 2 & 9 & . & . & . \\ 4 & 2 & 3 & 6 & 9 & . & . & . & . \\ 2 & 6 & 5 & 9 & . & . & . & . & . \\ 8 & 5 & 9 & . & . & . & . & . & . \\ 7 & 9 & . & . & . & . & . & . & . \\ 9 & . & . & . & . & . & . & . & . \end{vmatrix}$$

$$M = \begin{vmatrix} 1 & 8 & 7 & 2 & 3 & 4 & 5 & 6 & 9 \\ 6 & 4 & 1 & 8 & 5 & 3 & 7 & 9 & . \\ 5 & 3 & 6 & 1 & 7 & 8 & 9 & . & . \\ 3 & 5 & 8 & 4 & 2 & 9 & . & . & . \\ 4 & 2 & 3 & 6 & 9 & 1 & . & . & . \\ 2 & 6 & 5 & 9 & 8 & 7 & . & . & . \\ 8 & 5 & 9 & 7 & 4 & 2 & . & . & . \\ 7 & 9 & 4 & 5 & 1 & 6 & . & . & . \\ 9 & 1 & 2 & 3 & 6 & 5 & . & . & . \end{vmatrix}$$

M is a good extension of P(L).

$M_4 \setminus \{9\} \subseteq L_4$

$M_5 \setminus \{9\} \subseteq L5$

$M_6 \setminus \{9\} \subseteq L_6$

$M_7 \setminus \{9\} \subseteq L_7$

$M_8 \setminus \{9\} \subseteq L_8$

	$(r_1, r_2, \ldots, r_t)$	Starting row
1	(8,5,7)	7
2	(7,8)	8
3	(6)	6
4	-	9
5	(5)	5
6	(8,4)	4
7	-	9
8	-	9

Since 2 is the entry in cell (3,7) of L,
$(x = x_1, x_2, \ldots, x_m) = (2, 3, 4)$.

$$M^* = \begin{vmatrix} 1 & 8 & 7 & 2 & 3 & 4 & 5 & 6 & 9 \\ 6 & 4 & 1 & 8 & 5 & 3 & 7 & 9 & . \\ 5 & 3 & 6 & 1 & 7 & 8 & 9 & . & . \\ 3 & 5 & 8 & 4 & 2 & 9 & 1 & . & . \\ 4 & 2 & 3 & 6 & 9 & 1 & 8 & . & . \\ 2 & 6 & 5 & 9 & 8 & 7 & 3 & . & . \\ 8 & 5 & 9 & 7 & 4 & 2 & 6 & . & . \\ 7 & 9 & 4 & 5 & 1 & 6 & 2 & . & . \\ 9 & 1 & 2 & 3 & 6 & 5 & 4 & . & . \end{vmatrix}$$

M* is a good extension of P(L).

$M_3^* \setminus \{9\} \subseteq L_3$

$M_4^* \setminus \{9\} \subseteq L_4$

$M_5^* \setminus \{9\} \subseteq L_5$

$M_6^* \setminus \{9\} \subseteq L_6$

$M_7^* \setminus \{9\} \subseteq L_7$

$M_8^* \setminus \{9\} \subseteq L_8$

THEOREM 10.2 [C.C.Lindner(1970)]. The Evans' Conjecture is true provided that the occupied cells lie in at most n/2 of the rows.

Proof. Let P be a partial n×n latin square with n-1 cells occupied and such that the occupied cells occur in at most n/2 of the rows of P. Let α be a permutation of $\{1,2,3,\ldots,n\}$ such that row $\alpha(i)$ of P has at least as many occupied cells as row $\alpha(i+1)$. In C.C.Lindner(1970), it is shown that P can be completed to a latin square by filling in one row at a time provided that row $\alpha(i)$ is filled in before row $\alpha(i+1)$. Each row is completed by using a straightforward application of Philip Hall's Theorem. []

COROLLARY The Evans' Conjecture is true provided that at least one of the following is true: either the occupied cells lie in at most n/2 of the rows, or the occupied cells lie in at most n/2 of the columns, or at most n/2 distinct symbols occur in the occupied cells.

Proof. At least one of the conjugates of such a partial latin square has the property that the occupied cells lie in at most n/2 of the rows. []

THEOREM 10.3 An n×n partial latin square with n-1 cells occupied can always be completed to a latin square.

Proof. The proof is by induction on n. The Evans' Conjecture is obviously true for n = 1 or 2 and so we can assume that n ≥ 3 and that the Evans' Conjecture is true for all m < n. So, let P be a partial n×n latin square with n-1 cells occupied. By virtue of the Corollary of Theorem 10.2, we can assume that the number of distinct symbols occurring in the occupied cells is at most n/2. If so, at least one symbol, say n, occurs exactly once. It is a routine matter to permute the rows and columns of P so as to obtain a partial latin square P* with n-2 occupied cells above the back diagonal and with exactly one cell of the back diagonal occupied, and so that this cell is occupied by n.

Let Y be the part of P* above the back diagonal and form an (n-1)×(n-1) partial latin square Q such that the part of Q on and above the back diagonal is Y and all cells below the back diagonal are empty. (Note that the n-2 occupied cells of Y contain symbols from the set $\{1,2,\ldots,n-1\}$.) Thus, Q is an (n-1)×(n-1) partial latin square (based on

1,2,...,n-1) with at most n-2 cells occupied and so, by induction, can be completed to an (n-1)×(n-1) latin square Q*. Now form the n×n partial latin square P(Q*) and complete this to a latin square L by Smetaniuk's Theorem. This gives a completion of P*. If α and β were the permutations of the rows and columns of P used to obtain P*, then applying α^{-1} and β^{-1} to the rows and columns of L transforms L into an n×n latin square containing P. This completes the proof. []

The following example illustrates the technique of proof used in Theorem 10.3.

EXAMPLE.

P* =
```
. . 7 . . . . . .
. . . 8 . . . . .
. . . . . 9 . .
3 . . . 2 . . . .
. 2 . . . . . . .
. . 5 . . . . . .
. . . . . . . . .
7 . . . . . . . .
. . . . . . . . .
```

P* is a 9×9 partial latin square with 8 occupied cells: 7 of the occupied cells are above the back diagonal <u>and</u> exactly one cell on the back diagonal is occupied (P* has been obtained from P by permuting rows and columns and renaming symbols).

Q =
```
. . 7 . . . . .
. . . 8 . . . .
. . . . . . . .
3 . . . 2 . . .
. 2 . . . . . .
. . 5 . . . . .
. . . . . . . .
7 . . . . . . .
```

8×8 partial latin square with 7 cells occupied.

Q* =
```
1 8 7 2 3 4 5 6
6 4 1 8 5 3 7 2
5 3 6 1 7 8 2 4
3 5 8 4 2 6 1 5
4 2 3 6 1 5 8 7
2 6 5 3 8 7 4 1
8 5 2 7 4 1 6 3
7 1 4 5 6 2 3 8
```

Q* is the completion of Q which is guaranteed by induction.

$$P(Q^*) = \begin{bmatrix} 1 & 8 & 7 & 2 & 3 & 4 & 5 & 6 & 9 \\ 6 & 4 & 1 & 8 & 5 & 3 & 7 & 9 & . \\ 5 & 3 & 6 & 1 & 7 & 8 & 9 & . & . \\ 3 & 5 & 8 & 4 & 2 & 9 & . & . & . \\ 4 & 2 & 3 & 6 & 9 & . & . & . & . \\ 2 & 6 & 5 & 9 & . & . & . & . & . \\ 8 & 5 & 9 & . & . & . & . & . & . \\ 7 & 9 & . & . & . & . & . & . & . \\ 9 & . & . & . & . & . & . & . & . \end{bmatrix} \quad L = \begin{bmatrix} 1 & 8 & 7 & 2 & 3 & 4 & 5 & 6 & 9 \\ 6 & 4 & 1 & 8 & 5 & 3 & 7 & 9 & 2 \\ 5 & 3 & 6 & 1 & 7 & 8 & 9 & 2 & 4 \\ 3 & 5 & 8 & 4 & 2 & 9 & 1 & 6 & 7 \\ 4 & 2 & 3 & 6 & 9 & 1 & 8 & 7 & 5 \\ 2 & 6 & 5 & 9 & 8 & 7 & 3 & 4 & 1 \\ 8 & 5 & 9 & 7 & 4 & 2 & 6 & 1 & 3 \\ 7 & 9 & 4 & 5 & 1 & 6 & 2 & 3 & 8 \\ 9 & 1 & 2 & 3 & 6 & 5 & 4 & 8 & 7 \end{bmatrix}$$

L is the complettion of $P(Q^*)$ which is guaranteed by Smetaniuk's Theorem.

COMMENTS. The solution by L.D.Andersen and A.J.W.Hilton(1983) of the Evans' Conjecture actually shows which partial n×n latin squares with n cells occupied can be completed to a latin square. This was extended further by L.D.Andersen(1985) who showed which partial n×n latin squares with n+1 cells occupied can be completed. Quite recently, L.D.Andersen and A.J.W.Hilton(1987,199a) extended their method for completing partial latin squares to deal with the commutative case. They showed exactly which partial n×n commutative latin squares with up to n+1 cells occupied can be completed to a commutative latin square.

ACKNOWLEDGEMENT. The author wishes to thank Chris Rodger (Auburn University) and A.J.W.Hilton (University of Reading) for many helpful discussions during the preparation of this survey.

APPENDIX (1)   Alternative description of Smetaniuk's proof of the Evans' Conjecture.

For our alternative explanation of Smetaniuk's proof, we take Theorem 10.2 (due to C.C.Lindner) and its Corollary as our starting point. That is,

"An n×n incomplete latin square in which at most n-1 of the rows or at most n–1 of the columns contain occupied cells and in which at most $\lfloor \frac{1}{2}n \rfloor$ different symbols occur in the occupied cells can be completed to an n×n latin square".

It follows that the only case of T.Evans' conjecture which has to be proved is that in which more than $\lfloor \frac{1}{2}n \rfloor$ different symbols occur in the n-1 occupied cells. In the latter case, one of the symbols, say the symbol k, occurs only once. For, suppose that all of $\lfloor \frac{1}{2}n \rfloor + 1$ symbols occurred twice. There would then be at least $2(\lfloor \frac{1}{2}n \rfloor + 1) > n$ cells occupied all together.

Let the symbol k which occurs only once occur in row r_1 and suppose that all together n_1 of the cells of row r_1 are occupied. Suppose that, of the remaining cells of the incomplete latin square which are occupied, n_2 are in row r_2, n_3 are in row $r_3, \ldots$, n_s are in row r_s. Then $n_1 + n_2 + \ldots + n_s = n-1$.

We observe that if a given incomplete latin square can be completed, this property is unaffected by interchanges of rows or interchanges of columns.

Let us permute the rows of our incomplete square in such a way that row r_1 becomes row $n-n_1$, row r_2 becomes row $n-(n_1+n_2)$, row r_3 becomes row $n-(n_1+n_2+n_3), \ldots$, row r_s becomes row $n-(n_1+n_2+\ldots+n_s)$.

Next let us permute the columns in such a way that the columns which now contain the n_1 occupied cells of row $n-n_1$ become the first n_1 columns; the columns which now contain the remaining occupied cells (there are at most n_2 such columns) of row $n-(n_1+n_2)$ become the next columns counting from the left; the columns which contain the remaining occupied cells (there are at most n_3 such columns) of row $n-(n_1+n_2+n_3)$ become the next columns; and so on.

After completion of these interchanges, all occupied cells of the modified incomplete square M are above the main right-to-left diagonal and on or above the $(n-n_1)$-th row. Let us suppose that the symbol k which occurs only once occurs in the cell $(n-n_1, c)$. We now exchange the (n_1+1)-th and c-th columns, to yield an incomplete square M' in which the symbol k lies on the main right-to-left diagonal and all other occupied cells occur above this diagonal. If we delete the symbol k and also the n-th row and the n-th column of this square, we get an incomplete $(n-1)\times(n-1)$ latin square which has n-2 cells occupied with symbols from the set $\{1,2,\ldots,k-1,k+1,\ldots,n\}$.

Since T.Evans' conjecture is true for 2×2 incomplete latin squares with one cell occupied, let us take as induction hypothesis that it is true for all incomplete latin squares of size smaller than n×n. In that case, the above incomplete $(n-1)\times(n-1)$ latin square can be completed to an $(n-1)\times(n-1)$ latin square B on the n-1 symbols of the set $\{1,2,\ldots,k-1,k+1,\ldots,n\}$. We may use B.Smetaniuk's lemma below to extend B to an n×n latin square B* on the symbols $\{1,2,\ldots,k-1,k+1,\ldots,n\}$ in such a way that the portion A of B which occurs above and includes its main right-to-left diagonal is the portion of B* which is above the main right-to-left diagonal and so that B* has the symbol k in every cell along the main right-to-left diagonal. It is clear, therefore, that those n-1 cells of the modified incomplete n×n square M' which were occupied are occupied by the same symbols in the complete latin square B* because, with the exception of the cell which contained k and was on the main right-to-left diagonal of M', all other occupied cells were above that diagonal and so were on or above the main right-to-left diagonal of B. Consequently, B* is a completion of M' and so T.Evans' conjecture is true in all cases.

B.Smetaniuk's lemma. A given n×n latin square L_n can be extended to a latin square L_{n+1} in such a way that the portion A of L_n which occurs above and includes the main right-to-left diagonal is the portion of L_{n+1} which is above its main right-to-left diagonal and so that L_{n+1} has an assigned additional symbol.

Proof. We start with the portion A of L_n. We form a new main right-to-left diagonal for L_{n+1} by adjoining the additional element α at the end of every row of A and at the foot of the first column of A. We then complete the columns of L_{n+1} by an iterative process.

Whenever possible, we complete the c-th column of L_{n+1} by the following simple construction.
Simple construction of a column: "Put the below main right-to-left diagonal element of L_n in the last row of L_{n+1} provided that it does not already

occur in that row and leave all the other elements of the c-th column of L_n in the same rows as before".

When the simple construction of a column (say the c-th column of L_{n+1}) fails, we construct the c-th column as follows:
"If the element h of the (c+1,c)-th cell cannot be put in the new last row because that row already contains an element h, put it in the row from which that element h was removed during the formation of the new last row. However, if the element h in question has already been replaced in that row during the construction of an earlier column, put it in the row from which the replacement was taken. In any event, put the element k (of column c) which is consequently displaced in the new last row. If this also is impossible because the new last row already contains an element k, put the element k in the row from which it was removed during the formation of the new last row. However, if the element k in question has already been replaced in that row during the construction of an earlier column, put it in the row from which the replacement was taken. In any event, put the element ℓ (of column c) which is consequentially displaced in the new last row. If this again is impossible because the new last row already contains an element ℓ, put the element ℓ in the row from which it was removed during the formation of the new last row unless it has already been replaced in that row during the construction of an earlier column. In the latter case, put it in the row from which the replacement was taken. In any event, put the element m (of column c) which is consequentially displaced in the new last row."

Since the partially filled last row of L_{n+1} contains only c-1 elements, after at most c-1 iterations of this process an element to fill the c-th place of the new last row will have been found. Since all the elements of the c-th column which have not been affected by the iterative procedure are left in the same rows as before, no row of the enlarged square L_{n+1} can contain any element more than once bacause no element has been introduced into any row which had not previously been displaced from that row. Moreover, the elements in any column of L_{n+1} comprise the additional element α (in main right-to-left diagonal position) together with the elements which occur in that same column of L_n.

EXAMPLE.

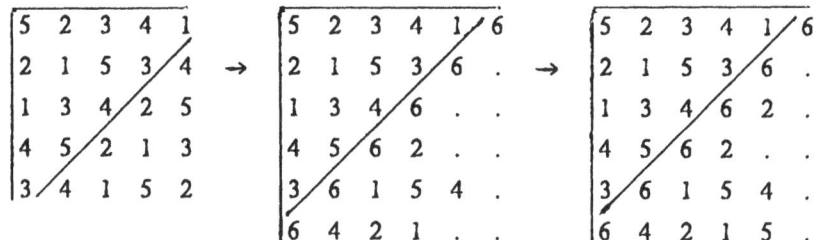

APPENDIX (2) Additional Bibliography

During the past fifteen years, a large number of papers concerned with embedding and extension of latin squares have been published. We conclude this Chapter by listing a number of those which are not explicitly mentioned in the above text, as follows: L.D.Andersen(1982); L.D. Andersen, R.Häggkvist, A.J.W.Hilton and W.B.Poucher(1980); A.B.Cruse (1974b,1974c); R.Dacič(1978); T.Evans and C.C.Lindner(1977); F.R. Giles, J.Oyama and L.E.Trotter(1977); A.J.W.Hilton(1974,1975,1980, 1981,1982); A.J.W.Hilton and C.A.Rodger(1982); D.G.Hoffman(1983); D.G.Hoffman and C.A.Rodger (1987); C.C.Lindner(1971b,1975a, 1976); C.A.Rodger(1983,1984); R.Tosič(1980).

We do not guarantee that the list is exhaustive. Also, some of the results obtained in the earlier papers have subsequently been improved to those listed by Lindner in this Chapter.

Finally, we draw attention to two further papers, A.B.Cruse(1975) and D.A.Drake and M.P.Hale(1974), which are somewhat related to the subject matter of this Chapter and are of interest in their own right.

CHAPTER 9

LATIN SQUARES AND CODES (J.Dénes and A.D.Keedwell)

The codes which we consider in this chapter are those used for communication of messages of all kinds. The messages will be encoded into digital form for transmission and may also be <u>ciphered</u> so as to render them unintelligible to unauthorized interceptors. The primary message may be one-dimensional (for example, the message may consist of English sentences) or two-dimensional (for example, it may be required to transmit a picture by radio). In any event, the message has to be decoded at the receiving end and any errors of transmission detected and, if possible, corrected. For this reason, an error-detecting and correcting code will usually be used.

Suppose that we wish to transmit a sequence of digits $a_1, a_2, \ldots$ across a noisy channel : for example, through a telegraph cable or from a space satellite. Occasionally (in the case of a satellite, frequently), the channel noise will cause a transmitted digit a_i to be mistakenly interpreted as a different digit a_j with the result that the message received at the destination will differ from that which was transmitted.

Although it is not possible to prevent the channel from causing such errors, we can reduce their undesirable effects by sending with each "word" $(a_1, a_2, \ldots, a_k)$ of k <u>information digits</u> a sequence of r additional <u>check digits</u> $a_{k+1}, \ldots, a_{k+r}$ so that the <u>codeword</u> $(a_1 \, a_2 \, \ldots \, a_{k+r})$ actually transmitted is of length k+r. At the receiver, we hope to be able to use these additional digits to enable us to detect, and preferably also correct, errors in <u>any</u> of the digits of the lengthened word which we transmitted. In this way, we are able to recover the original message of k digits.

In a mobile radio telephone system, the area to be covered will be divided into smaller regions, each serviced by a local transmitter. If two such transmitters are allocated the same frequency, then there may be interference and this is an additional form of noise which has to be avoided so far as is possible.

In the resolution of all the above problems of coding, decoding and transmission interference, latin squares play a part. Indeed, in the 1940's, when information theory was an entirely new idea, the pioneers [notably C.E.Shannon(1949), R.Schauffler(1956) and H.Rohrbach(1953)] regarded latin squares as an important tool in the construction of codes with good characteristics. Later, when electronic digital computers became widespread and when, for the most part, only binary codes were used, the important role which latin squares could play tended to become overlooked because the pioneers had used them mainly for the construction of non-binary codes. However, as we shall show in this chapter and as we also showed in Chapter 10 of [DK], latin squares have a role to play in the construction and ciphering of binary codes as well as non-binary ones. It is also the case that, for satellite communication where wideband channels are used, a non-binary code may provide a more efficient transmission medium than a binary one.

(1) Basic facts about error-detecting and correcting codes

As we remarked above, if a message of k information digits is to be sent we shall instead transmit a word of length $n = k+r$, where the magnitude of r depends on the amount of interference expected from the channel noise. We may regard each transmitted word as being a row matrix $\underline{a} = (a_1 \ a_2 \ \ldots \ a_n)$, as a vector of a vector space $V_n(F)$ over a field F, or as a point of an affine space of n dimensions. The digits $a_1, a_2, \ldots a_n$ belong to an alphabet of q symbols which are usually taken to be the elements of a finite Galois field $GF[q]$. In the case of a binary code, $q = 2$. Otherwise the code is called q-ary.

We define the Hamming distance $d(\underline{a},\underline{b})$ between two codewords of length n as equal to the number of positions in which the two words differ.

If a set of codewords is such that the minimum Hamming distance between any two of them is two then the code will be a single-error-detecting code, since any word which is transmitted with a single error will be received as a meaningless n-vector. However, a double error may go undetected.

A set of codewords whose minimum Hamming distance is three will be a single-error-correcting or, alternatively, a double-error-detecting code. If it is assumed that a received word will have at most one error, the received word will always represent a point (of affine space) which is closer to the code point representing the intended codeword than to any other code point. Consequently, it will be possible to correct the error. Double errors will be detected but may be miscorrected.

More generally, a set of codewords whose minimum Hamming distance is d will be a d-1 error-detecting or, alternatively, a $\lfloor (d-1)/2 \rfloor$ error-correcting code.

A good code is one for which, given an assigned minimum Hamming distance d and codeword length n, the number of codewords is as large as possible. This maximum number of codewords for assigned d, n and alphabet size q is evidently bounded above. Two well-known upper bounds are the Joshi-Singleton bound and the Plotkin bound, which we shall now obtain. [See D.D.Joshi(1958), R.C.Singleton(1964) and M.Plotkin(1960).] For later convenience, we shall obtain these bounds in generalized form applicable to so-called variable-order codes as well as to the fixed order ones so far discussed.

DEFINITION. Let $n \geq 1$ and let $2 \leq q_1 \leq q_2 \leq \ldots \leq q_n$ be integers. Let $F_1, F_2, \ldots, F_n$ be finite sets of cardinalities $q_1, q_2, \ldots, q_n$ respectively. The sets $F_1, F_2, \ldots, F_n$ are called alphabets and the cardinal numbers $q_1, q_2, \ldots, q_n$ are called orders. A variable-order block code C is a non-empty subset C of $F_1 \times F_2 \times \ldots \times F_n$.

When $F_1 = F_2 = \ldots = F_n = F$ say, C becomes an ordinary q-ary fixed order code of the kind previously discussed with alphabet comprising the elements of F.

THEOREM 1.1. Let $C \subset F_1 \times F_2 \times \ldots \times F_n$ be a variable order code with minimum Hamming distance d. Then $|C| \leq q_1 q_2 \ldots q_{n-d+1}$. (In particular, for a fixed order q-ary code, $|C| \leq q^{n-d+1}$. This is the

Joshibound given in Theorem 10.1.1 of [DK].)

Proof. Let $g_j = (\alpha_{j1} \; \alpha_{j2} \; \cdots \; \alpha_{j,n-d+1} \; 0 \; 0 \; \cdots \; 0)$, where $\alpha_{ji} \in F_i$ and the vector g_j has length n. Each codeword of C coincides in the first n−d+1 places with one of these $q_1 q_2 \cdots q_{n-d+1}$ vectors g_j. If $|C| > q_1 q_2 \cdots q_{n-d+1}$, then two vectors of C must coincide in the first n−d+1 places with the same g_j and these two codewords would then differ in at most n−(n−d+1) = d−1 places : a contradiction to the fact that the minimum Hamming distance of the code is d. []

THEOREM 1.2. Let $C \subset F_1 \times F_2 \times \cdots \times F_n$ be a variable order code with minimum Hamming distance d. Then $d \leq \frac{M}{M-1}(n - \frac{1}{q_1} - \frac{1}{q_2} - \cdots - \frac{1}{q_n})$, where $M = |C|$ denotes the total number of codewords.

Proof. We compute the sum $D = \sum_{x \in C} \sum_{y \in C} \rho(x,y)$, where $\rho(x,y)$ denotes the Hamming distance between the codewords x and y, in two ways. D is called the total distance of the code C and it includes both $\rho(x,y)$ and $\rho(y,x)$. Then $\sum_{y \in C} \rho(x,y) \geq (M-1)d$ since $\rho(x,x) = 0$. Thence, $D \geq M(M-1)$. We note that equality holds only if C is an equidistant code. That is, only if $\rho(x,y) = d$ for all pairs of codewords x and y.

Next, suppose that the codewords of C are listed as the rows of an M×n matrix A. We consider the contribution made by the i-th column of this matrix to the total distance D. If M_{ij} denotes the number of codewords which have $j \in F_i$ as entry in the i-th place, then $M - M_{ij}$ is the number of codewords which have an entry not equal to j in the i-th place. Hence, the contribution of the i-th column of A to the total distance D is

$$\sum_{j=1}^{q_i} M_{ij}(M - M_{ij}) = M \sum_{j=1}^{q_i} M_{ij} - \sum_{j=1}^{q_i} M_{ij}^2 = M^2 - \sum_{j=1}^{q_i} M_{ij}^2.$$

It follows that

$$D = \sum_{i=1}^{n} (M^2 - \sum_{j=1}^{q_i} M_{ij}^2) = nM^2 - \sum_{i=1}^{n} \sum_{j=1}^{q_i} M_{ij}^2.$$

Now, we have that the sum

$$(M_{i1}-M_{i2})^2 + (M_{i1}-M_{i3})^2 + \ldots + (M_{i1}-M_{i,q_i-1})^2 + (M_{i1}-M_{i,q_i})^2$$
$$+ (M_{i2}-M_{i3})^2 + \ldots + (M_{i2}-M_{i,q_i-1})^2 + (M_{i2}-M_{i,q_i})^2$$
$$+ \ldots + \ldots + \ldots$$
$$+ (M_{i,q_i-2}-M_{i,q_i-1})^2 + (M_{i,q_i-2}-M_{i,q_i})^2$$
$$+ (M_{i,q_i-1}-M_{i,q_i})^2$$

is positive or zero. That is, $\sum_{i \leq j < k \leq q_i} (M_{ij}-M_{ik})^2 \geq 0$.

So, $(q_i-1) \sum_{j=1}^{q_i} M_{ij}^2 \geq 2 \sum_{1 \leq j < k \leq q_i} M_{ij} M_{ik}$.

(We note that equality holds if and only if $M_{ij} = M_{ik}$ for all $j, k \in F_i$.)

Hence, $q_i \sum_{j=1}^{q_i} M_{ij}^2 \geq \sum_{j=1}^{q_i} M_{ij}^2 + 2 \sum_{1 \leq j < k \leq m} M_{ij} M_{ik} = (M_{i1}+M_{i2}+\ldots+M_{i,q_i})^2 = M^2$.

So, from above, $M(M-1)d \leq D \leq nM^2 - \sum_{i=1}^{n} (M^2/q_i)$

Thence, $d \leq \dfrac{M}{M-1} (n - \dfrac{1}{q_1} - \dfrac{1}{q_2} - \ldots - \dfrac{1}{q_n})$

This is called the Plotkin bound. []

Theorems 1.1 and 1.2 for variable order codes can be found in W. Heise and P. Quattrocchi (1981).

We shall need a few further definitions :

DEFINITIONS. Let C be a variable order block code which is a non-empty subset of $F_1 \times F_2 \times \ldots \times F_n$. A set of k indices $1 \leq i_1 < i_2 < \ldots < i_k \leq n$ is a set of total (or partial) information positions if, for any k symbols $f_1 \in F_{i_1}$, $f_2 \in F_{i_2}$, ..., $f_k \in F_{i_k}$, there exists exactly (or at most) one codeword $\underline{x} = (x_1 \; x_2 \; \ldots \; x_n)$ of C such that $x_{i_1} = f_1, x_{i_2} = f_2, \ldots, x_{i_k} = f_k$.

The code is an (n,k)-code if there exists a set of k total information positions. It is a systematic (n,k)-code if the first k positions are a set of total information positions. A systematic (n,k)-code is said to be <u>separable</u> if any set of k positions is a set of partial information positions.

It is a consequence of Theorem 1.1 above that

THEOREM 1.3 For any variable order (n,k)-code, we have $k \leq n-d+1$. (This is called the <u>Singleton bound</u>.)

<u>Proof</u>. Let C be a variable-order (n,k)-code whose total information positions are $i_1, i_2, \ldots, i_k$. Then $q_{i_1} q_{i_2} \cdots q_{i_k} = |C| \leq q_1 q_2 \cdots q_{n-d+1}$. Since $q_{i_1} \geq q_1$, $q_{i_2} \geq q_2$, $\ldots$, $q_{i_k} \geq q_k$, this inequality implies that $k \leq n-d+1$.

Equality is possible only if the code is a systematic (n,k)-code. []

DEFINITION. A variable order code C is said to be a <u>maximum distance code</u> if $|C| = q_1 q_2 \cdots q_{n-d+1}$.

From the above it is clear that a maximum distance code must be a systematic (n,k)-code with $k = n-d+1$ and, moreover, it must be separable (otherwise $|C| \geq q_{i_1} q_{i_2} \cdots q_{i_k}$ for some set of k information positions $i_1, i_2, \ldots, i_k$.)

On the other hand, if C is a separable (n,k)-code, then no two distinct codewords coincide in more than $k-1$ positions and so $d \geq n-(k-1)$, whence $k \geq n-d+1$. However, since, by definition, C is a systematic (n,k)-code, we also have $k \leq n-d+1$. So $k = n-d+1$ and $|C| = q_1 q_2 \cdots q_{n-d+1}$.

Thus, the concepts of maximal distance code (with Hamming distance $d = n-k+1$) and separable (n,k)-code coincide. A code which has either of these equivalent properties will be called an <u>MDS-code</u>.

(2) <u>Codes based on orthogonal latin squares and their generalizations</u>.

As we mentioned in the introduction, orthogonal latin squares have been used in the construction of both binary and non-binary error correcting codes. One of the earliest writers on this subject was G.B. Olderogge (1963) who used orthogonal latin squares for the construction of an ingenious binary code of word length n^2+4n+1 having n^2 information digits and capable of correcting all single and double error patterns and most triple error patterns. Olderogge's work was described in detail in [DK], page 356. Considerably later, a somewhat similar but more far-reaching concept was developed by M.Y. Hsiao, D.C. Bossen and R.T. Chien (1970) who constructed a class of rapidly decodable multiple error-correcting codes under the name of "Orthogonal latin square codes."

In the year following publication of Olderogge's work, S.W. Golomb and E.C. Posner (1964) established an important connection between the existence of sets of mutually orthogonal latin squares and the existence of efficient q-ary codes : namely they showed that the following three concepts are equivalent,

(1) a set of t mutually orthogonal latin squares of order q,

(2) a set of q^2 super rooks of power t on a q^{t+2} board (that is, a (t+2)-dimensional cube having q cells along each edge) such that no two super rooks attack each other,

(3) a code of word length t+2 and minimum Hamming distance t+1 having q^2 words constructed from an alphabet of q symbols.

If the t mutually orthogonal latin squares of order q are $A^{(k)} = (a_{ij}^{(k)})$, $k = 1, 2, \ldots, t$, then the words of the Golomb-Posner code are of the form $(i, j, a_{ij}^{(1)}, a_{ij}^{(2)}, \ldots, a_{ij}^{(t)})$, where $i, j = 0, 1, \ldots, q-1$. A description of this code was given also in R.C. Singleton (1964).

This code has subsequently been generalised in several ways. Firstly, E. Gergely and one of the present authors showed that the existence of a set of t pairwise orthogonal groupoids of order q is equivalent to the existence of a q-ary code with word length t, minimal Hamming distance t-1 and containing at most q^2 codewords. [See J. Dénes and E. Gergely (1975).] Secondly, analogous generalizations were made for which n-ary quasigroups were used in place of binary ones. Details will be found in J. Ušan (1978) and (1979), Z. Stojakovič and J. Ušan (1978) and in A. Surda, R. Tosič and J. Ušan (1979). Thirdly, one of the present authors considered the generalization of the Golomb-Posner result which arises when latin squares are replaced by latin rectangles. In the latter case, we get a variable-order code as we explain below. Fourthly, the original Golomb-Posner construction was rediscovered by A.M.Andrew(1975). The latter author pointed out that the known existence of pairs of mutually orthogonal latin squares of order 10 allows the construction of a decimal code having codewords of length 4 with two information digits and which is capable of correcting single errors. (There are 100 codewords and the minimum Hamming distance is 3.)

Let us now describe the generalization of the Golomb-Posner construction which makes use of latin rectangles and which is due to one of the present authors. [See J.Dénes (1977).] We remind the reader that two m×n latin rectangles (m<n) each defined on a set of n symbols are called <u>orthogonal</u> if, when one is superimposed on the other, each ordered pair of symbols occurs in at most one cell of the superimposed pair. The code construction makes use of the following theorem due to P.Quattrocchi (1968):

For every prime number p and every integer q($\geq$p) such that each prime divisor of q is not less than p, there exists at least one complete system of pq-1 mutually orthogonal latin rectangles of size pq×p. (This theorem is stated and proved in [DK] as Theorem 5.4.1.)

The (i+j)-th codeword is $(i,j,r_{ij}^{(1)},r_{ij}^{(2)},\ldots,r_{ij}^{(pq-1)})$, where $r_{ij}^{(h)}$ is the entry in the (i,j)-th cell of the h-th rectangle R_h, h = 1,2,...,pq-1. The resulting code has codewords of length pq+1, minimal Hamming distance pq and contains p^2q codewords. We note that the second component ranges over an alphabet of p symbols, while the remaining components belong to an alphabet of pq symbols. Thus, the code is a variable order code and the number of its codewords attains the Joshi-Singleton bound p(pq) given in Theorem 1.1.

In the book of W.Heise and P.Quattrocchi(1983), two further useful properties of this code were pointed out: (i) for any two integers i and j with 1 $\leq$ i $\leq$ pq and 1 $\leq$ j $\leq$ p, there is exactly one codeword $(x_1\ x_2\ \ldots\ x_{pq+1})$ for which x_1 = i and x_2 = j (and also, if 1 $\leq$ i,k $\leq$ pq, there is at most one codeword for which x_m = i and x_n = k if m,n are different from 2); (ii) the code is a maximal distance separable code (MDS code).

An example of the code for the case p = 2, q = 3 showing its construction is given in Figure 2.1. This example was originally given in J.Dénes (1977). In fact, because the proof of Quattrocchi's theorem is constructive, a code of p^2q codewords of the above kind can easily be constructed for any integers p and q which satisfy Quattrocchi's conditions.

The reader who is interested in variable order codes will find further constructions of such codes in P.Lancellotti and C.Pellegrino (1982), A.Bonisoli (1984), C.Fiori and P.Lancellotti (1984) and N.A.Malara and C.Pellegrino (1984). Some applications of variable order codes are listed in W.Heise and P.Quattrocchi (1981). In particular, they have been used in the construction of International Standard Book Numbers, personal identity numbers, licence numbers and Gray codes.

$$R_1 = \begin{vmatrix} 6 & 1 \\ 1 & 2 \\ 2 & 3 \\ 3 & 4 \\ 4 & 5 \\ 5 & 6 \end{vmatrix} \quad R_2 = \begin{vmatrix} 6 & 2 \\ 1 & 3 \\ 2 & 4 \\ 3 & 5 \\ 4 & 6 \\ 5 & 1 \end{vmatrix} \quad R_3 = \begin{vmatrix} 6 & 3 \\ 1 & 4 \\ 2 & 5 \\ 3 & 6 \\ 4 & 1 \\ 5 & 2 \end{vmatrix} \quad R_4 = \begin{vmatrix} 6 & 4 \\ 1 & 5 \\ 2 & 6 \\ 3 & 1 \\ 4 & 2 \\ 5 & 3 \end{vmatrix} \quad R_5 = \begin{vmatrix} 6 & 5 \\ 1 & 6 \\ 2 & 1 \\ 3 & 2 \\ 4 & 3 \\ 5 & 4 \end{vmatrix}$$

(1 1 6 6 6 6) (4 1 3 3 3 3) (1 2 1 2 3 4 5) (4 2 4 5 6 1 2)
(2 1 1 1 1 1) (5 1 4 4 4 4) (2 2 2 3 4 5 6) (5 2 5 6 1 2 3)
(3 1 2 2 2 2) (6 1 5 5 5 5) (3 2 3 4 5 6 1) (6 2 6 1 2 3 4)

Figure 2.1

A quite different application of orthogonal latin squares to coding is given in J.I.Lewandowski, J.W.S.Liu, C.L.Liu (1983). These authors have shown how orthogonal latin squares can be applied to the so-called Satellite-Switched Time Division Multiple Access Systems (SS/TDMA systems) in space communication. In such a system, the area to be scanned by the satellite is divided up into several geographical zones. A high-gain antenna on board the satellite provides spotbeam coverage of the zones and a system of switches permits the various beams directed to and from the satellite (called uplink and downlink beams) to be connected to ground receiving stations (which may be mobile or static) on a time sharing basis. A major problem faced by such systems is the traffic scheduling problem. This problem has been studied extensively by several authors among whom are J.J.Lewandowski, J.S.W.Liu and C.L.Liu. In (1983), these authors proposed a solution to the problem which uses orthogonal latin squares and which we shall now explain.

The current switching mode of the system is specified by an n×n matrix P of zeros and ones, called the switching mode matrix. A switching mode matrix $P = ||P_{ij}||$ is said to cover the (i,j)-th cell if $p_{ij} = 1$. Two switching mode matrices are said to intersect at the (i,j)-th place if they both cover the (i,j)-th cell. A set $S = \{P_1, P_2, \ldots, P_{2n}\}$ of 2n switching mode matrices (usually just called "switching modes") is a set of restricted switching modes if (i) each of the n^2 cells of an n×n array is covered by exactly two switching modes of the set S, (ii) the set S can be partitioned into two subsets S_A and S_B each containing n switching modes and such that no two switching modes in the same subset cover the same cell, and (iii) each switching mode of the subset S_A intersects each switching mode of the subset S_B in exactly one place.

Lewandowski et al observed that there is a one-to-one correspondence between sets of 2n switching modes of order n and pairs A,B of n×n orthogonal latin squares. This correspondence is illustrated in Figure 2.2, where the method of construction of the subsets S_A and S_B which correspond to the orthogonal latin squares exhibited there is easy to see. Lewandowski's construction implies that sets of 2n restricted switching modes exist for all $n \geq 3$ except $n = 6$. It also shows why, from a practical point of view, it is desirable to know how many isomorphically different pairs of orthogonal latin squares exist of a given order n. (For a graphical representation of this problem, see Part II of this book.)

$$A = \begin{vmatrix} 1 & 2 & 3 \\ 2 & 3 & 1 \\ 3 & 1 & 2 \end{vmatrix} \quad B = \begin{vmatrix} 1 & 2 & 3 \\ 3 & 1 & 2 \\ 2 & 3 & 1 \end{vmatrix} \quad S_A = \left\{ \begin{pmatrix} 1 & 0 & 0 \\ 0 & 0 & 1 \\ 0 & 1 & 0 \end{pmatrix}, \begin{pmatrix} 0 & 1 & 0 \\ 1 & 0 & 0 \\ 0 & 0 & 1 \end{pmatrix}, \begin{pmatrix} 0 & 0 & 1 \\ 0 & 1 & 0 \\ 1 & 0 & 0 \end{pmatrix} \right\}$$

$$S_B = \left\{ \begin{pmatrix} 1 & 0 & 0 \\ 0 & 1 & 0 \\ 0 & 0 & 1 \end{pmatrix}, \begin{pmatrix} 0 & 1 & 0 \\ 0 & 0 & 1 \\ 1 & 0 & 0 \end{pmatrix}, \begin{pmatrix} 0 & 0 & 1 \\ 1 & 0 & 0 \\ 0 & 1 & 0 \end{pmatrix} \right\}$$

Figure 2.2

In R.Mandl(1985), the author has recommended the use of orthogonal latin squares for the testing of a computer language compiler. Although in some sense this is a coding problem, it really amounts to a rediscovery of the role of orthogonal latin squares in experimental design (see Chapter 10 of [DK] and Chapter 10 of the present book) as the purpose of Mandl's paper is to show how, by judicious use of orthogonal latin squares, a set of n^3 compiler fault tests (that is, "experiments" on the efficiency of the compiler) can be reduced to n^2 such tests.

The generalized form of orthogonality known as r-orthogonality and discussed in detail in Chapter 6 is also very useful in coding theory. We remind the reader that two latin squares of order n, defined on the same set Σ of symbols, are called <u>r-orthogonal</u> if exactly r different ordered pairs of the set $\Sigma \times \Sigma$ are obtained when the two squares are superimposed.

From this concept, the following efficient encoding procedure can be derived:

Let $L_1, L_2, \ldots, L_k$ be a set of mutually r-orthogonal latin squares of order n defined on a set Σ of n symbols and suppose that the r-set of distinct ordered pairs obtained when the squares L_s and L_t are superimposed is different for each choice of the pair s,t (s<t). These $\frac{1}{2}k(k-1)$ r-sets may be regarded as representing messages which are to be transmitted in coded form. Let H_{st} denote the message which is defined by the latin squares L_s, L_t. Then, provided that the set of r-orthogonal squares is stored at each of the receiving stations as well as at the transmitting station, the message H_{st} can be transmitted as the pair s,t which is both compact and also an effective coded form. This can provide an efficient alarm code. (For example, if the nearest three of a number of tracking stations are to be alerted quickly when an airborne moving object has been detected then a set of mutually 3-orthogonal latin squares could be used.)

It will be clear from our description that, for the construction of such a code it is important to know what is the largest number k of n×n latin squares which can be constructed such that every pair of the collection is exactly r-orthogonal and so that, in addition, the r-sets defined by each of the $\frac{1}{2}k(k-1)$ pairs of squares in the collection are distinct. A weaker form of the same question is to ask for which values of n,k,r and m does a collection of k pairwise exactly r-orthogonal n×n latin squares exist such that at least m of the r-sets obtained by superimposition of the squares of the collection in pairs are distinct. (When such a collection of squares exists, it can be used to encode m messages. For given values of m and r, it is desirable to minimize k and n.)

In G.Quattrocchi(1984), the author has obtained some partial answers to the latter problem (which had earlier been published as a research problem by one of the present authors, J.Dénes(1979)). G.Quattrocchi's results are listed in Section 7 of Chapter 6.

In G.B.Belyavskaya(1976), one can find an example of a 15-orthogonal set of latin squares of order 5 containing 5 latin squares and with m=6 distinct subsets H_{st}. (See also Figure 6.1 of Chapter 6.) These six different subsets are as follows: H_{12}, H_{13}, $H_{14}=H_{24}=H_{45}$, H_{15}, $H_{23}=H_{35}$, $H_{25}=H_{34}$.

The main problem in the construction of alarm codes is that of syncronization. The above method permits the construction of alarm codes with perfect syncronization.

The idea which we have just used to construct an efficient alarm code can be used also to provide a ciphering system which can encode each of k(k-1) words into 2-character messages. For this purpose the r-sets defined by the various pairs L_s, L_t of r-orthogonal latin squares must be ordered. (The message alphabet comprises the n^2 ordered pairs of symbols of the Cartesian product $\Sigma \times \Sigma$).

We may do this in the following way:

If the encoded message is s,t with s<t, then the message comprises the distinct pairs which arise when the squares L_s, L_t are superimposed but taken in the order of cells which is such that cell (h,i) precedes cell (h,j) whenever i<j and cell (h,i) precedes cell (k,j) if h<k whatever the values of i,j. On the other hand, if s>t, the message is the same as that just described but in reverse. We note that now two pairs, say L_r, L_s and L_u, L_v, of exactly r-orthogonal latin squares which define the same r elements of $\Sigma \times \Sigma$ may yet define different messages and so we should now ask what is the largest number of messages for given r and n which can be encoded into a code with words of length 2. The answer to this question depends partly, but not entirely, on the question "What is the maximum number of pairwise exactly r-orthogonal latin squares of order n which can be constructed?"

A somewhat similar ciphering system which uses latin squares (and certain other combinatorial designs) has been proposed by D.G.Sarvate and J.Seberry(1986). These authors proposed to cipher a two digit n-ary message (i,j) by selecting two orthogonal n×n latin squares from a k-set of such squares and then transmitting the digits in the (i,j)-th cells of these two squares in place of the original message. It would also of course be necessary to inform the recipient of the message which pair of the set of orthogonal squares had been used in order that the message could be deciphered. Sarvate and Seberry called this the key. They also suggested that, in place of a pair of orthogonal latin squares, a skew Room square could be used in a similar way to cipher a two digit message.

The scheme proposed by Sarvate and Seberry seems inferior to the one described earlier in a number of respects. Firstly, the length of message which can be encoded is much shorter and secondly, in the scheme of the present authors, only the key has to be transmitted. The message itself is generated from the k-set of mutually r-orthogonal latin squares stored at the receiving end. Also, this means of storing messages is very economical in space and, moreover, k>n-1 is possible if r-orthogonal latin squares are used in place of n^2-orthogonal ones.

Next we remark that orthogonal latin squares can be used to provide an efficient method of generating pseudo-random permutations. Let H be a set of k mutually orthogonal latin squares of order n. We can use these to generate a set of k(k-1) permutations of degree n^2 in the following way. We first number the cells of each square as shown in Figure 2.3 counting from left to right along rows starting with the topmost row. We select two squares L_i and L_j from H and, after juxtaposing them in the order i,j, we regard each cell entry as being an n-ary number in the range 0 to n^2-1. This has the effect of replacing the pair of squares L_i, L_j by a nearly magic square M. (The diagonal sums may fail to fulfil the requirements for a magic square, cf. pages 208 to 212 of [DK]). We illustrate our construction in Figure 2.4. By reading the entries of the nearly magic square in the order determined by Figure 2.3 we obtain a permutation φ of the set $\{0,1,\ldots,n^2-1\}$, as shown in Figure 2.5. To each of the k(k-1) ordered pairs i,j, there corresponds just one such permutation. If we adjoin a square R (each of whose columns contains the integers $0, 1, 2, \ldots, n-1$ in natural order) and a square C (each of whose rows contains the integers $0, 1, 2, \ldots, n-1$, in natural order) our set of $k-1$ mutually orthogonal latin squares becomes a set of $k+1$ mutually orthogonal equi-n-squares (see Chapter 2) and the same construction then gives $k(k+1)$ permutations.

$$\begin{bmatrix} 0 & 1 & 2 & 3 & \ldots & n-1 \\ n & n+1 & n+2 & n+3 & \ldots & 2n-1 \\ 2n & 2n+1 & 2n+2 & 2n+3 & \ldots & 3n-1 \\ \cdot & \cdot & \cdot & \cdot & \ldots & \\ n^2-n & \cdot & \cdot & \cdot & \ldots & n^2-1 \end{bmatrix}$$

Figure 2.3

$$L_1 = \begin{bmatrix} 0 & 1 & 2 & 3 \\ 1 & 0 & 3 & 2 \\ 2 & 3 & 0 & 1 \\ 3 & 2 & 1 & 0 \end{bmatrix} \quad L_2 = \begin{bmatrix} 0 & 1 & 2 & 3 \\ 2 & 3 & 0 & 1 \\ 3 & 2 & 1 & 0 \\ 1 & 0 & 3 & 2 \end{bmatrix} \quad L_3 = \begin{bmatrix} 0 & 1 & 2 & 3 \\ 3 & 2 & 1 & 0 \\ 1 & 0 & 3 & 2 \\ 2 & 3 & 0 & 1 \end{bmatrix}$$

$$L_1/L_3 = \begin{bmatrix} 0_0 & 1_1 & 2_2 & 3_3 \\ 1_3 & 0_2 & 3_1 & 2_0 \\ 2_1 & 3_0 & 0_3 & 1_2 \\ 3_2 & 2_3 & 1_0 & 0_1 \end{bmatrix} \quad M = \begin{bmatrix} 0 & 5 & 10 & 15 \\ 7 & 2 & 13 & 8 \\ 9 & 12 & 3 & 6 \\ 14 & 11 & 4 & 1 \end{bmatrix}$$

Figure 2.4

$$\Phi = \begin{bmatrix} 0 & 1 & 2 & 3 & 4 & 5 & 6 & 7 & 8 & 9 & 10 & 11 & 12 & 13 & 14 & 15 \\ 0 & 5 & 10 & 15 & 7 & 2 & 13 & 8 & 9 & 12 & 3 & 6 & 14 & 11 & 4 & 1 \end{bmatrix}$$

Figure 2.5

Moreover, we have the following theorem which shows that the permutations which we generate are "random" provided that we randomly select our set of mutually orthogonal equi-n-squares.

THEOREM 2.1. Let $n (\geq 2)$ be any chosen positive integer. Then there exists at least one set of $N(n)+2$ mutually orthogonal equi-n-squares whose set of generated permutations includes any arbitarily chosen permutation ψ of degree n^2.

Proof. The integer $N(n)$ denotes the maximum number of mutually orthogonal latin squares of order n which exist. (See Chapter 5.) If we adjoin the squares R and C defined above to these, we get a set H^* of $N(n)+2$ mutually orthogonal equi-n-squares.

Now let $\psi = \begin{bmatrix} 0 & 1 & 2 & \ldots & n^2-1 \\ a_0 & a_1 & a_2 & \ldots & a_{n^2-1} \end{bmatrix}$ be an arbitrarily chosen permutation of degree n^2. We interpret the integers $a_0, a_1, \ldots, a_{n^2-1}$ as n-ary numbers and write them into the cells of an $n \times n$ square M^* in the order prescribed in Figure 2.3.

Then, by reordering the entries of the cells in all the members of the set H^* simultaneously, we can easily derive a set of $N(n)+2$ mutually orthogonal equi-n-squares which generate a set of permutations of the integers $0, 1, 2, \ldots, n^2-1$ which includes the permutation ψ. □

(We illustrate this for the case n=3 in Figure 2.6.)

$\psi = \begin{bmatrix} 0 & 1 & 2 & 3 & 4 & 5 & 6 & 7 & 8 \\ 1 & 5 & 2 & 4 & 3 & 6 & 8 & 0 & 7 \end{bmatrix}$ $M^* = \begin{bmatrix} 0_1 & 1_2 & 0_2 \\ 1_1 & 1_0 & 2_0 \\ 2_2 & 0_0 & 2_1 \end{bmatrix}$

$\begin{vmatrix} 0 & 0 & 0 \\ 1 & 1 & 1 \\ 2 & 2 & 2 \end{vmatrix}$ $\begin{vmatrix} 0 & 1 & 2 \\ 0 & 1 & 2 \\ 0 & 1 & 2 \end{vmatrix}$ $\begin{vmatrix} 0 & 1 & 2 \\ 1 & 2 & 0 \\ 2 & 0 & 1 \end{vmatrix}$ $\begin{vmatrix} 0 & 1 & 2 \\ 2 & 0 & 1 \\ 1 & 2 & 0 \end{vmatrix}$

The rearranged equi-n-squares are

$\begin{vmatrix} 0 & 1 & 0 \\ 1 & 1 & 2 \\ 2 & 0 & 2 \end{vmatrix}$ $\begin{vmatrix} 1 & 2 & 2 \\ 1 & 0 & 0 \\ 2 & 0 & 1 \end{vmatrix}$ $\begin{vmatrix} 1 & 0 & 2 \\ 2 & 1 & 2 \\ 1 & 0 & 0 \end{vmatrix}$ $\begin{vmatrix} 1 & 1 & 2 \\ 0 & 2 & 1 \\ 0 & 0 & 2 \end{vmatrix}$

Figure 2.6

In order to calculate the efficiency of this method of generation of permutations, let us look at the storage capacity which we use. The maximum efficiency can be obtained if n is a prime power, since in that case a complete set of orthogonal latin squares of order n exists. (See Theorem 5.2.2. of [DK].) This means that we can use an orthogonal system of equi-n-squares with the maximum cardinality of n+1 squares. To store this orthogonal system, we need a storage capacity of $(n+1)n^2\log n$ bytes and we are then able to generate $n(n+1)$ permutations of degree n^2 by the method described above. On the other hand, if these permutations were stored in the usual way we would need

$$(n+1)n(n^2\log n^2) = 2(n+1)n^3\log n$$

bytes of storage capacity.

We note also that our method is such that each ordered pair (i,j), $(1 \leq i,j \leq n)$, defines a unique permutation and so gives us a means of transmitting permutations in a reduced and coded form.

REMARKS. (1) H. Niederreiter (1987) made use of latin squares to generate random numbers and permutations in a quite different way from that described above.

(2) Readers of Chapter 10 will observe that random permutations are commonly used in the process of randomizing an experimental design.

We end this Section by stating a conjecture of one of the present authors as follows:

Suppose that a non-binary code on a q-symbol alphabet exists such that there are q^2 codewords of length t+2, q>t, and the minimum Hamming distance is t+1, then there exists a set of at least t mutually orthogonal latin squares of order q.

This conjecture "inverts" the Golomb-Posner construction described at the beginning of the Section.

ADDED IN PROOF. The construction of Lewandowski et al has recently been generalized. See J.L. Lewandowski and C.L. Liu (1987).

(3) Row and column complete latin squares in coding theory.

We discussed row and column latin squares and their construction in detail in Chapter 3. Like orthogonal latin squares, these squares have many useful applications in coding.

We start by mentioning an analogue of the Golomb-Posner construction which we described in the preceding section.

THEOREM 3.1 Let $A = ||a_{ij}||$ denote a row complete latin square of order q. Then the set of codewords $(i, j, a_{ij}, a_{i,j+1})$, $i = 0,1,\ldots,q-1$ and $j = 0,1,\ldots,q-2$ form a q-ary code having $q(q-1)$ codewords each of length 4 and with minimum Hamming distance 3.

The proof, which is simple, was given as Theorem 10.1.6 in [DK] and so we shall not repeat it here.

A much more important and useful application of row complete latin squares arises in connection with picture processing.

DEFINITION. A two-dimensional non-null array of zeros and ones is said to have the u×v horizontal window property (u,v⩾2) if every non-null pattern of zeros and ones is seen at most once when a window of u rows and v columns is moved horizontally across the array. (To avoid trouble at the ends, the array should be imagined as being written on a vertical cylinder). Similarly the u×v vertical window property holds if every nonzero view appears at most once when the window is moved vertically down the array (and the array is imagined to be written on a horizontal cylinder).

This concept has practical applications, in particular in connection with the coding and transmission of pictures, as has been pointed out in S.L.Ma(1984) and, more recently, also by G.Fazekas(1989).

Two-dimensional arrays of zeros and ones which have the stronger property that every non-null pattern of zeros and ones is seen exactly once when a window of u rows and v columns is moved horizontally and vertically across the array have been called perfect maps or pseudo-random arrays of a special type. Construction of such arrays has been discussed in J.S.Reed and I.M.Stewart(1962), B.Gordon(1966), T.Nomura, H.Miyakawa, H.Imai and A.Fukuda(1972), F.J.MacWilliams and N.J.A.Sloane(1976), J.H.vanLint, F.J.MacWilliams and N.J.A.Sloane(1979), S.L.Ma(1984), L.G.Khachatryan(1982) and, more recently, by C.T.Fan, S.M.Fan, S.L.Ma and M.K.Siu(1985), T.Etzion(1988) and A.Ivanyi(1987,1989). Many of the constructions used by these authors make use of sequences which have a one-dimensional window property (de Bruijn sequences).

DEFINITIONS. A non-binary m×n array is said to have the u×v window property if all its sub-arrays of size u×v are different.

A non-binary sequence (one-dimensional array) is said to have the window property for windows of width u if every subsequence of length u is different.

We shall show how row-complete latin squares and row-complete row-latin squares can be used to construct non-binary one and two-dimensional arrays which have the u×v horizontal and/or vertical window property for all sufficiently large windows and how binary arrays with a similar property can be obtained. (Since an array can only be a perfect map for one particular size of window, the arrays which we construct are not perfect but this is not always important in practice.)

For most of our constructions, it is not necessary that each symbol occurs exactly once in each column but it is necessary that each symbol occurs exactly once in each row and also that each ordered pair of symbols occurs exactly once among the rows. A square with these properties is a row-complete row-latin square or, in S.W.Golomb and H.Taylor(1985), it has been called a Tuscan square. (See also Chapter 3 of this book.)

In a Tuscan square, each symbol occurs exactly once in each row and is followed exactly once by each of the other symbols. Consequently, there is exactly one row in which a particular symbol has no successor. This means that each symbol occurs exactly once in the last column of the square. Similarly, each symbol is preceded exactly once by each of the other symbols and so each symbol occurs exactly once in the first column of the square.

Let us take an n×n Tuscan square T defined on the symbols $0, 1, 2, \ldots, n-1$ and adjoin to it two additional columns, the first being a duplicate of the last column of T and the second a column all of whose entries are the integer n, and then construct a sequence which consists of the successive rows of the augmented square written one after the other, the r-th following the $(r-1)$-th and the first following the last after the adjunction of an additional copy of the symbol n. Then it is easy to see that the sequence so obtained is a non-binary sequence which has the window property for a window of length 2 and consquently also for windows of all lengths greater than 2. An example for the case n = 4 is shown in Figure 3.1.

$$T = \begin{bmatrix} 0 & 3 & 1 & 2 \\ 1 & 0 & 2 & 3 \\ 2 & 1 & 3 & 0 \\ 3 & 2 & 0 & 1 \end{bmatrix} \quad \begin{bmatrix} 0 & 3 & 1 & 2 & 2 & 4 \\ 1 & 0 & 2 & 3 & 3 & 4 \\ 2 & 1 & 3 & 0 & 0 & 4 \\ 3 & 2 & 0 & 1 & 1 & 4 & 4 \end{bmatrix}$$

The sequence is 0 3 1 2 2 4 1 0 2 3 3 4 2 1 3 0 0 4 3 2 0 1 1 4 4 ... ...

Figure 3.1

We observe that such a sequence is perfect relative to a window of length 2 since every ordered pair (u, v) occurs exactly once in the sequence.

In order to be able to construct binary sequences which have the window property we need to make use of the concept of a comma-free code.

DEFINITION. A set C of codewords each of length m and defined over an alphabet of q symbols is said to form a <u>comma-free code</u>, which we shall denote by $CF(q, m)$, if for each pair of words $\underline{a} = (a_1 \ a_2 \ \ldots \ a_m)$ and $\underline{b} = (b_1 \ b_2 \ \ldots \ b_m)$ in C, none of the words $(a_{i+1} \ a_{i+2} \ \ldots \ a_m \ b_1 \ b_2 \ \ldots \ b_i)$, where $i = 1, 2, \ldots, m-1$, is in C.

It will be convenient to denote the set of all words of the form

$(a_{i+1}\ a_{i+2}\ \ldots\ a_m\ b_1\ b_2\ \ldots\ b_i)$ by $L_i(C)$ and to call it the i-th overlap of the code C.

Suppose now that the n+1 different integers which occur in the augmented Tuscan square of Figure 3.1 are replaced by n+1 different codewords of a binary comma-free code $CF(2,m)$. The resulting binary sequence S will consist of $[(n+2)n+1]m = (n+1)^2 m$ symbols and has the window property for all windows of width 3m-1 or greater. To see this, we note that a window of length $w \geq 3m-1$ spans at least one pair (c_u, c_v) of consecutive complete codewords of C which have replaced a corresponding pair (u,v) of consecutive integers in the non-binary sequence. Since the code $CF(2,m)$ is comma-free, no codeword of $CF(2,m)$ occurs as the first m binary digits of any subsequence of S except one starting at the (im+1)-th digit of S for some i, $i=0,1,\ldots,(n+2)n$. Consequently, the subsequence $t=c_u c_v$ of binary digits occurs only once in S and is in a different position in each window of length w which spans it.

As examples, we exhibit in Figure 3.2 the complete binary sequence which is constructed from the (unique) row-complete latin square L of order 2 using the three codewords c_0, c_1, c_2 of a comma-free code $CF(2,4)$. This sequence contains $(2+1)^2 \times 4 = 36$ binary digits and has the window property for all windows of length 11 or greater. (If the window width is 11, there are 26 different windows.)

$$L = \begin{bmatrix} 0 & 1 \\ 1 & 0 \end{bmatrix} \quad \begin{bmatrix} 0 & 1 & 1 & 2 \\ 1 & 0 & 0 & 2 & 2 \end{bmatrix} \quad \begin{matrix} c_0 = (1\ 0\ 0\ 0) \\ c_1 = (1\ 0\ 0\ 1) \\ c_2 = (1\ 0\ 1\ 1) \end{matrix}$$

The binary sequence is

1 0 0 0 1 0 0 1 1 0 0 1 1 0 1 1 1 0 0 1 1 0 0 0 1 0 0 0 1 0 1 1 1 0 1 1 ...

Figure 3.2

9:21 *Latin squares and codes* 287

$$T = \begin{bmatrix} 0 & 1 & 2 & 3 & 4 & 5 & 6 \\ 1 & 6 & 4 & 2 & 5 & 3 & 0 \\ 2 & 4 & 0 & 6 & 5 & 1 & 3 \\ 3 & 2 & 0 & 5 & 4 & 6 & 1 \\ 4 & 1 & 5 & 0 & 3 & 6 & 2 \\ 5 & 2 & 6 & 3 & 1 & 0 & 4 \\ 6 & 0 & 2 & 1 & 4 & 3 & 5 \end{bmatrix}$$

$$\begin{bmatrix} 0 & 1 & 2 & 3 & 4 & 5 & 6 & 6 & 7 \\ 1 & 6 & 4 & 2 & 5 & 3 & 0 & 0 & 7 \\ 2 & 4 & 0 & 6 & 5 & 1 & 3 & 3 & 7 \end{bmatrix}$$

$c_0 = (0\ 0\ 0\ 1\ 0\ 0)$
$c_1 = (0\ 0\ 1\ 1\ 0\ 0)$
$c_2 = (0\ 0\ 1\ 1\ 0\ 1)$
$c_3 = (0\ 1\ 0\ 1\ 0\ 0)$
$c_4 = (0\ 1\ 1\ 1\ 0\ 0)$
$c_5 = (0\ 1\ 1\ 1\ 0\ 1)$
$c_6 = (0\ 1\ 0\ 1\ 1\ 0)$
$c_7 = (0\ 1\ 1\ 1\ 1\ 0)$

The first part of the sequence is

0 0 0 1 0 0 0 0 1 1 0 0 0 0 1 1 0 1 0 1 0 1 0 0 0 1 1 1 0 0 0 1 1 1 0
1 0 1 0 1 1 0 0 1 0 1 1 0 0 1 1 1 1 0 0 0 1 1 0 0 0 1 0 1 1 0 0 1 1 1
0 0 0 0 1 1 0 1 0 1 1 1 0 1 ... ...

Figure 3.3

In Figure 3.3, we exhibit the first part of the binary sequence which is constructed from the Tuscan square T_7 of order 7 which is shown there. [The Tuscan square has been taken from a list given in S.W.Golomb and H.Taylor(1985)]. The non-binary digits are replaced by the first eight codewords of a comma-free code $CF(2,6)$. This binary sequence has the window property for all windows of length 17 or greater. For the window width 17, there are $[(7+1)^2 \times 6]-16 = 368$ different windows in the constructed sequence.

We turn next to two-dimensional arrays with the window property. We observe firstly that any Tuscan square or row complete latin square of order $n (n>2)$ or any $r \times n$ sub-rectangle of such a square ($r<n$ and $n>2$) is a non-binary array with the window property for all $u \times v$ windows with $v \geq 2$. The same statement remains true if two additional columns are adjoined to the Tuscan square as in Figure 3.1.

It follows immediately that, if the $h (n \leq h \leq n+1)$ different integers which occur in the square or sub-rectangle (or augmented square with n+2 columns) are replaced by h different codewords of a binary comma-free code $CF(2,m)$, then we shall obtain a zero-one array of size $r \times (n+2)m$ (in the case of the augmented square) which has the $u \times v$ window property for all $v \geqslant 3m-1$. (If and only if $v \geqslant 3m-1$, each $1 \times v$ window will span completely the binary representation of some two consecutive non-binary digits and, since the binary representation of every two consecutive digits is different, the result follows as for sequences.)

As an illustration, we exhibit in Figure 3.4 a row-complete latin square of order 6 to which an additional column has been adjoined and, in Figure 3.5, we exhibit the 3×35 rectangle of zeros and ones derived from its first three rows using a comma-free code $CF(2,5)$ with codewords $c_0, c_1, c_2, c_3, c_4, c_5$. The latter rectangle has the $u \times v$ window property for all $v \geqslant 14$ and for all values of u.

$$\begin{vmatrix} 0 & 1 & 5 & 2 & 4 & 3 & 3 \\ 1 & 2 & 0 & 3 & 5 & 4 & 4 \\ 2 & 3 & 1 & 4 & 0 & 5 & 5 \\ 3 & 4 & 2 & 5 & 1 & 0 & 0 \\ 4 & 5 & 3 & 0 & 2 & 1 & 1 \\ 5 & 0 & 4 & 1 & 3 & 2 & 2 \end{vmatrix}$$

$c_0 = (1\ 0\ 1\ 0\ 0)$ $c_3 = (1\ 0\ 0\ 0\ 1)$
$c_1 = (1\ 0\ 1\ 0\ 1)$ $c_4 = (1\ 0\ 0\ 1\ 1)$
$c_2 = (1\ 0\ 0\ 0\ 0)$ $c_5 = (1\ 0\ 1\ 1\ 1)$

```
10100101011011110000100111000110001
10101100001010010001101111001110011
10000100011010110011101001011110111
```

Figure 3.4

Figure 3.5

For the implementation of the above constructions, it is necessary to be able to construct binary comma-free codes $CF(2,m)$ and to know how many codewords each contains. A summary of what is known about this question will be found in J.Dénes and A.D.Keedwell(1990b), in which paper the above results were first published. It is also shown there that the proportion of the total number of different binary $u \times (2m-1)$ windows which is given by the above construction varies with n.

R.D.Yates and G.R.Cooper(1966) have proposed making use of row complete latin squares or Tuscan squares to solve some of the problems which arise in connection with frequency hopping.

A frequency hopping system may be specified as follows:
There are n available frequencies and up to n communicators in a communication network. The communicators each use the given set of frequencies for transmission but they hop from one frequency to another at specified time intervals in order to avoid signal jamming or other interference. Each communicator operates independently of the others and they employ no common time reference but the signal duration between hops is effectively the same for each of them. The order in which a particular communicator uses the n frequencies is decided by some pre-determined permutation of them. It is required to so choose these permutations that any two of the communicators will be transmitting on the same frequency at most once per cycle of the n frequencies, no matter what the displacement between their starting times. (We might regard this requirement as being that of a special kind of multiple access coding.)

S.W.Golomb and H.Taylor(1985) have remarked that a general solution to this problem can be given by assigning to each communicator one of the rows of a so-called Florentine square of order n as his permutation of the n frequencies. This concept was defined in Chapter 3 but, for convenience, we repeat the definition here.

A Tuscan k-square is a Tuscan square which has distinct ordered pairs of symbols $2, 3, \ldots, k$ apart in its rows as well as 1-apart.

A Florentine square of order n is a Tuscan n-square for which each ordered pair of symbols (i,j) from the set $\{1, 2, \ldots, n\}$ has a different signed distance in each of the n rows of the array, where columns are read cyclically. (It differs from a Tuscan n-square in the fact that columns are read cyclically, as is necessary for a solution of the frequency hopping problem described above.)

A Florentine square of each even order p-1 for which p is an odd prime can be obtained immediately from the multiplication table of the field Z_p of residue classes modulo p. (Such a square is of course a latin square.) This was first pointed out in R.D.Yates and G.R.Cooper(1966). Golomb and Taylor were unsuccessful in finding any other examples.

The former authors pointed out, however, that if a Tuscan square is used to assign frequency permutations to the communicators, then each will use a different pattern of frequency transitions. But, use of such a square does not solve the original problem.

One of the present authors has observed that a pseudo-random sequence on n symbols with good properties is obtained if the entries of a row-complete latin square (or Tuscan square) are written down row by row. The sequence obtained has even better properties if a Tuscan n-square is used since then almost all k-sequences (for k = 1,...,n) in the sequence will be different.

(4) Two-dimensional coding problems

The coding and transmission of pictures, one aspect of which we discussed in the previous section, may be regarded as one kind of two-dimensional coding problem in which latin squares have a role to play. Another concerns the allocation of frequencies in a mobile radio telephone system and a third arises in connection with the so-called two dimensional codes (especially cyclic codes) which have recently become a subject of attention.

In a mobile radio telephone system, each region is served by a local transmitter and it is required to allocate frequencies to these transmitters in such a way that there is no interference between neighbouring transmitters. Consequently, adjacent transmitters must be allocated different frequencies. In practice, each transmitter will require several frequencies for sending messages and further frequencies will be required for receiving back messages from the outstations. Since the total number of frequencies which are available for the system is usually limited, it is important to be able to allocate frequencies as efficiently as possible.

In a recent paper [J.Dénes and A.D.Keedwell(1988)], the present authors have suggested several ways in which the frequency allocation may be carried out. Two of these make use of latin squares and we shall now describe them.

If we suppose that the transmitters are arranged in the form of a rectangular grid, an optimal solution may be obtained as follows:

Since each transmitter is surrounded by eight others, of which no two adjacent ones may be assigned the same frequency, at least five different frequencies are necessary, one for the central transmitter of the group and at least four for the eight transmitters which surround it. We show that this minimal number of frequencies is achievable by repetition (in both horizontal and vertical directions) of a certain 5×5 latin square as many times as is required for coverage of the entire area of the mobile radio system. Our construction is illustrated in Figure 4.1. In this schema, there is at least a knight's move separation between any two transmitters which are allocated the same frequency.

The 5×5 latin square which we use is one which has appeared many times in the literature of Mathematics. It has the remarkable properties of being both a strict knight's move square (as defined by P.J.Owens(1987), see also Part II of the present book) and also a totally diagonal latin square or Knut Vik design (as defined by A.Hedayat and W.T.Federer(1975)).

```
        2 3 4 0 | 1 2 3 4 0 | 1 2 3 4 0 | 1
      3 4 0 1 2 | 3 4 0 1 2 | 3 4 0 1 2 |
    ─────────────────────────────────────
    4 | 0 1 2 3 4 | 0 1 2 3 4 | 0 1 2 3
    1 | 2 3 4 0 1 | 2 3 4 0 1 | 2 3 4
    3 | 4 0 1 2 3 | 4 0 1 2 3 | 4 0
    0 | 1 2 3 4 0 | 1 2 3 4 0 | 1
    2 | 3 4 0 1 2 | 3 4 0 1 2 |
    ─────────────────────────────
    4 | 0 1 2 3 4 | 0 1 2 3
    1 | 2 3 4 0 1 | 2 3 4
    3 | 4 0 1 2 3 | 4 0           Figure 4.1
```

If a schema which provides at least a knight's move separation between any two transmitters which are allocated the same frequency is insufficient, a considerable improvement can be obtained if a minimum of seven frequencies is available. In this case, we regard the area which lies within the range of each transmitter as being circular and the various circular regions as being close-packed so that each region is adjacent to six others. It is then possible to distribute the available seven frequencies in such a way that each transmitter is surrounded by six others all of which are transmitting on frequencies which are distinct both from that of the transmitter in their centre and also from each other. The array which has to be repeated is a skewed latin square. The required repeating pattern is shown in Figure 4.2 and it is clear that the solution is again optimal. Transmitters which use the same frequency are an extended knight's move apart.

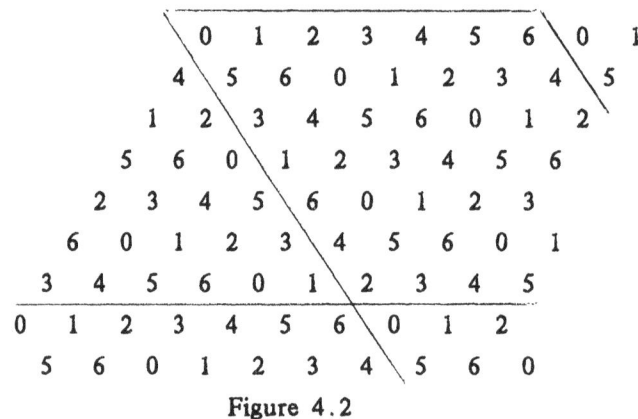

Figure 4.2

In his paper (1984), H.Imai has defined a <u>two-dimensional binary cyclic code of area m×n</u> as a set C of m×n matrices with symbols from GF[2] which satisfies the following conditions: (i) C is a linear code; (ii) the matrix obtained by shifting the columns of any code array in C cyclically one step to the right is also in C; and (iii) the matrix obtained by shifting the rows of any code array in C one step downwards is also in C. Imai has pointed out that one way of constructing such codes is by forming the direct product of two one-dimensional cyclic codes. As an example, he has given the code whose arrays are shown in Figure 4.3, which is the direct product of the cyclic code 0 0 0 0 1 1 1 0 1 1 1 0 of four words with itself. In general, it is not easy to find the minimum Hamming distance between the code arrays of a two-dimensional binary cyclic code.

However, we should like to draw attention here to the fact that a two-dimensional non-binary cyclic code of constant weight and whose minimum Hamming distance is easy to state can be derived by means of latin squares. The arrays of the code comprise all possible distinct Cayley tables of the cyclic group of order n and then, by virtue of theorem 3.2.1

```
0 0 0      1 0 1      1 1 0      0 1 1
0 0 0      0 1 1      1 0 1      1 1 0
0 0 0      1 1 0      0 1 1      1 0 1

0 1 1      1 0 1      0 1 1      1 1 0
1 0 1      0 0 0      0 0 0      0 0 0
1 1 0      1 0 1      0 1 1      1 1 0

1 1 0      0 0 0      0 0 0      0 0 0
0 1 1      1 0 1      0 1 1      1 1 0
1 0 1      1 0 1      0 1 1      1 1 0

1 0 1      1 0 1      0 1 1      1 1 0
1 1 0      1 0 1      0 1 1      1 1 0
0 1 1      0 0 0      0 0 0      0 0 0
```

Figure 4.3

of [DK], the minimum Hamming distance between any two arrays of the code is 2n. For convenience, we repeat our theorem here and we also illustrate our construction by giving in Figure 4.4 the two-dimensional non-binary cyclic code obtained when n = 3.

```
1 2 3    3 1 2    2 3 1    1 2 3    2 3 1    3 1 2
2 3 1    1 2 3    3 1 2    3 1 2    1 2 3    2 3 1
3 1 2    2 3 1    1 2 3    2 3 1    3 1 2    1 2 3

1 3 2    3 2 1    2 1 3    1 3 2    2 1 3    3 2 1
2 1 3    1 3 2    3 2 1    3 2 1    1 3 2    2 1 3
3 2 1    2 1 3    1 3 2    2 1 3    3 2 1    1 3 2
```

Figure 4.4

THEOREM 4.1 Two different Cayley tables, A and B, of a given group G of order n differ from each other in at least 2n places.

Proof. If no two corresponding rows of the two Cayley tables are the same then every row of A differs from the corresponding row of B in at least two places. Likewise, if no two corresponding columns of the two tables are the same, then every column of A differs from the corresponding column of B in at least two places. In either event, B differs from A in at least 2n places.

For the remaining part of the proof, we may suppose that at least one row and at least one column of B are the same as the corresponding row and column of A. Let us suppose that the equal rows are the u-th rows and that the equal columns are the v-th columns. Then, we have $a_{uv} = b_{uv}$, $a_{uj} = b_{uj}$ and $a_{iv} = b_{iv}$ whence, by the quadrangle criterion, $a_{ij} = b_{ij}$ for all pairs of indices i and j, and so A = B. Consequently, this case cannot occur. []

It was shown by E.N. Gilbert (1965) that the number of different Cayley tables of a given cyclic group G of order n is $n!(n-1)!(n-1)!/\varphi(n)$, where $\varphi(n)$ is Euler's function. For example, when n = 3, this number is 12, as in Figure 4.4.

Another class of two-dimensional arrays which arise in coding theory and which have connections with latin squares are the so-called Costas arrays.

A Costas array of order n is an n×n array of blanks and ones with the property that the $\frac{1}{2}n(n-1)$ vectors which connect pairs of ones in the matrix are all distinct as vectors. (For an example of order six, see Figure 4.5). Thus, a translation of the array without rotation produces at most one pair of superimposed cells both of which contain an entry one. Such arrays are of value in determining the range and velocity of a moving object by means of radar or sonar signals. A detailed account of the history of these arrays and of their applications will be found in J.P. Costas (1984).

A Costas array is called vertically singly-periodic (horizontally singly-periodic) if all its vertical translates (horizontal translates), when read cyclically as if on a horizontal (vertical) cylinder, are also Costas arrays. It is known that singly-periodic Costas arrays exist for all orders p-1, where p is a prime number, and the question has been raised as to whether they exist for orders not of this form. Several attempts have been made to solve this problem with the aid of latin squares. In order to be able to show the connection, let us first remark that if we replace the blanks of a Costas array by zeros, we may regard it as a permutation matrix. If the corresponding permutation of the integers $1,2,\ldots,n$ maps i to $\theta(i)$ then the cells of the array which contain the entry one have co-ordinates $[i,\theta(i)]$ and the Costas property requires that the vectors $[i+k,\theta(i+k)]-[i,\theta(i)]$ and $[j+k,\theta(j+k)]-[j,\theta(j)]$ be different whenever $i \neq j$. So θ must be a permutation of $1,2,\ldots,n$ such that

$\theta(i+k)-\theta(i) = \theta(j+k)-\theta(j) \Rightarrow i = j$ for all choices of k, $k = 1,2,\ldots,n-2$.

We shall call this <u>the first condition on θ</u>.

As an example, the permutation $\theta = \begin{bmatrix} 1 & 2 & 3 & 4 & 5 & 6 \\ 4 & 1 & 2 & 6 & 5 & 3 \end{bmatrix}$ defines the Costas array given in Figure 4.5. We observe that the differences $\theta(i+1)-\theta(i) = -3,1,4,-1,-2$ are all different, the differences $\theta(i+2)-\theta(i) = -2,5,3,-3$ are all different, the differences $\theta(i+3)-\theta(i) = 2,4,1$ are all different and, finally, the differences $\theta(i+4)-\theta(i) = 1,2$ are different. We can exhibit the differences compactly in the form of a triangle as in Figure 4.5.

$$\theta = \begin{bmatrix} 1 & 2 & 3 & 4 & 5 & 6 \\ 4 & 1 & 2 & 6 & 5 & 3 \end{bmatrix}$$

```
         4    1    2    6    5    3
           -3    1    4   -1   -2
             -2    5    3   -3
                2    4    1
                  1    2
```

0	0	0	1	0	0
1	0	0	0	0	0
0	1	0	0	0	0
0	0	0	0	0	1
0	0	0	0	1	0
0	0	1	0	0	0

Figure 4.5

In terms of the permutation θ, a Costas array is vertically singly-periodic if, in addition to the permutation θ, each of the permutations $i \to \theta(i+h)$, $h = 1,2,\ldots,n-1$, defines a Costas array. It is horizontally singly-periodic if, in addition to the permutation θ, each of the permutations i $\longrightarrow \phi_h(i)$, where $\phi_h(i) = \theta(i)-h \pmod{n}$ and h = 1,2,...,n-1, defines a Costas array. For the example of Figure 4.5, we note that i $\longrightarrow \theta(i+2)$ defines a Costas array (shown in Figure 4.6) but i $\longrightarrow \theta(i+1)$ does not. Also, i $\longrightarrow \phi_1(i) = \theta(i)-1 \pmod{6}$ gives a Costas

i $\longrightarrow \theta(i+2)$

```
2   6   5   3   4   1
  4  -1  -2   1  -3
    3  -3  -1  -2
      1  -2  -4
        2  -5
```

0	0	0	1	0	0
1	0	0	0	0	0
0	1	0	0	0	0
0	0	0	0	0	1
0	0	0	0	1	0
0	0	1	0	0	0
0	0	0	1	0	0
1	0	0	0	0	0

Figure 4.6

i $\longrightarrow \theta(i)-1 \pmod{6}$

```
3   6   1   5   4   2
  3  -5   4  -1  -2
   -2  -1   3  -3
      2  -2   1
        1  -4
```

0:	0	0	1	0	0	0
1:	0	0	0	0	0	1
0:	1	0	0	0	0	0
0:	0	0	0	0	1	0
0:	0	0	0	1	0	0
0:	0	1	0	0	0	0

Figure 4.7

array (shown in Figure 4.7) but i $\longrightarrow \phi_2(i)$ does not.

T. Etzion (1989) has shown that when and only when a Costas array is horizontally singly-periodic, the first condition on θ is replaced by the stronger condition

$\theta(i+k)-\theta(i) \equiv \theta(j+k)-\theta(j) \pmod{n} \Rightarrow i=j$ for all choices of k, k=1,2,...,n-2. We shall call this the second condition on θ.

When the second condition on θ holds, the latin square whose (i,j)-th cell contains $\theta(j)+(i-1)$ is a Vatican square: that is, a Tuscan-(n-1) square which is a latin square. (For a discussion of these concepts, see Section 5 of Chapter 3.) A Vatican square which takes this particular form (in which the i-th row is obtained by adding i–1 to each of the elements of the first row) is said to be constructed by the polygonal path construction [S.W.Golomb, T.Etzion and H.Taylor(1990)] because, when the vertices of a regular n-gon are numbered $1, 2, \ldots, n$ in order, the joins $\theta(i+1)-\theta(i)$ form n–1 directed chords no two of which coincide when the polygon is rotated.

Thus we have the following theorems:

THEOREM 4.2. A Costas array is horizontally singly-periodic if and only if its defining permutation θ satisfies the second conditon above.

Proof. It is easy to see that the second condition on θ inplies that each of the mappings $i \longrightarrow \phi_h(i) = \theta(i)-h \pmod{n}$ satisfies the first condition on θ.

Conversely, suppose if possible that θ defines a Costas array which is horizontally singly-periodic and that, for some set of values i, j, k, the first condition on θ holds but the second does not. We can suppose without loss of generality that $\theta(i+k)-\theta(i) = \theta(j+k)-\theta(j)+n$, $i \neq j$. Then, $\theta(i+k)-\theta(j) > 0$ and $\theta(j+k)-\theta(j) < 0$ since all differences lie between $\pm(n-1)$. Let $m = \min[\theta(i), \theta(j+k)]$. Then $\phi_m(i+k) = \theta(i+k)-m$ since $\theta(i+k) > \theta(i) \geq m$, and $\phi_m(j) = \theta(j)-m$ since $\theta(j) > \theta(j+k) \geq m$.

If $m = \theta(i)$, we have $\phi_m(i) = n$ (since all permutation entries lie between 1 and n) and $\phi_m(j+k) = \theta(j+k)-m$. Then $\phi_m(i+k)-\phi_m(i) = \theta(i+k)-m-n = \theta(i+k)-\theta(i)-n = \theta(j+k)-\theta(j) = \phi_m(j+k)-\phi_m(j)$ which cannot occur with $i \neq j$ because $i \longrightarrow \phi_m(i)$ defines a Costas array.

If $m = \theta(j+k)$, we have $\phi_m(i) = \theta(i)-m$ and $\phi_m(j+k) = n$. This gives $\phi_m(j+k)-\phi_m(j) = n-[\theta(j)-m] = n+\theta(j+k)-\theta(j) = \theta(i+k)-\theta(i) = \phi_m(i+k)-\phi_m(i)$ which again cannot occur with $i \neq j$.

From these contradictions, we conclude that the second condition on θ holds when θ defines a Costas array which is horizontally singly-periodic. []

THEOREM 4.3. [T.Etzion, S.W.Golomb and H.Taylor(1990)]. A Costas array of order n which is horizontally singly-periodic exists if and only if a Vatican square of order n constructed by the polygonal path construction exists.

Proof. The permutation θ which defines the Costas array satisfies the second condition on θ and consequently the latin square whose i-th row is θ(1)+(i-1), θ(2)+(i-1), ..., θ(n)+(i-1), where arithmetic is mod n, is a Vatican square. []

An illustration of Theorem 4.3 is given in Figure 4.8, where both the Costas array and the Vatican square to which it is equivalent are exhibited.

$$\theta = \begin{bmatrix} 1 & 2 & 3 & 4 & 5 & 6 \\ 6 & 4 & 5 & 2 & 1 & 3 \end{bmatrix}$$

$$\begin{array}{|cccccc|} \hline 0 & 0 & 0 & 0 & 0 & 1 \\ 0 & 0 & 0 & 1 & 0 & 0 \\ 0 & 0 & 0 & 0 & 1 & 0 \\ 0 & 1 & 0 & 0 & 0 & 0 \\ 1 & 0 & 0 & 0 & 0 & 0 \\ 0 & 0 & 1 & 0 & 0 & 0 \\ \hline \end{array}$$

$$\begin{array}{|cccccc|} \hline 6 & 4 & 5 & 2 & 1 & 3 \\ 1 & 5 & 6 & 3 & 2 & 4 \\ 2 & 6 & 1 & 4 & 3 & 5 \\ 3 & 1 & 2 & 5 & 4 & 6 \\ 4 & 2 & 3 & 6 & 5 & 1 \\ 5 & 3 & 4 & 1 & 6 & 2 \\ \hline \end{array}$$

Figure 4.8

Since the permutation $i \longrightarrow \theta(i)$ in Figure 4.8 defines a horizontally singly-periodic Costas array whose non-empty cells have co-ordinates $[i,\theta(i)]$, it follows that the Costas array whose non-empty cells have co-ordinates $[\theta(i),i]$ or $[j,\theta^{-1}(j)]$ will be vertically singly-periodic. Consequently, each of the permutations $j \longrightarrow \theta^{-1}(j+h)$ defines a Costas array. In particular, this means that, when the six differences $\theta^{-1}(j+1)-\theta^{-1}(j)$, where $j = 1,2,3,4,5,6$ and $6+1 = 1$, are read cyclically, every consecutive five of them are distinct and define a proper difference triangle. Moreover, when the six differences $\theta^{-1}(j+h) - \theta^{-1}(j)$, $h = 2,3,4$, are read cyclically, every consecutive $6-h$ of them are distinct. Now, it is easy to see that if a sequence of n symbols is arranged in a circle and if every subset of $r \geq 1 + \lfloor n/2 \rfloor$ consecutive symbols has all its members distinct, then the n symbols must be distinct.

Generalizing our example, we have:

THEOREM 4.4. If the Costas array with defining permutation θ is vertically singly-periodic then the Costas array with defining permutation

θ^{-1} is horizontally singly-periodic. A necessary condition that a Costas array of order n should be vertically singly-periodic is that its defining permutation θ satisfies the condition $\theta(i+k)-\theta(i) = \theta(j+k)-\theta(j) \Rightarrow i = j$ for all $k \geq 1+\lfloor n/2 \rfloor$, where j+k is computed modulo n.

In fact, only one systematic construction for vertically singly-periodic Costas arrays is known. This is due to L.R.Welch [see S.W.Golomb and H.Taylor(1982)] and is as follows:

Let p be a prime and α a generating integer of the multiplicative group of the Galois field GF[p]. Then a Costas array of order p-1 is obtained by defining $\theta(i) = \alpha^i$, $i = 1, 2, \ldots, p-1$.

Since $\theta(i+k)-\theta(i) = \alpha^i(\alpha^k-1)$, the defining permutation of a Costas array which is constructed by the Welch method has a stronger property than that required by Theorem 4.4: namely, it satisfies the condition

$\theta(i+k)-\theta(i) \equiv \theta(j+k)-\theta(j) \mod(n+1) \Rightarrow i = j$ for all values of k, $1 \leq k \leq n$,

where j+k is computed modulo n. We shall call this <u>the third condition on θ</u>.

It follows then that the (n+1)×n array whose (i,j)-th cell is $\theta(j)+(i-1)$ and where arithmetic is modulo n+1, is a circular Vatican array.

T.Etzion(198γ) has shown further that if and only if a permutation θ satisfies the above third condition then θ^{-1} satisfies the condition

$\theta^{-1}(i+k)-\theta^{-1}(i) \equiv \theta^{-1}(j+k)-\theta^{-1}(j) \mod n \Rightarrow i = j$ for all values of k, $1 \leq k \leq n$,

where j+k is computed modulo n+1. Consequently, the integers $\theta^{-1}(1), \theta^{-1}(2), \ldots, \theta^{-1}(n)$ form a polygonal path for an n×(n+1) circular Vatican array whose (i,n+1)-th entry is 0 and whose (i,j)-th entry is $\theta^{-1}(j)+(i-1)$, for $1 \leq i, j \leq n$, where arithmetic is modulo n. It has been proved that such a circular Vatican array can only exist if n+1 is prime. [See T.Etzion, S.W.Golomb and H.Taylor(1989).] Also, we have:

THEOREM 4.5. [T.Etzion(198α)] An n×(n+1) circular Vatican array exists if and only if there exists a complete set $\{A_1, A_2, \ldots, A_n\}$ of mutually orthogonal latin squares of order n+1 for which all the rows are cyclic permutations of $0, 1, 2, \ldots, n$ and the top row of each square begins with 0.

Proof. For $2 \leq i, j \leq n+1$, we put k in the (i-1,j-1)-th cell of the Vatican array if and only if the (i,j)-th cell of A_k contains 0. The last column of the Vatican array consists entirely of zeros. []

[Compare Theorem 5.1 of Chapter 3 and the remarks which follow it and also Theorem 7.4.1 of [DK]. The latter theorem gives a direct construction of a complete set of mutually orthogonal latin squares from an (n+1)×n circular Vatican array of the type whose existence is guaranteed by the third condition on θ.]

We provide illustrations of some of the above-mentioned results in Figures 4.9, 4.10 and 4.11. Thus, the Costas array given in Figure 4.9 is constructed by the Welch construction taking p = 7 and α = 5. It is vertically singly-periodic and its defining permutation $\psi = \begin{bmatrix} 1 & 2 & 3 & 4 & 5 & 6 \\ 5 & 4 & 6 & 2 & 3 & 1 \end{bmatrix}$ is the inverse of that used in Figure 4.8. So the integers 5,4,6,2,3,1 provide a polygonal path for an n×(n+1) circular Vatican array as shown in Figure 4.10 and the integers 6,4,5,2,1,3 form a polygonal path for an (n+1)×n Vatican array as in Figure 4.11. A further observation is that a vertically (horizontally) singly-periodic Costas array of order n defines an n×n latin square with the property that all those cells which contain the same symbol represent a vertical (horizontal) translate of the Costas array defined by the cells which contain the entry one. This fact also is illustrated in Figure 4.9. We may think of it as a means of filling the plane with Costas arrays each of which is a translate of each other.

0	0	0	0	1	0
0	0	0	1	0	0
0	0	0	0	0	1
0	1	0	0	0	0
0	0	1	0	0	0
1	0	0	0	0	0

2	4	3	6	1	5
3	5	4	1	2	6
4	6	5	2	3	1
5	1	6	3	4	2
6	2	1	4	5	3
1	3	2	5	6	4

Figure 4.9

Since it appears likely that singly-periodic Costas arrays only exist for orders of the form p−1, we may ask the more interesting (but more difficult) question whether the plane can be filled with Costas arrays in such a way that no one is the translate of any other. In other words, is it possible to construct a latin square of order n such that all those cells which contain the same symbol form a Costas array and no two of these n Costas arrays are translates one of another?

```
5 4 6 2 3 1                      5 4 6 2 3 1
6 2 3 1 5 4                      6 5 7 3 4 2
1 5 4 6 2 3   differences,       7 6 1 4 5 3
5 4 6 2 3 1      mod 7           1 7 2 5 6 4
1 5 4 6 2 3                      2 1 3 6 7 5
6 2 3 1 5 4                      3 2 4 7 1 6
                                 4 3 5 1 2 7
```

Figure 4.10

```
6 4 5 2 1 3 0                    6 4 5 2 1 3 0
. 4 1 3 5 2 .                    1 5 6 3 2 4 0
. 5 4 2 1 . 3   differences,     2 6 1 4 3 5 0
1 . 2 3 4 . 5      mod 6.        3 1 2 5 4 6 0
2 . 1 5 . 4 3                    4 2 3 6 5 1 0
4 5 . 3 . 1 2                    5 3 4 1 6 2 0
1 4 . . 2 5 3
```

Figure 4.11

To construct such a latin square, we require n permutations $i \longrightarrow \theta_u(i)$, $u = 1, 2, \ldots, n$, such that (i) each permutation has the difference properties required for it to define a Costas array, (ii) $\theta_u(i) \neq \theta_v(i)$ for $u \neq v$, (iii) $\theta_v(i) \neq \theta_u(i+h)$ and $\theta_v(i) \neq \theta_u(i)-h$ mod n for any fixed h; where i varies from 1 to n.

In an unpublished preprint, the present authors called the Costas arrays constructed from such a set of permutations a set of plane-filling Costas arrays. They showed that such a set exists when n = 6 (as in Figure 4.12) but were unable to find a set for n = 5. More recently, T. Etzion (198β) has obtained several general constructions for sets of plane-filling Costas arrays to which he has given the alternative name of latin Costas arrays. He has also shown the existence of plane-filling Costas arrays with a stronger property: namely that the latin square which they fill is a Vatican square. He has called such a design a <u>Vatican Costas square</u>. All these squares have orders of the form p-1, p prime.

i =	1	2	3	4	5	6						
$\theta_1(i) =$	2	6	4	5	1	3	6	1	5	2	4	3
$\theta_2(i) =$	4	2	6	3	5	1	4	2	6	3	5	1
$\theta_3(i) =$	6	4	5	1	3	2	5	6	4	1	3	2
$\theta_4(i) =$	5	1	3	4	2	6	3	5	2	4	1	6
$\theta_5(i) =$	3	5	1	2	6	4	1	4	3	6	2	5
$\theta_6(i) =$	1	3	2	6	4	5	2	3	1	5	6	4

Figure 4.12

The following two constructions are claimed to give Vatican Costas squares:

(1) Let α be a primitive root modulo the prime number p. We put k in the (i,j)-th cell of a (p-1)×(p-1) square if and only of $\alpha^{k-i} \equiv j \pmod{p}$, where $1 \leq i,j,k \leq p-1$.

(2) Let $p = 2^r - 1$ be a Mersenne prime and let α and β be primitive elements of the field GF[p+1]. We put k in the (i,j)-th cell of a (p-1)×(p-1) square if and only if $\alpha^i + \beta^{jk} = 1$, where $1 \leq i,j,k \leq p-1$.

However, the first of these constructions does <u>not</u> give a set of plane-filling Costas arrays according to the above definition since each Costas array which it constructs is a vertical translate of each other and so is vertically singly-periodic. On the other hand, the second construction does appear to give a plane-filling set, although Etzion does not prove this.

The following two constructions from T. Etzion (198y) give latin Costas arrays which partially satisfy our definition of plane-filling. (A few of the Costas arrays are translates of others, according to Etzion, private communication.)

(3) Let p be a prime of the form $p = (2^n \cdot t) + 1$, where t is odd and $n \geq 1$, and let α be a primitive root modulo p. We put k in the (i,j)-th cell of a (p-1)×(p-1) square if and only if $\alpha^{(-k+j)(2kt+1)} \equiv i \pmod{p}$, where $1 \leq i,j,k \leq p-1$.

(4) Let α and β be primitive roots modulo the prime p and let us construct a (p-1)×(p-1) square by putting k in the (i,j)-th cell if (i) when k is odd, $\alpha^{(-k+j)} \equiv i \pmod{p}$, and (ii) when k is even, $\beta^{(-k+j)} \equiv i \pmod{p}$.

Finally, in this Section, we draw attention to the use of a single latin square in connection with the allocation of the pixels of a two-dimensional raster-graphics display to the M memory chips which store the display in such a way that it is possible to copy or alter small rectangles of N pixels very rapidly whatever their shape. Essentially, the requirement is that the N pixels of a randomly chosen rectangle should all be allocated to distinct memory chips. In B.Chor, C.E.Leiserson and R.L.Rivest(1982), the authors have shown that this requirement can be met using only approximately $\sqrt{5N}$ memory chips if the allocation of memory chips to the pixels of the complete array is described by a certain cyclic latin square with rows rearranged in a prescribed way. The prescription for rearranging the rows is described in terms of Fibonacci numbers.

(5) Secret-sharing systems

In his book W.W.Wu(1985), the author stated (see pages 163 to 167) that all the secret-sharing systems known at the time his book was written are connected with latin squares and he provided some constructions of such systems using orthogonal latin squares. In fact, his examples all construct secret-sharing systems in which only two keys are needed to unlock the secret. He also made what the present authors consider to be a more general observation to the effect that all these systems can be constructed with the aid of Reed-Solomon codes. In this section, we describe those three of the constructions of secret-sharing systems which most closely fit in with the pattern observed by Wu and elucidate more precisely the role played by latin squares.

In a secret-sharing system of the kind which we are about to describe, it is supposed that there are n keys and that we require that any k of them should unlock the secret. In the first two of the systems which we describe, the secret is assumed to consist of a single number which we shall denote by a_0.

The first system is due to A.Shamir(1979).

We define a polynomial $q(x) = a_0 + a_1 x + \ldots + a_{k-1} x^{k-1}$, modulo p, where p is a prime greater than n and where the polynomial is so chosen that it has distinct values modulo p for n different values $x_1, x_2, \ldots, x_n$ of x. (For some further remarks about such polynomials, see the end of this Section.) The keys to the secret are the n different ordered pairs of integers $(x_i, q(x_i))$ for $i = 1, 2, \ldots, n$. By Lagrange's interpolation formula for polynomials

$$q(x) = \sum_{i=1}^{k} \frac{q(x_i)(x-x_1)(x-x_2)\ldots(x-x_{i-1})(x-x_{i+1})\ldots(x-x_k)}{(x_i-x_1)(x_i-x_2)\ldots(x_i-x_{i-1})(x_i-x_{i+1})\ldots(x_i-x_k)}$$

where $x_1, x_2, \ldots, x_k$ are any k of the n keys. Since $q(x)$ can be calculated using any k of the keys, such a set of k keys unlocks the secret.

The second system is due to R.J.McEliece and D.V.Sarwate(1981).

Let $\alpha_0 = 1$, $\alpha_1 = 1$, $\alpha_2, \ldots, \alpha_{q-1}$ be the elements of a finite field GF[q] with $q = p^m$ (p prime) elements, where $q \geq n+1$. Then the matrix

$$G = \begin{bmatrix} 1 & 1 & \ldots & 1 \\ \alpha_1 & \alpha_2 & \ldots & \alpha_{q-1} \\ \alpha_1^2 & \alpha_2^2 & \ldots & \alpha_{q-1}^2 \\ \cdot & \cdot & & \cdot \\ \cdot & \cdot & & \cdot \\ \alpha_1^{k-1} & \alpha_2^{k-1} & & \alpha_{q-1}^{k-1} \end{bmatrix}$$

is the generator matrix of a Reed-Solomon code of word length q-1 which expresses each k-tuple $\underline{a} = (a_0\ a_1\ \ldots\ a_{k-1})$ in coded form as a (q-1)-tuple $\underline{b} = (b_1\ b_2\ \ldots\ b_{q-1})$, where $\underline{b} = \underline{a}G$. Since a Reed-Solomon code is an MDS code, the minimum Hamming distance between codewords $\underline{b}$ is $(q-1)-k+1$. Consequently, no two codewords $\underline{b}$, $\underline{b}^*$ agree in more than k-1 places. In other words, any assignment of k of the elements $b_1, b_2, \ldots, b_{q-1}$ determines a unique codeword $\underline{b}$ and consequently determines $\underline{a}$ and the secret a_0 associated with $\underline{a}$ uniquely. The $n(\leq q-1)$ keys, any k of which unlock the secret, are n of the elements $b_1, b_2, \ldots b_{q-1}$. We note that $b_i = q(\alpha_i)$, where $q(x) = a_0 + a_1 x + a_2 x + \ldots + a_{k-1} x^{k-1}$, so this method is a generalization of that of Shamir.

The third system is due to E.D.Karnin, J.W.Greene and M.E.Hellman(1983).

For this system the secret may have m components. Let $A_0, A_1, A_2, \ldots, A_n$ be a set of n+1 matrices with elements in the finite field GF[q], each of size km×m and with the property that the km×km matrix formed by adjoining any set of k of the A_i is non-singular. Let μ be a vector of length km such that the secret is the 1×m matrix μA_0. Then the n keys are $\mu A_1, \mu A_2, \ldots, \mu A_n$ and any k of these unlock the secret in the following way.

Suppose that $\mu A_{i_1}, \mu A_{i_2}, \ldots, \mu A_{i_k}$ are given. We suppose that the inverses of the non-singular matrices formed by summing any k of the matrices $A_1, A_2, \ldots, A_n$ are stored in the unlocking device and that the matrix A_0 also is stored there. Then, by postmultiplying $(\mu A_{i_1} \mu A_{i_2} \ldots \mu A_{i_k})$ by the inverse of the matrix $(A_{i_1} A_{i_2} \ldots A_{i_k})$, we obtain the vector μ from the k given keys. By postmultiplying this again by the matrix A_0, the secret is revealed.

Before showing how this scheme is related to the two schemes previously described, we illustrate it with an example.

EXAMPLE. Over GF[2], the following five 4×2 matrices meet the requirements of the scheme:

$$A_0 = \begin{bmatrix} 1 & 0 \\ 0 & 1 \\ 0 & 0 \\ 0 & 0 \end{bmatrix} \quad A_1 = \begin{bmatrix} 0 & 0 \\ 0 & 0 \\ 1 & 0 \\ 0 & 1 \end{bmatrix} \quad A_2 = \begin{bmatrix} 1 & 0 \\ 1 & 1 \\ 1 & 0 \\ 0 & 1 \end{bmatrix} \quad A_3 = \begin{bmatrix} 0 & 1 \\ 1 & 0 \\ 1 & 0 \\ 0 & 1 \end{bmatrix} \quad A_4 = \begin{bmatrix} 1 & 0 \\ 0 & 1 \\ 1 & 1 \\ 0 & 1 \end{bmatrix}$$

If $\mu = (u_1 \ u_2 \ u_3 \ u_4)$, the secret is $\mu A_0 = (u_1 \ u_2)$.
The keys are $\mu A_1 = (u_3 \ u_4)$ $\quad \mu A_2 = (u_1+u_2+u_3 \ u_2+u_4)$
$\mu A_3 = (u_2+u_3 \ u_1+u_4)$ $\quad \mu A_4 = (u_1+u_3 \ u_2+u_3+u_4)$

Suppose that the keys μA_2 and μA_3 are given.
Then $\mu (A_2 \ A_3) = (u_1+u_2+u_3 \ u_2+u_4 \ u_2+u_3 \ u_1+u_4)$
and so $\mu = (u_1+u_2+u_3 \ u_2+u_4 \ u_2+u_3 \ u_1+u_4)(A_2 \ A_3)^{-1}$.

The matrix A_0 and the six matrices $(A_i \ A_j)^{-1}$ are all stored in the decoder.

The above example is one of those given by Karnin et al. These authors pointed out that a simple method by which a set of $km \times m$ matrices satisfying the requirements may be constructed over $GF[q]$ is by means of the matrix

$$\begin{bmatrix} 1 & 0 & 1 & 1 & \ldots & 1 \\ 0 & 0 & \alpha_1 & \alpha_2 & \ldots & \alpha_{q-1} \\ . & . & . & . & \ldots & . \\ . & . & . & . & \ldots & . \\ 0 & 1 & \alpha_1^{h-1} & \alpha_2^{h-1} & \ldots & \alpha_{q-1}^{h-1} \end{bmatrix}$$

where $h = km (\leq q)$ and $\alpha_1, \alpha_2, \ldots, \alpha_{q-1}$ are the non-zero elements of the finite field $GF[q]$, as before. The determinant formed by any h of the columns of this matrix is non-singular. Consequently, any decomposition of the matrix into pairwise disjoint sets of m columns defines a set of $km \times m$ matrices suitable for the secret sharing scheme.

Consider now the special case when $m = 1$ (and $h = k$). Then the matrices $A_0, A_1, \ldots, A_n$ ($n \leq q-1$) are of size $k \times 1$ and we may take them to be the first n columns of the matrix

$$\begin{bmatrix} 1 & 1 & 1 & \ldots & 1 \\ 0 & \alpha_1 & \alpha_2 & \ldots & \alpha_{q-1} \\ 0 & \alpha_1^2 & \alpha_2^2 & \ldots & \alpha_{q-1}^2 \\ . & . & . & \ldots & . \\ . & . & . & \ldots & . \\ 0 & \alpha_1^{k-1} & \alpha_2^{k-1} & \ldots & \alpha_{q-1}^{k-1} \end{bmatrix}$$

If $\mathbf{u} = (a_0\ a_1\ a_2\ \ldots\ a_{k-1})$, the secret is $\mathbf{u}A_0 = a_0$ and the n keys are $b_i = a_0 + a_1 \alpha_i + a_2 \alpha_i^2 + \ldots + a_{k-1} \alpha_i^{k-1}$ any k of which unlock the secret. Thus, we are back to the method of McEliece and Sarwate.

In order to substantiate his claim that all of the above secret-sharing systems are connected with latin squares, W.W.Wu(1985) showed that the codewords of a Reed-Solomon code of word length 3 and with minimum Hamming distance 2 can be constructed from a single latin square of order 4 by taking triples of the form (i, j, a_{ij}), where a_{ij} is the entry in the i-th row and j-th column of the latin square, as the codewords. He further showed that an extended Reed-Solomon code of word length 4 and with minimum Hamming distance 3 can be constructed from two orthogonal latin squares of order 4 by taking as the codewords quadruples of the form (i,j,a_{ij},b_{ij}), where a_{ij} and b_{ij} are the entries in the i-th row and j-th column of the two latin squares. In the latter case, the generator matrix of the code takes the form $G = \begin{bmatrix} 1 & 0 & 1 & 1 \\ 0 & 1 & \alpha_2 & \alpha_3 \end{bmatrix}$ where $0, 1, \alpha_2$ and α_3 are the elements of GF[4] and it can be used to construct a secret sharing system with four keys any two of which unlock the secret by either the method of McEliece and Sarwate or the method of Karnin, Greene and Hellman. In fact both these examples of Reed-Solomon codes are Golomb-Posner codes (which we discussed in Section 2 of this chapter). We have the following general theorem (see J. Dénes and A.D. Keedwell (1990a):

THEOREM 5.1. Every extended Reed-Solomon code with a generator matrix of two rows and which is defined over the Galois field GF[q] can be constructed from a complete set of orthogonal latin squares of order q and is identical to the Golomb-Posner code of that order.

Proof Let $\alpha_0 = 0$, $\alpha_1 = 1$, $\alpha_2, \alpha_3, \ldots, \alpha_{q-1}$ be the elements of GF[q]. Then the generator matrix of the extended Reed-Solomon code corresponding to a secret-sharing system in which k keys are needed to unlock the secret is

$$\begin{bmatrix} 1 & 0 & 1 & 1 & \ldots & 1 \\ 0 & 0 & \alpha_1 & \alpha_2 & \ldots & \alpha_{q-1} \\ 0 & 0 & \alpha_1^2 & \alpha_2^2 & \ldots & \alpha_{q-1}^2 \\ . & . & . & . & \ldots & . \\ . & . & . & . & \ldots & . \\ 0 & 1 & \alpha_1^{k-1} & \alpha_2^{k-1} & \ldots & \alpha_{q-1}^{k-1} \end{bmatrix}$$

If there are only two rows (k = 2), the matrix reduces to

$$\begin{bmatrix} 1 & 0 & 1 & 1 & \ldots & 1 \\ 0 & 1 & \alpha_1 & \alpha_2 & \ldots & \alpha_{q-1} \end{bmatrix}$$

The codeword which is obtained by adding α_i times row 1 to α_j times row 2 is $(\alpha_i \; \alpha_j \; \alpha_i + \alpha_1\alpha_j \; \alpha_i + \alpha_2\alpha_j \; \ldots \; \alpha_i + \alpha_{q-1}\alpha_j)$. But this is exactly the codeword which the Golomb-Posner construction obtains by using the entries of the i-th row and j-th column of the members of the complete set of q-1 mutually orthogonal latin squares constructed from GF[q] by the method originally described by R.C.Bose(1938) and independently by W.L.Stevens(1939). (See also Theorem 5.2.4 of [DK].) The prescription for filling the k-th square L_k of the set $L_1, L_2, \ldots, L_{q-1}$ of mutually orthogonal latin squares of order q as given by these authors is that the cell of the i-th row and j-th column should be filled by the entry $\alpha_i + \alpha_k \alpha_j$. This demonstrates the truth of the theorem. []

REMARK. The polynomials q(x) used by A.Shamir (see above) are examples of permutation polynomials (also called Redei polynomials because they were first studied by L.Redei(1946)). A polynomial with coefficients in a finite field GF[q] is called a permutation polynomial if the associated polynomial function $f: x \longrightarrow f(x)$ from GF[q] into itself is a permutation of GF[q]. Such polynomials are discussed in detail in R.Lidl and H.Niederreiter(1983).

The following interesting connection between permutation polynomials and latin squares is due to G.L.Mullen(1981). A polynomial f(x,y) in two variables over a finite field GF[q] is called a local permutation polynomial if the one variable polynomials f(x,a) and f(b,y) are permutation polynomials in x,y respectively for all choices of the constants a,b in GF[q]. Mullen has proved that the number of local permutation polynomials in two variables is equal to the number of latin squares of order q.

(6) Miscellaneous results

In A.Ecker and G.Poch(1986), the authors have given a survey of error-detecting decimal and alphanumeric codes. They have shown that in most systems each codeword $(a_0 \ a_1 \ a_2 \ .. \ a_n)$ incorporates a single check digit a_0 which is computed from an equation of the form

$\sum_{i=0}^{n} w_i a_i \equiv c \mod m$, where c is usually taken to be zero, m is usually 10 or 11 for a decimal code or 36 or 37 for an alphanumeric code, and the w_i are weights whose judicious choice determines the proportion of errors which can be detected.

For example, in a decimal code with codewords of length 7, the check equation might be $-a_0+a_1+3a_2+7a_3+a_4+3a_5+7a_6 \equiv 0$ mod 10, where a_0 is the check digit and $a_1, a_2, \ldots, a_6$ are information digits. This code will detect all single errors but only a proportion of other types of error, such as transposition of symbols. For example, if $a_2 = 2$ and $a_3 = 7$, transposition of these symbols will not be detected.

In practice, the most important errors are transcription errors a → b which comprise about 80% of all errors and transpositions ab → ba which form about 10%. Less common errors are jump transpositions acb → bca, twin errors aa → bb and jump twin errors aca → bcb. In order that all single errors shall be detected, it is necessary that each weight w_i is relatively prime to the modulus m and, in order that all transpositions shall be detected, it is necessary that $w_{i+1}-w_i$ be relatively prime to m for $i = 0, 1, \ldots, n-1$. If m is 10, these two requirements are incompatible because the first requires that each w_i is odd and then $w_{i+1}-w_i$ is even and not relatively prime to 10. This is a special case of a more general result, as we shall show. Ecker and Poch also list the requirements for the remaining three types of error listed above to be detected. These are that $w_{i+2}-w_i$, w_i+w_{i+1}, w_i+w_{i+2} respectively be relatively prime to m.

As Ecker and Poch point out, a particularly effective code for detecting errors of all but one of these types is that used for the International Standard Book Numbers. Each book published by a Western Publishing House is assigned a ten digit number b_9-$b_8b_7b_6$-$b_5b_4b_3b_2b_1$-b_0, where the information digits $b_9, b_8, \ldots, b_1$ belong to the alphabet $\{0, 1, \ldots, 9\}$ while the check digit b_0 belongs to the alphabet $\{0, 1, \ldots, 9, X\}$ and is calculated from the equation

$$10b_9+9b_8+8b_7+\ldots+2b_1+b_0 \equiv 0 \text{ mod } 11.$$

It is easy to check that the conditions for detection of single errors of transcription, simple transpositions, jump transpositions, and jump twin errors are all satisfied by this choice of weights. The majority of adjacent twin errors aa → bb will also be detected, an exception being, for example, a twin error in the digits b_5 and b_6.

We may regard multiplying the i-th digit a_i of a codeword $(a_0 \ a_1 \ \ldots \ a_n)$ by a weight w_i as being a special case of applying a permutation δ_i of the codeword alphabet to that digit. For example, if $m = 10$, the effect of premultiplying the digit a_i by the weight 3 is the same as that of applying the permutation

$$\delta_i = \begin{bmatrix} 0 & 1 & 2 & 3 & 4 & 5 & 6 & 7 & 8 & 9 \\ 0 & 3 & 6 & 9 & 2 & 5 & 8 & 1 & 4 & 7 \end{bmatrix}$$

to it. Also, addition modulo m is a particular example of a group structure imposed on the codeword alphabet. This leads us to look at the following more general situation.

We suppose that the codeword alphabet A forms a group G under an operation (+) and that $\delta_0, \delta_1, \ldots, \delta_n$ are mappings of the set A into itself. Then our check equation becomes $\sum_{i=0}^{n} \delta_i(a_i) = c$, where $c \in A$.

In order that all single errors be detected, it is necessary and sufficient that $\delta_i(a) \neq \delta_i(b)$ for $a, b \in A$. That is, δ_i must be a permutation of A. To be able to detect all transpositions, it is necessary and sufficient that $\delta_i(a) + \delta_{i+1}(b) \neq \delta_i(b) + \delta_{i+1}(a)$. If $\delta_i(a) = x$ and $\delta_i(b) = y$, this requirement can be re-written in the form $x + \delta_{i+1}\delta_i^{-1}(y) \neq y + \delta_{i+1}\delta_i^{-1}(x)$. In the case when the group defined on A is abelian, the requirement is met if and only if $\phi_i(x) = \delta_{i+1}\delta_i^{-1}(x) - x$ is a <u>complete mapping</u> of G: that is, if and only if $\phi_i(x)$ and $x + \phi_i(x)$ are both permutations of G. (See Chapter 2 of this book).

Suppose now that $\theta(x)$ is a complete mapping of G and that $\eta(x) = x + \theta(x)$. Then $\eta(x)$ is a permutation of A and so we may define $\delta_i = \eta^i$ for $i = 0, 1, \ldots, n$, where η^0 is the identity mapping. Then $\phi_i(x) = \eta(x) - x = \theta(x)$ and so ϕ_i is a permutation of A which is a complete mapping of G, as required.

As an illustration of the latter construction, consider the case when $A = \{0, 1, \ldots, 9, X\}$ and the group is Z_{11}, the group formed by carrying out addition in A modulo 11. The identity mapping $\theta(x) = x$ is then a complete mapping and $\eta(x) = 2x$. The check equation for computing the check digit a_0 becomes

$a_0 + 2a_1 + 4a_2 + 8a_3 + 5a_4 + 10a_5 + 9a_6 + 7a_7 + 3a_8 + 6a_9 + a_{10}$
$\qquad + 2a_{11} + 4a_{12} + \ldots \ldots + 2^n a_n \equiv 0 \bmod 11.$

We see that we have obtained the following theorem of Ecker and Poch:

THEOREM 6.1. Let $G = (A,+)$ be an abelian group defined on the codeword alphabet A. Then a necessary and sufficient condition that all single errors and all transposition errors be detected in a set of codewords $(a_0\ a_1\ \ldots\ a_n)$, where a_0 is a check digit determined by an equation of the form $\sum_{i=0}^{n} \delta_i(a_i) = c$, is that G has a complete mapping.

COROLLARY. No decimal code whose check equation uses addition modulo 10 can check all single errors and all transposition errors.

Proof. The abelian group $(Z_{10},+)$ has no complete mapping because it has a unique element of order 2. (See Chapter 2 of the present book.) []

We consider next the possibility of using a non-abelian group G defined on the codeword alphabet A. As is customary, we now use multiplicative notation for the group. If $\delta_0, \delta_1, \ldots, \delta_n$ denote mappings of the set A into itself as before, the check equation for a codeword $(a_0\ a_1\ \ldots\ a_n)$ now becomes $\prod_{i=0}^{n} \delta_i(a_i) = c$, where $c \in A$. In order that all single errors be detected, it is again necessary and sufficient that each δ_i be a permutation of A. In order that all transpositions be detected, it is necessary and sufficient that $\delta_i(a) \cdot \delta_{i+1}(b) \neq \delta_i(b) \delta_{i+1}(a)$ for $i = 0, 1, \ldots, n$. This requirement can be written in the form $x \cdot \delta_{i+1}\delta_i^{-1}(y) \neq y \cdot \delta_{i+1}\delta_i^{-1}(x)$, where $x = \delta_i(a)$ and $y = \delta_i(b)$. If $\psi(x)$ is a permutation of the non-abelian group G such that $x \cdot \psi(y) \neq y \cdot \psi(x)$ for all $x, y \in G$ then the assignment $\delta_i(x) = \psi^i(x)$ will provide a checking scheme with a single check digit a_0 which will detect both all single errors and all transposition errors since then $\delta_{i+1}\delta_i^{-1}(x) = \psi(x)$ for each value of i.

In their survey paper, Ecker and Poch have shown that for a dihedral group of singly-even order such a permutation $\psi(x)$ always exists. They have proved the following theorem:

THEOREM 6.2. Let $D_n = \langle a, b : a^n = b^2 = e,\ ba = a^{-1}b \rangle$ be the dihedral group of order 2n, where $n \geq 3$ is odd. Then the permutation ψ defined by $\psi(a^m) = a^{n-1-m}$ and $\psi(a^m b) = a^m b$ has the property $x \cdot \psi(y) \neq y \cdot \psi(x)$ for all $x \neq y$.

If we represent the element a^m by the integer m, the element $a^m b$ by the integer n+m and the group operation by the symbol (∗), we have

h∗k = (h+k) mod n for 0 ⩽ h,k ⩽ n-1;
h∗k = n+[(h+k) mod n] for 0 ⩽ h ⩽ n-1, n ⩽ k ⩽ 2n-1;
h∗k = n+[(h-k) mod n] for n ⩽ h ⩽ 2n-1, 0 ⩽ k ⩽ n-1;
h∗k = (h-k) mod n for n ⩽ h,k ⩽ 2n-1.

Then, $\psi(m)$ = n-1-m for 0 ⩽ m ⩽ n-1 and $\psi(m)$ = m otherwise.

In particular, when n = 5, we have

$$\psi = \begin{bmatrix} 0 & 1 & 2 & 3 & 4 & 5 & 6 & 7 & 8 & 9 \\ 4 & 3 & 2 & 1 & 0 & 5 & 6 & 7 & 8 & 9 \end{bmatrix}$$

and our code is then the decimal code. Thus, we have obtained a means of checking all single and transposition errors in a decimal code using a single check digit. In fact, permutations more advantageous than ψ have been found such that not only are all single and transposition errors detected but also almost all twin errors, jump transpositions and jump twin errors. For further details, see A.Ecker and G.Poch(1986), page 295.

Finally, in their paper, Ecker and Poch have considered the possiblility of using an arbitrary quasigroup (that is, latin square) for the purpose of computing the check digit. They have noted that, for some codeword alphabet sizes, it is then possible to do without any permutation mapping of the symbols and still be able to detect all single errors and all simple transposition errors. If (Q,∗) denotes the quasigroup defined on the codeword alphabet A, then the check digit a_0 of the codeword (a_0 a_1 ... a_n) is computed directly from the product

$$a_0 = \{..[(a_1 * a_2) * a_3] * a_4 ...\} * a_n.$$

The quasigroup cancellation laws a∗x = a∗y => x = y and x∗a = y∗a => x = y alone ensure that all single errors are detected. (Of course, this is true for groups as well.) To be able to detect simple transposition errors, it is necessary that (a∗b)∗c ≠ (a∗c)∗b unless b = c. If the alphabet A is the set {0,1,...,n-1} and the quasigroup operation (∗) is defined by x∗y = y-kx mod n, where k is a fixed integer relatively prime to n such that k+1 also is relatively prime to n, then (a∗b)∗c = c-k(b-ka) = c-kb+k^2a. Consequently, (a∗b)∗c = (a∗c)∗b would imply c-kb ≡ b-kc mod n: that is, b = c when k+1 is relatively prime to n. If in addition k-1 is relatively prime to n then all jump transpositions as well will be detected.

For example, if $n = 11$, we could choose $k = 4$. There still remains the possibility of introducing a permutation mapping into the check equation as an additional feature (as we did for groups). Ecker and Poch discuss briefly this further possibility in their paper.

Much earlier than the above work of Ecker and Poch, R.Schauffler(1956) discussed using not only one quasigroup but several in the construction of parity check equations of a non-binary code. A short summary of this will be found on page 365 of [DK].

An authentication scheme for messages which uses a quasigroup in a somewhat similar way to that just described has recently been devised by the present authors. [See J.Dénes and A.D.Keedwell(1990d).]

In an authentication scheme, additional digits are appended to each message in such a way that the recipient of the message can check that the message is "authentic": that is, that it has not been altered or replaced by a substitute message by some intercepter in the course of transmission. The additional digits are said to form a signature for the message. They are independent of any arrangements for error-detecting or error-correcting which may subsequently be applied to the authenticated message.

In the scheme proposed by the present authors, the message digits are separated into subsets of equal, or nearly equal, length in a well-defined way (for example, by means of a latin square) and each of these subsets is used to compute an additional signature digit by means of a quasigroup. At the receiving end, a corresponding computation is made (the recipient having a copy of the relevant quasigroup and also of the scheme of computation) and each signature digit is checked. If all signature digits are correct, it is assumed that the message received is authentic. In such a scheme, the same signature digit in a particular position in the signature may arise from several messages. It is then important for the scheme to be effective that, in each particular position in the signature each of the digits of the available alphabet should be "equally likely" to appear: that is, among the totality of possible messages, each digit should occur equally often in that position. The scheme proposed in J.Dénes and A.D.Keedwell(1990d) satisfies this requirement. For further details and for the proof of the last-mentioned fact, the reader is referred to the original paper.

Latin squares have sometimes been used in ciphering. In particular, J.L.Massey, U.Maurer and M.Wang(1988) have recently extended the pioneering work of Shannon on this topic which we referred to at the beginning of this Chapter.

In S.R.Jacobsen(1988), latin squares and geometric k-nets have found a somewhat suprising and unexpected application to the resolution of a problem concerning re-arrangement of the connections in a two-stage multiplication fan-out switching network of a type which had earlier been proposed and introduced by G.W.Richards and F.H.Hwang(1985).

Another surprising and interesting connection between latin squares and a notion related to coding has recently been made by G.L.Mullen(1989). This author has exhibited a linkage between non-singular feedback shift registers, latin squares and permutation polynomials. (Linkages between the latter two concepts are, of course, already well known, see, for example, R.Lidl and H.Niederreiter(1983) for details.)

ADDITIONAL REMARKS

The authors recently came across what they believe to be one of the very earliest articles on error-detecting which makes use of a latin square: namely, W.F.Friedmann and C.J.Mendelsohn(1932). By contrast, a very recent article which provides a connection between MDS codes, discussed in Section 1, and latin rectangles is A.Bonisoli and G.Fiori(199α). These authors show that an (m+1,2)–MDS code of order q exists if and only if there exist q-1 mutually column orthogonal m×q latin rectangles.

An additional and very recent paper on the subject of generalizing the Golomb-Posner codes, discussed in Section 2, is J.Dénes, G.L.Mullen and S.J.Suchower(1989). In this paper, use is made of orthogonal frequency squares which latter topic we consider in detail in Chapter 12 of this book.

With regard to Section 4, we should like to draw attention to some historical remarks in A.D.Keedwell(1976a) concerning the so-called polygonal path construction method mentioned above Theorem 4.2. Also, we wish to point out with regret that the statement and proof of Theorem 4.1 are not correct. In the second paragraph of the proof, it is assumed that, when two latin squares represent the same group (and so satisfy the quadrangle criterion), the fourth members of a particular quadrangle will be the same in both squares. Miss S.Frische has shown by means of the counter-example exhibited below that this assumption is false. The counter-example exhibits two latin squares which are Cayley tables of the cyclic group of order 6 and which differ in only nine places. In her Diplome (Lateinische Quadrate, Vienna, 1988), Miss Frische has shown that Theorem 4.1 fails to be valid when, and only when, n=4 or 6.

```
| 0 1 2 3 4 5     (0) (1) (2) (3) (4) (5)      | 0 1 2 3 4 5     (0) (1) (2) (3) (4) (5)
| 1 2 0 4 5 3     (0 1 2)(3 4 5)               | 1 2 0 4 5 3     (0 1 2)(3 4 5)
| 2 0 1 5 3 4     (0 2 1)(3 5 4)               | 2 0 1 5 3 4     (0 2 1)(3 5 4)
| 3 4 5 1 2 0     (0 3 1 4 2 5)                | 3 4 5 0 1 2     (0 3)(1 4)(2 5)
| 4 5 3 2 0 1     (0 4)(1 5)(2 3)              | 4 5 3 1 2 0     (0 4 2 3 1 5)
| 5 3 4 0 1 2     (0 5 2 4 1 3)                | 5 3 4 2 0 1     (0 5 1 3 2 4)
```
  Generating permutation (0 3 1 4 2 5)   Generating permutation (0 4 2 3 1 5)

On the topic of error-detecting and error-correcting of decimal and alphanumeric codes, discussed in Section 6, there are several papers not mentioned in that Section to which attention should be drawn: namely H.P.Gumm(1985); R.-H.Schulz(1989,199α); A.S.Sethi, V.Ragaraman and P.S.Kenjale(1978); and J.Verhoeff(1969).

CHAPTER 10

LATIN SQUARES AS EXPERIMENTAL DESIGNS (D.A.Preece)

The main theme of this chapter is the use of latin-square designs in quantitative research in agriculture, medicine, industry and elsewhere. Its author, D.A.Preece, has had wide experience of experimental design in agricultural research, mainly as a statistician at Rothamsted Experimental Station, but more recently as a biometrician at East Malling Research Station (now part of the Institute of Horticultural Research). During the greater part of the preparation of this chapter and before taking up his present post, he held a post funded by the Overseas Development Administration. He is a well-known authority on combinatorial aspects of experimental design.

The content of the chapter is explained in detail in its introductory section. It includes a description of the basic concepts involved in the design and analysis of comparative experiments, illustrative examples of different types of practical application, explanation of some of the simpler methods of statistical analysis of the numerical results, and discussion of principles upon which such analysis is based.

(1) Introduction

This Chapter discusses the use of latin squares in quantitative research work in agriculture, forestry, horticulture, psychology, medicine, industry, etc. The principles of their use come from the statistical subject of Experimental Design, which is itself an inseparable component of the wider statistical subject usually known as the Design and Analysis of Experiments. These subjects are mathematical, in that they make heavy demands on mathematical theory and argument, but they also depend greatly on the practical needs and circumstances of experimenters. Practical considerations are very important in determining whether, for example, a latin square is appropriate for specifying the layout or structure of a proposed experiment. A statistician advising an experimenter needs to be alert to both mathematical and practical realities, notable amongst the latter being the various components of the variability inherent in the experimental material, whether that material be land, plants, animals, people or industrial environments.

A single Chapter can outline only parts of all this. The material chosen here starts with details of the basic concepts of the Design and Analysis of Experiments. Then practical examples - some of them of special historical interest - are given of the use of latin squares in experimental work. Next, other "latin" experimental designs are briefly mentioned. Then come sections on the statistical analysis of data from latin square designs, and on the tricky statistical subject of Randomization. The emphasis throughout is on the use of latin squares as designs in their own right; the mathematical use of latin squares to construct other statistically useful designs such as balanced incomplete block designs is not covered, nor are other combinatorial matters. The final Section moves outside the statistical boundaries of the Design and Analysis of Experiments to outline the use of complete latin squares (that is, latin squares each of which is both row-complete and column-complete) as "polycross designs" in plant-breeding research on clones of a self-fertilising crop.

(2) The design and analysis of experiments

The statistical subject of the Design and Analysis of Experiments is concerned primarily with <u>comparative</u> experiments [see F.J.Anscombe(1948); D.R.Cox(1958a), Section 1.1; D.J.Finney(1960), Section 1.2] which compare different <u>treatments</u> or, more precisely, the quantifiable <u>effects</u> of different treatments. In agricultural research there are, for example, field trials for comparing different varieties of barley, and animal-feeding trials for comparing how different diets affect the quality of meat obtained from the animals. In medical research, there are experiments to compare the effects of different drugs. In industry, an experiment might be run to permit comparison of the effects of different acid concentrations used during a manufacturing process.

Many comparative experiments are factorial [see F.Yates(1937); D.R.Cox(1958a), Chapter 6], permitting simultaneous study of different treatment factors (often called just factors), each at different levels (sometimes called factor levels, for clarity). Thus in a field experiment, three varieties of wheat (factor A, with 3 levels) might be grown each at four rates of nitrogen fertiliser (factor B, with 4 levels); in this 3×4 factorial experiment the twelve combinations of variety and nitrogen-rate are the twelve treatment combinations. A factor may be qualitative (for example, different varieties of wheat, as above, or different colours in a psychological experiment) or quantitative (for example, different nitrogen rates, as above, or different exposure times in a photographic experiment).

Confused terminology often arises in discussions of factorial experiments, as "treatment" is sometimes used for "treatment combination", sometimes for "factor level", and sometimes for "factor". A further complication arises because an experiment might, for example, include an extra treatment as well as all combinations of a 2-level factor A and a 3-level factor B; one might then say that there are seven treatments, with a factorial structure on six of them, or that the experiment is a "2×3 factorial with one extra treatment". The present account tries to avoid ambiguity by using the phrase "treatment or treatment combination" to mean "treatment" if there is no factorial structure in the experiment, to mean "treatment combination" if the experiment is factorial, and to mean (for example) one or other of the seven possibilities in the above experiment with factorial structure plus extra treatment.

Every comparative experiment comprises a set of experimental units (often called plots, because of agricultural precedent, and sometimes called runs, because of how some industrial experiments are organised). Different treatments or treatment combinations are applied to different experimental units. Plots may be patches of land, pens of animals, individual animals or plants, individual leaves on growing plants, etc. Commonly, treatments are replicated: that is, there is more than one plot per treatment or treatment combination.

The comparisons between treatments, between factor levels, and between treatment combinations are based on data recorded after the treatments have had a chance to affect the experimental material. These data comprise one or more variates, each consisting of one value for each plot (except that there may be missing values arising from mishaps). Variates may be continuous (for example, weights, heights, temperatures, pressures, times) or discrete (counts, scores); they may be primary (obtained by direct reading of a scale, by counting, or by scoring) or derived (obtained from one or more other variates by arithmetical calculation). In statistical usage, the phrase analysis of an experiment connotes the analysis of a variate or variates; such an analysis may also involve other relevant data.

The design of an experiment is harder to define. D.J.Finney[(1955), Section 1.2 and (1960), Section 1.5] defined it to mean

(i) the set of treatments selected for comparison;

(ii) the specification of the plots;

(iii) the rules for allocating the treatments to the plots;

(iv) the specification of the measurements or other records to be made on each plot.

The need for a statistical subject of Design and Analysis of Experiments arises from variability inherent in the experimental material, in the experimental environment, and in how the experiment is managed. This variability leads to uncontrolled - indeed uncontrollable - variability in the observed variate values: that is, to what experimenters and statisticians call experimental error (where "error" does not mean "mistake"). A variate thus includes, not only any variability from one treatment or treatment combination to another, but also the error variability (a sort of background noise) that would have been present even if a single treatment (or none) had been applied throughout.

In a properly designed comparative experiment, the treatments or treatment combinations are allocated to plots in accordance with the three basic principles (or Three R's) of Fisherian experimental design: Replication, Randomization and Blocking. The three were explicitly formulated by R.A.Fisher, and discussed by him in R.A.Fisher(1925) and later publications.

Replication, as stated above, involves a treatment or treatment combination being applied to more than one plot; the number of plots carrying the treatment or treatment combination is the number of replications or replicates of the treatment or treatment combination. Replication enables estimates of experimental error to be made.

Randomization [see D.R.Cox(1958a), Chapter 5; D.J.Finney(1960), Section 1.6] is a procedure for allocating the treatments or treatment combinations to the plots. The allocation is taken at random from a set of possible allocations, the set being such that the experiment shall provide unbiased estimates of treatment differences and of experimental error. When such a random allocation is used, the treatments or treatment combinations have been randomized. In many experiments, an acceptable randomization gives all the plots an equal chance of receiving any particular treatment or treatment combination. This is the usual requirement where the allocation has the pattern of a latin square, with the treatments or treatment combinations represented by the symbols of the square; a particularly simple randomization procedure consists of taking the n rows of a latin square in random order and the n columns in another independently determined random order, where each random order is taken at random from all n! possible orders (see Section 8 of this Chapter).

Blocking is a grouping of the experimental units into blocks (often all containing the same number of plots) with the aim of reducing the experimental error in the variates of interest. The grouping is done before the treatments or treatment combinations are allocated to the plots. The aim is to capture any large components of the inherent variability as between-blocks variability which can then be separated out in the statistical analysis: plots thought likely to give the most similar results in the absence of treatment effects are, as far as possible, grouped together within blocks. Perhaps the commonest experimental design is the randomized complete block design, in which each block contains exactly one plot for each treatment or treatment combination, the allocation of the t treatments or treatment combinations within each block being taken at random from all the t! possible allocations for that block. However, some experiments, notably on fruit-trees in orchards, take the concept of blocking further, in having a rectangular array of plots arranged in a row-and-column design with two overlaid systems of blocks, namely rows and columns. The rows and columns in the orchard experiments, or in similar agricultural field experiments, are strips of land, with the row strips at right angles to the column strips, the plots being the intersections of strips in different directions; each strip may comprise a set of adjacent lines of trees. The simplest row-and-column design is the n×n latin square design with n rows, n columns and n treatments or treatment combinations, each row and each column being a replicate of the treatments or treatment combinations represented by the symbols in the square.

(3) Some practical examples of latin squares used as row-and-column designs

Latin square designs feature prominently in textbook accounts of the design and analysis of comparative experiments, and are often said to be of particular value in agricultural research. This emphasis has led to their uncritical use, especially by agricultural experimenters, who have sometimes preferred them to randomized complete block designs on the mistaken grounds that "two blocking systems must be better than one!" This over-simple argument overlooks the possibility that compactly arranged arrays of the available plots (perhaps 2×2 or 3×2 arrays) might constitute the most appropriate blocks.

The latin square used as a row-and-column design is nevertheless important in the history of agricultural experimentation, particularly before randomization and the dependence of the statistical analysis on randomization were fully understood. Until then, special interest attached to the 5×5 knight's-move squares in which (a) no diagonally adjacent cells have the same symbols, and (b) all cells with the same symbol can be traversed by a series of knight's moves, as on a chessboard, without visiting cells with any other treatments. Examples in the agricultural research literature were provided by E.Lindhard(1909), who used the term *Springertraek* ("knight's move"), by H.Kryger-Larsen(1913,page 109), by N.A.Hansen(1915,page 510), and by K.Vik(1924), after whom they were named Knut Vik squares even though R.A.Fisher found that they had been known in Denmark since about 1872. The 4×4 <u>balanced chessboard</u> of E.J.Maskell(1925,page 379) is also a latin square with property (b), but necessarily it does not have property (a); another such 4×4 square had been given earlier by N.A.Hansen(1915,page 510).

R.A.Fisher(1925,Section 49,pages 229-232) devoted the last Section of the First Edition of his "Statistical Methods for Research Workers" to the use of the latin square as a row-and-column design, and it was use for field experiments that he mentioned specifically.

The value of row-and-column designs for orchard experiments was mentioned in Section 2 above. However, trees available for an experiment often cannot be grouped into an n×n arrangement of plots, and even if that grouping is possible there may be no way of making the number of rows and columns agree with the number of treatments or treatment combinations. The row-and-column designs for orchard experiments are thus commonly more elaborate than latin square designs. A similar difficulty arises with glasshouse experiments where the plots are pots on a bench: lighting and heating patterns in a glasshouse often suggest the use of a row-and-column design, but the configuration of the bench is unlikely to lend itself to a suitable latin square.

Latin square designs have had a place in forestry research too. Indeed R.A.Fisher is said to have produced a randomized latin square as the design for an experiment in an English forest nursery as early as 1924 [see J.F.Box(1978),page 156]. H.M.Steven(1928,pages 23,31,37,etc.) referred to various English forest nursery experiments in 1925 and 1926 as having had <u>latin square chessboard</u> layouts with 6,5 and 4 replications. Steven gave a 6×6 example which looks very unlikely to have been randomized and which he described on page 23 of his paper as "somewhat regular"; however he stipulated that randomization should be done.

In experimentation on the land, an n×n latin square design with one of the blocking systems used in a non-standard way has been recommended for occasional use when <u>all</u> the n^2 experimental plots are side by side [see R.A.Fisher(1925), Section 49, page 231; W.G.Cochran and G.M.Cox(1957), page 118]. In this specialised application, which could be used if an inherent yield gradient were suspected that ran from one plot to the next, throughout them all, the rows are n blocks of land each consisting of n consecutive plots, whereas the columns specify the order that the treatments have within each block.

Latin square designs have also been used for experimentation on animals, both small and large. For example, C.G.Butler, D.J.Finney and P.Schiele(1943) described an experiment on the response of honey-bees to differing concentrations of lime-sulphur (a component of orchard sprays) in sucrose solution imbibed. The "plots" were individual cells of dry brood comb each filled with one of the eight solutions tested. Eight cells of each solution were put in a chamber where about 100 bees were released for about two hours; a latin square arrangement in an 8×8 layout was used so that, as far as possible, any positional effects within the chamber could be eliminated from the data as row effects and column effects [see D.J.Finney(1952),pages 188-189]. The weights of the bees' uptakes of liquid from each cell did indeed provide indications of positional effects. For this experiment, there are statistical reasons (see Section 7 below) for rejecting an analysis of the weights of liquid in favour of an analysis of the logarithms of the weights. Whether or not this is done, elmination of the row and column effects produces a modest reduction in the calculated experimental error.

Larger animals were involved in a 7×7 latin square design used for a diet experiment on 49 male rats, seven of which came from each of seven litters [see J.A.John and M.H.Quenouille(1977),pages 39-40]. The rows of the square represented litters and the columns represented the order, by weight, of the rats in each litter. Here again, the double blocking led to a modest reduction in the calculated experimental error.

Amongst industrial applications of latin squares used as designs with two blocking systems are examples described by V.R.Main and L.H.C.Tippett(1941), who were concerned with assessing different sizing treatments applied to the warp used in cloth-weaving.

(4) Some other uses of latin squares in experimental design

Sometimes a set of two or more latin squares of the same size is used to make up an experimental design. The simplest example is where two n×n latin squares are combined into a design with n rows and 2n columns; if this is to be used as a row-and-column design without further restriction, the 2n columns will be taken in random order and thus the separate identities of the two component squares will be lost. Similar 2n×n designs that cannot be rearranged into two latin squares are also easy to write down; an example is the 8×4 design in Figure 4.1. This design, like one

```
A B C D A B C D
B C A A D D B C
C D B C B A D A
D A D B C C A B
```

Figure 4.1

made up from two 4×4 latin squares, has each letter once in each column and twice in each row. As an unrestricted row-and-column design, the 8×4 design derives neither more nor less merit from not being rearrangeable as two latin squares.

Both of the above types of 2n×n design provide examples of F-rectangles (see Chapter 12).

Another use of a set of n×n latin squares arises when the individual squares are kept separate and constitute a blocking system of their own. If the rows within one square have a similar role to those within another, and the same is true of the columns, then the design has three blocking systems, namely squares, rows within squares, and columns within squares. This seems to have been the sort of design referred to by T.N.Hoblyn(1930,page 53), when he mentioned a blackcurrant variety trial consisting of six 4×4 latin squares, in each of which each plot had four bushes.

H.W.Norton(1939,page 270) described a latin square as having three constraints, namely rows, columns and symbols. In each of the examples in Section 3 above, two of the constraints, namely rows and columns, were used for systems of blocks (that is, for block factors) and the remaining constraint (namely symbols) was used for treatments. However, sometimes only one constraint is used for a block factor (the other two being used for treatment factors) and sometimes all three are used for treatment factors. An example of the latter, provided by O.L.Davies(1956),pages 166-169, consists of a 7×7 latin square used to study the merits of various methods of preparation of an insecticidal dust: the rows of the square were used for seven methods of mixing the ingredients, the columns for seven forms of the active ingredient, and the letters for seven inert diluents (powders). This design could be used because it "was reasonable to suppose that the three factors would act independently, to a first approximation at least" [O.L.Davies(1956),page 166]. More generally, an n×n latin square used for two or three treatment factors, each with n levels, is appropriate only if there is little or no interaction between those factors: that is, if the relative merits of the different levels of one of the factors are the same, whatever levels of the other factor or factors they are combined with. If there are no interactions between treatment factors, we say that these factors have only main effects, and thus an n×n latin square used for two or three n-level treatment factors can be described as a main-effect plan.

Sometimes it is hard to say whether a given factor is a block factor, a treatment factor, or perhaps a factor of some other sort. Particular conceptual difficulty seems to attach to factors such as sex and breed in livestock experiments: sex or breed cannot be allocated randomly to existing piglets, so Sex and Breed here are unlike other treatment factors, yet sex and breed effects may well have an interest for us that block effects do not, and interactions with sex and breed can of course arise. This makes it hard to categorise the earliest known latin-square experimental design, a 4×4 latin square used by Cretté de Palluel(1788) for an experiment on sixteen sheep: the factors were breed, time of food, and date of slaughter.

Another historically important design that is hard to categorise is the 5×5 latin square design supplied by R.A.Fisher for a forestry experiment laid out on a steep slope in Beddgelert Forest in Wales in 1929 [see R.F.Wood(1974), Plates 12 and 13; J.F.Box(1978),page 156 and Plates 6 and 7]. The five "treatments" represented by the symbols of the square were two different species of conifer and three mixes of species. However, the experiment is recorded as having been designed to study "exposure" as well as species. The rows of the latin square were at different altitudes up the slope, which suggests particular interest in possible differences between rows. The complex local topography and wind-patterns suggest interest in column effects too.

(5) The use of latin squares in experiments with changing treatments

Changes of treatments often occur in orchard experimentation, in animal and medical trials, and in long-term agricultural field experiments. Fruit-trees in research-station orchards often have to be used in successive years for testing different sets of treatments; removal and replacement of a plantation of trees after just a single year's experimentation on them would be too costly. The single year's treatments (for example, different sprays, different methods of managing the surrounding sward) may, of course, leave residual effects detectable in the trees during the following year or even later, but these residual effects may sometimes reasonably be supposed not to interact, or to interact only negligibly, with the effects of new treatments. Then a method of superimposing the allocation of the new treatments on that of the old is required, so that residual and new effects can be disentangled. A simple such superimposition can be determined by a latin square, and this can be illustrated by a trial on the control of Apple Mildew [see T.N.Hoblyn, S.C.Pearce and G.H.Freeman(1954), page 509]. The trees of this trial were in a row 240 feet long, which was divided into 36 plots of equal length. The plots were grouped into 6 blocks, each of 6 adjacent plots. In 1951, six spraying treatments, here represented by capital letters, were applied in a randomized complete block design; in 1952, six different spraying treatments, represented by lower case letters, were applied instead, as follows:

```
DFCABE   BFEADC   AFDEBC   ABFECD   EBDAFC   CADEBF
edabcf   afecdb   abfdec   edcafb   cbafed   edcbfa
```

The allocation here is that of a latin square whose three constraints are used for the factors "Blocks", "1951 treatments" and "1952 treatments".

In dietary experiments on animals such as bacon pigs and dairy cows, and in medical experiments on human subjects, a succession of treatments is often given to each animal or patient, so that within-patient information on treatment differences is available. If an animal experiment is divided up into periods, and the treatments change from period to period for each animal, the design might simply be a row-and-column design, possibly made up from one or more latin squares, with rows for periods (as in the weaving experiment mentioned at the end of Section 3) and columns for animals. However, if the treatments - say different diets - leave residual effects detectable when observations are made for periods after their application, then a special design is needed whose structure will depend on what assumptions can reasonably be made about the residual effects. The simplest situation is where there are only first residuals (that is, when each treatment has a residual effect only in the one period immediately after its application) and the magnitude of these first residuals is independent of the current treatment. Then a member of one of the simpler classes of change-over design may be appropriate. These designs were introduced by E.J.Williams(1949), and include column-complete latin squares (discussed in Chapter 3) and also those pairs of latin squares where, as in the column-complete single square, each treatment is preceded by each other treatment the same number of times within columns. (See Figure 5.1 for an example.)

```
Period
  I      A B C D E      A B C D E
  II     B C D E A      C D E A B
  III    D E A B C      B C D E A
  IV     E A B C D      E A B C D
  V      C D E A B      D E A B C
```

Figure 5.1

E.J.Williams(1949) and (1950) also used latin squares in the construction of designs for dealing simultaneously with current effects, first residuals and second residuals. The use of column-complete latin squares in psychological experimentation was discussed by B.R.Bugelski(1949), whose motivation for using them was similar to that of Williams.

(6) Other "latin" experimental designs

Apart from latin squares, other designs with "latin" in their names are graeco-latin squares (see Chapter 1 for the definition), semi-latin squares, latin rectangles, quasi-latin squares and latin cubes.

Graeco-latin squares, despite their appearance in many statistics textbooks, seem to be infrequently used as experimental designs. Their use in some psychological experimentation was discussed by D.A.Grant(1948). Graeco-latin squares are appropriate for use as experimental designs only if there are no interactions, or merely negligible interactions, between the factors represented by the rows, by the columns, by the Roman letters and by the Greek letters.

A semi-latin square with $n = kp$ symbols is a rectangular arrangement with p rows and n columns, these latter being grouped into p sets each containing k consecutive columns; each symbol occurs just once in each row and just once in each set of columns [see F.Yates(1935)], as in Figure 6.1. The name "semi-latin square" seems to have first been used in print

```
A  B    C  D    E  F    G  H
C  D    A  B    G  H    E  F
E  F    G  H    A  D    C  B
G  H    E  F    C  B    A  D
       (n = 8, p = 4, k = 2)
```

Figure 6.1

by R.A.Scott(1932, pages 3, 7, 11 and 13), but has been attributed to E.J.G.Pitman. The semi-latin square was proposed - and indeed used - as an experimental design with two blocking systems, namely (i) rows and (ii) sets of columns; the individual columns were not taken as the blocks of any blocking system. However, Yates [see F.Yates(1935), Section 6, last paragraph and Discussion] showed that this design, if analysed analogously to a latin square design, is statistically defective. The design has nevertheless reappeared from time to time subsequently [see D.A.Preece and G.H.Freeman(1983)]. O.C.Riddle and G.A.Baker(1944) used the name "modified latin square" for it, and this latter became "cuadro latino modificado" in agronomic work in Mexico. In Russian, the name ЛАТИНСКИЙ КВАДРАТ С РАЗДЕЛЕННЫМИ ЭЛЕМЕНТАМИ (latin square with separated elements) has been used. German authors have called the design a "lateinisches Rechteck", despite the greater generality of the two usual combinatorial definitions of <u>latin rectangle</u>, namely (a) a latin square with some row or rows omitted, and (b) any r×s array using n symbols (r < n, s ≤ n) none of which is repeated in any row or any column.

Latin rectangles consisting of latin squares with rows omitted are of little statistical interest except for those designs where the symbols are to some extent "balanced" with respect to the incomplete columns. Foremost amongst the exceptions are <u>Youden 'squares'</u> (which despite their name, are <u>not</u> square); in these designs the assignment of treatments to columns (ignoring order within them) is the same as that to the blocks of a symmetrical balanced incomplete block design. An example of a Youden 'square' is shown in Figure 6.2. Youden 'squares' arose first as possible

```
A  B  C  D  E  F  G
B  C  D  E  F  G  A
C  D  E  F  G  A  B
E  F  G  A  B  C  D
```

Figure 6.2

designs for greenhouse experiments [see W.J.Youden(1937) and (1940)].

For explanations of <u>quasi-latin square</u> and <u>half-plaid latin square</u> the reader should see pages 35-36 and 78-80 respectively of F.Yates(1937). Statistically, these designs now seem to be of hardly more than historic interest.

<u>Latin cubes</u> and related designs have suffered from a tiresome confusion of definitions, reviewed by D.A.Preece, S.C.Pearce and J.R.Kerr(1973), who proposed use of these designs for experiments in three space dimensions, for example in an incubator or refrigerated fruit store. The use of latin cubes, etc., to produce main-effect plans has had much attention from Soviet authors, notably E.V.Markova; the literature was reviewed in D.A.Preece(1975) and (1979).

(7) Statistical analysis of latin square designs

The standard method for statistical analysis of a variate from an n×n latin square design with a single non-factorial set of n treatments includes an <u>analysis of variance</u> of the form given in Table 1. In the Table, y_{ijk}

Standard analysis of variance for a latin square

Source of variation	Number of degrees of freedom	Sum of squares	Mean square
Rows	n-1	$S_R = \frac{1}{n}\Sigma R_i^2 - \frac{G^2}{n^2}$	$S_R/(n-1)$
Columns	n-1	$S_C = \frac{1}{n}\Sigma C_j^2 - \frac{G^2}{n^2}$	$S_C/(n-1)$
Treatments	n-1	$S_T = \frac{1}{n}\Sigma T_k^2 - \frac{G^2}{n^2}$	$S_T/(n-1)$
Error	(n-1)(n-2)	$S_E = S_{TOTAL} - S_R - S_C - S_T$	$S_E/(n-1)(n-2)$
Total	n^2-1	$S_{TOTAL} = \Sigma y_{ijk}^2 - \frac{G^2}{n^2}$	—

Table 1

denotes the variate value (for example, yield of grain) for the plot in row i and column j, this plot having been assigned treatment k; the suffices i,j and k customarily all take the values 1 to n inclusive. Also, R_i denotes the total of the yields of the n plots in row i; likewise the quantities C_j and T_k are the column and treatment totals, and G is the grand total of all n yields. The Table displays a partitioning of the total sum of squares S_{TOTAL} into four parts, one for each of the four listed sources of variation. The total number of degrees of freedom n^2-1 (which is obtained as 1 less than the total number of plots) is partitioned likewise (the number of degrees of freedom for rows being obtained as 1 less than the number of rows, and so on). Each mean square is calculated as the corresponding sum of squares divided by the corresponding number of degrees of freedom.

A measure of the variability of the variate values, after the row, column and treatment effects have been eliminated from them, is the standard error per plot, calculated - as for other designs - as the square root of the error mean square M_E (= $S_E/(n-1)(n-2)$). The standard error (S.E.) per plot is sometimes expressed as a percentage of the general mean $\bar{y} = G/n^2$, to give the coefficient of variation C.V. = $100\sqrt{M_E}/\bar{y}$. This over-used dimensionless quantity was first introduced (in a different context) by K. Pearson (1896), pages 271 and 276-277.

After an analysis of variance has been completed, the quantities of most interest are the treatment means T_k/n and either the S.E. of a treatment mean

$$S.E.(mean) = \sqrt{M_E/n}$$

or the S.E. of the difference between any two treatment means

$$S.E.D. = \sqrt{2M_E/n}.$$

The S.E. of a mean is often given simply in the form ±h, where h is its numerical value. However, the "±" notation has several other uses than as a label for standard errors, so its use unexplained should be avoided.

If the treatments mean square M_T in the analysis of variance is much larger than the error mean square M_E, there are clear differences in effect between at least some of the treatments. If however the treatments do not differ in their effects, M_T and M_E are both estimates of experimental error; then, so long as some standard statistical assumptions about the data are satisfied, the value of the ratio M_T/M_E (a variance ratio) is a random variable whose statistical distribution is an example of the so-called F-distribution. This distribution thus provides a statistical test - if such a test be desired - of the null hypothesis that the treatments do not differ in their effects. The test is made by comparing the value of M_T/M_E with a percentage point for the F-distribution having (n-1) and (n-1)(n-2) degrees of freedom, these being the numbers of degrees of freedom for the numerator and denominator of the ratio. The most commonly used percentage points are the 5% and 1% points (sometimes called the 95% and 99% points); these values, to be found tabulated in all standard sets of statistical tables, are such that just 5% and just 1% of the F-distribution comprises values greater than them. If M_T/M_E is greater than the tabulated 5% point, the ratio is said to be "significant at the 5% level".

Many statisticians use such "significance" as a criterion for "rejecting the null hypothesis at the 5% level" and for "accepting the alternative hypothesis (that there are indeed differences in effect) at the 5% level". A variance ratio that is significant at the 5% testing level is often marked with a single asterisk, whereas two asterisks are used for significance at the 1% level.

The statistical analysis of Table 1 is appropriate only if there are no interactions between rows, columns and treatments. Indeed, the latin square has insufficient degrees of freedom for the proper identification of any interactions between these three factors. If interaction is suspected, confirmation can sometimes be gained by separating one or more degrees of freedom for non-additivity from the error, and comparing the non-additivity mean square with the new residual mean square. The simplest such procedure involves a form of Tukey's one degree of freedom for non-additivity [see J.W.Tukey(1955)]; elaborations were given by G.A.Milliken and F.A.Graybill(1972). Effects of non-additivities in latin squares were considered by M.B.Wilk and O.Kempthorne(1957), whose work was discussed critically by D.R.Cox(1958b) and by J.A.Nelder(1977, page 62).

The assumptions that the variate values y_{ijk} must satisfy if the analysis of variance in Table 1 is to be valid and appropriate are

(i) the row, column and treatment effects are additive throughout the variate: that is, there are no interactions between the row, column and treatment factors;

(ii) the magnitude of the error variability is the same throughout;

(iii) the error components of the individual variate values have a normal (that is Gaussian) distribution; and

(iv) the error components of any two of the variate values are statistically independent of one another.

Readers must turn to specialised statistical texts for information on what to do if these assumptions are neither satisfied nor nearly satisfied. Only one possibility can be mentioned here, namely that help might come from a transformation of the variate, by which statisticians mean a monotonic, continuous, one-one transformation of the variate. Thus, for example, if the values y_{ijk} do not satisfy the assumptions, then the transformed values $z_{ijk} = \sqrt{y_{ijk}}$ may satisfy or sufficiently nearly satisfy them. This square root transformation and the logarithmic transformation $z_{ijk} = \log y_{ijk}$ are commonly used; use of the latter is mentioned in Section 3 above.

If a set of s latin squares all of size n×n is used as a design with three blocking systems, these being squares, rows within squares, and columns within squares, then an analysis of variance can easily be obtained, with the total number of degrees of freedom partitioned as in Table 2, given on the next page.

To avoid confusion, we should note that the partition in Table 2 has one component fewer than there is in the partition given by P.W.M.John(1971), page 115, Table 6.2, and reproduced here in Table 3. John's version has the extra component denoted "Squares×Treatments", which is for interaction between squares and treatments. There may well be circumstances in which such interaction occurs, but then difficulty in interpreting the results of the experiment must arise - as it does whenever there is interaction between blocks and treatments: reporting the treatment effects averaged over the blocks could be very misleading.

The form of the analysis of variance in Table 1 is appropriate not only for a latin square used as a row-and-column design, but also for a latin square used as a main-effect plan with one blocking system or none (see Section 4 above). However, if a column-complete latin square such as that shown in Figure 7.1 is used as a change-over design for estimation

		SUBJECT					
		1	2	3	4	5	6
	I	A	B	C	D	E	F
	II	B	C	D	E	F	A
PERIOD	III	F	A	B	C	D	E
	IV	C	D	E	F	A	B
	V	E	F	A	B	C	D
	VI	D	E	F	A	B	C

Figure 7.1

Form of the analysis of variance for a set of s
latin squares used as a design with the 3 blocking systems listed

Source of variation	Number of degrees of freedom
Squares	$s-1$
Rows within squares	$s(n-1)$
Columns within squares	$s(n-1)$
Treatments	$n-1$
Error	$(n-1)(sn-s-1)$
Total	sn^2-1

Table 2

Form of the analysis of variance given by P.W.M.John (1971)
for an experiment involving s latin squares

Source of variation	Number of degrees of freedom
Squares	$s-1$
Treatments	$n-1$
Squares×Treatments	$(s-1)(n-1)$
Rows within squares	$s(n-1)$
Columns within squares	$s(n-1)$
Error	$s(n-1)(n-2)$
Total	sn^2-1

Table 3

of current effects of the treatments and first residuals (Section 5 above), then an extra n-1 degrees of freedom must be separated out for the first residuals. The calculation of sums of squares and mean squares is now no longer straightforward, however. The difficulties arise because period I - unlike the others - can provide no information on first residuals, and because no treatment ever follows itself. The effects of the rows and the columns, and the currrent and residual treatment effects, are thus partly entangled; for example, straightforward differences between column totals include differences between residual treatment effects as well as true between-columns differences. In technical statistical language, the entanglement is non-orthogonality between the row effects, the column effects, the current treatment effects and the residual treatment effects. The statistical analysis is therefore non-orthogonal, and a straightforward partition of the total sum of squares is no longer possible, even though there is no difficulty over partitioning the total number of degrees of freedom. Without going into details, we can say that the total sum of squares with n^2-1 degrees of freedom can be partitioned into an appropriate five components in two different ways that are of interest; for these, the components are as follows:

Partition (i)
Periods, ignoring first residuals
Subjects, ignoring first residuals
Current treatment effects, ignoring first residuals
+First residuals, after allowing for periods, subjects and current effects
Error

Partition (ii)
Periods, ignoring first residuals
Subjects, ignoring first residuals
*Current treatment effects, after allowing for first residuals
First residuals, ignoring current effects, but after allowing for periods
    and subjects
Error

The sums of squares for components + and * have elaborate formulae, but corresponding mean squares are obtained merely by dividing the sums of squares by n-1. The mean square for component + can be used to assess whether residual effects are indeed present in the results of the experiment. The mean square for component * can be used to assess current effects when allowance is made for residual effects. Partition (i) is the basis of the illustrative Table 11.5.3 to be found on page 205 of J.A.John and M.H.Quenouille(1977), whose discussion continues to calculation of the sum of squares for *.

If a latin square is used as a row-and-column design for a factorial experiment, then a more detailed analysis of variance table is called for than that in Table 1. Suppose, for example, that a 6×6 latin square is used as a row-and-column design for a 3×2 factorial experiment with factors A (having 3 levels) and B (2 levels), each of the six treatment combinations occurring once per row and once per column. Then the 5 degrees of freedom for the treatment combinations need to be partitioned into 2 for the main effect of factor A (2 being 1 less than the number of levels of A), 1 for the main effect of factor B, and the remaining 2 for interaction between factors A and B. The interaction between A and B is often denoted merely by A×B or indeed by AB; likewise, in an analysis of variance table, the main effects are often written merely as A and B. Thus the form of the analysis of variance for our 6×6 latin square design is as in Table 4. Similar partitioning for A, B and AB is required for most other factorial experimental designs with two treatment factors.

Form of the analysis of variance for a 6×6 latin Square used for a 3×2 factorial experiment

Source of variation	Number of degrees of freedom
Rows	5
Columns	5
A	2
B	1
AB	2
Error	20
Total	35

Table 4

Even if a latin square is used as a row-and-column design for a non-factorial experiment, a partitioning of the degrees of freedom for treatments (see Table 1), and of the sum of squares for treatments, may be useful. Most importantly, special interest may attach to a particular contrast between the treatment means T_k/n. Such a treatment contrast is defined as a linear combination

$$\sum_{k=1}^{n} \lambda_k (T_k/n)$$

where the coefficients λ_k sum to zero. Any such contrast accounts for precisely one degree of freedom from amongst the n-1 degrees of freedom for treatments. The treatments sum of squares with n-1 degrees of freedom can therefore be partitioned into a sum of squares for a chosen contrast and a residual treatments sum of squares with n-2 degrees of freedom. The mean square for the treatment constrast is of course the same as the corresponding sum of squares, as division of the sum of squares by its number of degrees of freedom is just division by 1; its value is

$$(\sum_{k=1}^{n} \lambda_k T_k)^2 / n \sum_{k=1}^{n} \lambda_k^2 \ .$$

The concept of a mean square for a treatment contrast is important in the randomization theory discussed in the next Section of this Chapter.

(8) Randomization of latin square designs

One of the purposes of randomization is to try to ensure that assumptions associated with a proposed method of statistical analysis are satisfied or nearly satisfied. Criteria are therefore needed for assessing proposed randomization procedures.

R.A.Fisher(1925, Sections 48 and 49) and F.Yates(1933) proposed an adequacy criterion for randomization of a latin square used as a row-and-column design, and more generally for randomization of any design with a single error term in the analysis of variance. This so-called weak criterion requires that, in the absence of treatment effects, the expectation of the error mean square shall be equal to that of the treatments mean square (see Table 1), the two expectations being obtained by averaging over all possible outcomes of the randomization. The more stringent <u>strong criterion</u>, foreshadowed by F.Yates(1933) and given more formally by P.M.Grundy and M.J.R.Healy(1950,page 290), requires the error mean square to have the same expectation as the mean square for <u>any</u> treatment contrast in the absence of treatment effects. The strong criterion was generalized by J.A.Nelder(1965a,1965b), who provided detailed supporting theory for it.

S.C.Pearce(1975,pages 73-74) showed that the weak criterion is satisfied when both the rows and columns of any row-and-column design are <u>permuted at random</u>: that is, when the rows are taken in random order and the columns are taken in an independently obtained random order. However, when the n×n latin square is used as a row-and-column design this design is so special that the weak criterion is satisfied if merely n-1 of its rows are permuted at random [see F.Yates(1933)].

Permuting all rows at random, instead of just n-1 of them, satisfies the weak criterion, of course, and is recommended as it increases the number of designs open for selection.

For the strong criterion to be satisfied, the columns or the letters must be permuted at random, in addition to n-1 of the rows. Permuting both letters <u>and</u> columns at random (as well as all the rows) is generally recommended, as this too may increase the number of accessible designs. Random choice of a permutation of the letters is, of course, done in practice by allocating treatments to letters at random.

R.A.Fisher(1935) showed that selecting an n×n latin square at random from a complete set of n-1 mutually orthogonal n×n latin squares satisfies the weak validity criterion; the strong criterion is satisfied if treatments are randomly permuted too.

F.Yates(1933) added that, although appropriate random permutation within the constraints of a single square will satisfy the strong criterion of validity for a latin square used as a row-and-column design, "it would seem theoretically preferable to choose a square at random from all the possible squares of given size". Such a wide choice requires in practice that the latin squares of the size in hand should have been enumerated and representative squares tabulated. For most practical purposes likely to arise in experimental design, Table XV of R.A.Fisher and F.Yates(1963) provides all that is needed.

The randomization of latin designs other than straightforward row-and-column designs has received little attention in the literature. The problem of converting a randomized complete block design with n blocks and n treatments to an n×n latin square (see Section 5 above) was however considered by D.A.Preece, R.A.Bailey and H.D.Patterson(1978). They pointed out that J.A.Nelder's theory can be used to justify the following procedure for allocating the new set of treatments: select any n×n latin square; label its rows and columns with random permutations of the original experiment's block labels and treatment lables respectively; use the letters of the square to determine the allocation of the new treatments.

D.A.Preece, R.A.Bailey and H.D.Patterson(1978) also considered randomization schemes for superimposing one latin square orthogonally on another to form a graeco-latin square design. Here they found themselves in much deeper water; even a summary of the difficulties that they encountered cannot be included in the present book.

Deep waters become stormy seas when we turn to thoughts of randomizing change-over designs. Consider for example the change-over design of Figure 7.1. The design as printed is balanced in the combinatorial sense that each treatment is succeeded by each other within columns, and in a consequential statistical sense. The balance would however almost certainly be lost if the rows were randomized. Of the 5! = 120 latin squares obtained by permuting the last 5 rows of the printed square, only 4 have the balance of that square; these are the latin squares whose initial columns have their elements in the following orders:- A,B,F,C,E,D; A,F,B,E,C,D; A,C,B,E,F,D; and A,E,F,C,B,D. The columns and the letters can of course be randomized, but the randomization theory already quoted does not validate this simple randomization, as there are both the residual and main effects to be estimated. This sort of problem, ignored by statistical textbooks and experimenters alike, has hardly been touched on in the statistical literature, but was considered very briefly - along with other difficulties pertaining to change-over designs - by H.D.Patterson(1971).

(9) Polycross designs

Outside the statistical subject of Design and Analysis of Experiments, but nevertheless of importance in plant-breeding research on a cross-fertilising crop, are polycross designs. These are used for simultaneously flowering clones of the crop, when the aim is to produce offspring from all possible crosses between the clones and to ensure, as far as is possible, that each clone has an equal chance of pollinating, or of being pollinated by, any of the others. C.E.Wright(1962,1965) and K.Olesen and O.J.Olesen(1973) discussed the use of complete latin squares (latin squares each of which is both column-complete and row-complete; see Chapter 3) to determine crop-planting layouts to meet this aim: each symbol of a complete latin square is used for a different clone, and the planting layout - in rows and columns of plots - is then a straight copy of the layout of the latin square. No statistical analysis of any results is envisaged here, so the lack of randomization in the design is irrelevant. K.Olesen(1976) described these complete-latin-square designs as "balanced with respect to nearest neighbours ... in any of the four main directions" but "not balanced with respect to nearest neighbours in the intermediate (that is, diagonal) directions". K.Olesen(1976) went on to give a completely balanced polycross design consisting of a set of n complete n×n latin squares (n+1 being prime) such that, within the squares, there is balance (equal frequency of the other clones as neighbours) in the four intermediate directions too. Olesen's example with n = 4 is given in Figure 9.1, where the clones are represented by digits.

```
1 2 3 4      2 4 1 3      3 1 4 2      4 3 2 1
2 4 1 3      4 3 2 1      1 2 3 4      3 1 4 2
3 1 4 2      1 2 3 4      4 3 2 1      2 4 1 3
4 3 2 1      3 1 4 2      2 4 1 3      1 2 3 4
```

Figure 9.1

If we imagine North to be at the top of the page and East to the right, we can, for example, consider neighbours to the North-East. In each square, no element from the first row or last column has such a neighbour, but every other element has one. Thus, throughout the four squares in the example, 36 elements have North-Eastern neighbours; these 36 consist of 9 occurrences of each of the 4 clones. Each clone has itself as a North-Eastern neighbour three times and has each other clone as a North-Eastern neighbour twice. The same result is true of North-Western, South-Western and South-Eastern neighbours.

CHAPTER 11

LATIN SQUARES AND GEOMETRY (J.Dénes and A.D.Keedwell)

It is well-known that a set of k mutually orthogonal latin squares of order n is equivalent to a (k+2)-net of order n and, in particular, that a set of n-1 such squares (called a complete set) is equivalent to a projective plane of order n. In this chapter, we discuss a number of latin square and quasigroup problems which can be investigated by making use of this relationship.

(1) Complete sets of mutually orthogonal latin squares and projective planes.

For convenience of reference, we first re-establish the equivalence between a set of mutually orthogonal latin squares and a geometric net (cf. Chapter 8 of [DK] and Sections 2,3 of Chapter 5 in the present book).

DEFINITION. A geometric net is a set of objects called "points" together with certain designated subsets called "lines". The lines are partitioned into classes called "parallel classes", such that (a) each point belongs to exactly one line of each parallel class; (b) if ℓ_1 and ℓ_2 are lines of different parallel classes, then ℓ_1 and ℓ_2 have exactly one point in common; (c) there are at least three parallel classes and at least two points on a line.

A net possessing k parallel classes is called a k-net.

If the net is finite, then it is characterized by a parameter n, called the order of the net, such that (i) each line contains exactly n points; (ii) each parallel class consists of exactly n lines; and (iii) the total number of points is n^2.

An (n+1)-net of order n is called an affine plane. An affine plane can be uniquely completed to a <u>projective plane</u> by adjoining one additional "line at infinity" containing n+1 points, each of these additional points being incident with all the lines of one parallel class.

THEOREM 1.1. A geometric (k+2)-net N of order n is equivalent to a set of k mutually orthogonal latin squares of order n.

Proof. We designate the various parallel classes of the (k+2)-net N by capital letters $A, E, B_1, B_2, \ldots, B_k$. (They can also be thought of as pencils of lines with the points of a "line at infinity" as vertices.) Let the lines of these parallel classes be labelled as follows: $a_1, a_2, \ldots, a_n$ are the lines of the class A; $e_1, e_2, \ldots, e_n$ are the lines of the class E; $b_{j1}, b_{j2}, \ldots, b_{jn}$ are the lines of the class B_j. Every point P(h,i) of N can then be identified with a set of k+2 numbers $(h, i, \ell_1, \ell_2, \ldots, \ell_k)$ describing the k+2 lines $e_h, a_i, b_{1\ell_1}, b_{2\ell_2}, \ldots, b_{k\ell_k}$ with which it is incident, one from each of the k+2 parallel classes, and a set of k mutually orthogonal latin squares can be formed in the following way: In the j-th square, put ℓ_j in the (h,i)-th place. Each square is latin since, as h varies with i fixed, so does ℓ_j, and as i varies with h fixed, so does ℓ_j. Each two squares L_p and L_q are orthogonal: for, if not, we would have two lines belonging to distinct parallel classes with more than one point in common.

Conversely, from a given set of k mutually orthogonal latin squares, we may construct a (k+2)-net N. We define a set of n^2 points (h,i), h = 1, 2, ..., n; i = 1, 2, ..., n; where the point (h,i) is to be identified with the (k+2)-tuple of numbers $(h, i, \ell_1, \ell_2, \ldots, \ell_k)$, ℓ_j being the entry in the h-th row and the i-th column of the j-th latin square L_j. We form (k+2)n lines $b_{j\ell}$, j = -1, 0, 1, 2, ..., k; ℓ = 1, 2, ..., n; where $b_{j\ell}$ is the set of all points whose (j+2)-th entry is ℓ and $b_{-1\ell} = e_\ell$, $b_{0\ell} = a_\ell$. Thus, we obtain k+2 sets of parallel lines. (Two lines are <u>parallel</u> if they have no point in common). Also, from the orthogonality of the latin squares, it follows that two lines of distinct parallel classes intersect in one and only one point, so we have a (k+2)-net. []

COROLLARY. A complete set of mutually orthogonal latin squares is equivalent to a projective (or affine) plane of order n.

Because desarguesian projective planes exist of all prime power orders, it is known that complete sets of mutually orthogonal latin squares exist of all such orders. Moreover, it is known that, up to isomorphism, only one complete set exists for each of the orders 2,3,4,5,7 and 8. (See [DK], page 169). It is conjectured that only one complete set exists for each prime order: that is, it is conjectured that the only projective plane of any given prime order p which can exist is the desarguesian plane of that order. Two recent attempts to prove or disprove this conjecture are described in Section 3 below.

Not even a pair of orthogonal latin squares exists of order 6. While, for order 9, four isomorphically distinct complete sets of m.o.l.s. (mutually orthogonal latin squares) are known corresponding to the four known projective planes of that order: namely, the desarguesian plane, the translation plane, its dual, and the Hughes plane (see also [DK], page 280 and Section 6 below.)

For the order 10, numerous pairs of orthogonal latin squares have been constructed, some with special properties as, for example, those of L.Weisner(1963), one of which is the transpose of the other. (These provide a so-called self-orthogonal latin square of order 10, see also Part II of the present book.) However, no triad of m.o.l.s. of order 10 has so far been constructed although one of the present authors [A.D.Keedwell(1980)] found some triads which were close to being orthogonal. The best result to date is the construction by A.Brouwer(1984) of a set of four latin squares of order 10 which fail to be orthogonal only in respect of the fact that all four share a common 2×2 latin subsquare.

Despite this limited amount of success in constructing sets of m.o.l.s. of order 10, a very large amount of effort has been expended in recent years in trying to show existence or non-existence of a projective plane of order 10. Parallel efforts have been made in respect of analysing the properties that a projective plane of order 12 would need to have if it existed and in trying to find a new (fifth) projective plane of order 9. We shall now describe some of this work.

(2) Projective planes of orders 9, 10, 12 and 15.

As long ago as 1957, D.R.Hughes(1957a,1957b) showed that, for a projective plane of order n where n ≡ 2 mod 4 and n>2, the collineation group is of odd order. He also showed that for a plane of order 10 the only primes dividing the order of the collineation group of the plane could be 3, 5 or 11 and that a collineation of order 3 would have 3 or 9 fixed points while a collineation of order 5 would have exactly one fixed point and one fixed line. In (1976), Sue Whitesides showed that a collineation of order 11 would be impossible and, in (1979a,1979b), she further showed that collineation groups of orders 9, 25 or 15 would be impossible. This work left only 1, 3 or 5 as possible orders for the collineation group of a projective plane of order 10. The possibility of order 5 was eliminated by R.P.Anstee, M.Hall and J.G.Thompson(1980) using an ingenious argument involving coding theory and the possibility of order 3 was eliminated by Z.Janko and T.van Trung(1981a) using a mixture of algebraic argument and computer elimination of specific cases. This leaves us with the conclusion that a projective plane of order 10, if it exists, is wholly irregular in that it can have no collineation other than the identity mapping. (See also the Addendum to this Chapter.)

Subsidiary to this main result are a number of other properties most of which were obtained in the early 1970's. L.Baumert and M.Hall(1973) showed that no projective plane of order 10 can be co-ordinatized by a double loop with the property that its multiplication loop is a group. A.Bruen and J.C.Fisher(1973) showed that a blocking set in a projective plane of order 10 would have to comprise at least 16 points. (A blocking set S of a projective plane π is a set of points such that each line of π contains at least one point which is in S and one point which is not in S.) They also showed that a net of order 10 which contains 6 or more parallel classes can be completed to a plane of order 10 in at most one way. (This is an improvement on Theorem 3.1 of R.H.Bruck(1963) for the case n = 10.) F.J.MacWilliams, N.J.A.Sloane and W.Thompson(1973) showed that, if a projective plane of order 10 exists and C denotes the (111,56) binary error-correcting code generated by the rows of its incidence matrix then the number of codewords of weight 15 in C must be zero. These authors themselves gave a lengthy direct proof of the result but they also showed in the course of their argument that a codeword of weight 15 would give rise to a blocking set of 15 points in the plane. The Bruen-Fisher result to the effect that no such blocking set can exist permits a much shorter proof of the MacWilliams et al result. A further investigation of the code C and its relation to the incidence matrix of a plane of order 10 was made by M.Hall(1980).

A recent result due to S.S.Sane(1985) states that if a projective plane of order ten were to exist then it would be pointwise extendible to a $3-(n^2+n+2,n+2,1)$ design if and only if a quasi-symmetric $2-(111,12,10)$ design exists.

A somewhat similar investigation of the properties which a projective plane of order 12 would have to possess if it existed has recently been carried out. In particular, in a long series of papers published from 1980 onwards, Z.Janko and T.van Trung have obtained the following properties of the collineation group of such a plane:

(a) Let π be a projective plane of order 12 which has a subplane π_0 of order 3 and suppose that σ is an automorphism of π_0 which has order 13. Then σ cannot be extended to an automorphism of π. [See(1980a,1981d).]

(b) A projective plane π of order 12 cannot have any collineation of order 5, 11 or 13. [See(1982a,1982b).]

(c) The full collineation group of a projective plane π of order 12 is a $\{2,3\}$-group. [See(1982b).] The argument for this important result goes as follows. Suppose that σ is an automorphism of π which has prime order p and suppose that, if possible, p>13. If σ has no fixed points then p divides 157 (=12^2+12+1) which is the number of points in the plane. But 157 is a prime, so p = 157. This implies that σ is a cyclic collineation of Singer type. However, it is known that such a collineation cannot exist in a projective plane whose order is divisible both by 2 and by 3 [D.R.Hughes and F.Piper(1973), page 266]. Suppose next that σ has exactly one fixed point. Then p divides 156 = 4.3.13 which contradicts p>13. Consequently σ, if it exists, has at least two fixed points A and B. If ℓ is the join AB then AB has 13 points, so p>13 implies that σ fixes AB pointwise. Thus, σ must be a perspectivity. If σ is a homology with vertex V, then σ permutes the eleven non-fixed points of any line through V so p = 11. If σ is an elation with vertex V then σ permutes the twelve non-fixed points of any line through V (except ℓ) so p divides 12. Since this contradicts p>13 in both cases, we deduce that any automorphism σ of π of prime order p has p≤13. By (b) above, p ≠ 5, 11 or 13. If p = 7 then σ has at least three fixed points: for each point which is not fixed is permuted in an orbit of length which is a multiple of 7 and the total number of points in the plane is 157 = (7×22)+3. Let A,B,C be three points which are left fixed by σ. σ is not a perspectivity (otherwise p = 11 or p divides 12, as already shown) so if ℓ = AB, σ must permute a set of 7 points on ℓ and leave the remaining 6 points, say A,B,D,E,F,G fixed. But, if σ fixes 6 points on ℓ, it fixes at least 10 points all together (because 157 = (7×21)+10) and so it fixes at least 4 points not on ℓ. Let H be one such point. The fixed line HD has fixed points distinct from H and D (since it has at least 6 fixed points). Let K be one of these points. Then the proper quadrangle ABHK generates a subplane π_0 which is fixed by σ and which contains at least the 8 fixed points A,B,D,E,F,G,H,K. Thus, the order of this subplane must be at least 3. Since the order m of any subplane of π satisfies $m^2 \leq 12$ (by a theorem due to R.H.Bruck(1955), see page 454 of [DK]), the order of the subplane must be 3 and it must be fixed pointwise. It follows that there are 157-13 = 144 points of π which are not fixed by σ and, because this number is not divisible by 7, we have a contradiction. We are left with the possibilities that p = 2 or 3. This completes Janko and van Trang's proof that the full collineation group is a {2,3}-group

In a subsequent paper, Z.Janko(1984), that author has shown that the {2,3}-group cannot have a subgroup of order 27.

(d) A projective plane π of order 12 cannot have a non-abelian group of order 6 as a collineation group nor can it possess an elation of order 3 or 6. [See(1981b,1981c).]

(e) A projective plane π of order 12 cannot possess a collineation group of order 4 consisting of elations with a fixed centre and axis. Moreover, the collineation group G of π cannot contain a Klein four-group and so any Sylow 2-subgroup of G is either cyclic or is a generalized quaternion group. [See (1982c, 1982d).]

The fact that a collineation group of a projective plane of order 12 cannot be of order 27 [see (c) above and Z.Janko(1984)] has also been obtained independently by J.M.Nowlin Brown(1983b) (without the use of a computer) who obtained it as a corollary to some more general results. In the first place, Nowlin Brown showed that if a projective plane of order pt with a collineation group of order p^x exists, where p is odd, p does not divide t and t≤p+1, then x≤3. Next, she showed that if p is odd and a projective plane of order pt with a collineation group of order p^3 exists then 1<t<p is impossible. If t = p+1, then a number of conditions must be met and these are sufficient to show that no plane of order 3(3+1) = 12 with a collineation group of order 3^3 can exist. See also J.M.Nowlin Brown(1983a) for an earlier somewhat weaker result.

In a much older paper(1973) of L.Baumert and M.Hall, it was shown that there is no projective plane of order 12 which possesses a collineation group of order 12 consisting of elations with a fixed point as centre and a fixed line as axis.

We mention two further results on existence of projective planes of order 12. In R.H.Bruck(1973) the author conjectured that, for a prime power q, the three dimensional geometry PG(3,q) of points and lines might be extended to a projective plane of order q(q+1) by adding more points and lines but retaining the point-line incidences of PG(3,q). However, M.Hall and R.Roth(1984) have shown that in the case q = 3 such an extension (to a plane of order 12) is impossible. A projective plane is <u>singly-generated</u> if, by successive free extensions from one of its quadrangles (which may include some forced coincidences of points or lines at intermediate stages), the entire plane can be generated. R.B.Killgrove and Ed.Milne(1974) have shown that any projective plane of order 12 which may exist is necessarily singly-generated. (By contrast, a desarguesian projective plane is singly-generated if and only if it is of prime order since in such a plane every quadrangle generates a subplane of order p.)

As we mentioned at the beginning of this section, much work in recent years has been done on trying to determine whether any projective planes of order 9 distinct from the four known ones exist. Almost all this work has to a larger or smaller extent made use of computers. In (1976), R.B.Killgrove, E.T.Parker and Ed.Milne proved that if a plane of order 9 has any subplane of order 3, then it is one of the four known planes. A year later, R.B.Killgrove and E.T.Parker(1977) showed that if a projective plane of order 9 distinct from the four known ones does exist then it does not have a multiplication loop isomorphic to any one of the 21 loops which occur as multiplication loops in the various possible co-ordinatizing systems which can be used for the four known planes. These authors (1980) also showed that such a plane does not have a collineation of order 13. More recently, Z.Janko and T.van Trung(1981e,1982e) have shown that, among the planes of order 9, only the four known ones can contain non-trivial perspectivities or collineations of order 2 while S.H.Whitesides(1985) has proved that, of the planes of order 9 known and unknown, only the desarguesian plane has a collineation of order 7 with no fixed points. R.Shull(1984) has made use of both these results in showing that the order of the full collineation group of any projective plane of order 9 other than one of the four known ones is either a power of three or else five times a power of three. The latter possibility is ruled out by a later result in R.Shull(1985) to the effect that, if a projective plane of order 9 has a collineation of order 5, then it must be either the desarguesian plane or the known translation plane (Hall plane) or its dual.

A detailed description of the structure and properties of the four known projective planes of order 9 is in the book of T.G.Room and P.B.Kirkpatrick(1971). (See also Section 6 of this chapter). A further interesting result due to R.B.Killgrove and D.I.Kiel(1980) is that the Hughes plane (self-dual plane) of order 9 contains the fifth free extension π_5 of a quadrangle as a configuration. The possibility that the same configuration may also exist in the other two known non-desarguesian planes of order 9 is not ruled out.

The next order after 12 for which it is not known whether existence of a projective plane (that is, a complete set of mutually orthogonal latin squares) is possible is 15. Here, the best result to date is that there exist sets of four mutually orthogonal latin squares. This result was obtained by P.J.Schellenberg, G.H.T.Van Rees and S.A.Vanstone(1978). (For the order 12, sets of five mutually orthogonal latin squares are known to exist. See Chapter 7 of [DK].) As regards the collineation group of a projective plane of order 15 (should one exist), it is known that the order of the full collineation group cannot be divisible by any prime number except possibly 2,3,5 or 7. [See V.Cigić(1983,1984).]

Some further investigations of the collineation groups of projective planes of small orders will be found in V.Cigić(1985), Z.Janko and V.Cigić(1985), C.Y.Ho(1986a,1986b), as well as those to be mentioned in the next Section of this Chapter.

(3) Non-desarguesian projective planes of prime order.

Attempts to construct non-desarguesian planes of prime order have been made in two recent papers: namely, in N.S.Mendelsohn and B.Wolk(1985) and in A.B.Evans and R.L.McFarland(1984). Both papers employ the same general idea: that of trying to construct a complete set of mutually orthogonal latin squares using orthogonal mappings of the cyclic group C_p of order p.

Let $x \to xM_i$ be a permutation mapping of the cyclic group C_p written additively. M_i is called an <u>orthogonal mapping</u> or <u>orthomorphism</u> of the group C_p if the elements $x - xM_i$, $x \in C_p$, are all distinct: that is, if the mapping $x \to x - xM_i$ is again a permutation of the elements of C_p. When this is the case, the latin squares L_0 and L_i whose (x,y)-th entries are x-y and $x - yM_i$ respectively are orthogonal. If M_i and M_j are two orthogonal mappings and if, in addition, the elements $xM_i - xM_j$, $x \in C_p$, are all distinct, then the latin squares L_0, L_i, L_j form a mutually orthogonal triple. When this occurs, we shall call the pair of orthogonal mappings <u>special</u>. Consequently, the existence of p-2 orthogonal mappings $M_1, M_2, \ldots, M_{p-2}$, each pair of which is special, guarantees the existence of a complete set of m.o.l.s. and so also of a projective plane of order p. (This same idea was used much earlier to construct sets of five m.o.l.s. of order 12. See D.M.Johnson, A.G.Dulmage and N.S.Mendelsohn(1961) and R.C.Bose,

J.M.Chakravarti and D.E.Knuth(1960,1961) or pages 232 and 233 of [DK].) If at least one of the orthogonal mappings M_i is different from a multiplication mapping of the form $xM_i = xi \bmod p$, then the plane constructed will be non-desarguesian.

Using these ideas, N.S.Mendelsohn and B.Wolk(1985) obtained, for $p = 13$, sets of five pairwise special orthogonal mappings which include at least one non-multiplication mapping and they showed by a computer search that no larger set fulfilling the latter requirement exists. For $p = 17$, they obtained sets of seven pairwise special orthogonal mappings which include at least one non-multiplication mapping but, for this value of p, their investigations were nowhere near being exhaustive.

Using the same ideas in a different way (which included a graph theoretical interpretation of the requirements), A.B.Evans and R.L.McFarland(1984) re-established that a non-desarguesian projective plane of order 11 which admits a translation group among its collineations does not exist. [This same result can be deduced from the work of D.M.Johnson, A.G.Dulmage and N.S.Mendelsohn described in their paper(1961).]

Two further papers which provide information on the collineation group of projective planes of order 11 are C.Y.Ho and G.E.Moorhouse(1985) and I.Matulić-Bedenić(1986), wherein it is shown that the plane will be desarguesian if its collineation group contains the alternating group A_4 of order 12 or if it contains a homology of order 5.

The general question of existence of non-desarguesian projective planes of prime order remains completely open. However, a novel new approach to the problem has been presented in C.Y.Ho(1986a) and the reader's attention is also drawn to C.Y.Ho and A.Gonclaves(1986).

(4) Digraph complete sets of latin squares and incidence matrices.

Every finite projective plane of order n can be described by means of an $(n^2+n+1) \times (n^2+n+1)$ incidence matrix obtained as follows. Number the points and lines arbitrarily from 1 to n^2+n+1 and form a matrix in which the rows are identified with the points and the columns are identified with the lines. The entry in the (i,j)-th position is either 1 or 0 according as the point P_i is incident with the line ℓ_j or not.

Clearly, such an incidence matrix must be equivalent to a complete set of n-1 m.o.l.s. of order n. L.J.Paige and C.Wexler investigated this relationship in (1953). They derived a canonical form for the incidence matrix N and suggested a means of constructing a complete set of m.o.l.s. from it by first obtaining what they called a digraph complete set of latin squares as an intermediate construct. A set $D_1, D_2, \ldots, D_{n-1}$ of latin squares of order n is digraph complete if, for any given pair of columns, say the r-th and s-th, s≠r, the n(n-1) number pairs (h,k) obtained by picking out the entries which occur in these columns for each row of each of the n-1 latin squares in turn are all distinct. The squares of a digraph complete set need not be orthogonal.

More recently, S.Bourn(1983) has shown that, by modifying the Paige-Wexler canonical form of the incidence matrix to what he calls the ordered canonical form it is possible to go directly from the matrix to a complete set of m.o.l.s. or vice versa. He has also shown that there is a one-to-one correspondence between ordered canonical form incidence matrices of a given projective plane and planar ternary rings co-ordinatizing that plane. We shall now explain these relationships. (An account of the earlier work of Paige and Wexler will be found in Section 8.5 of [DK]. There, rows of the incidence matrix represent lines and columns represent points).

We note that the incidence matrix N of a projective plane cannot contain a quadruple of cells (h,j), (i,j), (h,k), (i,k) all of which contain 1's (we call this a sign-rectangle) otherwise there would be two lines incident with the same two points. Remembering this fact, we first describe how to obtain the Paige-Wexler canonical form.

We partition the last n^2+n rows of N into n+1 "bands" comprising n rows each. We label these n+1 bands by the symbols $\infty, 0, 1, \ldots, n-1$ and we label the rows within each band by the integers $0, 1, \ldots, n-1$. Then ⟨s,t⟩ denotes the t-th row of the s-th band: that is, row (s+1)n+(t+2) of the matrix N. We likewise partition the last n^2+n columns of N into n+1 bands and use a similar labelling system for these columns. The intersection of the ℓ-th row band with the m-th column band is an n×n submatrix of N which we denote by $C_{\ell m}$.

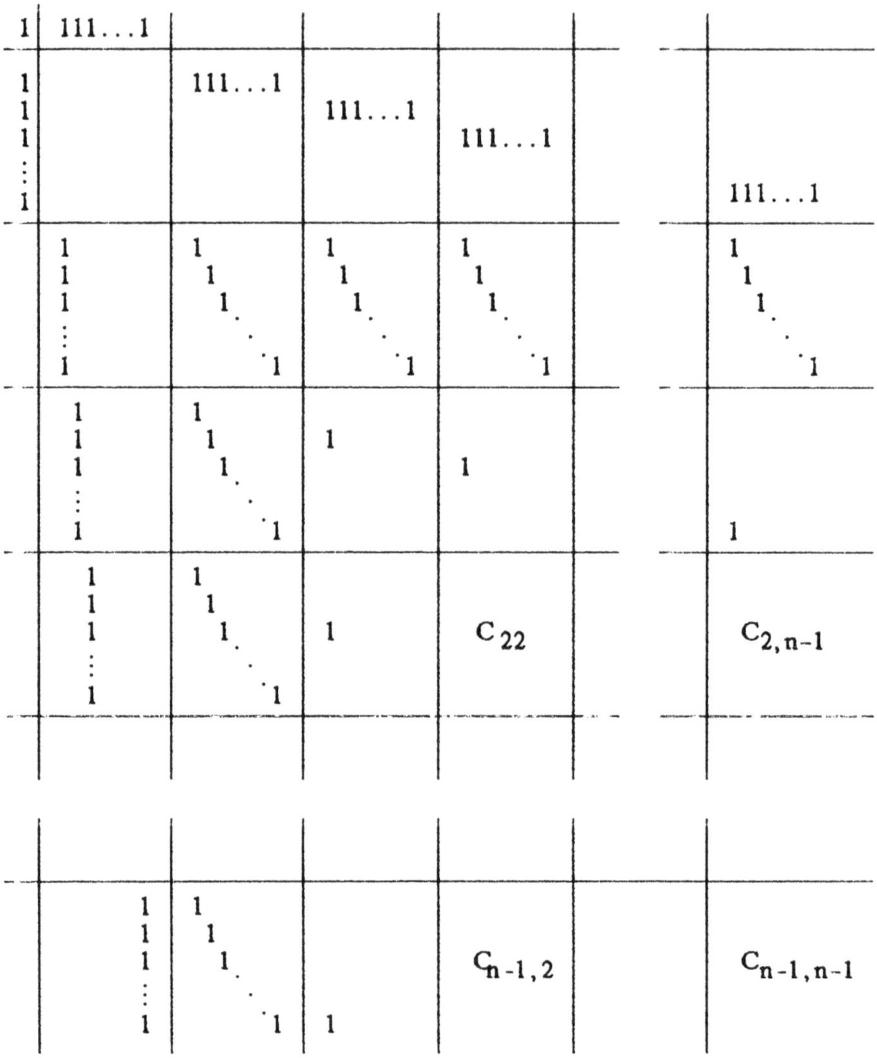

Figure 4.1

The following sequence of row permutations and column permutations is now applied to the matrix N so as to get the first 3n+1 rows and 2n+2 columns into the form shown in Figure 4.1. Because there cannot be any sign-rectangles, the submatrices $C_{\ell m}$ are then all permutation matrices and, in particular, those for which $\ell = 0$ or $m = 0$ are n×n identity matrices.

(1) Put the first row of N into the required form by appropriate column interchanges. Put the first column of N into the required form by appropriate row interchanges which do not affect the first row.

(2) Obtain the required pattern in the n rows of the ∞-th row band by interchanging columns and then the required pattern in the n columns of the ∞-th column band by interchanging rows. When the latter interchanges have been completed, the submatrices $C_{\ell m}$ will all be permutation matrices.

(3) In order to transform C_{om} into the identity matrix, make suitable interchanges of the columns of the m-th column band, $m = 0,1,2,\ldots,n-1$. Finally, in order to transform $C_{\ell 0}$ into the identity matrix, make suitable interchanges of the rows of the ℓ-th row band, $\ell = 1,2,\ldots,n-1$. The result is the Paige-Wexler canonical form.

In order to change this canonical form to the Bourn ordered canonical form, we carry out the following further step:

(4) Put the row band labelled 1 into the form shown in Figure 4.1 by rearranging the last n-1 column bands. (This is possible in view of the fact that the Paige-Wexler canonical form contains no sign-rectangles and therefore none of the submatrices $C_{11}, C_{12}, \ldots, C_{1,n-1}$ can have a one in its top left corner.) Then rearrange the rows of the ∞-th row band so as to restore its required form. Next, put the column band labelled 1 into the form shown in Figure 4.1 by rearranging the last n-2 row bands. Finally, rearrange the columns of the ∞-th column band so as to restore its required form.

DEFINITION. A planar ternary ring with ternary operation T is said to <u>correspond naturally</u> to an incidence matrix N in ordered canonical form if and only if the relation $T(x,m,c) = y$ implies that there is a 1 in the cell of the ⟨x,y⟩-th row and ⟨m,c⟩-th column of N.

Using this definition, Bourn has proved:

		∞			0			1			2		
	1	1	1	1	0	0	0	0	0	0	0	0	0
	1				1	1	1						
∞	1							1	1	1			
	1										1	1	1
	0	1			1			1			1		
0	0	1				1			1			1	
	0	1					1			1			1
	0		1		1					*1*		*1*	
1	0		1			1		1					*1*
	0		1				1		*1*		1		
	0			1	1				**1**				**1**
2	0			1		1				**1**	**1**		
	0			1			1	1				**1**	

$$D_1 = \begin{vmatrix} 0 & \mathit{2} & \mathit{1} \\ 1 & 0 & \mathit{2} \\ 2 & \mathit{1} & 0 \end{vmatrix} \quad D_2 = \begin{vmatrix} 0 & \mathbf{1} & \mathbf{2} \\ 1 & \mathbf{2} & \mathbf{0} \\ 2 & 0 & \mathbf{1} \end{vmatrix} \quad L_1 = \begin{vmatrix} 0 & 1 & 2 \\ \mathit{2} & 0 & \mathit{1} \\ \mathbf{1} & \mathbf{2} & 0 \end{vmatrix} \quad L_2 = \begin{vmatrix} 0 & 1 & 2 \\ \mathit{1} & \mathit{2} & 0 \\ \mathbf{2} & \mathbf{0} & \mathbf{1} \end{vmatrix}$$

Figure 4.2

Latin squares and geometry

THEOREM 4.1. There is a one-to-one natural correspondence between the planar ternary rings and the ordered canonical form incidence matrices associated with any finite projective plane.

Bourn has also shown that a complete set of m.o.l.s. of order n may be constructed directly from an $(n^2+n+1) \times (n^2+n+1)$ incidence matrix N in ordered canonical form as follows:

If there is a 1 in the cell of the $\langle x,y \rangle$-th row and $\langle m,c \rangle$-th column of N, then the entry in row x and column y of the latin square L_m is c. Here, x, y and c are restricted to the symbol set $\{0,1,2,\ldots,n-1\}$ and m to the symbol set $\{1,2,\ldots,n-1\}$.

The fact that this construction gives a complete set of m.o.l.s. follows from Theorem 4.1 and the fact that each line of the associated projective plane (except ℓ_∞) has an equation of the form $y = T(x,m,c)$ or $x = k$. Alternatively, we may make a direct verification from the properties of the incidence matrix.

A digraph complete set of m.o.l.s. of order n may also be constructed directly from N. For these squares, the entry in row y and column m of the latin square D_x is c if and only if there is a 1 in the cell of the $\langle x,y \rangle$-th row and $\langle m,c \rangle$-th column of N, where m, y and c are restricted to the symbol set $\{0,1,2,\ldots,n-1\}$ and x to the symbol set $\{1,2,\ldots,n-1\}$.

Finally, Bourn has pointed out that, when a complete set of m.o.l.s. is constructed by the method described above, the canonical form obtained is not standard form (with first row and column of the latin square L_1 in natural order) but is the so-called <u>normal form</u> obtained by G.E.Martin(1968).

The complete construction of N and the two associated sets of latin squares is illustrated in Figure 4.2 for the case n = 3.

(5) <u>Complete sets of column orthogonal latin squares and affine planes</u>.

K. Vedder (1983) has defined two latin squares $A = [a_{ij}]$ and $B = [b_{ij}]$ of the same order n to be <u>column orthogonal</u> if, for each fixed pair of integers j,k, we have $a_{ij} = b_{ik}$ for at most one value of i. He has given the following example of a set of four mutually column-orthogonal latin squares of order 3 defined on the symbol set $\{1,2,3,4\}$:

$$A_1 = \begin{vmatrix} 2 & 3 & 4 \\ 3 & 4 & 2 \\ 4 & 2 & 3 \end{vmatrix} \quad A_2 = \begin{vmatrix} 1 & 3 & 4 \\ 4 & 1 & 3 \\ 3 & 4 & 1 \end{vmatrix} \quad A_3 = \begin{vmatrix} 1 & 2 & 4 \\ 2 & 4 & 1 \\ 4 & 1 & 2 \end{vmatrix} \quad A_4 = \begin{vmatrix} 1 & 2 & 3 \\ 3 & 1 & 2 \\ 2 & 3 & 1 \end{vmatrix}$$

Such a set of n squares of order n-1 based on a set of n symbols and with one of the symbols omitted from each square of the set, he has called a <u>complete set of mutually column orthogonal latin squares of type n-1</u>. He has proved the following theorem. (Compare Theorem 1.1 of this Chapter and its corollary.)

THEOREM 5.1. A complete set of mutually column orthogonal latin squares of type n-1 is equivalent to a projective (or affine) plane of order n.

Proof. We designate the various parallel classes of the affine plane by the capital letters $A, E, B_1, B_2, \ldots, B_{n-1}$ as in Theorem 1.1 of this Chapter. Let the lines of these parallel classes be labelled as follows: $a_1, a_2, \ldots, a_n$ are the lines of the class A; $e_0, e_1, \ldots, e_{n-1}$ are the lines of the class E; $b_{j1}, b_{j2}, \ldots, b_{jn}$ are the lines of the class B_j. Every point $P(h,i)$ of the plane can then be identified with a set of n+1 numbers $(h, i, \ell_1, \ell_2, \ldots, \ell_{n-1})$ describing the n+1 lines $e_h, a_i, b_{1\ell_1}, b_{2\ell_2}, \ldots, b_{n-1\ell_{n-1}}$ with which it is incident, one from each of the n+1 parallel classes.

We construct a set of column orthogonal latin squares $A_1, A_2, \ldots, A_n$ in the following way. The columns of the i-th square A_i are defined by the n lines through the point $(0,i)$. Let $b_{j\ell}$ be one of these lines and suppose that it contains the points $(1, i_1), (2, i_2), \ldots, (n-1, i_{n-1})$. Then the j-th column of the square A_i is the column vector $(i_1, i_2, \ldots, i_{n-1})^T$.

Since each of the n points on the line $b_{j\ell}$ is incident with a different line of the parallel class A, the set $\{i, i_1, i_2, \ldots, i_{n-1}\}$ is the set of natural numbers $\{1, 2, \ldots, n\}$. Also, since two lines $b_{j\ell}$ and b_{km} of distinct parallel classes have exactly one point in common which is the point $(0, i)$ if and only if $b_{j\ell}$ and b_{km} define two columns of the same square A_i, each square is latin and columns of distinct squares agree in at most one row. That is, the squares form a complete set of mutually column orthogonal latin squares of type n-1, as required.

From the method of construction, it is clear that if j≠k then the j-th column of square A_i agrees in exactly one place with the k-th column of each other square, while on the other hand the j-th columns of each pair of the squares do not agree in any place. A complete set of mutually column orthogonal latin squares which has this property will be called normalized. It is easy to see that any complete set can be so normalized and then, by reversing the construction described above, we complete the proof of our theorem. □

Vedder has shown that, from a normalized complete set of mutually column orthogonal latin squares of type n-1, we can construct a set of n-1 mutually column orthogonal n×n latin squares and conversely. He has called a set of the latter type a complete set of mutually column orthogonal latin squares of type n.

Let $A_1, A_2, \ldots, A_n$ be a normalized complete set of mutually column orthogonal latin squares of type n-1. We form an n×n latin square B_j (j = 1, 2, ..., n-1) by taking as the columns of B_j, the j-th columns of $A_1, A_2, \ldots, A_n$ respectively each headed by its missing symbol. The fact that these missing symbols are all different follows from the construction of Theorem 5.1. The fact that each of the squares B_j is latin follows from the fact that no two j-th columns agree in any place. Hence, the squares $B_1, B_2, \ldots, B_{n-1}$ form a complete set of mutually column orthogonal latin squares of type n. This construction is illustrated for the case n = 4 in Figure 5.1.

In general, complete sets of mutually orthogonal latin squares need not be mutually column orthogonal (the complete set of m.o.l.s. which

$$A_1 = \begin{bmatrix} 2 & 3 & 4 \\ 3 & 4 & 2 \\ 4 & 2 & 3 \end{bmatrix} \quad A_2 = \begin{bmatrix} 1 & 4 & 3 \\ 4 & 3 & 1 \\ 3 & 1 & 4 \end{bmatrix} \quad A_3 = \begin{bmatrix} 4 & 1 & 2 \\ 1 & 2 & 4 \\ 2 & 4 & 1 \end{bmatrix} \quad A_4 = \begin{bmatrix} 3 & 2 & 1 \\ 2 & 1 & 3 \\ 1 & 3 & 2 \end{bmatrix}$$

$$B_1 = \begin{bmatrix} 1 & 2 & 3 & 4 \\ 2 & 1 & 4 & 3 \\ 3 & 4 & 1 & 2 \\ 4 & 3 & 2 & 1 \end{bmatrix} \quad B_2 = \begin{bmatrix} 1 & 2 & 3 & 4 \\ 3 & 4 & 1 & 2 \\ 4 & 3 & 2 & 1 \\ 2 & 1 & 4 & 3 \end{bmatrix} \quad B_3 = \begin{bmatrix} 1 & 2 & 3 & 4 \\ 4 & 3 & 2 & 1 \\ 2 & 1 & 4 & 3 \\ 3 & 4 & 1 & 2 \end{bmatrix}$$

Figure 5.1

represents the Hughes Plane of order 9 given in the next section is a counterexample). However, Vedder has shown that orthogonal latin squares which are constructed by the automorphism method of H.B.Mann (see H.B.Mann(1942) or page 234 of [DK]) are necessarily column orthogonal also. In particular:

THEOREM 5.2. A complete set of m.o.l.s. based on the addition group of a nearfield and constructed by the automorphism method (using multiplications) is a complete set of mutually column orthogonal latin squares of type n.

(6) The Paige-Wexler latin squares.

Throughout this Chapter, we have made use of the fact that a complete set of m.o.l.s. is equivalent to a projective plane. A question which then arises is: "Given a complete set of m.o.l.s. of order n, can we determine to which projective plane of order n it is equivalent?" The question first becomes of significance when n=9 and was brought to the attention of one of the present authors when an amateur mathematician, B.K.Kelly(1975), wrote him a letter showing how the complete set of m.o.l.s. given by L.J.Paige and C.Wexler(1953) and reproduced on page

285 of [DK] could be transformed in a series of steps to a set in which all the squares have the same rows but these rows appear in a different order in each of the squares. This implies closure of certain Desargues' configurations and so casts doubt on the assumption (made as an assertion on page 284 of [DK]) that these squares represent the Hughes plane. (In fact, Paige and Wexler stated only that "they correspond to the smallest known non-Desarguesian projective geometry".)

In this section, we first describe the Kelly transformations and also some other transformations of the same squares due to D.A.Preece and then explain how recent work of P.J.Owens has shown that the squares actually represent the dual of a translation plane. (In fact, it has been shown by computer search that the known translation plane of order 9 is the only translation plane of that order. See, for example, page 483 of [DK] for details.)

In his letter, Kelly first pointed out that Figure 6.1 provides a means by which all the squares of the Paige-Wexler set can be described in terms of the eight non-identity permutations α_i, $i=1,2,\ldots,8$, which define the second, third, ..., ninth rows of the square L_i of the set as permutations of its first row. Here, L_i is the latin square exhibited in Figure 6.2 and α_2, for example, is the permutation $(0\ 2\ 8)(1\ 6)(3\ 7)(4\ 5)$.

For the purpose of transforming the squares, Kelly used the following two simple Lemmas.

LEMMA 6.1. If the latin squares L_h and L_k whose row permutations are $\alpha_0, \alpha_1, \ldots, \alpha_{n-1}$ and $\beta_0, \beta_1, \ldots, \beta_{n-1}$ respectively are orthogonal, then so are the latin squares L_h' and L_k' whose row permutations are $\alpha_0\alpha, \alpha_1\alpha, \ldots, \alpha_{n-1}\alpha$ and $\beta_0\beta, \beta_1\beta, \ldots, \beta_{n-1}\beta$ respectively, where α and β are arbitrary permutations of the n symbols.

Proof. Because the squares L_h and L_k are orthogonal, the permutations $\alpha_0^{-1}\beta_0, \alpha_1^{-1}\beta_1, \ldots, \alpha_{n-1}^{-1}\beta_{n-1}$ form a sharply transitive set. [H.B.Mann(1942).] Consequently, the permutations $\alpha^{-1}(\alpha_0^{-1}\beta_0)\beta$, $\alpha^{-1}(\alpha_1^{-1}\beta_1)\beta, \ldots, \alpha^{-1}(\alpha_{n-1}^{-1}\beta_{n-1})\beta$ also form a sharply transitive set. The result follows. □

LEMMA 6.2 (B.K.Kelly's lemma). If the latin squares L_h, $h=1,2,\ldots,n$, whose row permutations are $\epsilon, \alpha_{h1}, \alpha_{h2}, \ldots, \alpha_{hn}$, form a complete set of mutually orthogonal squares of order n+1 then so also do the latin squares $L'_h, h=1,2,\ldots,n$, where L'_1 has row permutations $\epsilon, \alpha_{11}^{-1}, \alpha_{12}^{-1}, \ldots, \alpha_{1n}^{-1}$ and $L'_h (h \neq 1)$ has row permutations $\epsilon, \alpha_{11}^{-1}\alpha_{h1}, \alpha_{12}^{-1}\alpha_{h2}, \ldots, \alpha_{1n}^{-1}\alpha_{hn}$.

Proof. The condition for the latin square L'_1 to be orthogonal to the latin square $L'_h (h \neq 1)$ is that the permutations $\epsilon, (\alpha_{11}^{-1})^{-1}\alpha_{11}^{-1}\alpha_{h1}, (\alpha_{12}^{-1})^{-1}\alpha_{12}^{-1}\alpha_{h2}, \ldots, (\alpha_{1n}^{-1})^{-1}\alpha_{1n}^{-1}\alpha_{hn}$ form a sharply transitive set and this is true because L_h is a latin square. The condition for the latin square L'_h to be orthogonal to the latin square L'_k (h,k≠1) is that the permutations $\epsilon, \alpha_{h1}^{-1}\alpha_{k1}, \alpha_{h2}^{-1}\alpha_{k2}, \ldots, \alpha_{hn}^{-1}\alpha_{kn}$ form a sharply transitive set and this is true because the latin squares L_h and L_k are orthogonal. □

L_1	L_2	L_3	L_4	L_5	L_6	L_7	L_8
ϵ	ϵ	ϵ	ϵ	ϵ	ϵ	ϵ	ϵ
α_1	$\alpha_1\alpha_7^{-1}\alpha_5$	$\alpha_1\alpha_4^{-1}\alpha_7$	$\alpha_1\alpha_3^{-1}\alpha_4$	$\alpha_1\alpha_5^{-1}\alpha_3$	$\alpha_1\alpha_6^{-1}\alpha_8$	$\alpha_1\alpha_6$	$\alpha_1\alpha_8^{-1}\alpha_1$
α_2	α_5	α_7	α_4	α_3	α_8	α_6	α_1
α_3	$\alpha_3\alpha_1^{-1}\alpha_5$	$\alpha_3\alpha_7$	$\alpha_3\alpha_6^{-1}\alpha_4$	$\alpha_3\alpha_8^{-1}\alpha_3$	$\alpha_3\alpha_7^{-1}\alpha_8$	$\alpha_3\alpha_5^{-1}\alpha_6$	$\alpha_3\alpha_4^{-1}\alpha_1$
α_4	$\alpha_4\alpha_5$	$\alpha_4\alpha_6^{-1}\alpha_7$	$\alpha_4\alpha_8^{-1}\alpha_4$	$\alpha_4\alpha_1^{-1}\alpha_3$	$\alpha_4\alpha_5^{-1}\alpha_8$	$\alpha_4\alpha_3^{-1}\alpha_6$	$\alpha_4\alpha_7^{-1}\alpha_1$
α_5	$\alpha_5\alpha_8^{-1}\alpha_5$	$\alpha_5\alpha_1^{-1}\alpha_7$	$\alpha_5\alpha_4$	$\alpha_5\alpha_6^{-1}\alpha_3$	$\alpha_5\alpha_4^{-1}\alpha_8$	$\alpha_5\alpha_7^{-1}\alpha_6$	$\alpha_5\alpha_3^{-1}\alpha_1$
α_6	$\alpha_6\alpha_3^{-1}\alpha_5$	$\alpha_6\alpha_5^{-1}\alpha_7$	$\alpha_6\alpha_7^{-1}\alpha_4$	$\alpha_6\alpha_4^{-1}\alpha_3$	$\alpha_6\alpha_1^{-1}\alpha_8$	$\alpha_6\alpha_8^{-1}\alpha_6$	$\alpha_6\alpha_1$
α_7	$\alpha_7\alpha_6^{-1}\alpha_5$	$\alpha_7\alpha_8^{-1}\alpha_7$	$\alpha_7\alpha_1^{-1}\alpha_4$	$\alpha_7\alpha_3$	$\alpha_7\alpha_3^{-1}\alpha_8$	$\alpha_7\alpha_4^{-1}\alpha_6$	$\alpha_7\alpha_5^{-1}\alpha_1$
α_8	$\alpha_8\alpha_4^{-1}\alpha_5$	$\alpha_8\alpha_3^{-1}\alpha_7$	$\alpha_8\alpha_5^{-1}\alpha_4$	$\alpha_8\alpha_7^{-1}\alpha_3$	$\alpha_8\alpha_8$	$\alpha_8\alpha_1^{-1}\alpha_6$	$\alpha_8\alpha_6^{-1}\alpha_1$

Figure 6.1

0	1	2	3	4	5	6	7	8
1	4	5	8	7	6	2	0	3
2	6	8	7	5	4	1	3	0
3	7	1	6	0	8	4	2	5
4	3	6	0	1	2	5	8	7
5	8	3	4	2	7	0	6	1
6	2	7	5	8	0	3	1	4
7	0	4	2	3	1	8	5	6
8	5	0	1	6	3	7	4	2

Figure 6.2

Using Lemma 6.2, the latin squares of Figure 6.1 can be transformed to the form shown in Figure 6.3 and thence, using Lemma 6.1, to the form shown in Figure 6.4 in which all the squares have the same rows. Finally, the squares of Figure 6.4 may be transformed into a standardized set by first pre-multiplying every permutation by α_2 and then rearranging the rows of all the squares simultaneously so that the first square is put into standard form as in Figure 6.5. Kelly ended his discussion by remarking that when the squares have been transformed to the form shown in Figure 6.5, the permutations which describe the rearrangements of the rows by which the squares $L_2, L_3, \ldots, L_8$ are obtained from the square L_1 are the non-identity elements of the quaternion group.

L_1	L_2	L_3	L_4	L_5	L_6	L_7	L_8
ϵ	ϵ	ϵ	ϵ	ϵ	ϵ	ϵ	ϵ
α_1^{-1}	$\alpha_7^{-1}\alpha_5$	$\alpha_4^{-1}\alpha_7$	$\alpha_3^{-1}\alpha_4$	$\alpha_5^{-1}\alpha_3$	$\alpha_6^{-1}\alpha_8$	α_6	$\alpha_8^{-1}\alpha_1$
α_2^{-1}	$\alpha_2^{-1}\alpha_5$	$\alpha_2^{-1}\alpha_7$	$\alpha_2^{-1}\alpha_4$	$\alpha_2^{-1}\alpha_3$	$\alpha_2^{-1}\alpha_8$	$\alpha_2^{-1}\alpha_6$	$\alpha_2^{-1}\alpha_1$
α_3^{-1}	$\alpha_1^{-1}\alpha_5$	α_7	$\alpha_6^{-1}\alpha_4$	$\alpha_8^{-1}\alpha_3$	$\alpha_7^{-1}\alpha_8$	$\alpha_5^{-1}\alpha_6$	$\alpha_4^{-1}\alpha_1$
α_4^{-1}	α_5	$\alpha_6^{-1}\alpha_7$	$\alpha_8^{-1}\alpha_4$	$\alpha_1^{-1}\alpha_3$	$\alpha_5^{-1}\alpha_8$	$\alpha_3^{-1}\alpha_6$	$\alpha_7^{-1}\alpha_1$
α_5^{-1}	$\alpha_8^{-1}\alpha_5$	$\alpha_1^{-1}\alpha_7$	α_4	$\alpha_6^{-1}\alpha_3$	$\alpha_4^{-1}\alpha_8$	$\alpha_7^{-1}\alpha_6$	$\alpha_3^{-1}\alpha_1$
α_6^{-1}	$\alpha_3^{-1}\alpha_5$	$\alpha_5^{-1}\alpha_7$	$\alpha_7^{-1}\alpha_4$	$\alpha_4^{-1}\alpha_3$	$\alpha_1^{-1}\alpha_8$	$\alpha_8^{-1}\alpha_6$	α_1
α_7^{-1}	$\alpha_6^{-1}\alpha_5$	$\alpha_8^{-1}\alpha_7$	$\alpha_1^{-1}\alpha_4$	α_3	$\alpha_3^{-1}\alpha_8$	$\alpha_4^{-1}\alpha_6$	$\alpha_5^{-1}\alpha_1$
α_8^{-1}	$\alpha_4^{-1}\alpha_5$	$\alpha_3^{-1}\alpha_7$	$\alpha_5^{-1}\alpha_4$	$\alpha_7^{-1}\alpha_3$	α_8	$\alpha_1^{-1}\alpha_6$	$\alpha_6^{-1}\alpha_1$

Figure 6.3

A quite different transformation of the squares of the Paige-Wexler complete set which turned out to be useful in P.J. Owens' recent investigations has been given by D.A. Preece. Preece's objective was to discover which main classes of squares are represented and also to count intercalates: that is, 2×2 latin subsquares. He observed that, if the rows of all the squares are simultaneously permuted according to the permutation $\theta = (0\ 5\ 8\ 1\ 2)(3)(4)(6)(7)$, then the columns according to the permutation $\phi = (0\ 1\ 8\ 5\ 3\ 2)(4\ 7\ 6)$ and finally the symbols according to the permutation $\psi = (0\ 5\ 6\ 8)(1\ 7\ 2)(3\ 4)$, the squares L_1, L_3, L_5, L_6 of

L_1	L_2	L_3	L_4	L_5	L_6	L_7	L_8
ϵ	α_5^{-1}	α_7^{-1}	α_4^{-1}	α_3^{-1}	α_8^{-1}	α_6^{-1}	α_1^{-1}
α_1^{-1}	α_7^{-1}	α_4^{-1}	α_3^{-1}	α_5^{-1}	α_6^{-1}	ϵ	α_8^{-1}
α_2^{-1}	α_2^{-1}	α_2^{-1}	α_2^{-1}	α_2^{-1}	α_2^{-1}	α_2^{-1}	α_2^{-1}
α_3^{-1}	α_1^{-1}	ϵ	α_6^{-1}	α_8^{-1}	α_7^{-1}	α_5^{-1}	α_4^{-1}
α_4^{-1}	ϵ	α_6^{-1}	α_8^{-1}	α_1^{-1}	α_5^{-1}	α_3^{-1}	α_7^{-1}
α_5^{-1}	α_8^{-1}	α_1^{-1}	ϵ	α_6^{-1}	α_4^{-1}	α_7^{-1}	α_3^{-1}
α_6^{-1}	α_3^{-1}	α_5^{-1}	α_7^{-1}	α_4^{-1}	α_1^{-1}	α_8^{-1}	ϵ
α_7^{-1}	α_6^{-1}	α_8^{-1}	α_1^{-1}	ϵ	α_3^{-1}	α_4^{-1}	α_5^{-1}
α_8^{-1}	α_4^{-1}	α_3^{-1}	α_5^{-1}	α_7^{-1}	ϵ	α_1^{-1}	α_6^{-1}

Figure 6.4

L_1	L_2	L_3	L_4	L_5	L_6	L_7	L_8
ϵ	ϵ	ϵ	ϵ	ϵ	ϵ	ϵ	ϵ
$\alpha_2\alpha_6^{-1}$	$\alpha_2\alpha_3^{-1}$	$\alpha_2\alpha_5^{-1}$	$\alpha_2\alpha_7^{-1}$	$\alpha_2\alpha_4^{-1}$	$\alpha_2\alpha_1^{-1}$	$\alpha_2\alpha_8^{-1}$	α_2
α_2	$\alpha_2\alpha_5^{-1}$	$\alpha_2\alpha_7^{-1}$	$\alpha_2\alpha_4^{-1}$	$\alpha_2\alpha_3^{-1}$	$\alpha_2\alpha_8^{-1}$	$\alpha_2\alpha_6^{-1}$	$\alpha_2\alpha_1^{-1}$
$\alpha_2\alpha_7^{-1}$	$\alpha_2\alpha_6^{-1}$	$\alpha_2\alpha_8^{-1}$	$\alpha_2\alpha_1^{-1}$	α_2	$\alpha_2\alpha_3^{-1}$	$\alpha_2\alpha_4^{-1}$	$\alpha_2\alpha_5^{-1}$
$\alpha_2\alpha_5^{-1}$	$\alpha_2\alpha_8^{-1}$	$\alpha_2\alpha_1^{-1}$	α_2	$\alpha_2\alpha_6^{-1}$	$\alpha_2\alpha_4^{-1}$	$\alpha_2\alpha_7^{-1}$	$\alpha_2\alpha_3^{-1}$
$\alpha_2\alpha_4^{-1}$	α_2	$\alpha_2\alpha_6^{-1}$	$\alpha_2\alpha_8^{-1}$	$\alpha_2\alpha_1^{-1}$	$\alpha_2\alpha_5^{-1}$	$\alpha_2\alpha_3^{-1}$	$\alpha_2\alpha_7^{-1}$
$\alpha_2\alpha_1^{-1}$	$\alpha_2\alpha_7^{-1}$	$\alpha_2\alpha_4^{-1}$	$\alpha_2\alpha_3^{-1}$	$\alpha_2\alpha_5^{-1}$	$\alpha_2\alpha_6^{-1}$	α_2	$\alpha_2\alpha_8^{-1}$
$\alpha_2\alpha_3^{-1}$	$\alpha_2\alpha_1^{-1}$	α_2	$\alpha_2\alpha_6^{-1}$	$\alpha_2\alpha_8^{-1}$	$\alpha_2\alpha_7^{-1}$	$\alpha_2\alpha_5^{-1}$	$\alpha_2\alpha_4^{-1}$
$\alpha_2\alpha_8^{-1}$	$\alpha_2\alpha_4^{-1}$	$\alpha_2\alpha_3^{-1}$	$\alpha_2\alpha_5^{-1}$	$\alpha_2\alpha_7^{-1}$	α_2	$\alpha_2\alpha_1^{-1}$	$\alpha_2\alpha_6^{-1}$

Figure 6.5

the set (as given in Figure 6.1) take the semi-cyclic form shown in Figure 6.6. When the first row and column are disregarded, each broken left-to-right diagonal either consists entirely of zeros or else contains the integers 1,2,...,8 in natural order but read cyclically, 1 following 8. The first column contains a further cyclic permutation of the integers 1,2,...,8. [It is interesting to note that the sets of three mutually orthogonal latin squares of order 14 obtained by D.T.Todorov(1985) have this same structure.]

With the aid of the isotopism (θ,ϕ,ψ), Preece deduced that just two of the main classes of 9×9 latin squares are represented, one by the squares L_1 and L_6 and another by the remaining six squares. Squares of the first of these main classes have 48 intercalates and squares of the second have 24 intercalates: the semi-cyclic representations providing a demonstration of the fact that the number of intercalates must be a multiple of 8 in each case. The second main class comprises a single isotopy class while the first comprises three isotopy classes of which one is represented by the square L_1 and another by the square L_6 (which is a parastrophe of L_1).

$L_6 = $
0	1	2	3	4	5	6	7	8
1	5	8	2	6	0	7	4	3
2	4	6	1	3	7	0	8	5
3	6	5	7	2	4	8	0	1
4	2	7	6	8	3	5	1	0
5	0	3	8	7	1	4	6	2
6	3	0	4	1	8	2	5	7
7	8	4	0	5	2	1	3	6
8	7	1	5	0	6	3	2	4

$L_1 = $
0	1	2	3	4	5	6	7	8
5	0	7	4	3	1	8	2	6
6	7	0	8	5	4	2	1	3
7	4	8	0	1	6	5	3	2
8	3	5	1	0	2	7	6	4
1	5	4	6	2	0	3	8	7
2	8	6	5	7	3	0	4	1
3	2	1	7	6	8	4	0	5
4	6	3	2	8	7	1	5	0

$L_5 = $
0	1	2	3	4	5	6	7	8
8	7	3	6	0	2	5	1	4
1	5	8	4	7	0	3	6	2
2	3	6	1	5	8	0	4	7
3	8	4	7	2	6	1	0	5
4	6	1	5	8	3	7	2	0
5	0	7	2	6	1	4	8	3
6	4	0	8	3	7	2	5	1
7	2	5	0	1	4	8	3	6

$L_3 = $
0	1	2	3	4	5	6	7	8
6	3	0	1	7	4	2	8	5
7	6	4	0	2	8	5	3	1
8	2	7	5	0	3	1	6	4
1	5	3	8	6	0	4	2	7
2	8	6	4	1	7	0	5	3
3	4	1	7	5	2	8	0	6
4	7	5	2	8	6	3	1	0
5	0	8	6	3	1	7	4	2

Figure 6.6

Let us return now to the recent investigation of P.J.Owens into the problem of deciding which projective plane of a particular order is represented by a given complete set of m.o.l.s.

Owens has used the relationship described in Theorem 1.1 of this

Chapter as his main tool. Let π_n be a given projective plane of order n. We construct a set of n pairwise orthogonal squares (of which the first is not a latin square) as follows. Choose a line of π_n and denote it by ℓ_∞. Label the n+1 points of ℓ_∞ as $E=E_0$, $E_1=A$, E_2, $E_3,\ldots,E_n$. Label the n lines, other than ℓ_∞, that pass through E as $e_1,e_2,\ldots,e_n$. Label the n lines, other than ℓ_∞, that pass through A as $a_1,a_2,\ldots,a_n$. Denote the point $e_1 \cap a_j$ by A_j, $1 \leq j \leq n$. For each of the points E_h, $1 \leq h \leq n$, construct an associated n×n square L_h by putting the symbol ℓ in the (i,j)-th cell of L_h if the line joining E_h to the point $e_i \cap a_j$ meets e_1 at A_ℓ. The square which corresponds to the point $E_1(=A)$ is then a row-latin square each row of which contains the symbols $1,2,\ldots,n$ in natural order. The remaining squares form a complete set of m.o.l.s. in canonical form: that is, each has 1 2...n as its first row. (Figure 6.7 illustrates these facts). We shall call the set of n squares a <u>complete canonical set</u>.

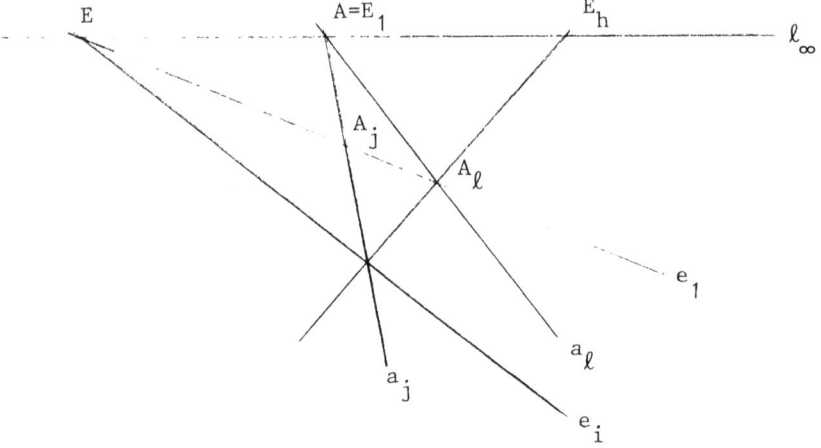

Figure 6.7

In general, if we change our choice of ℓ_∞ in π_n or if we keep ℓ_∞ fixed but change the labelling of the points on it or if we change the order in which we label the lines through E or the lines through A, we shall alter the complete canonical set of squares. For example, if we relabel the lines other than ℓ_∞ which pass through A so that the line a_j becomes the line $a_{j\theta}$, where θ is some permutation of the symbols, the effect is to re-order the columns in each of the squares L_1, $L_2,\ldots,L_n$ according to the permutation θ. If we then relabel the points on the line e_1 so that they are re-matched with the labels of the lines through A (this implies that the point A_j is relabelled as $A_{j\theta}$), the effect is to permute the symbols in each of the squares $L_1, L_2, \ldots, L_n$ according to the permutation θ so that the new first rows again contain the symbols $1,2,\ldots,n$ in natural order. In other words, applying the isotopism $(1,\theta,\theta)$ to all the squares has the geometrical effect of relabelling the lines through A and changing the labels of the points on e_1 accordingly.

Owens has defined five types of transformation on the complete canonical set of squares which correspond to a given projective plane π_n and has described the geometrical effect of each such transformation. Let us denote these five types of transformation by Tm, $m=1,2,\ldots,5$. We may list their geometrical effects as follows:

T1 relabels the lines other than ℓ_∞ which pass through E, so that e_i becomes $e_{i\theta}$, where θ is a permutation of the symbols $1,2,\ldots,n$.

T2 relabels the lines other than ℓ_∞ which pass through A, so that a_j becomes $a_{j\theta}$ and relables the points of e_1 so that A_j becomes $A_{j\theta}$. (This transformation amounts to applying the isotopism $(1,\theta,\theta)$ to all the squares, as we explained above.)

T3 interchanges the points E and A and also relables the line e_i as a_i and the line a_j as e_j, for $1 \leq i \leq n$ and $1 \leq j \leq n$.

T4 interchanges the points $A \equiv E_1$ and E_r on ℓ_∞ but without changing the labelling of the points A_j on e_1.

T5 replaces the given projective plane π_n by its dual (with interchange of ℓ_∞ and E, $E_1 \equiv A$ and e_1, E_h and e_h, A_j and a_j).

T4 and T5 are the most interesting of the transformations and, in order to describe them, Owens has introduced an n×n matrix of permutations called a representational array according to the following definition:

DEFINITION. Suppose that $L_1, L_2, \ldots, L_n$ is the complete canonical set of squares which correspond to a given projective plane π_n, where L_1 is row-latin and the remaining squares are latin (as described above), and let ρ_{ih} be the permutation which converts the first row of the square L_h into its i-th row. Then the n×n matrix $R = (\rho_{ih})$ is called the <u>representational</u>

array of the squares $L_1, L_2, \ldots, L_n$. (Note that each entry in the first row and column of R is the identity permutation.)

Transformation T5 is shown to consist of transposing the representational array R and then replacing each of the entries by its inverse, while transformation T4 corresponds to replacing the representational array $R = (\rho_{ih})$ by a new representational array in which each permutation ρ_{ih} is replaced by $\rho_{ir}^{-1}\rho_{ih}$.

It is quite easy to see that suitable combinations of the above transformations will transform an arbitrary given complete canonical set of squares into any other complete canonical set which represents either the same plane or its dual.

Owens has used the above transformations to prove a strengthened form of Theorem 8.4.2 of [DK] as follows:

THEOREM 6.3. Let S be a complete canonical set of squares which represents a given projective plane π_n (of order $n \geq 3$) relative to co-ordinatizing points $E, A, E_2, E_3, \ldots, E_n$ on a line ℓ_∞ of π_n. Then the plane is (E, ℓ_∞)-desarguesian if and only if (i) every latin square in S has the same set of n row permutations and (ii) this set of n row permutations forms a group.

Proof. Suppose firstly that the plane is (E, ℓ_∞)-desarguesian. This means that every pair of triangles which are in perspective from E and which have two pairs of corresponding sides meeting on ℓ_∞ have the third pair of corresponding sides also meeting on ℓ_∞.

Consider the latin squares L_u and L_v of S. Suppose that the entry ℓ appears in the j-th column of the r-th row of L_u and that ℓ also appears in the j-th column of the s-th row of L_v. Then the join of the point $e_r \cap a_j$ (B_1 say) to E_u meets the line e_1 in the point A_ℓ and the join of the point $e_s \cap a_j$ (C_1 say) to E_r also meets the line e_1 at the point A_ℓ. (See Figure 6.8.) If the entry in the k-th column of the r-th row of L_u is m, then the join of the point $e_r \cap a_k$ (B_2 say) to E_u meets the line e_1 at A_m. Let us denote the point $e_s \cap a_k$ by C_2. Then, the triangles $A_\ell B_1 C_1$ and $A_m B_2 C_2$ are in perspective from A with the sides $B_1 A_\ell$ and $B_2 A_m$ meeting at E_u and the sides $C_1B_1=a_j$ and $C_2B_2=a_k$ meeting at A so, since the plane is (E,ℓ_∞)-desarguesian, the third sides C_1A_ℓ and C_2A_m meet on ℓ_∞ (at E_v). That is, the join of the point $e_s \cap a_k$ to E_u meets the line e_1 at A_m and so m appears in the k-th column of the s-th row of L_v. Since this argument can be used for every value of k, the s-th row of L_v must be the same as the r-th row of L_u.

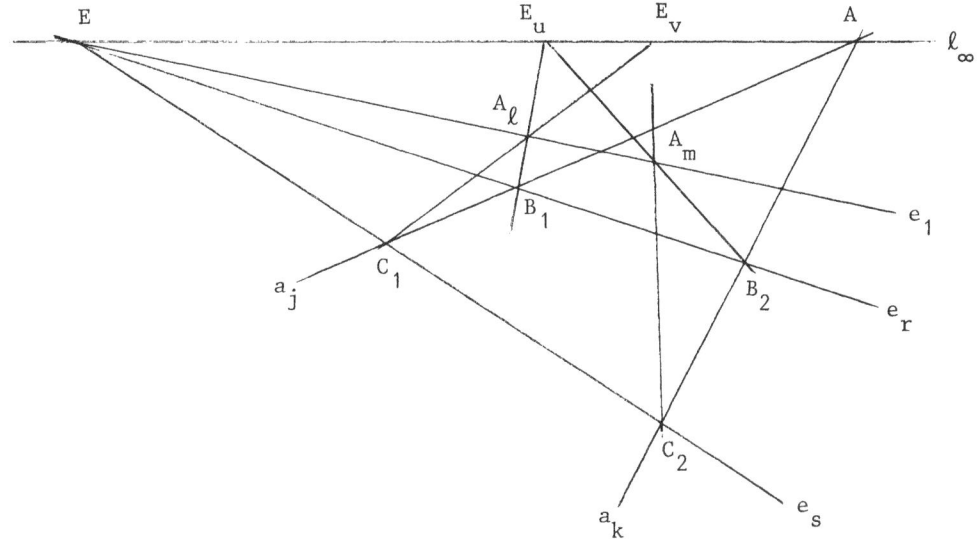

Figure 6.8

Let $R=(\rho_{ih})$ be the representational array for the complete canonical set S of latin squares. The above argument shows that each column of R contains the same set G of n permutations in some order. Suppose now that we interchange the points A and E_r of π_n. This will have the effect of replacing S by a new complete canonical set of latin squares whose representational array is $R'=(\rho_{ir}^{-1}\rho_{ih})$. Since π_n is (E,ℓ_∞)-desarguesian, the new representational array will again have the property that each of its columns contains the same n permutations in some order. However, since $\rho_{i1}=\epsilon$ (the identity permutation) for $1 \leq i \leq n$, the first column of R' consists of the n permutations $\rho_{1r}^{-1}, \rho_{2r}^{-1}, \ldots, \rho_{nr}^{-1}$. Consequently, each column of R' consists of the inverses of the n permutations of the set G. Because $n \geq 3$, there are at least two points E_2 and E_3 on ℓ_∞ different from E and

$A \equiv E_1$ so we can first move A to E_2 (by interchanging E_1 and E_2), then move A to E_3 by a second interchange and finally move A back to E_1 by a third interchange. Each transformation inverts the n distinct permutations which occur in the representational array. However, the final representational array is the same as the first and there have been three inversions, so $G = G^{-1}$. Now let α, β be any two permutations of G. Since $\alpha \in G^{-1}$, $\alpha = \rho_{ar}^{-1}$ for some a. Since $\beta \in G$, and since the permutations in the a-th row of R are all different (because the squares of S are orthogonal), $\beta = \rho_{ab}$ for some b. Thus, $\alpha\beta = \rho_{ar}^{-1}\rho_{ab}$ occurs in R' and so $\alpha\beta \in G$. It now follows that G is a group.

For the converse, suppose that every latin square S has the same set of n row permutations and that these form a group G. By reversing the first part of the argument above, we can show that, if two triangles $A_\ell B_1 C_1$ and $A_m B_2 C_2$ are in perspective from E, have the vertices A_ℓ and A_m incident with the line e_1 and have the two pairs of sides $B_1 A_\ell, B_2 A_m$ and $C_1 B_1, C_2 B_2$ meeting on ℓ_∞ at the points A and E_u respectively, then the third pair of sides will meet on ℓ_∞. This is by virtue of the fact that every pair of latin squares in S has the same n rows. Note that the requirements that the lines $C_1 B_1$ and $C_2 B_2$ be incident with A and that the points A_ℓ and A_m be incident with e_1 are essential. The latter requirement was omitted from the statement of Theorem 8.4.2 in [DK] although it was used implicitly in the proof.

Now let $A_1 B_1 C_1$ and $A_2 B_2 C_2$ be two triangles of π_n which are in perspective from E with A_1, A_2 on the line e_a, B_1, B_2 on the line e_b and C_1, C_2 on the line e_c. Suppose further that $B_1 A_1$, $B_2 A_2$ meet at E_u on ℓ_∞ and that $C_1 A_1$, $C_2 A_2$ meet at E_w on ℓ_∞. An Owens transformation of type T4 will interchange E_w and $E_1 = A$ and an Owens transformation of type T1 will interchange the lines e_a and e_1. The effects on the representational array $R = (\rho_{ih})$ of the complete canonical set of squares S are to replace each permutation ρ_{ih} by the permutation $\rho_{iw}^{-1}\rho_{ih}$ and then to re-order the row permutations in each of the squares. By virtue of the fact that the permutations ρ_{ih}, for $i = 1, 2, \ldots, n$ and for each fixed h, are the elements of the same group G of order n, it follows that the new representational array has the same properties as R and so can be used to show that the third sides of triangles $A_1 B_1 C_1$ and $A_2 B_2 C_2$ meet at a point on ℓ_∞. Hence, the plane π_n is (E, ℓ_∞)-transitive and the theorem is proved. □

We are now in a position to explain how Owens applied his transformations and the above Theorem to the case of the squares given in L.J.Paige and C.Wexler(1953) and displayed on page 285 of [DK]. He first observed that the presence of an intercalate in one of the latin squares which represent a projective plane π_n shows the presence of a Fano subplane in π_n. He also noted that in the Preece representation of the Paige-Wexler squares which we described earlier and which we partially exhibited in Figure 6.6, the cell of the third row and third column is the only one which does not form part of an intercalate in any of the eight squares. He was aware of the fact that every point of the dual translation plane of order 9 except the translation point occurs in some Fano subplane, whereas in the remaining known planes of order 9, no single point is distinguished in this particular way. [For more details, see T.G.Room and P.B.Kirkpatrick(1971)].

His next step was to move the special cell to leading position in first row and column of the squares by applying transformations of types T1 and T2 to the Paige-Wexler squares (augmented by the usual row-latin square L_1). He then applied transformations of types T5,T3,T5,T3,T5 successively to obtain a complete canonical set of squares which satisfy the two conditions of Theorem 6.3 above. He was thus able to show that the plane represented by this latter set of squares is (E, ℓ_∞)-desarguesian. By applying a further transformation of type T3, he was able to show that this latter plane is also (A, ℓ_∞)-desarguesian and hence that it is a translation plane with ℓ_∞ as translation line [using, for example, Theorems 4.19 and 4.29 in D.R.Hughes and F.Piper(1973)]. Since each transformation of type T5 changes the plane represented into its dual, Owens was finally able to deduce that the Paige-Wexler squares represent the dual of a translation plane.

Full details of Owens' work will be found in P.J.Owens(199α).

Since it is now clear that the squares reproduced on page 285 of [DK] do not represent the Hughes plane, we provide in Figure 6.9 a complete set of m.o.l.s. which do represent it. This particular set of squares has previously been given in R.C.Bose and K.R.Nair(1941) and in the Ph.D. thesis of R.Kamber (Technische Hochschule, Zürich, 1976).

L_1

0	1	2	3	4	5	6	7	8
1	2	0	4	5	3	7	8	6
2	0	1	5	3	4	8	6	7
3	4	5	6	7	8	0	1	2
4	5	3	7	8	6	1	2	0
5	3	4	8	6	7	2	0	1
6	7	8	0	1	2	3	4	5
7	8	6	1	2	0	4	5	3
8	6	7	2	0	1	5	3	4

L_2

0	1	2	3	4	5	6	7	8
2	0	1	5	3	4	8	6	7
1	2	0	4	5	3	7	8	6
6	7	8	0	1	2	3	4	5
8	6	7	2	0	1	5	3	4
7	8	6	1	2	0	4	5	3
3	4	5	6	7	8	0	1	2
5	3	4	8	6	7	2	0	1
4	5	3	7	8	6	1	2	0

L_3

0	1	2	3	4	5	6	7	8
3	4	5	6	7	8	0	1	2
6	7	8	0	1	2	3	4	5
2	0	1	7	8	6	4	5	3
5	3	4	1	2	0	7	8	6
8	6	7	4	5	3	1	2	0
1	2	0	8	6	7	5	3	4
4	5	3	2	0	1	8	6	7
7	8	6	5	3	4	2	0	1

L_4

0	1	2	3	4	5	6	7	8
4	5	3	7	8	6	1	2	0
8	6	7	2	0	1	5	3	4
7	8	6	4	5	3	2	0	1
2	0	1	8	6	7	3	4	5
3	4	5	0	1	2	7	8	6
5	3	4	1	2	0	8	6	7
6	7	8	5	3	4	0	1	2
1	2	0	6	7	8	4	5	3

L_5

0	1	2	3	4	5	6	7	8
5	3	4	8	6	7	2	0	1
7	8	6	1	2	0	4	5	3
4	5	3	2	0	1	7	8	6
6	7	8	4	5	3	0	1	2
2	0	1	6	7	8	5	3	4
8	6	7	5	3	4	1	2	0
1	2	0	7	8	6	3	4	5
3	4	5	0	1	2	8	6	7

L_6

0	1	2	3	4	5	6	7	8
6	7	8	0	1	2	3	4	5
3	4	5	6	7	8	0	1	2
1	2	0	8	6	7	5	3	4
7	8	6	5	3	4	2	0	1
4	5	3	2	0	1	8	6	7
2	0	1	7	8	6	4	5	3
8	6	7	4	5	3	1	2	0
5	3	4	1	2	0	7	8	6

L_7

0	1	2	3	4	5	6	7	8
7	8	6	1	2	0	4	5	3
5	3	4	8	6	7	2	0	1
8	6	7	5	3	4	1	2	0
3	4	5	0	1	2	8	6	7
1	2	0	7	8	6	3	4	5
4	5	3	2	0	1	7	8	6
2	0	1	6	7	8	5	3	4
6	7	8	4	5	3	0	1	2

L_8

0	1	2	3	4	5	6	7	8
8	6	7	2	0	1	5	3	4
4	5	3	7	8	6	1	2	0
5	3	4	1	2	0	8	6	7
1	2	0	6	7	8	4	5	3
6	7	8	5	3	4	0	1	2
7	8	6	4	5	3	2	0	1
3	4	5	0	1	2	7	8	6
2	0	1	8	6	7	3	4	5

Figure 6.9

(7) <u>Miscellanea</u>.

In the previous Section, we discussed how the plane represented by a particular complete set of m.o.l.s. can be determined. We begin this Section by drawing attention to a simple construction which provides a set of m.o.l.s. which represent the non-desarguesian translation plane of order 9. The construction consists of taking $x_r x_i^\epsilon + x_j$ as the entry in the (i,j)-th cell of the r-th latin square L_r, $r = 1, 2, \ldots, 8$, where $x_0 = 0$, $x_1 = 1$, $x_2, \ldots, x_8$ are the elements of the Galois field $GF[3^2]$ and $\epsilon = 1$ or 3 according as x_i is a square or a non-square of the field.

The fact that this prescription gives a set of eight m.o.l.s. depends on the fact that the cube of an element of $GF[3^2]$ is a square or a non-square according as the element itself is a square or a non-square.

Suppose that the squares L_r and L_s were not orthogonal. There would then be cells (i,j) and (k, ℓ) such that $u = x_r x_i^{\epsilon_1} + x_j = x_r x_k^{\epsilon_2} + x_\ell$ and $v = x_s x_i^{\epsilon_1} + x_j = x_s x_k^{\epsilon_2} + x_\ell$. Then $u-v = (x_r - x_s) x_i^{\epsilon_1} = (x_r - x_s) x_k^{\epsilon_2}$. Since $x_r \neq x_s$, $x_i^{\epsilon_1} = x_k^{\epsilon_2}$. This equality can only hold if $\epsilon_1 = \epsilon_2$ and $x_i = x_k$ since $\epsilon_2 = 3 \Rightarrow x_k$ is a non-square $\Rightarrow x_k^3$ is a non-square $\Rightarrow x_i^{\epsilon_1}$ is a non-square $\Rightarrow \epsilon_1 = 3$. It follows that $x_j = x_\ell$ and that $i = k$, $j = \ell$ whence L_r and L_s are orthogonal.

In effect, the above construction is equivalent to replacing the field $GF[3^2]$ which co-ordinatizes the desarguesian plane of order 9 by a ring R which has the same addition as $GF[3^2]$ but has multiplication re-defined as follows:

(i) if $x_u x_v = x_w$ in $GF[3^2]$ and x_v is a square, then $x_u x_v = x_w$ in R;
(ii) if $x_u x_v^3 = x_w$ in $GF[3^2]$ and x_v is a non-square, then $x_u x_v = x_w$ in R.

The construction was given explicitly in R.D.Carmichael(1937) and in R.C.Bose and K.R.Nair(1941) but is implicit in the much earlier paper of O.Veblen and J.H.M.Wedderburn(1907).

In Chapter 2, Section 5 of this book, we discussed sets of m.o.l.s. which are or are not extendible to a larger set. A related question is to ask what can be said about the structure of a latin square which can be embedded in a complete set of m.o.l.s. and thus form part of a projective plane. J.Bierbrauer(1985c) has considered this question using the concept of a blocking set of a projective plane as a main tool.

DEFINITION. Let B be a subset of the points of a projective plane π such that every line of π is incident both with points of B and with points not in B. Then B is called a blocking set of π.

It is known that the cardinal $|B|$ of B is not less than $n+\sqrt{n}+1$ [A.A.Bruen(1971)]. On the other hand, not more than $|B|-n$ points of B can be collinear. When π contains a line g such that g is incident with exactly $|B|-n$ points of B, B is said to be of Rédei type [A.A.Bruen and R.Silverman(1981)]. Blocking sets of Rédei type arise from common transversals of a complete set of m.o.l.s. which do not represent lines of the associated projective plane [A.A.Bruen and J.C.Fisher(1973)].

$$L = \begin{vmatrix} A & B \\ C & D \end{vmatrix}$$

Figure 7.1

Bierbrauer has proved the following results:

Let L be a latin square of order n>2 with a subsquare A of size a×a, where a>0 and n-a is odd. Let X denote any subset of the entries of L of cardinal $|X| = n-a$. Further, let L be partitioned as shown in Figure 7.1 and let K_A, K_D denote the sets of cells of the subsquares A, D respectively which contain elements of the set X. Let $\bar{K}_B, \bar{K}_C$ similarly denote the sets of cells of the subrectangles B, C which contain elements of the complement $\bar{X}$ of X. Then every transversal of L interests the union $U = K_A \cup K_D \cup \bar{K}_B \cup \bar{K}_C$ non-trivially. Also, $|K_A| = |\bar{K}_B| = |\bar{K}_C|$ and $|K_D| = |K_A| - (n-a)(n-2a)$.

Suppose further that L is the Cayley table of the additive loop of the ternary ring of some projective plane π of order n. Let E, A, B be the points on ℓ_∞ which correspond to the rows, columns and entries of L respectively (cf. Theorem 1.1 of this Chapter). Then each cell of L is associated with a finite point of π. Bierbrauer has shown that the set of points of π which correspond to the cells of U together with the points E,A,B of ℓ_∞ form a blocking set of π provided that $|K_D| < a$. Thence he has proved the following two theorems. (We use the same notation as above.)

Theorem 7.1. Let L be a latin square of order n which is a member of a complete set of m.o.l.s. Further, let A be a subsquare of size a×a and X a subset of the entries of L of cardinal $|X| = n-a$. Then,

(i) if $n = 4m+2 > 2$, $a = 2m+1$, we have $K_A \geq m+\frac{1}{4}(\lfloor\sqrt{n}\rfloor+2)$;

(ii) if $n = 4m+1 \geq 9$, $a = 2m$, we have $K_A \geq \frac{1}{2}m+\frac{1}{4}\lfloor\sqrt{n}\rfloor$;

(iii) if $n = 4m$, $a = 2m-1$, we have $K_A \geq \frac{1}{4}(\lfloor\sqrt{n}\rfloor-2)$; also $K_A \neq 0$ for $n=8$;

(iv) if $n = 4m+3 \geq 7$, $a = 2m+1$, we have $K_A \geq \frac{1}{2}m+\frac{3}{8}$.

Theorem 7.2. Let L be a latin square of order $n = 4m+2 \geq 10$ which is a member of a complete set of m.o.l.s. Then every $(2m+1)\times(2m+1)$ subsquare of L contains at least $2m+3$ different entries.

Bierbrauer has used these two theorems to obtain restrictions on the extent (defined below) of a blocking set of Rédei type in a projective plane and, in particular, in a projective plane of order $4m+2$ should one exist.

DEFINITION. Let B be a blocking set in a projective plane of order n. If there is a configuration $G = \{g_1,g_2,g_3\}$ of three lines of π such that all points of B are incident with G and such that the points $g_i \cap g_j$ belong to B, then B is called 3-concurrent or triangular according as the lines of G are concurrent or form a proper triangle. If $k_i = |g_i \cap B|$, i=1,2,3, and $k_1 \geq k_2 \geq k_3$, then the extent of B is the triple (k_1-1,k_2-1,k_3-1) if B is 3-concurrent or the triple (k_1-2,k_2-2,k_3-2) if B is triangular.

If, in addition, B is of Rédei type then $k_1 = |g_1 \cap B| = |B|-n$.

Bierbrauer has made an analogous definition for latin squares:

DEFINITION. A latin square L of order n has extent (x,y,z) if L has a y×z subrectangle which contains exactly x distinct elements, x<n.

Suppose now that B is a 3-concurrent blocking set of extent (k_1-1, k_2-1, k_3-1) and of Rédei type whose points lie on the lines g_1, g_2, g_3, where g_1, g_2, g_3 are concurrent at the point P. Denote the points of g_i which belong to B by G_{ij}, where $j=1,2,\ldots,k_i$ and $i=1,2,3$. Then it can be shown that each join $G_{2u}G_{3v}$ meets g_1 at a point G_{1w} of B [J. Bierbrauer (198∝)]. Let L be a latin square of order $n = (k_2-1)+(k_3-1)$ whose (u,v)-th entry is w for $1 \leq u \leq k_2$ and $1 \leq v \leq k_3$. Then L has extent (k_1-1, k_2-1, k_3-1).

Conversely, Bierbrauer has shown that if L is a latin square of order $n = y+z$ with extent (x,y,z) and if there is a projective plane π of order n with three concurrent lines g_1, g_2, g_3 such that, after suitable choice of notation, L is the latin square determined in the above way by the configuration $G = \{g_1, g_2, g_3\}$, then π has a 3-concurrent blocking set of Rédei type and with extent (x,y,z).

Combining this result with Theorems 7.1 and 7.2, Bierbrauer has deduced:

<u>Theorem 7.3</u>. If π is a finite projective plane of order n and B a 3-concurrent blocking set of π, then

(i) B cannot have extent $(n/2, n/2, n/2)$ or $((n/2)+1, n/2, n/2)$ if $n \equiv 2 \pmod 4$, $n \geq 10$;

(ii) B cannot have extent $((n+1)/2, (n+1)/2, (n-1)/2)$ if $n \equiv 1 \pmod 2$, $n \geq 7$;

(iii) B cannot have extent $((n/2)+1, (n/2)+1, (n/2)-1)$ if $n \equiv 0 \pmod 2$, $n \geq 8$.

Bierbrauer used this Theorem to rule out existence in a projective plane of order 10 of 3-concurrent blocking sets of extent $(6,5,5)$, so augmenting an earlier study of blocking sets in such a plane [J. Bierbrauer (1985a, 1985b)].

ADDENDUM

Soon after this Chapter was completed, the Authors received news of the completion of exhaustive computer searches for new projective planes of order 9 and for a projective plane of order 10. Both searches proved negative but there remains a not-quite-negligible possibility of computer error.

The search for projective planes of order 9 different from the four known ones was carried out by G.Kolesova, C.W.H.Lam and L.Thiel(1988a,b) and was dependent on first generating all non-isomorphic latin squares of order 8. In carrying out the latter generation, Kolesova et al calculated the total number of reduced 8×8 latin squares to be 535,281,401,856, the number of isotopy classes to be 1,676,267 and the number of main classes to be 283,657. The first number agrees with that obtained by M.B.Wells(1967), reported on page 144 of [DK] and later verified by S.E.Bammel and J.Rothstein(1975), but the number of isotopy classes is ten more than the number obtained by J.W.Brown(1968). Also, the number of main classes is different from the number computed by V.L.Arlazarov, A.M.Baraev, J.U.Golfand and I.A.Faradzhov(1978). The latter authors found only 283,640 such classes.

After a representative of each main class of 8×8 latin squares had been obtained, it was used to construct part of the incidence matrix of a projective plane of order 9. Kolesova et al have shown that the incidence matrix of a finite projective plane can be put into a canonical form (which is somewhat similar to that of S.Bourn(1983) described in Section 4 of this Chapter) and which is such that the first 27 columns of the matrix are completely determined by a particular 8×8 latin square. Latin squares of order 8 which belong to the same main class are associated with partial incidence matrices which represent isomorphic projective planes. The computer programme attempted to complete each of the 283,657 canonical partial incidence matrices which was constructed from the catalogue of main class representatives of 8×8 latin squares to a projective plane of order 9. 325 complete incidence matrices were obtained. Each corresponds to one of the known planes.

The search for a projective plane of order 10 was carried out by C.W.H.Lam, L.Thiel and S.Swiercz(1989). Again, a partial incidence matrix of the supposed plane was first constructed. For this purpose, the earlier work of F.J.MacWilliams, N.J.A.Sloane and W.Thompson(1973), C.W.H.Lam, L.Thiel, S.Swiercz and J.McKay(1983), and C.W.H.Lam, L.Thiel and S.Swiercz(1986), which together showed that the (111,56) binary error-correcting code generated by the rows of the incidence matrix would contain no codewords of weights 12, 15 or 16, was used and also a paper of M.Hall(1980) which characterized codewords of weight 19 in terms of geometrical properties of the supposed plane. (See also Section 2 of this Chapter.) An attempt to complete each partial incidence matrix was then made. For details, readers should consult the original paper. The search was exhaustive but required a very large amount of computing time which the authors estimate as the equivalent of 800 days of CPU time on a VAX-11/780 computer together with 2000 hours on a CRAY-1A supercomputer.

In their paper, C.W.H.Lam, L.Thiel and S.Swiercz(1989), the authors have given a frank and careful assessment of the possibility that their computer programmes were in error or that errors arose in the actual computation. We quote the first paragraph of this assessment: "Because of the use of a computer, one should not consider these results as a "proof", in the traditional sense, that a plane of order 10 does not exist. They are experimental results and there is always the possibility of mistakes. Despite all these reservations, we are going to present reasons that the probability of the existence of an undiscovered plane of order 10 is very small."

Additional Remarks: In a private communication to the Authors, C.W.H.Lam has raised the question as to who first asked about the existence of a projective plane of order 10. It appears that the probable answer is that it dates back to O.Veblen and J.H.M.Wedderburn(1907). These authors were the first to construct non-desarguesian projective planes and the last sentence of their paper reads: "The problem of determining a non-desarguesian geometry of minimum number of points per line remains unsolved." This is followed by a footnote to say that C.R.MacInnes(1907) had shown that the only plane of order five is the desarguesian one and that no projective plane of order six exists, also that the latter result had been obtained independently by F.H.Safford(1907).

It is interesting to note that the connection between non-existence of a projective plane and non-existence of a complete set of mutually orthogonal latin squares was not appreciated at this time so that the relevance of G.Tarry's work (1900) on non-existence of a pair of orthogonal 6×6 latin squares was not realized.

In a recent letter, G.L.Mullen has drawn the attention of the present Authors to the fact that, although the connection with projective planes was not realized, E.H.Moore had constructed complete sets of m.o.l.s. by a method analogous to that of R.C.Bose(1938) and W.L.Stevens(1939) as early as 1896 under the title of "two-fold School-Girl-System of class m−1, index 2 and degree m." In the same paper, Moore also proved what is usually known as MacNeish's theorem (see, for example, page 390 of [DK]). For the details see E.H.Moore(1896), page 282, paragraph(ℓ); page 286, paragraph(c); and page 288, paragraph(k). Mullen states that Moore's work was mentioned to him by R.D.Baker.

We conclude by remarking that, because of limitations of space, a number of important papers connecting latin squares with geometry via quasigroups and nets have not been discussed at all in the foregoing pages. We call attention especially to A.Barlotti(1976); A.Barlotti and K.Strambach(1983); V.D.Belousov(1971,1979); D.Jungnickel(1984b,1988); K.Repphun(1965); and K.Strambach(1981).

Also there have been a large number of papers (additional to those already mentioned in this Chapter) which have discussed the properties of latin squares which represent the various projective planes of order 9, or which have tried to construct new planes of that order, and a similarly large number which have tried to get as close as possible to constructing a pair of orthogonal latin squares of order 6 or a triple of m.o.l.s. of order 10. On the first of these topics, we mention M.Hall, J.D.Swift and R.B.Killgrove(1959); L.J.Hararen(1968); R.B.Killgrove(1960); R.B.Killgrove and E.T.Parker(1964,1979); A.D.Lumpov(1972,1974); A.E.Malih (1972,1976); and L.J.Panteleeva(1968). On the second, we mention J.W.Brown and E.T.Parker(1982,1984,1985); M.F.Franklin(1984); K.Heinrich(1977e); K.Heinrich and W.D.Wallis(1980); A.D.Keedwell (1984b); and E.T.Parker(1988), in addition to A.E.Brouwer(1984) and

A.D.Keedwell(1980) which were cited in Section 11.1. [For the record, the uniqueness of the projective plane of order 7 follows from the work of R.C.Bose and K.R.Nair(1941), W.A.Pierce(1953) and M.Hall(1953,1954) and that of order 8 from M.Hall, J.D.Swift and R.Walker(1956). See also Section 11.1 and [DK], page 169.]

A further paper on the existence of projective planes of orders 12 and 20 not mentioned in Section 11.2 is Z.Janko and T.van Trung(1980b).

Earlier in this Chapter, we mentioned the work of S.Bourn(1983) on finding a canonical form for the incidence matrix of a finite projective plane. Another paper on the same topic is B.Montaron(1985). Also, earlier in this Chapter, we mentioned the self-orthogonal latin square of order 10 obtained by L.Weisner(1963) but we omitted to mention that the same square was obtained independently by A.J.Lyamzin(1963). See also [DK], page 235.

CHAPTER 12

FREQUENCY SQUARES (J.Dénes and A.D.Keedwell)

The idea of generalizing the concept of latin square to that of frequency square has its origins in three papers of D.J.Finney(1945,1946a, 1946b). In the late 1960's, the same idea was discussed anew in S.Addelman(1967) and G.H.Freeman(1966). The formal definition of frequency square was first given in A.Hedayat's Ph.D. thesis of 1969 and the properties of such squares were developed in A.Hedayat and E.Seiden (1970) and in A.Hedayat, D.Raghavaro and E.Seiden(1975). Since the concept became more widely known after the publication of these papers and also of [DK] in 1974, quite a spate of papers on the subject has appeared and the purpose of this short chapter is to summarize the results so far obtained.

(1) F-squares and orthogonal F-squares.

We begin with the necessary definitions and notation.

DEFINITION. Let $A = ||a_{ij}||$ be an n×n matrix and let $\Sigma = \{c_1, c_2, \ldots, c_m\}$, $m \leq n$, be the set of distinct elements of A. Suppose further that, for each i, where $i = 1, 2, \ldots, m$, the element c_i appears precisely λ_i times ($\lambda_i \geq 1$) in each row and column of A. Then A is called an F-square of order n on the set Σ with frequency vector $(\lambda_1, \lambda_2, \ldots, \lambda_m)$. Note that, by virtue of the definition, $n = \lambda_1 + \lambda_2 + \ldots + \lambda_m$.

Briefly, we say that the matrix A is an $F(n; \lambda_1, \lambda_2, \ldots, \lambda_m)$ square. We may abbreviate the notation further by allowing powers of the λ_i to stand for repeated occurrences of the same symbol. Thus, an $F(n; \lambda_1, \lambda_1, \lambda_3, \lambda_4, \lambda_4, \lambda_6)$ square may be denoted by $F(n; \lambda_1^2, \lambda_3, \lambda_4^2, \lambda_6)$.

In their paper (1970), Hedayat and Seiden defined orthogonality of F-squares as follows.

DEFINITION. The two F-squares $F_1(n;\lambda_1,\lambda_2,\ldots,\lambda_\ell)$ defined on the set $\Sigma_1 = \{a_1,a_2,\ldots,a_\ell\}$ and $F_2(n;\mu_1,\mu_2,\ldots,\mu_m)$ defined on the set $\Sigma_2 = \{b_1,b_2,\ldots,b_m\}$ are <u>orthogonal</u> if each ordered pair $a_i b_j$ appears $\lambda_i \mu_j$ times when the squares F_1 and F_2 are placed in juxtaposition.

$$F_1(5;2^2,1) \qquad F_2(5;1^2,3) \qquad F_3(5;1^3,2)$$

1	2	3	1	2		1	2	3	3	3		1	2	3	4	4
2	1	2	3	1		3	3	1	2	3		3	4	4	1	2
1	2	1	2	3		2	3	3	3	1		4	1	2	3	4
3	1	2	1	2		3	1	2	3	3		2	3	4	4	1
2	1	1	2	1		3	3	3	1	2		4	4	1	2	3

Figure 1.1

An illustrative example of a set of three mutually orthogonal F-squares of order 5 which comes from A.Hedayat and E.Seiden(1970) is given in Figure 1.1.

We note that a latin square $F(n;1^n)$ is a special case of a frequency square and that the concept of orthogonal latin squares is embraced by the above definition of orthogonal F-squares.

The maximum number of orthogonal latin squares of order n which can be contained in a pairwise orthogonal set is n-1 and this result is a particular case of a more general theorem proved in A.Hedayat, D.Raghavaro and E.Seiden(1975). In view of its importance, we both state the theorem and give these authors' proof of it.

<u>THEOREM 1.1</u>. The maximum number t of pairwise orthogonal F-squares of type $F(n;\lambda^m)$, where $n = \lambda m$, satisfies the inequality $t \leq (n-1)^2/(m-1)$.

<u>Proof</u> Let $F_1, F_2, \ldots, F_t$ be a set of t pairwise orthogonal F-squares of type $F(n;\lambda^m)$. We may suppose without loss of generality that these squares are all based on the same m elements $1,2,\ldots,m$. For each square

F_h, we define an $n^2 \times m$ matrix $N_h = ||n^h_{(i,j),k}||$, where $n^h_{(i,j),k} = 1$ if the element k occurs in the (i,j)-th cell of square F_h and is zero otherwise $(i,j = 1,2,\ldots,n)$.

Let M denote the $n^2 \times tm$ matrix $||N_1:N_2:\ldots:N_t||$ and let the row of M whose entries are $n^h_{(i,j),k}$ for $h = 1,2,\ldots,t$ and $k = 1,2,\ldots,m$ be denoted by r_{ij}. Since each element k occurs λ times in each row of each square, $\sum_{j=1}^{n} r_{ij} = (\lambda \; \lambda \; \ldots \; \lambda)$ for each value of i, $i = 1,2,\ldots,n$. Also, since each element k occurs λ times in each column of each square $\sum_{i=1}^{n} r_{ij} = (\lambda \; \lambda \; \ldots \; \lambda)$ for each value of j, $j = 1,2,\ldots n$. Thus, we have at least $2n-2$ independent dependence relations between the rows of M: namely, $n-1$ of the form $\sum_{j=1}^{n} r_{ij} = \sum_{j=1}^{n} r_{nj}$ and $n-1$ of the form $\sum_{i=1}^{n} r_{ij} = \sum_{i=1}^{n} r_{in}$. It follows that the row rank of M is at most $n^2 - (2n-2) = (n-1)^2 + 1$. Hence, $\text{rank}(M) \leq \min((n-1)^2+1, tm)$.

Also, using these same properties of an F-square together with the facts that in each row of each submatrix N_h all but one of the entries is zero and that, because the F-squares are orthogonal, two columns of M belonging to distinct submatrices N_h and N_i both have 1's as entries in just λ^2 of the n^2 rows, we find that

$$M^T M = \begin{bmatrix} n\lambda I_m & \lambda^2 J_m & \lambda^2 J_m & \ldots & \lambda^2 J_m \\ \lambda^2 J_m & n\lambda I_m & \lambda^2 J_m & \ldots & \lambda^2 J_m \\ \cdot & \cdot & \cdot & \ldots & \cdot \\ \cdot & \cdot & \cdot & \ldots & \cdot \\ \cdot & \cdot & \cdot & \ldots & \cdot \\ \lambda^2 J_m & \lambda^2 J_m & \lambda^2 J_m & \ldots & n\lambda I_m \end{bmatrix}$$

where M^T denotes the transpose of M, I_m is the $m \times m$ identity matrix, J_m is the $m \times m$ matrix all of whose entries are ones, and there are t rows and t columns in the above expression for $M^T M$.

Next, we find the eigenvalues of $M^T M$. To do this it is simplest to consider the nature of its eigenvectors. We note first that every row of $M^T M$ has the same sum $n\lambda + m(t-1)\lambda^2 = nt\lambda$ and so $(1 \; 1 \; \ldots \; 1)^T$ is a column eigenvector with eigenvalue $nt\lambda$. Since M^TM is symmetric, we can span each of its eigenspaces with an orthogonal set of eigenvectors. In particular, we observe that $(\theta\ \theta\ \ldots\ \alpha_h\ \ldots\ \theta)^T$, where θ is a row of m zeros and α_h is a row of elements (not all zero) occurring in the (hm+1)-th, (hm+2)-th,..., (hm+m)-th places whose sum is zero, is an eigenvector for $h = 0, 1, 2, \ldots, t-1$ and that these eigenvectors are orthogonal to each other and to the eigenvector of all ones. Also, for each h, m-1 mutually orthogonal vectors of this type exist. Each of these $t(m-1)$ eigenvectors corresponds to an eigenvalue $n\lambda$. Finally, there also exist $tm-1-t(m-1) = t-1$ independent eigenvectors of the type $(-J\ -J\ \ldots\ J_h^*\ \ldots\ -J)^T$ where $-J$ denotes a row of m (-1)'s and J_h^* denotes a row of m $(t-1)$'s occurring in the (hm+1)-th, (hm+2)-th,..., (hm+m)-th places. Each of these latter eigenvectors corresponds to the eigenvalue 0.

Since all together this accounts for tm eigenvalues of M^TM and since M^TM is similar to a diagonal matrix because it is symmetric, we deduce that $\text{rank}(M^TM) = tm-(t-1) = 1+t(m-1)$.

Since $\text{rank}(M^TM) \leq \text{rank}(M) \leq \min((n-1)^2+1, tm)$ from above, we have $1+t(m-1) \leq (n-1)^2+1$ from which $t \leq (n-1)^2/(m-1)$ as required. []

In Figure 1.2 we illustrate the above theorem by means of a set of three pairwise orthogonal F-squares of type $F(4; 2^2)$ and the corresponding matrix M.

As a consequence of the above theorem, A. Hedayat, D. Raghavaro and E. Seiden (1975) have given the following definition.

DEFINITION. A set of t pairwise orthogonal F-squares of type $F(n; \lambda^m)$, where $n = \lambda m$, is said to be a <u>complete set</u> if $t = (n-1)^2/(m-1)$.

$$F_1 = \begin{bmatrix} 1 & 2 & 1 & 2 \\ 2 & 1 & 2 & 1 \\ 1 & 2 & 1 & 2 \\ 2 & 1 & 2 & 1 \end{bmatrix} \quad F_2 = \begin{bmatrix} 1 & 2 & 1 & 2 \\ 1 & 2 & 1 & 2 \\ 2 & 1 & 2 & 1 \\ 2 & 1 & 2 & 1 \end{bmatrix} \quad F_3 = \begin{bmatrix} 1 & 1 & 2 & 2 \\ 2 & 2 & 1 & 1 \\ 2 & 2 & 1 & 1 \\ 1 & 1 & 2 & 2 \end{bmatrix}$$

$$M = \begin{bmatrix} 1 & 0 & 1 & 0 & 1 & 0 \\ 0 & 1 & 0 & 1 & 1 & 0 \\ 1 & 0 & 1 & 0 & 0 & 1 \\ 0 & 1 & 0 & 1 & 0 & 1 \\ \hline 0 & 1 & 1 & 0 & 0 & 1 \\ 1 & 0 & 0 & 1 & 0 & 1 \\ 0 & 1 & 1 & 0 & 1 & 0 \\ 1 & 0 & 0 & 1 & 1 & 0 \\ \hline 1 & 0 & 0 & 1 & 0 & 1 \\ 0 & 1 & 1 & 0 & 0 & 1 \\ 1 & 0 & 0 & 1 & 1 & 0 \\ 0 & 1 & 1 & 0 & 1 & 0 \\ \hline 0 & 1 & 0 & 1 & 1 & 0 \\ 1 & 0 & 1 & 0 & 1 & 0 \\ 0 & 1 & 0 & 1 & 0 & 1 \\ 1 & 0 & 1 & 0 & 0 & 1 \end{bmatrix} \qquad M^T M = \begin{bmatrix} 8 & 0 & 4 & 4 & 4 & 4 \\ 0 & 8 & 4 & 4 & 4 & 4 \\ 4 & 4 & 8 & 0 & 4 & 4 \\ 4 & 4 & 0 & 8 & 4 & 4 \\ 4 & 4 & 4 & 4 & 8 & 0 \\ 4 & 4 & 4 & 4 & 0 & 8 \end{bmatrix}$$

Figure 1.2.

These authors have further proved:

<u>THEOREM 1.2</u>. A complete set of pairwise orthogonal $F(n; \lambda^m)$ squares exists whenever λ is a power of m and m is a prime power.

The proof is constructive and makes use of a symmetrical factorial design. The authors have illustrated their method by constructing a complete set of nine F-squares of type $F(4; 2^2)$. (The squares given in our Figure 1.2 are a subset of three of these nine F-squares.)

These authors have also given a Kronecker product construction as follows:

THEOREM 1.3. Let $F_i \otimes F_j$ denote the Kronecker product of the F-squares F_i and F_j. Then, if F_1 is orthogonal to F_3 and F_2 is orthogonal to F_4, it follows that $F_1 \otimes F_2$ is orthogonal to $F_3 \otimes F_4$. If F_1 is of type $F(m; \lambda_1, \lambda_2, \ldots, \lambda_s)$, where $m = \sum_{i=1}^{s} \lambda_i$, and F_2 is of type $F(n; \mu_1, \mu_2, \ldots, \mu_t)$, where $n = \sum_{j=1}^{t} \mu_j$, then $F_1 \otimes F_2$ is of type $F(mn; \nu_{11}, \nu_{12}, \ldots, \nu_{st})$, where $\nu_{ij} = \lambda_i \mu_j$.

J.P. Mandeli, F.C.H. Lee and W.T. Federer (1981) have generalized Theorem 1.1 above to the following:

THEOREM 1.4 Let $F_1, F_2, \ldots, F_t$ be pairwise orthogonal F-squares of types $F(n; \lambda_i^{m_i})$, $i = 1, 2, \ldots, t$, where F_i is defined on a set of cardinality m_i and $n = \lambda_i m_i$. Then t satisfies the inequatlity $(\sum_{i=1}^{t} m_i) - t \leq (n-1)^2$.

C. Pellegrino and N.A. Malara (1986) have shown that the same result holds in a more general situation:

THEOREM 1.5. Let $F_1, F_2, \ldots, F_t$ be pairwise orthogonal F-squares of types $F(n; \lambda_{i1}, \lambda_{i2}, \ldots, \lambda_{im_i})$, $i = 1, 2, \ldots, t$, where F_i is defined on a set of cardinality m_i and $n = \sum_{j=1}^{m_i} \lambda_{ij}$. Then t satisfies the inequality $(\sum_{i=1}^{t} m_i) - t \leq (n-1)^2$.

Each of these theorems can be proved by a simple extension of the method used to prove Theorem 1.1.

J.P. Mandeli and W.T. Federer (1983) have proved the following extension of MacNeish's theorem (see, for example, page 390 of [DK]) concerning the minimum number of mutually orthogonal latin squares of a given composite order n which can be guaranteed to exist.

THEOREM 1.6 If the prime decomposition of the number n is given by $n = p_1^{e_1} p_2^{e_2} \ldots p_m^{e_m}$ with $p_1^{e_1} < p_2^{e_2} < \ldots < p_m^{e_m}$ then there exists a set of $p_m^{e_m}-1$ pairwise orthogonal F-squares of order n. This set contains $p_{i+1}^{e_{i+1}}-p_i^{e_i}$ pairwise orthogonal F-squares of type $F(n; \lambda_i^{t_i})$, where $\lambda_i = p_1^{e_1} p_2^{e_2} \ldots p_i^{e_i}$ and $t_i = p_{i+1}^{e_{i+1}} p_{i+2}^{e_{i+2}} \ldots p_m^{e_m}$ for $i = 0, 1, \ldots, m-1$ (with the convention that $\lambda_0 = 1$).

P. Lancellotti and C. Pellegrino (1986) have generalized Theorem 1.6 to the case of systems of orthogonal F-squares of composite order having a variable number of symbols.

These authors and a number of others have given constructions for complete sets of orthogonal F-squares of various kinds. Space does not permit descriptions of all the results which have been obtained. Instead we conclude this section with a list of further papers on the subject.

In W.T. Federer (1977), a complete set of $(4h-1)^2$ F-squares of type $F(4h; 2h, 2h)$ is constructed. This paper is the first to use Hadamard matrices for this purpose. In S.J. Schwager, J.P. Mandeli and W.T. Federer (1984), complete sets of F-squares of order 2^n with a cyclic latin square as one member are obtained. (Readers will recall from Chapter 2 that such a latin square has no orthogonal latin square mate.) Further constructions for complete sets of orthogonal F-squares will be found in B.L. Raktoe and W.T. Federer (1985), in J.P. Mandeli and W.T. Federer (1984b) and in G.L. Mullen (1988), see also Section 5 below.

S.J. Schwager, W.T. Federer and B.L. Raktoe (1984) have defined two sets of orthogonal frequency squares to be isomorphic if the squares of the second set can be obtained from the squares of the first set by applying the same sequence of row and column permutations to all of the squares of the first set. With the aid of Hadamard matrices, these authors have been able to construct three non-isomorphic complete sets of nine mutually orthogonal $F(4; 2^2)$ squares. A related paper is J. Seberry (1980).

(2) Enumeration and classification of F-squares.

The major paper on the topic of the title of this section is that of L.J.Brant and G.L.Mullen(1986). These authors have attempted for orders up to 6 an enumeration of F-squares analogous to that made in the now distant past for latin squares. (Details of the latter are in Chapter 4 of [DK].)

They have defined an F-square of type $F(n;\lambda_1,\lambda_2,\ldots,\lambda_m)$ with symbols from the set $\{1,2,\ldots,m\}$ to be reduced (or in standard form) if in both the first row and the first column of the square, the symbol 1 occurs in the first λ_1 positions, the symbol 2 occurs in the next λ_2 positions, ..., and the symbol m occurs in the last λ_m positions. They have proved:

__THEOREM 2.1__ Let $F(n;\lambda_1,\lambda_2,\ldots,\lambda_m)$ and $f(n;\lambda_1,\lambda_2,\ldots,\lambda_m)$ respectively denote the total number and the number of reduced F-squares of type $F(n;\lambda_1,\lambda_2,\ldots,\lambda_m)$. Then,

$$F(n;\lambda_1,\lambda_2,\ldots,\lambda_m) = \left(\frac{n!}{\lambda_1!\lambda_2!\ldots\lambda_m!}\right)\left(\frac{(n-1)!}{(\lambda_1-1)!\lambda_2!\ldots\lambda_m!}\right)f(n;\lambda_1,\lambda_2,\ldots,\lambda_m).$$

They have also defined isotopy between F-squares by saying that two such squares of the same type are isotopic if one can be obtained from the other by permutations σ_r, σ_c and $\sigma_\#$ of the rows, columns and symbol set respectively.

Their main achievement is the enumeration of the reduced F-squares of all possible types up to order 6 and the classification of them into isotopy classes. Also, a representative square for each class is given, analogous to those given for each isomorphism class of latin squares of order six or less in [DK], pages 129-137.

Other (earlier) papers relevant to the topic of this Section are D.J.Finney(1982) and H.C.Kirton(1985).

(3) Completion of partial F-squares.

We devoted a large part of Chapter 8 to a discussion of partial latin squares and their completion. We included B.P.Smetaniuk's proof of the Evans' conjecture. In B.P.Smetaniuk(1983), this author has shown that the analogous problem of embedding a partial F-square in a complete F-square can be solved by reducing it to a corresponding problem about latin squares. To achieve this, Smetaniuk has shown that, to any given partial F-square A, there corresponds at least one derived partial latin square. Then he has proved that A can be completed if and only if at least one of its derived partial latin squares can be completed. We shall summarize the main steps.

DEFINITION. A partial $F(n;\lambda_1,\lambda_2,\ldots,\lambda_m)$ square defined on the symbol set $\Sigma = \{c_1,c_2,\ldots,c_m\}$ is an n×n matrix $A = ||a_{ij}||$ in which (i) each cell is either empty or else contains an element of Σ, (ii) the element c_i occurs at most λ_i times in each row and column of A.

DEFINITION. Let A be an n×n matrix and let s be an integer such that $1 \leq s \leq n$ and suppose that M is a set whose elements are cells of A. If M has the properties that each row and column of A has at most s cells which are elements of M and that at least one row or column of A has exactly s cells which are elements of M, we shall say that M is a partial s-region of A. If each row and column of A has s cells which are elements of M, we shall say that M is an s-region of A.

In particular, if $F(n;\lambda_1,\lambda_2,\ldots,\lambda_m)$ is an F-square defined on the symbol set $\Sigma = \{c_1,c_2,\ldots,c_m\}$ then those cells which contain the symbol c_i form a λ_i-region of F. If $F(n;\lambda_1,\lambda_2,\ldots,\lambda_m)$ is a partial F-square then those cells which contain the symbol c_i form a partial μ_i-region of F, where μ_i, $1 \leq \mu_i \leq \lambda_i$, is the maximum number of occurrences of c_i in any row or column of F.

For example, in the partial $F(6;3,3)$ square A defined on the symbol set $\{1,2\}$ which is shown in Figure 3.1, the cells which contain the symbol 1 form a partial 2-region of A while the cells which contain the symbol 2 form a partial 3-region of A.

$$\begin{bmatrix} 1 & 1 & 2 & . & . & . \\ 2 & . & 1 & 2 & 2 & . \\ 2 & 2 & . & . & . & . \\ 1 & 2 & . & . & . & 1 \\ . & . & 2 & . & 2 & 2 \\ . & 1 & 1 & . & 2 & . \end{bmatrix}$$

$$\begin{array}{|cccccc|} \hline 0+1 & 0+2 & 3+1 & . & . & . \\ 3+1 & . & 0+1 & 3+2 & 3+3 & . \\ 3+2 & 3+1 & . & . & . & . \\ 0+2 & 3+2 & . & . & . & 0+1 \\ . & . & 3+2 & . & 3+1 & 3+3 \\ . & 0+1 & 0+2 & . & 3+2 & . \\ \hline \end{array}$$

Figure 3.1   Figure 3.2

Let A be a partial $F(n;\lambda_1,\lambda_2,\ldots,\lambda_m)$ square on the symbol set $\Sigma = \{c_1,c_2,\ldots,c_m\}$ and let M^i denote the partial μ_i-region of A comprising those cells of A which contain the symbol c_i, where $1 \leq \mu_i \leq \lambda_i$. We can express M^i as a union of μ_i mutually disjoint partial 1-regions M^i_1, M^i_2, ..., $M^i_{\mu_i}$ by virtue of the following argument. First form an n×n matrix N of 0's and 1's by inserting a one in each cell which belongs to M^i and a zero in each other cell. By a theorem due to L. Mirsky (1971), N can be written as the sum of μ_i (0,1)-matrices $N_1, N_2, \ldots, N_{\mu_i}$ each of which contains at most one non-zero entry in any row or column. Those cells of N_j which contain 1 as entry define a partial 1-region M^i_j (j = 1,2,...,μ_i) and these partial 1-regions have the properties stated above. As an illustration, we carry out the decomposition for the partial $F(6;3,3)$ square which we gave in Figure 3.1. The decomposition is not unqiuely defined but one possible outcome is as shown in Figure 3.3.

Having separated the partial F-square A into partial 1-regions M^i_j in the manner just described, we can form from it an n×n partial latin square B based on the symbol set $\{1,2,\ldots,n\}$ by the following two construction rules:

(i) if cell (i,j) of A is empty, then cell (i,j) of B is empty;

(ii) if cell (i,j) of A contains the symbol c_ℓ and cell (i,j) of A belongs to the partial 1-region M^ℓ_k then insert the integer $k + \sum_{h<\ell} \lambda_h$ in cell (i,j) of B.

For example, if we carry out this construction for the partial F-square of Figure 3.1, we get the partial latin square illustrated in Figure 3.2.

The partial latin square B constructed in this way from the partial F-square A is called a <u>derived partial latin square</u> of A.

12:11 Frequency squares

$$\begin{bmatrix} 1 & 1 & \cdot & \cdot & \cdot & \cdot \\ \cdot & \cdot & 1 & \cdot & \cdot & \cdot \\ \cdot & \cdot & \cdot & \cdot & \cdot & \cdot \\ 1 & \cdot & \cdot & \cdot & \cdot & 1 \\ \cdot & \cdot & \cdot & \cdot & \cdot & \cdot \\ \cdot & 1 & 1 & \cdot & \cdot & \cdot \end{bmatrix} = \begin{bmatrix} 1 & \cdot & \cdot & \cdot & \cdot & \cdot \\ \cdot & \cdot & 1 & \cdot & \cdot & \cdot \\ \cdot & \cdot & \cdot & \cdot & \cdot & \cdot \\ \cdot & \cdot & \cdot & \cdot & \cdot & 1 \\ \cdot & \cdot & \cdot & \cdot & \cdot & \cdot \\ \cdot & 1 & \cdot & \cdot & \cdot & \cdot \end{bmatrix} + \begin{bmatrix} \cdot & 1 & \cdot & \cdot & \cdot & \cdot \\ \cdot & \cdot & \cdot & \cdot & \cdot & \cdot \\ \cdot & \cdot & \cdot & \cdot & \cdot & \cdot \\ 1 & \cdot & \cdot & \cdot & \cdot & \cdot \\ \cdot & \cdot & \cdot & \cdot & \cdot & \cdot \\ \cdot & \cdot & 1 & \cdot & \cdot & \cdot \end{bmatrix}$$

$\qquad\qquad M^1 \qquad\qquad\qquad M^1_1 \qquad\qquad\qquad M^1_2$

$$\begin{bmatrix} \cdot & \cdot & 2 & \cdot & \cdot & \cdot \\ 2 & \cdot & \cdot & 2 & 2 & \cdot \\ 2 & 2 & \cdot & \cdot & \cdot & \cdot \\ \cdot & 2 & \cdot & \cdot & \cdot & \cdot \\ \cdot & \cdot & 2 & \cdot & 2 & 2 \\ \cdot & \cdot & \cdot & \cdot & 2 & \cdot \end{bmatrix} = \begin{bmatrix} \cdot & \cdot & 2 & \cdot & \cdot & \cdot \\ 2 & \cdot & \cdot & \cdot & \cdot & \cdot \\ \cdot & 2 & \cdot & \cdot & \cdot & \cdot \\ \cdot & \cdot & \cdot & \cdot & \cdot & \cdot \\ \cdot & \cdot & \cdot & \cdot & 2 & \cdot \\ \cdot & \cdot & \cdot & \cdot & \cdot & \cdot \end{bmatrix} +$$

$\qquad\qquad M^2 \qquad\qquad\qquad M^2_1$

$$+ \begin{bmatrix} \cdot & \cdot & \cdot & \cdot & \cdot & \cdot \\ \cdot & \cdot & \cdot & 2 & \cdot & \cdot \\ 2 & \cdot & \cdot & \cdot & \cdot & \cdot \\ \cdot & 2 & \cdot & \cdot & \cdot & \cdot \\ \cdot & \cdot & 2 & \cdot & \cdot & \cdot \\ \cdot & \cdot & \cdot & \cdot & 2 & \cdot \end{bmatrix} + \begin{bmatrix} \cdot & \cdot & \cdot & \cdot & \cdot & \cdot \\ \cdot & \cdot & \cdot & \cdot & 2 & \cdot \\ \cdot & \cdot & \cdot & \cdot & \cdot & \cdot \\ \cdot & \cdot & \cdot & \cdot & \cdot & \cdot \\ \cdot & \cdot & \cdot & \cdot & \cdot & 2 \\ \cdot & \cdot & \cdot & \cdot & \cdot & \cdot \end{bmatrix}$$

$\qquad\qquad\qquad M^2_2 \qquad\qquad\qquad M^2_3$

Figure 3.3

We are now in a position to state Smetaniuk's main theorem:

<u>THEOREM 3.1</u> Let A be a partial $F(n; \lambda_1, \lambda_2, \ldots, \lambda_m)$ square defined on the set $\Sigma = \{c_1, c_2, \ldots, c_m\}$. Then A can be completed to an $F(n; \lambda_1, \lambda_2, \ldots, \lambda_m)$ square A^* if and only if there exists a derived partial latin square B of A such that B can be completed to a latin square.

Proof. Suppose first that B is a derived partial latin square of A which can be completed to a latin square D. We can use D to define an $F(n;\lambda_1,\lambda_2,\ldots,\lambda_m)$ square B^* as follows:

If d_{ij} denotes the (i,j)-th cell of D, then the (i,j)-th cell of A^* contains c_1 if $0 < d_{ij} \leq \lambda_1$, c_2 if $\lambda_1 < d_{ij} \leq \lambda_1+\lambda_2$, c_3 if $\lambda_1+\lambda_2 < d_{ij} \leq \lambda_1+\lambda_2+\lambda_3$, ..., c_m if $\sum_{k=1}^{m-1}\lambda_k < d_{ij} \leq \sum_{k=1}^{m}\lambda_k$. It is immediately evident that A^* is a completion of the partial F-square A.

Conversely, suppose that A can be completed to an $F(n;\lambda_1,\lambda_2,\ldots,\lambda_m)$ square A^*. We may use A^* to construct a latin square D in exactly the same way as we described above for constructing a partial latin square from a partial F-square. Let M_k^ℓ ($k = 1,2,\ldots,\lambda_\ell$; $\ell = 1,2,\ldots,m$) denote the 1-regions which arise in this construction. Now define a partial latin square B as follows: (i) if cell (i,j) of A is empty, then cell (i,j) of B is empty; (ii) if cell (i,j) of A contains the symbol c_ℓ and cell (i,j) of A^* belongs to the 1-region M_k^ℓ then insert the integer $k + \sum_{h<\ell} \lambda_h$ in cell (i,j) of B. Then evidently B is a derived partial latin square of A which can be completed to the latin square D. []

COROLLARY. Let A be a partial $F(n;\lambda_1,\lambda_2,\ldots,\lambda_m)$ square with n-1 cells occupied. Then A can be completed to an $F(n;\lambda_1,\lambda_2,\ldots,\lambda_m)$ square.

Proof. This result follows immediately from Trevor Evans' theorem described in Section 10 of Chapter 8. []

(4) F-rectangles and other generalizations

The generalized concepts of F-rectangle and F-hyper-rectangle have crept into the literature quite naturally and it is difficult to attribute priority. However, the following definitions were given in A.S.Hedayat and W.T.Federer(1984).

Frequency squares

DEFINITIONS. Let $V = \{1,2,\ldots,v\}$ be a set of v symbols. Then an r×c matrix whose elements belong to the set V is called an F-rectangle if (i) every element of V occurs the same number of times, h = rc/v, in the matrix, and (ii) the number of occurrences of each symbol in each row and column is as uniform as possible.

Two r×c F-rectangles F_1 and F_2 are called orthogonal if, when they are juxtaposed, each ordered pair of elements of V occurs equally often in the cells of the juxtaposed pair.

A consequence of the requirement (ii) in the above definitions is that if r ≤ v, no element of V appears more than once in each column, and if c = tv, each element occurs t times in each row.

In A.S.Hedayat and W.T.Federer(1984), it has been proved that:

THEOREM 4.1. (i) There exists a pair of orthogonal (v/2)×2v F-rectangles for all even integers v; (ii) For even values of v, the existence of t mutually orthogonal v×v latin squares of order v implies the existence of t mutually orthogonal (v/2)×2v F-rectangles.

Proof of (i). We construct three (v/2)×v F-rectangles A,B,C with cyclic permutations of the elements $1,2,3,\ldots,v$ as their rows in the following way. The rectangle A has $1,2,\ldots,v/2$ as entries in its first column, the rectangle B has $(v/2)+1, (v/2)+2,\ldots,v$ as entries in its first column and the rectangle C has the odd integers $1,3,\ldots,v-1$ as entries in its first column. It is then easy to see that the F-rectangles $F_1 = ||A:B||$ and $F_2 = ||C:C||$ are orthogonal. The case when v = 6 is illustrated in Figure 4.1.

Proof of (ii). Let $L_1, L_2, \ldots, L_t$ be a set of t mutually orthogonal latin squares of order v and suppose that A_i and B_i respectively denote the matrices formed of the first v/2 rows and the last v/2 rows of L_i. Then, if $F_i = ||A_i:B_i||$, the F-rectangles $F_1, F_2, \ldots, F_t$ are mutually orthogonal. []

```
1 2 3 4 5 6 4 5 6 1 2 3      1 2 3 4 5 6 1 2 3 4 5 6
2 3 4 5 6 1 5 6 1 2 3 4      3 4 5 6 1 2 3 4 5 6 1 2
3 4 5 6 1 2 6 1 2 3 4 5      5 6 1 2 3 4 5 6 1 2 3 4
```

Figure 4.1

In W.T.Federer, A.S.Hedayat and J.P.Mandeli(1984), it has been further proved that:

THEOREM 4.2. For all positive integers v, a pair of v×2v F-rectangle designs exists.

Proof. If L_1 and L_2 are two orthogonal latin squares of order v, we form the two F-rectangles $F_1 = ||L_1:L_1||$ and $F_2 = ||L_2:L_2||$. The cases v = 2,6 are treated separately. []

COROLLARY. A pair of orthogonal r×c F-rectangles which involve v distinct symbols exist whenever v ≠ 2,6 and both r and c are multiples of v or whenever r is a multiple of v and c is a multiple of 2v.

By extending their constructions slightly further, the same authors have proved:

THEOREM 4.3. A set of t mutually orthogonal pv×qv F-rectangles involving v distinct symbols exists whenever a set of t mutually orthogonal latin squares of order v exists, p and q being arbitrary positive integers.

They have shown that, in the above theorem, t is bounded above by the integer part of $(r-1)(c-1)/(v-1)$, where r = pv and c = qv. They have called a set of orthogonal F-rectangles of the above kind complete when t = $(pv-1)(qv-1)/(v-1)$ and have constructed such complete sets for some particular sets of values of p,q and v.

A more far-reaching generalization of the F-square was introduced by C.S.Cheng(1980). This author gave the following definitions of F-hyper-rectangle and orthogonal F-hyper-rectangles which embrace the corresponding definitions for latin squares, F-squares and latin cubes.

DEFINITIONS. Let H be an $N_1 \times N_2 \times \ldots \times N_n$ array involving v distinct symbols and suppose that v divides $\prod_{j \neq i} N_j$ for each i = 1,2,...,n. Then H is an F-hyper-rectangle if each of the v symbols occurs $v^{-1} (\prod_{j \neq i} N_j)$ times in the set of all cells $(j_1, j_2, \ldots, j_n)$ whose i-th component is a fixed integer:

that is, the set of all cells of every particular (n-1)-dimensional layer.

Two F-hyper-rectangles of the same size are <u>orthogonal</u> if, when they are placed in juxtaposition, every ordered pair of the symbols occurs the same number of times (that is: in the same number of pairs of corresponding cells).

Cheng showed that, for a set of t mutually orthogonal F-hyper-rectangles of size $N_1 \times N_2 \times \ldots \times N_n$ each containing the same number v of distinct symbols, we must have $t \leq (\prod_{j=1}^{n} N_j - \sum_{j=1}^{n}(N_j-1) -1)/(v-1)$. When this bound is attained the set of mutually orthogonal F-hyper-rectangles is a <u>complete set</u>. Cheng proved that:

<u>THEOREM 4.4</u>. If v is a prime power and each N_j is a power of v, then a complete set of mutually orthogonal $N_1 \times N_2 \times \ldots \times N_n$ F-hyper-rectangles, each involving v distinct symbols, exists.

J.P.Mandeli and W.T.Federer(1984a) have generalized Theorem 4.4 to the case when the F-hyper-rectangles in the pairwise orthogonal set have differing numbers of distinct symbols. They have drawn attention to the value of such sets in the design of multistage experiments. Indeed, many recent developments in the subject have been strongly influenced by its relevance to fractional factorial designs but we have not thought it appropriate to discuss this aspect further here.

Instead, we wish to point out that the same dichotomy occurs in regard to the definition of frequency cube as has occurred in defining the concept of latin cube. (See, for example, pages 181-189 of [DK] for a discussion of the latter.) The concept of frequency cube obtained by specializing Cheng's definition of F-hyper-rectangle given above is most appropriate for factorial design application but a more natural one from a mathematical point of view is that given in L.J.Brant and G.L.Mullen(1988), which is as follows:

DEFINITION. A three-dimensional <u>frequency cube</u> $C(n;\lambda_1,\lambda_2,\ldots,\lambda_m)$ is an n×n×n cube $C = (c_{ijk})$ having n^2 rows, n^2 columns and n^2 files and with the properties that (i) each of the elements c_{ijk} is an element of the set $\Sigma = \{1,2,\ldots,m\}$; and (ii) as one of the indices i,j,k varies from 1 to n while the other two remain fixed, the corresponding set $\{c_{ijk}\}$ of n elements contains the element h of Σ exactly λ_h times, h = 1,2,...,m. (It follows that $\lambda_1 + \lambda_2 + \ldots + \lambda_m = n$, as in the case of F-squares.)

Brant and Mullen have carried out an enumeration of such cubes for small values of n and have discussed their classification into isotopy classes.

(5) A generalized Bose construction for orthogonal F-squares

Since the above sections were written, G.L.Mullen(1988) has devised a method for constructing complete sets of mutually orthogonal F-squares which is quite independent of statistical designs and may be regarded as a natural generalization of the Bose construction [see R.C.Bose(1938)] for complete sets of mutually orthogonal latin squares.

The tool used is that of orthogonal permutation polynomials.

DEFINITIONS. A polynomial $f(x_1, x_2, \ldots, x_n)$ in n variables and with coefficients in the finite field GF[q] is called a <u>permutation polynomial</u> if the equation $f(x_1, x_2, \ldots, x_n) = \alpha$ has exactly q^{n-1} solutions for the n-tuple $(x_1, x_2, \ldots, x_n)$ in GF[q] for each choice of α in GF[q].

More generally, a set $f_1(x_1, x_2, \ldots, x_n)$, $f_2(x_1, x_2, \ldots, x_n)$, ..., $f_m(x_1, x_2, \ldots, x_n)$ of permutation polynomials with $1 \leq m \leq n$ is an <u>orthogonal set</u> in GF[q] if the system of equations $f_i(x_1, x_2, \ldots, x_n) = \alpha_i$ (i = 1,2,...,m) has exactly q^{n-m} solutions for the n-tuple $(x_1, x_2, \ldots, x_n)$ for each choice of constants $\alpha_1, \alpha_2, \ldots, \alpha_m$.

As an illustration of these definitions, suppose that a, b, α, β are elements of GF[q]. Then the pair of equations $au+v = \alpha$ and $bu+v = \beta$ have $q^{2-2} = 1$ solutions for u,v in GF[q], so the polynomials $f_a(u,v) \equiv au+v$ and $f_b(u,v) \equiv bu+v$ are orthogonal permutation polynomials in the variables u and v. We note that a consequence of this fact is that the latin squares L_a, L_b whose (u,v)-th cells contain $f_a(u,v)$ and $f_b(u,v)$ respectively are orthogonal. More generally, the $q-1$ latin squares L_{a_i} ($i = 1, 2, \ldots, q-1$), where $a_1, a_2, \ldots, a_{q-1}$ are the non-zero elements of GF[q] and where the (u,v)-th cell of L_{a_i} contains $a_i u + v$, are pairwise orthogonal. This is the Bose construction and is the one which has been generalized by Mullen.

Let $n = p^s$, where p is a prime, and let $m = p^{s/i}$. Then, by Theoerem 1.1 of this Chapter, the maximum possible number of mutually orthogonal F-squares of type $F(n; \lambda^m)$, where $\lambda = p^{(i-1)s/i}$, is $(p^s-1)^2/(p^{s/i}-1)$. Moreover, by Theorem 1.2, such a complete set always exists. Mullen describes the entries of each square of the set by means of a set of orthogonal permutation polynomials in $2i$ variables. If $f_h(u_0, u_1, \ldots, u_{i-1}, v_0, v_1, \ldots, v_{i-1})$ is one of these permutation polynomials, then $u = u_0 + u_1 p^{s/i} + u_2 p^{2s/i} + \ldots + u_{i-1} p^{(i-1)s/i}$ and
$v = v_0 + v_1 p^{s/i} + v_2 p^{2s/i} + \ldots + u_{i-1} p^{(i-1)s/i}$
are integers lying between 0 and p^s-1. In Mullen's construction, the (u,v)-th cell of the h-th F-square of type $F(n; \lambda^m)$ contains $f_h(u_0, u_1, \ldots, u_{i-1}, v_0, v_1, \ldots, v_{i-1})$.

Mullen's main theorem is as follows:

THEOREM 5.1. The $(p^s-1)^2/(p^{s/i}-1)$ permutation polynomials
$$f_{(a_1, a_2, \ldots, a_{2i})}(x_1, x_2, \ldots, x_{2i}) = a_1 x_1 + a_2 x_2 + \ldots + a_{2i} x_{2i}$$
over GF[q], where neither of the sets of coefficients $a_1, a_2, \ldots, a_i$ or $a_{i+1}, a_{i+2}, \ldots, a_{2i}$ are simultaneously all zeros and where no two of the polynomials are (non-zero) multiples one of the other in $GF[p^{s/i}]$, represent (in the way described above) a complete set of mutually orthogonal frequency squares of type $F(p^s; \lambda^m)$, where $\lambda = p^{(i-1)s/i}$ and $m = p^{s/i}$.

The three conditions on the coefficients of the permutation polynomials may be stated as follows:

(i) $(a_1, a_2, \ldots, a_i) \neq (0, 0, \ldots, 0)$;
(ii) $(a_{i+1}, a_{i+2}, \ldots, a_{2i}) \neq (0, 0, \ldots, 0)$;
(iii) $(a'_1, a'_2, \ldots, a'_{2i}) \neq t(a_1, a_2, \ldots, a_{2i})$ for any non-zero $t \in GF[p^{s/i}]$.

As an example of the application of the theorem, let us take $p = 2$ and $s = i = 2$. Then the three permutation polynomials

$$f_1(x_1,x_2,x_3,x_4) = x_1 +x_3$$
$$f_2(x_1,x_2,x_3,x_4) = x_2+x_3$$
$$f_3(x_1,x_2,x_3,x_4) = x_1+x_2 +x_4$$

are orthogonal and represent the three frequency squares F_1, F_2, F_3 of type $F(4; 2^2)$ which are illustrated in Figure 1.2 (with the symbols 1,2 replaced by 0,1 respectively). They form part of a set of nine such permutation polynomials (constructed as in Theorem 5.1) and of nine corresponding mutually orthogonal frequency squares of type $F(4; 2^2)$.

[The reader should note that, in the example of a set of nine mutually orthogonal frequency squares of type $F(4; 2^2)$ whose permutation polynomials are given in G.L.Mullen(1988), the roles of x_1 and x_2 and also those of x_3 and x_4 have been interchanged.]

In a private communication to the authors, G.L.Mullen has conjectured that the existence of a complete set of mutually orthogonal frequency squares of the type described in Theorem 5.1 is equivalent to the existence of an affine geometry $AG(2i, m)$: that is, an affine geometry of dimension 2i over the finite field $GF[m]$, where $m = p^{s/i}$. He has provided strong supporting evidence for the conjecture which, when $i = 1$, reduces to the well-known equivalence between a complete set of mutually orthogonal latin squares and an affine plane of the same order.

ADDITIONAL REMARKS

Since the above Chapter was written, the Authors have discovered that they were mistaken in attributing the origins of the concept of frequency square to D.J.Finney and other authors of the present century. The concept was in fact introduced by P.A.MacMahon(1898) under the name of quasi-latin square. MacMahon defined a reduced frequency square in exactly the way which we have attributed to L.J.Brant and G.L.Mullen and he also proved Theorem 2.1. He enumerated the reduced frequency squares of order four and separated them into isotopy classes. For the details, see P.A.MacMahon(1898), pages 276-280.

BIBLIOGRAPHY

Notes on the Bibliography.

(1) The italicized numbers which follow each bibliographic item indicate the Chapter and Section in which that item is referred to in the text. For example, the number *6.2* indicates that the item is referred to in Section 2 of Chapter 6.

(2) Because of limitations on space, only items which are cited in the text are included in the Bibliography.

(3) A $ following the date of an item indicates that the paper has been reprinted in the series "Advances in Discrete Mathematics and Computer Science", Hadronic Press, Nonantum, Massachusetts, U.S.A.

(4) For ease of reference, the names of authors of papers which have joint authorships have been re-arranged into alphabetical order and the paper is listed under the author whose initial letter comes earliest in the alphabet.

Addelman S.
   (1967) Equal and proportional frequency squares. J. Amer. Statist. Assoc. 62, 226-240. Not reviewed in MR. *12.0*

Afsarinejed K.
   (1986) Self-orthogonal Knut Vik designs. Statist. Prob. Letters 4, 289. MR 87j:62136. *5A*
   (1987) On mutually orthogonal Knut Vik designs. Statist. Prob. Letters 5, 323-324. MR88i:62038. *5A*

Afsarinejed K. and Hedayat A.
   (1975) Repeated measurements designs I. Proc. Internat. Sympos. Colorado State Univ., 1973, pp. 229-242. North Holland, Amsterdam. MR51(1976)#11869. *3A*
   (1978) Repeated measurements designs II. Ann. Statist. 6, 619-628. MR58(1979)#8058. *3.1, 3A*

Alimena B.S.
   (1962) A method of determining unbiased distribution in the Latin square. Psychometrika 27, 315-317. MR26(1963)#865. *3A*

Andersen L.D.
   (1982) Embedding latin squares with prescribed diagonal. Annals of Discrete Math. 15, 9-26. MR86a:05021. *8A*

(1985) Completing partial latin squares. Mat. Fys. Medd. Dan. Vid. Selsk. 41(1), 23-69. MR87k:05035. *8.10*

Andersen L.D., Häggkvist R., Hilton A.J.W. and Poucher W.B.
(1980) Embedding incomplete latin squares in latin squares whose diagonal is almost completely prescribed. European J. Combinatorics 1, 5-7. MR81i:05020b. *8A*

Andersen L.D. and Hilton A.J.W.
(1983) Thank Evans! Proc. London Math. Soc. (3) 47, 507-522. MR 85g:05034a. *8.5, 8.10*
(1987) The existence of symmetric latin squares with one prescribed symbol in each row and column. Annals of Discrete Math. 34, 1-26. MR 89a:05035. *8.10*
(199α) Symmetric latin square and complete graph analogues of the Evans conjecture. To be submitted for publication. *8.10*

Andersen L.D., Hilton A.J.W. and Mendelsohn E.
(1980) Embedding partial Steiner triple systems. Proc. London Math. Soc. (3) 41, 557-576. MR82a:05010. *8.8, 8.9*

Andersen L.D., Hilton A.J.W. and Rodger C.A.
(1982) A solution to the problem of embedding partial idempotent latin squares. J. London Math. Soc. (2) 26, 21-27. MR83i:05016. *8.2, 8.5, 8.9*

Andersen L.D. and Mendelsohn E.
(1982) A direct construction for latin squares without proper subsquares. Annals of Discrete Math. 15, 27-53. MR85m:05021. *4.2*

Anderson B.A.
(1973) A perfectly arranged Room square. Proc. Fourth S.E. Conf. on Combinatorics, Graph Theory and Computing, 1973. Congressus Numerantium 8, 141-150. MR50(1975)#6918. *4.2*
(1974) A class of starter induced one-factorizations. In "Graphs and Combinatorics", pp. 180-185. Lecture Notes in Mathematics No. 406, Springer Verlag, Berlin. MR51(1976)#268. *4.2*
(1975)$ Sequencings of certain dihedral groups. Proc. Sixth S.E. Conf. on Combinatorics, Graph Theory and Computing, 1975. Congressus Numerantium 14, 65-76. MR52(1976)#13437. *3.1*
(1976)$ Sequencings and starters. Pacific J. Math. 64, 17-24. MR55(1978) #161. *3.6*
(1987a) A fast method for sequencing low order non-abelian groups. Annals of Discrete Math. 34, 27-42. MR89f:20033. *3.1, 3.2, 3.4, 3.6*
(1987b) Sequencings of dicyclic groups. Ars Combinatoria 23, 131-142. MR88j:20022. *3.2, 3.6*
(1987c) S_5, A_5 and all non-abelian groups of order 32 are sequenceable. Congressus Numerantium 58, 53-68. MR? *3.2, 3.6*
(198α) Sequencings of dicyclic groups II. J. Combinatorial Math. and Combinatorial Comput. 3(1988), 5-27. *3.2, 3.6*
(198β) All dicyclic groups of order at least 12 have symmetric sequencings. Contemporary Math. (AMS). To appear. *3.2, 3.6*
(198γ) A product theorem for 2-sequencings. Discrete Math. To appear. *3.2, 3.6*

Anderson B.A. and Leonard P.A.
(1981) Sequencings and Howell designs. Pacific J. Math. 92, 249-256. MR 82g:05032. *3.1*
(1988) Symmetric sequencings of finite Hamiltonian groups with a unique element of order 2. Congressus Numerantium 65, 147-158. MR? *3.6*

Anderson B.A. and Morse D.
(1974) Some observations on starters. Proc. Fifth S.E. Conf. on Combinatorics, Graph Theory and Computing, 1974. Congressus Numerantium 10, 229-235. MR51(1976)#2975. *4.2*

Andrew A.M.
(1975) Decimal error-correction: a solution. Computer J. 18, 284-285. Not reviewed in MR. *9.2*

Anscombe F.J.
(1948) The validity of comparative experiments (with discussion). J. Royal Statist. Soc. 111, 181-211. MR10(1949), page 724. *10.2*

Anstee R.P., Hall M. and Thompson J.G.
(1980) Planes of order 10 do not have a collineation of order five. J. Combinatorial Theory A29, 39-58. MR82c:51012. *11.2*

Archdeacon D.A., Dinitz J.H., Stinson D.R. and Tillson T.W.
(1980) Some new row-complete latin squares. J. Combinatorial Theory A29, 395-398 and A30(1981), 116. MR82d:05031. *3.1*

Arlazarov V.L., Baraev A.M., Faradzev I.A. and Golfand Ja.Ju.
(1978) The construction with the aid of a computer of all latin squares of order eight. (Russian) In "Algorithmic Studies in Combinatorics", pp. 129-141 and 187. Izdat. "Nauka", Moscow. MR80a:05031. *11A*

Atkin A.O.L., Hay L. and Larson R.G.
(1977) Construction of Knut Vik designs. J. Statist. Planning and Inference 1, 289-297. MR58(1979)#10526. *5A*

Bailey R.A.
(1982) Latin squares with highly transitive automorphism group. J. Austral. Math. Soc. A33, 18-22. MR83g:05021. *4.3*
(1984) Quasi-complete latin squares: construction and randomization. J. Royal Statist. Soc. B46, 323-334. MR86i:62161. *3.2*
(1986) Private communication to the Authors. *3.6*

Bailey R.A., Patterson H.D. and Preece D.A.
(1978) A randomization problem in forming designs with superimposed treatments. Austral. J. Statist. 20(2), 111-125. MR82k:62153. *10.8*

Bailey R.A. and Praeger C.E.
(1988) Directed terraces for direct product groups. Ars Combinatoria 25A, 73-76. MR89i:20041. *3.6*

Baker G.A. and Riddle O.C.
(1944) Biasses encountered in large-scale yield trials. Hilgardia 16, 1-14.

Not reviewed in MR. *10.6*

Bammel S.E. and Rothstein J.
(1975) The number of 9×9 latin squares. Discrete Math. 11, 93-95. MR 51(1976)#7882. *11A*

Barlotti A.
(1976) Alcune questioni combinatorie nello studio delle strutture geometriche finite. Colloquio Internazionale sulle Teorie Combinatorie (Roma 1973), Tomo II, pp. 423-431. Atti dei Convegni Lincei, No.17, Accad. Naz Lincei, Rome. MR57(1979)#149. *11A*

Barlotti A. and Strambach K.
(1983) The geometry of binary systems. Advances in Math. 49, 1-105. MR 84k:57005. *11A*

Bartlett M.S.
(1978) Nearest neighbour models in the analysis of field experiments (with discussion). J. Royal Statist. Soc. B40, 147-174. Not reviewed in MR. *3A*

Bates G.E.
(1947) Free loops and their generalizations. Amer. J. Math. 69, 499-550. MR9(1948), page 8. *7.5*

Batten L.M.
(1986) Combinatorics of Finite Geometries. Cambridge Univ. Press. MR 87g:51030. *5.2*

Baumert L.D. and Hall M.
(1973)$ Non-existence of certain planes of order 10 and 12. J. Combinatorial Theory A14, 273-280. MR47(1974)#3206. *11.2*

Belousov V.D.
(1965) Systems of quasigroups with generalized identities. (Russian) Usp. Mat. Nauk. 20, No.1(121), 75-146. Translated as Russian Math. Surveys 20(1965), 73-143. MR30(1965)#3934. *1.3*
(1966) Balanced identities in quasigroups. (Russian) Mat. Sb. (N.S.) 70(112), 55-97. MR34(1967)#2757. *1.3*
(1967a) Non-associative binary systems. (Russian) Algebra, Topology, Geometry 1965, 63-81. Akad. Nauk SSSR Inst. Naučn. Tehn. Informacii, Moscow, 1967. MR35(1968)#5537. *(Preface)*
(1967b) Foundations of the Theory of Quasigroups and Loops. (Russian) Izdat. "Nauka", Moscow, 1967. MR36(1968)#1569. *6.3, 6.4*
(1968) Systems of orthogonal operations. (Russian) Mat. Sb. (N.S.) 77(119), 38-58. Translated as Mathematics of the USSR: Sbornik 5(1968), 33-52. MR38(1969)#1200. *6.2*
(1971) Algebraic Nets and Quasigroups. (Russian) Izdat. "Štiince" Kishinev, 1971. MR49(1975)#5214. *11A*
(1972) n-ary Quasigroups. (Russian) Izdat. "Stiince" Kishinev, 1972. MR 50(1975)#7396. *2.6*
(1979) Configurations in Algebraic Nets. (Russian) Izdat. "Štiince" Kishinev, 1979. MR80k:20074. *11A*

Belousov V.D. and Belyavskaya G.B.
   (1989) Latin Squares, Quasigroups and their Applications. Izdat. "Štiince" Kishinev, 1989. *(Preface)*

Belyavskaya G.B.
   (1976) r-orthogonal quasigroups I. (Russian) Mat. Issled. 39, 32-39. MR55(1978)#8283. *4.4, 6.0, 6.2, 9.2*
   (1977) r-orthogonal quasigroups II. (Russian) Mat. Issled. 43, 39-49. MR56(1978)#12162. *4.4, 6.0, 6.3*
   (1979) Construction of (n^2-2)-orthogonal quasigroups of even order, where $n-1 \neq 0 \pmod 3$. (Russian) Mat. Issled. 51, 23-26. MR80m:20053. *4.4, 6.5*
   (1982a) On spectra of partial admissibility of finite quasigroups (latin squares). (Russian) Mat. Zametki 32, 777-788. [English translation: Math. Notes 32(1982), 874-880(1983).] MR84j:05024. *6.3*
   (1982b) On spectra of partial orthogonality of quasigroups of small orders. (Russian) Mat. Issled. 66, 7-14. MR83j:20067. *6.4*
   (1983) On partially orthogonal quasigroups and systems of quasigroups. (Russian) Mat. Issled. 71, 25-33. MR85h:20087. *6.4, 6.6*

Belyavskaya G.B. and Nazarok A.V.
   (1987) Completion of groups and construction of orthogonal quasigroups of order $3t+i$, $i=0,1,2$, $t \neq 2,6$, with orthogonal subquasigroups of order t. (Russian) Mat. Issled. 95, 39-52. MR88i:20103. *4.4*

Belyavskaya G.B. and Russu A.F.
   (1975) On the admissibility of quasigroups. (Russian) Mat. Issled. 10, No.1(35), 45-57. MR52(1976)#3407. *2.3*
   (1976) On partial admissibility of quasigroups. (Russian) In "Quasigroups and Combinatorics". Mat. Issled. 43, 50-58. MR58(1979)#969. *2.2, 6.3*

Bennett F.E. and Mendelsohn N.S.
   (1980) On the spectrum of Stein quasigroups. Bull. Austral. Math. Soc. 21, 47-63. MR81g:20129. *4.4*

Bennett F.E. and Zhu L.
   (1987) Incomplete conjugate orthogonal idempotent latin squares. Discrete Math. 65, 5-21. MR88f:05025. *4.4*

Beth T.
   (1983) Einige Bemerkungen zur Abschätzung der Anzahl der orthogonalen lateinischen Quadrate mittels Siebverfahren. Abh. Math. Sem. Univ. Hamburg 53, 284-288. MR86f:05032. *5.6, 5A*

Betten D.
   (1983) Zum Satz von Euler-Tarry. Math. Nat. Unt. 36, 449-453. Not reviewed in MR. *1.2*
   (1984) Die 12 lateinischen Quadrate der Ordnung 6. Mit. Mat. Sem. Giessen No. 163, 181-188. MR86j:05038. *1.2*

Beutelspracher A.
   (1984) Universal algebra and combinatorics: a series of problems. Proc. Conf. on "Universal algebra and its links with logic, algebra,

combinatorics and computer science" (Darmstadt, 1983), pp.168-172. Heldermann, Berlin. MR86i:05026. *7.5*

Bierbrauer J.
(1985a) Blocking sets of 16 points in projective planes of order 10, Part II. Quart. J. Math. (2)36, 383-391. MR88a:51014a. *11.7*
(1985b) Blocking sets of 16 points in projective planes of order 10, Part III. Rend. Sem. Mat. Univ. Padova 74, 163-174. MR88a:51014b. *11.7*
(1985c) Necessary conditions for additive loops of finite projective planes. Geometriae Dedicata 19, 207-216. MR87a:05036. *11.7*
(198α) On projective planes whose binary code contains vectors of small weight. Combinatorica, submitted. *11.7*

Birkhoff G.
(1935) On the structure of abstract algebras. Proc. Cambridge Phil. Soc. 31, 433-454. *7.1*

Bonisoli A.
(1984) A class of variable order MDS codes. Proc. Conf. on combinatorial and incidence geometry: principles and applications (La Mendola, 1982), pp. 151-159. Rend. Sem. Mat. Brescia 7, Milan. MR84g:51001. *9.2*

Bonisoli A. and Fiori C.
(199α) MDS-codes, nets and column orthogonal latin rectangles. Preprint. *9 A*

Bosák I.
(1976) Latinske Stvorce. Vydalu v matematičke olympiade nakladatesvi. Mlada Fronta, Praha. (Slovak) Not reviewed in MR. *(Preface)*

Bose R.C.
(1938) On the application of the properties of Galois fields to the construction of hyper-Graeco-Latin squares. Sankhya 3, 323-338. Not reviewed in MR. *9.5, 11A, 12.5*
(1939) On the construction of balanced incomplete block designs. Ann. Eugenics 9, 353-399. MR1(1940), page 199. *8.8*

Bose R.C., Chakravarti I.M. and Knuth D.E.
(1960)$ On methods of constructing sets of mutually orthogonal latin squares using a computer I. Technometrics 2, 507-516. MR 23(1962)#A3099. *11.3*
(1961)$ On methods of constructing sets of mutually orthogonal latin squares using a computer II. Technometrics 3, 111-117. MR 23(1962)#A3100. (See also next entry.) *11.3*
(1978) Errata: On methods of constructing sets of mutually orthogonal latin squares using a computer II (Technometrics 3(1961), 111-117). Technometrics 20, 219. MR80a:05033.

Bose R.C. and Nair K.R.
(1941) On complete sets of latin squares. Sankhya 5, 361-382. MR4(1943), page 33. *11.6, 11.7, 11A*

Bose R.C., Parker E.T. and Shrikhande S.S.
(1960) Further results on the construction of mutually orthogonal latin squares and the falsity of Euler's conjecture. Canad. J. Math. 12, 189-203. MR23(1962)#A69.   *1.2, 4.4, 5.6*

Bose R.C. and Shrikhande S.S.
(1959) On the falsity of Euler's conjecture about the nonexistence of two orthogonal latin squares of order 4t+2. Proc. Nat. Acad. Sci. USA 45, 734-737. MR21(1960)#3343. (See also next entry.)   *4.1, 4.4*
(1960) On the construction of sets of mutually orthogonal latin squares and the falsity of a conjecture of Euler. Trans. Amer. Math. Soc. 95, 191-209. MR22(1961)#2557.   *5.3*

Bossen D.C., Chien R.T. and Hsiao M.Y.
(1970) Orthogonal latin square codes. IBM J. Res. Develop. 14, 390-394. MR 43(1972)#4548.   *9.2*

Bourn S.
(1983) A canonical form for incidence matrices of finite projective planes and their associated latin squares and planar ternary rings. In "Combinatorial Mathematics X" (Proceedings of the Conference held in Adelaide, Australia, 1982), pp. 111-120. Lecture Notes in Math. No.1036. Springer Verlag, Berlin. MR85f:05035.   *11.4, 11A.*

Box J.F.
(1978) R.A. Fisher: The Life of a Scientist. Wiley, New York. Not reviewed in MR.   *10.3, 10.4*

Bradley J.V.
(1958)$ Complete counterbalancing of immediate sequential effects in a latin square design. J. Amer. Statist. Assoc. 53, 525-528. Not reviewed in MR.   *3A*

Brant L.J. and Mullen G.L.
(1986) Some results on enumeration and isotopic classification of frequency squares. Utilitas Math. 29, 231-244. MR88a:05024.   *12.2, 12A*
(1988) A study of frequency cubes. Discrete Math. 69, 115-121. MR 89b:05050.   *12.4*

Brayton R.K., Coppersmith D. and Hoffman A.J.
(1974) Self-orthogonal latin squares of all orders $n \neq 2,3,6$. Bull. Amer. Math. Soc. 80, 116-118. MR48(1974)#5886.   *4.1*
(1976) Self-orthogonal latin squares. Colloq. Internat. sulle Teorie Combinatorie (Rome 1973), Tomo II, pp. 509-517. Atti dei Convegni Lincei, Rome. MR57(1979)#16101.   *4.1*

Brouwer A.E.
(1978) Mutually orthogonal latin squares. Report ZN81. Math. Centrum, Amsterdam. Not reviewed in MR.   *5.4*, 5.5*
(1979) The number of mutually orthogonal latin squares, a table up to order 10,000. Report ZW123. Math. Centrum, Amsterdam. MR80f:05013.   *4.1, 5.4*, 5.5*
(1980a) A series of separable designs with application to pairwise

orthogonal latin squares. European J. Combinatorics 1, 39-41. MR 81e:05035. *5.3**
(1980b) On the existence of 30 mutually orthogonal latin squares. Report ZW 136. Math. Centrum, Amsterdam. MR81d:05013. *5.5*
(1984) Four MOLS of order 10 with a hole of order 2. J. Statist. Planning and Inference 10, 203-205. MR86c:62091. *4A, 5.0, 6.6, 11.1, 11A*

Brouwer A.E. and van Rees G.H.J.
(1982) More mutually orthogonal latin squares. Discrete Math. 39, 263-282. MR84c:05019. *4.1, 5.5, 5.6*

Brouwer A.E., de Vries A.J. and Wieninga R.M.A.
(1978) A lower bound for the length of partial transversals in a latin square. Nieuw Archief voor Wiskunde (3)26, 330-332. MR58(1979)#282. *2.2*

Brown J. M. Nowlin
(1983a) Elations of order 3 in projective planes of order 12. In "Finite Geometries", Proc. Conf. at Pullman, Washington, pp.61-65. Lecture Notes in Pure and Applied Math. 82, Marcel Dekker AG, Basel. MR84f:51019. *11.2*
(1983b) On planes of order pt with collineation group of order p^3. (Lecture at Westfield College, Univ. London, 15th March, 1983.) *11.2*

Brown J.W.
(1968) Enumeration theory of latin squares with application to order 8. J. Combinatorial Theory 5, 177-184. MR37(1969)#5111. *11A*

Brown J.W., Cherry F., Most L., Most M., Parker E.T. and Wallis W.D.
(199α) The spectrum of orthogonal diagonal latin squares. Preprint. *5A*

Brown J.W. and Parker E.T.
(1982) A try for three order-10 orthogonal latin squares. Proc. Thirteenth S.E. Conf. on combinatorics, graph theory and computing, Boca Raton. Congressus Numerantium 36, 143-144. MR85f:05028. *11A*
(1984) Some attempts to construct orthogonal latin squares. Congressus Numerantium 43, 201-202. MR86e:05016. *11A*
(1985) An attempt to construct three MOLS of order ten with a common transversal. Proc. Conf. on Groups and Geometry, Part A (Madison, Wisconsin, 1985). Algebras, Groups and Geometries 2, 258-262. MR 87g:05045. *11A*

Bruck R.H.
(1951) Finite nets I. Numerical invariants. Canad. J. Math. 3, 94-107. MR 12(1951), page 580. *2.4*
(1955) Difference sets in a finite group. Trans. Amer. Math. Soc. 78, 464-481. MR16(1955), page 1081. *11.2*
(1958) A Survey of Binary Systems. Springer Verlag, Berlin. MR20(1959) #76. *(Preface)*
(1963) Finite nets II. Uniqueness and embedding. Pacific J. Math. 13, 421-457. MR27(1964)#4768. *2.5, 11.2*
(1973) Construction problems in finite projective spaces. In "Finite Geometric Structures and their Applications" (CIME, Bressanone, 1972), pp. 105-188. Edizioni Cremonese, Rome. MR49(1975)#7159. *11.2*

Bruen A.A.
(1971) Blocking sets in finite projective planes. SIAM J. Appl. Math. 21, 380-392. MR46(1973)#2543.   *11.7*
(1972) Unimbeddable nets of small deficiency. Pacific J. Math. 43, 51-54. MR52(1976)#1511.   *2.5*

Bruen A.A. and Fisher J.C.
(1973) Blocking sets, k-arcs and nets of order ten. Advances in Math. 10, 317-320. MR47(1974)#6515.   *2.5, 6.7, 11.2, 11.7*
(1974) Blocking sets and complete k-arcs. Pacific J. Math. 53, 73-84. MR 51(1976)#10127.   *2.5*

Bruen A.A. and Silverman R.
(1981) Arcs and blocking sets. In "Finite geometries and designs". (Proc. Conf., Chelwood Gate, 1980), pp. 52-60. London Math. Soc. Lecture Notes No. 49, Cambridge Univ. Press. MR83h:05024.   *11.7*

Bryant V.W.
(1984) Extending latin rectangles with restraints. European J. Combinatorics 5, 17-21. MR86a:05022.   *8.5*

Bugelski B.R.
(1949) A note on Grant's discussion of the latin square principle in the design of experiments. Psychological Bulletin 46, 49-50. Not reviewed in MR.   *3.0, 10.5*

Bush K.A.
(1952) A generalization of a theorem due to MacNeish. Ann. Math. Statist. 23, 293-295. MR14(1953), page 125.   *5.3*

Butler C.G., Finney D.J. and Schiele P.
(1943) Experiments on the poisoning of honeybees by insecticidal and fungicidal sprays used in orchards. Ann. Appl. Biology 30, 143-150. Not reviewed in MR.   *10.3*

Campbell G. and Geller S.
(1980) Balanced Latin Squares. University of Purdue preprint, Dept. of Statistics, Mimeoseries No. 80-26.   *3.2*

Carlitz L.
(1953) A note on abelian groups. Proc. Amer. Math. Soc. 4, 937-939. MR 15(1954), page 503.   *3.3*

Carmichael R.D.
(1937) Introduction to the Theory of Groups of Finite Order. Ginn and Co., Boston; Dover Reprint, 1956. MR17(1956), page 823.   *11.7*

Cheng C.S.
(1980) Orthogonal arrays with variable numbers of symbols. Ann. Statist. 8, 447-453. MR81d:05014.   *12.4*

Chor B., Leiserson C.E. and Rivest R.L.
(1982) An application of number theory to the organization of a raster-

graphics memory. 23rd Annual Symposium on foundations of Computer Science (Chicago III, 1982), pp. 92-99. IEEE, New York. MR85k:68007. *9.4*

Chowla S., Erdös P. and Straus E.G.
(1960) On the maximal number of pairwise orthogonal latin squares of a given order. Canad. J. Math. 12, 204-208. MR23(1962)#A70. *5.6*

Cigić V.
(1983) A theorem on finite projective planes of odd order and an application to planes of order 15. Arch. Math. (Basel) 41, 280-288. MR 85b:51010. (See also next entry.) *11.2*
(1984) Correction to the preceding paper. Arch. Math. (Basel) 43, 576. MR 86c:51011. *11.2*
(1985) Some possibilities of a collineation of prime order p on projective planes of order p+2 and p+3. Akad. Nauk. Umjet. Bosne Hercegov. Rad. Odjelj. Prirod. Mat. Nauka No. 24, 45-51. MR87j:51015. *11.2*

Cigić V. and Janko Z.
(1985) On planar collineations of order 13 acting on projective planes of order 16. Rad. Mat. 1, 163-172. MR87g:51012. *11.2*

Cochran W.G. and Cox G.M.
(1950) Experimental Designs. Chapman and Hall, London; Wiley, New York. MR11(1950), page 607. 2nd Edition 1957. MR19(1958), page 75. *10.3*

Cohen D. and Etzion T.
(1989) Row-complete latin squares which are not column-complete. Tech. Report No.588, Technion, Haifa, Israel. Ars Combinatoria. To appear. *3A*

Cooper G.R. and Yates R.D.
(1966) Design of large signal sets with good aperiodic correlation properties. Tech. Report TR-EE 66-13. Purdue University, West Lafayette, Indiana. *9.3*

Costas J.P.
(1984) A study of a class of detection waveforms having nearly ideal range-Doppler ambiguity properties. Proc. IEEE 72, 996-1009. Not reviewed in MR. *9.4*

Cox D.R.
(1958a) Planning of Experiments. Wiley, New York. MR20(1959)#2063. *10.2*
(1958b) The interpretation of the effects of non-additivity in the latin square. Biometrika 45, 69-73. Not reviewed in MR. *10.7*

Crampin D.J. and Hilton A.J.W.
(1975a) Remarks on Sade's disproof of the Euler conjecture with an application to latin squares orthogonal to their transposes. J. Combinatorial Theory A18, 47-59. MR51(1976)#197. *1.2, 4.1*
(1975b) On the spectra of certain types of latin square. J. Combinatorial Theory A19, 84-94. MR51(1976)#12563. *4.4*

Cretté de Palluel F.
(1788) Sur les avantages et l'économie que procurent les racines employées à l'engrais des moutons à l'étable. Mém. d'Agric., Trimestre d'Été, 17-23. English translation: Ann. Agric. 14(1790), 133-139. *10.4*

Cruse A.B.
(1974a) On embedding incomplete symmetric latin squares. J. Combinatorial Theory A16, 18-27. MR48(1974)#8265. *8.2, 8.4, 8.9*
(1974b) On the finite completion of partial latin cubes. J. Combinatorial Theory A17, 112-119. MR49(1975)#10584. *8A*
(1974c) On extending incomplete latin rectangles. Proc. Fifth S.E. Conf. on Combinatorics, Graph Theory and Computing, 1974. Congressus Numerantium 10, 333-348. MR50(1975)#6886. *8A*
(1975) A number theoretic function related to latin squares. J. Combinatorial Theory A19, 264-277. MR52(1976)#13427. *8A*

Cruse A.B. and Lindner C.C.
(1975) Small embeddings for partial semisymmetric and totally symmetric quasigroups. J. London Math. Soc. (2)12(1975/1976), 479-484. MR55(1978)#5779. *8.7, 8.8, 8.9*

Dacić R.
(1978) On the completion of incomplete latin squares. Publ. Inst. Math. (Beograd.) N.S. 23(37), 75-80. MR80a:05034. *8A*

Damerell R.M.
(1983) On Smetaniuk's construction for latin squares and the Andersen-Hilton theorem. Proc. London Math. Soc. (3)47, 523-526. MR85g:05034b. *8.10*

D'Angelo M. and Turgeon J.M.
(1982) Finite groups and homogeneous latin squares. University of Montreal Preprint No.82-19. *4.3*

Davies O.L. ed.
(1956) The Design and Analysis of Industrial Experiments. (2nd. Ed.) Oliver and Boyd, Edinburgh. Not reviewed in MR. *10.4*

Daykin D.E. and Häggkvist R.
(1984) Completion of sparse partial latin squares. In "Graph Theory and Combinatorics", Proc. Combinatorial Conf. in honour of Paul Erdős (Cambridge, 1983), pp. 127-132. Academic Press, New York. MR86e:05018. *4.3*

Dembowski P.
(1968) Finite Geometries. Springer Verlag, Berlin. MR38(1969)#1597. *4.3*

Dénes J.
(1977) Latin squares and non-binary encodings. Proc. Conf. Information Theory (Cachan, France, 1977), pp. 215-221. CNRS, Paris, 1979. Not reviewed in MR. *6.7, 9.2*
(1979) Research problems. Periodica Math. Hungar. 10, 311-312. Not reviewed in MR. *6.7, 9.2*
(1983) On a problem of A. Kotzig. Annals of Discrete Math. 18, 283-290.

MR84f:05046. *4.1*
(1984) Is it possible to avoid problems over attributing priority of discovery correctly? Publ. Recherches de Math. Pures de l'Université de Neuchatel Serie IV(11), 1-2. Not reviewed in MR. *2.1*
(1986) Research Problem No.40. Period. Math. Hungar. 17(3), 245-246. Not reviewed in MR. *2A*

Dénes J. and Gergely E.
(1975) Groupoids and Codes. Topics in Information Theory (Second Colloq., Keszthely, 1975), pp. 155-162. North Holland, Amsterdam. MR 56(1978)#18100. *9.2*

Dénes J. and Hermann P.
(1982) On the product of all the elements of a finite group. Annals of Discrete Math. 15, 105-109. MR86c:20024. *2.1, 2A*

Dénes J. and Keedwell A.D.
[DK] Latin Squares and their Applications. (Akadémiai Kiadó, Budapest; Academic Press, New York; English Universities Press, London, 1974). MR50(1975)#4338.
(1988) Frequency allocation for a mobile radio telephone system. IEEE Trans. on Communications COM-36, 765-767. MR? *9.4*
(1989) A new conjecture concerning admissibility of groups. European J. Combinatorics 10, 171-174. *2.1, 2A*
(1990a) On Golomb-Posner codes and a remark of W.W.Wu about secret-sharing systems. IEEE Trans. on Communications COM-38, 261-262. *9.5*
(1990b) A new construction of two-dimensional arrays with the window property. IEEE Trans. on Information Theory IT-36, 873-876. *9.3*
(1990c) On two conjectures related to admissible groups and quasigroups. Proc. Second Internat. Math. Mini-Conf. held in Budapest, August, 1988. Periodica Polytechnica, Budapest. To appear. *2.1*
(1990d) A new authentication system based on latin squares. Proc. Fifth Conf. of Program Designers, Budapest, 28 Aug.-1 Sept., 1989. Vol. 2, pp. 269-274. Edited by A. Iványi and S. Nagy. Eötvös Lorand Univ., Budapest. *9.6*

Dénes J., Mullen G.L. and Suchower S.J.
(1989) Another generalized Golomb-Posner code. Report No. PM9, Pennsylvania State Univ., U.S.A. To appear in IEEE Trans. on Information Theory IT-36(1990), 408-411. *9A*
(199α) On cyclic subgroups of the group of row-latin squares. Graphs and Combinatorics, submitted. *2A, 5A*

Dénes J. and Pásztor E.-né
(1963) A kvázicsoportok néhány problémájáról. Magyar Tud. Akad. Mat. Fiz. Oszt. Közl. 13, 109-118. (Hungarian). [Some problems on quasigroups.] MR29(1965)#180. *4.3*

Dénes J. and Török E.
(1970) Groups and graphs. Combinatorial Theory and its Applications, pp. 257-289. North Holland, Amsterdam, 1970. MR46(1973)#91. *3.1*

Denniston R.H.F.
(1978) Remarks on latin squares with no subsquares of order two. Utilitas

Math. 13, 299-302. MR58(1979)#21688.   *4.2*

Derienko J.J.
(1988) On the conjecture of Brualdi. (Russian) In "Operational Research and Quasigroups". Mat. Issled. 102, 53-65. MR89g:20108.   *2A*

Dinitz J.H. and Stinson D.R.
(1982) Private communication.   *3.1*
(1983) MOLS with holes. Discrete Math. 44, 145-154. MR84d:05044.   *4.1, 4.4, 5A*

Di Vincenzo O.M.
(1989) On the existence of complete mappings of finite groups. Rend. Mat. e Appl. (7) 9, 189-198. MR?   *2A*

Dow S.
(1983a) An improved bound for extending partial projective planes. Discrete Math. 45, 199-207. MR84m:51003.   *2.5*
(1983b) Transversal-free nets of small deficiency. Archiv der Math. 41, 472-474. MR85b:51014.   *2.5*

Doyen J. and Wilson R.M.
(1973) Embeddings of Steiner triple systems. Discrete Math. 5, 229-239. MR48(1976)#5881.   *8.8, 8.9*

Drake D.A.
(1977) Maximal sets of latin squares and partial transversals. J. Statist. Planning and Inference 1, 143-149. MR58(1979)#5272.   *2.2, 2.4, 2.5*

Drake D.A. and Hale M.P.
(1974) Half counts in latin quarters. Discrete Math. 8, 257-268. MR 54(1977)#5011.   *8A*

Drake D.A. and Larson J.A.
(1983) Pairwise balanced designs whose line sizes do not divide six. J. Combinatorial Theory A34, 266-300. MR85d:05038.   *4.4*

Drake D.A. and Lenz H.
(1980) Orthogonal latin squares with orthogonal subsquares. Archiv der Math. 34, 565-576. MR82b:05031.   *4.4*

Dulmage A.L., Johnson D.M. and Mendelsohn N.S.
(1961)$ Orthomorphisms of groups and orthogonal latin squares I. Canad. J. Math. 13, 356-372. MR23(1962)#A1544.   *2A, 5A, 11.3*

Ecker A. and Poch G.
(1986) Check character systems. Computing 37, 277-301. MR88a:94035.   *9.6*

Eckert S.R., Hancock T.W., Mayo O. and Wilkinson G.N.
(1983) Nearest neighbour analysis of field experiments. J. Royal Statist. Soc. B45, 151-211. Not reviewed in MR.   *3A*

Erdös P. and Evans A.B.
(1989) Representations of graphs and orthogonal latin square graphs. J. Graph Theory 13, 593-595. MR?   *2 A*

Erdös P., Hickerson D.R., Norton D.A. and Stein S.K.
(1988) Has every latin square of order n a partial latin transversal of size n-1? Amer. Math. Monthly 95, 428-430. Not reviewed in MR.   *2 A*

Erdös P. and Spencer J.
(199α) Lopsided Lovasz local lemma and latin transversals. Discrete Math. To appear.   *2 A*

Etzion T.
(1988) Constructions for perfect maps and pseudo-random arrays. IEEE Trans. on Information Theory IT-34, 1308-1316. MR?   *9.3*
(1989) Letter to the editors, dated 15/2/89.   *9.4*
(198α) Untitled part-manuscript.   *9.4*
(198β) Combinatorial designs with Costas arrays' properties. Annals of Discrete Math. To appear.   *9.4*
(198γ) Combinatorial designs derived from Costas arrays. Workshop on Sequences, Positano, Italy. To appear.   *9.4*
(198δ) On Hamiltonian decomposition of K_n^*, patterns with distinct differences and Tuscan squares. Discrete Math. To appear.   *3.5*

Etzion T., Golomb S.W., Taylor H.
(1989) Tuscan-k squares. Advances in Applied Math. 10, 164-174.   *9.4*
(1990) Polygonal path constructions for Tuscan-k squares. Ars Combinatoria. To appear.   *3.5, 9.4*

Euler L.
(1779) Recherches sur une nouvelle espèce de quarrés magiques. [Memoir presented to the Academy of Sciences of St. Petersburg on 8th March, 1779.] Published as (a) Verh. Zeeuwsch. Genootsch. Wetensch. Vlissengen 9(1782), 85-239; (b) Mémoire de la Société de Flessingue, Commentationes arithmetica collectae (elogé St. Petersburg 1783), 2(1849), 302-361; (c) Leonardi Euleri Opera Omnia, Serie 1, 7(1923), 291-392.   *1.2, 2.1, 4.4, 6.1*

Evans A.B.
(1987a)$ Generating orthomorphisms of GF[q]$^+$. Discrete Math. 63, 21-26. MR88f:20050.   *2.6, 2A*
(1987b)$ Orthomorphisms of Z_p. Discrete Math. 64, 147-156. MR88k:05037.   *2.6, 2A*
(1988) Orthomorphism graphs of Z_p. Ars Combinatoria 25B, 141-152. MR 89e:05104.   *2.6, 2A*
(1989a) Orthomorphisms of GF[q]$^+$. Ars Combinatoria 27, 121-132.   *2.6, 2A*
(1989b) Orthomorphisms of groups. In "Combinatorial Mathematics": Proc. Third Internat. Conf., 1985: Annals of the New York Academy of Sciences 555, 187-191.   *2 A*
(1989c) On planes of prime order with translations and homologies. J. Geometry 34, 36-41.   *2 A*
(1989d) Orthomorphism graphs of groups. J. Geometry 35, 67-74.   *2 A*

Evans A.B. and McFarland R.L.
  (1984)$ Planes of prime order with translations. Proc. Fifteenth S.E. Conf. on Combinatorics, Graph Theory and Computing, 1984. Congressus Numerantium 44, 41-46. MR86d:51007. *11.3*

Evans T.
  (1951) On multiplicative systems defined by generators and relations. I. Normal Form Theorems. Proc. Cambridge Phil. Soc. 47, 637-649. MR 13(1952), page 312. *7.5*
  (1953) On multiplicative systems defined by generators and relations. II. Monogenic Loops. Proc. Cambridge Phil. Soc. 49, 579-589. MR15(1954), page 283. *7.5*
  (1960) Embedding incomplete latin squares. Amer. Math. Monthly 67, 958-961. MR23(1962)#A68. *4.1, 8.1, 8.3, 8.9, 8.10*
  (1973) Latin cubes orthogonal to their transposes. A ternary analogue of Stein quasigroups. Aequationes Math. 9, 296-297. MR48(1974)#3763. *7.5*
  (1975) Algebraic structures associated with latin squares and orthogonal arrays. Proc. of the Conference on Algebraic aspects of Combinatorics (University of Toronto, 1974). Congressus Numerantium 13, 31-52. MR 52(1976)#13429. *7.5*
  (1976) The construction of orthogonal k-skeins and latin k-cubes. Aequationes Math. 14(1976), 485-491. MR54(1977)#2498. *7.5*
  (1979) Universal algebra and Euler's officer problem. Amer. Math. Monthly 86, 466-473. MR83c:05025. *7.5*
  (1982a) Universal-algebraic aspects of combinatorics. (Universal Algebra, Esztergom, 1977) Colloq. Math. Soc. János Bölyai 29, 241-266. MR 84d:08004. *7.5*
  (1982b) Some connections between Steiner systems and self-conjugate sets of MOLS. Algebraic and geometric combinatorics, pp.143-159. North Holland Math. Studies 65. MR86f:05029. 7.5

Evans T. and Lindner C.C.
  (1977) Finite Embedding Theorems for Partial Quasigroups. Les Presses de l'Université de Montréal. MR57(1979)#208. *8A*

Eynden C.V.
  (1978)$ Countable sequenceable groups. Discrete Math. 23, 317-318. MR 80b:20045. *3.1*

Fan C.T., Fan S.M., Ma S.L. and Siu M.K.
  (1985) On de Bruijn arrays. Ars Combinatoria 19A, 205-213. MR86i:05032. *9.3*

Fazekas G.
  (1989) On the coding of digitized pictures. Periodica Polytechnica, Budapest. To appear in 1990. *9.3*

Federer W.T.
  (1977) On the existence and construction of a complete set of orthogonal F(4t;2t,2t)-squares designs. Ann. Statist. 5, 561-564. MR55(1978)#2608. *12.1*

Federer W.T. and Hedayat A.S.
  (1975) On the non-existence of Knut-Vik designs for all even orders.

Ann. Statist. 3, 445-447. MR51(1976)#4577.   *5A, 9.4*
(1984) Orthogonal F-rectangles for all even v. Calcutta Statist. Assoc. Bull. 33, 85-92. MR86c:05039.   *12.4*

Federer W.T., Hedayat A.S. and Mandeli J.P.
(1984) Pairwise orthogonal F-rectangle designs. J. Statist. Planning and Inference 10, 365-374. MR86i:62166.   *12.4*

Federer W.T., Lee F.-C. and Mandeli J.P.
(1981) On the construction of orthogonal F-squares of order n from an orthogonal array (n,k,s,2) and an OL(s,t) set. J. Statist. Planning and Inference 5, 267-272. MR83c:62125.   *12.1*

Federer W.T. and Mandeli J.P.
(1983) An extension of MacNeish's theorem to the construction of sets of pairwise orthogonal F-squares of composite order. Utilitas Math. 24, 87-96. MR85e:62157.   *12.1*
(1984a) On the construction of mutually orthogonal F-hyper-rectangles. Utilitas Math. 25, 315-324. MR85i:05054.   *12.4*
(1984b) Complete sets of orthogonal F-squares of prime power order with differing numbers of symbols. In "Experimental Design, Statistical Models and Genetic Statistics" (Essays in honour of Oscar Kempthorne), edited K. Hinkelmann, pp. 45-59. MR86h:62115.   *12.1*

Federer W.T., Mandeli J.P. and Schwager F.J.
(1984) Embedding cyclic latin squares of order 2^n in a complete set of orthogonal F-squares. J. Statist. Planning and Inference 10, 207-218. MR86e:62107.   *12.1*

Federer W.T. and Raktoe B.L.
(1985) Lattice square approach to constructing mutually orthogonal F-squares. Ann. Inst. Statist. Math. 37, 329-336. MR87a:05040.   *12.1*

Federer W.T., Raktoe B.L. and Schwager S.J.
(1984) Nonisomorphic complete sets of orthogonal F-squares and Hadamard matrices. Communications in Statist. A13(11), 1391-1406. MR 85c:62201.   *12.1*

Finney D.J.
(1945) Some orthogonal properties of the 4×4 and 6×6 latin squares. Ann. Eugenics 12, 213-219. MR7(1946), page 107.   *12.0*
(1946a) Orthogonal partitions of the 5×5 latin squares. Ann. Eugenics 13, 1-3. MR7(1946), page 407.   *12.0*
(1946b) Orthogonal partitions of the 6×6 latin squares. Ann. Eugenics. 13, 184-196. MR8(1947), page 247.   *12.0*
(1952) Probit analysis: A Statistical Treatment of the Sigmoid Response Curve. Cambridge Univ. Press, 2nd Ed. MR14(1954), page 66.   *10.3*
(1955) Experimental design and its Statistical Basis. University of Chicago Press. Not reviewed in MR.   *10.2*
(1960) An Introduction to the Theory of Experimental Design. University of Chicago Press. MR22(1961)#11493.   *10.2*
(1982) Some enumerations for the 6×6 latin squares. Utilitas Math. 21, 137-153. MR 84g:05033.   *12.2*

Fiori C. and Lancelotti P.
(1984) Su una classe di codici di ordine variabile. Proc. Conf. on combinatorial and incidence geometry: principles and applications (La Mendola, 1982), pp. 307-322. Rend. Sem. Mat. Brescia 7, Milan. MR 86j:94058. *9.2*

Fisher R.A.
(1925) Statistical Methods for Research Workers. Oliver and Boyd, Edinburgh. [Latest Ed., 1958.] Not reviewed in MR. *10.2, 10.3, 10.8*
(1935) Contribution to "Discussion on Mr. Neyman's paper". J. Royal Statist. Soc. Suppl. 2, 154-157. Not reviewed in MR. *10.8*

Fisher R.A. and Yates F.
(1934) The 6×6 latin squares. Math. Proc. Cambridge Phil. Soc. 30, 492-507. *1.2*
(1963) Statistical Tables for Biological, Agricultural and Medical Research. Oliver and Boyd, Edinburgh, 6th edition. Not reviewed in MR, but see MR5(1945), page 207. *10.8*

Ford L.R. and Fulkerson D.R.
(1962) Flows in Networks. Princeton Univ. Press. MR28(1964)#291. *8.2*

Franklin M.F.
(1984) Triples of almost orthogonal 10×10 latin squares useful in experimental design. Ars Combinatoria 17, 141-146. MR86f:62127. *11A*

Freeman G.H.
(1966) Some non-orthogonal partitions of 4×4, 5×5 and 6×6 latin squares. Ann. Math. Statist. 37(1966), 666-681. MR33(1967)#839. *12.0*
(1979a) Some two-dimensional designs balanced for nearest neighbours. J. Royal Statist. Soc. B41(1979), 88-95. MR81j:62153. *3.2*
(1979b) Complete latin squares and related experimental designs. J. Royal Statist. Soc. B41(1979), 253-262. MR81m:65032. *3.2*
(1981) Further results on quasi-complete latin squares. J. Royal Statist. Soc. B43(1981), 314-320. MR83a:05037. *3.2*
(1985) Duplexes of 4×4, 5×5 and 6×6 latin squares. Utilitas Math. 27, 5-24. MR87d:05043. *2.6*
(1988) Nearest neighbour designs for three or four treatments in rows and columns. Utilitas Math. 34, 117-130. MR89k:05021. *3A*

Freeman G.H. and Preece D.A.
(1983) Semi-latin squares and related designs. J. Royal Statist. Soc. B45, 267-277. Not reviewed in MR. *10.6*

Freeman G.R., Hoblyn T.N. and Pearce S.C.
(1954) Some considerations in the design of successive experiments in fruit plantations. Biometrics 10, 503-515. Not reviewed in MR. *10.5*

Friedlander R.J.
(1976)$ Sequences in non-abelian groups with distinct partial products. Aequationes Math. 14, 59-66. MR53(1977)#8252. *3.1*
(1980) Minimal sequencings of groups. Proc. Eleventh S.E. Conf. on Combinatorics, Graph Theory and Computing, 1980. Congressus Numerantium

29, 461-478. MR82e:05032. *3.1*

Friedlander R.J., Gordon B. and Miller M.D.
 (1978)$ On a group sequencing problem of Ringel. Proc. Ninth S.E. Conf. on Combinatorics, Graph Theory and Computing, 1978. Congressus Numerantium 21, 307-321. MR80c:05079. *3.3*

Friedlander R.J., Gordon B. and Tannenbaum P.
 (1981)$ Partitions of groups and complete mappings. Pacific J. Math. 92, 283-293. MR83a:20079. *3.3*

Friedmann W.F. and Mendelsohn C.J.
 (1932) Notes on codewords. Amer. Math. Monthly 39, 394-409. *9A*

Fukuda A., Imai H., Miyakowa H. and Nomura T.
 (1972) A theory of two dimensional linear recurring arrays. IEEE Trans. on Information Theory IT-18, 775-785. MR52(1976)#2725. *9.3*

Gibbons P.B. and Mendelsohn E.
 (1987) The existence of a subsquare free latin square of side 12. SIAM J. Algebraic Discrete Methods 8, 93-99. MR? *4A*

Gilbert E.N.
 (1965) Latin squares which contain no repeated digrams. SIAM Rev. 7, 189-198. MR31(1966)#3346. *9.4*

Giles F.R., Oyama T. and Trotter L.E.
 (1977) On completing partial latin squares. Proc. Eighth S.E. Conf. on Combinatorics, Graph Theory and Computing, 1977. Congressus Numerantium 19, 523-543. MR57(1979)#16102. *8A*

Golomb S.W.
 (1970) Wilsonian products in groups. Bull. Amer. Math. Soc. 76, 973-974. Not reviewed in MR. *2A*

Golomb S.W. and Posner E.C.
 (1964) Rook domains, latin squares, affine planes, and error-distributing codes. IEEE Trans. Information Theory IT-10, 196-208. MR29(1965)#5657. *9.2*

Golomb S.W. and Taylor H.
 (1982) Two-dimensional synchronization patterns for maximum ambiguity. IEEE Trans. Information Theory IT-28, 600-604. MR83k:05037. *9.4*
 (1985) Tuscan squares - a new family of combinatorial designs. Ars Combinatoria 20B, 115-132. MR87i:05058. *3.5, 9.3*

Gordon B.
 (1961)$ Sequences in groups with distinct partial products. Pacific J. Math. 11, 1309-1313. MR24(1962)#A3193. *3.0, 3.1, 3.3, 3.6*
 (1966) On the existence of perfect maps. IEEE Trans. Information Theory IT-12, 486-487. MR35(1968)#2763. *9.3*

Grams G. and Jungnickel J.
(1986)$ Maximal difference matrices of order ≤ 10. Discrete Math. 58, 199-203. MR87e:05030. *2.5*

Grant D.A.
(1948) The latin square principle in the design and analysis of psychological experiments. Psychological Bull. 45, 427-442. Not reviewed in MR. *10.6*

Graybill F.A. and Milliken G.A.
(1972) Interaction models for the latin square. Austr. J. Statist. 14, 129-138. MR50(1975)#8860. *10.7*

Greene J.W., Hellman M.E. and Karnin E.D.
(1983) On secret sharing systems. IEEE Trans. on Information Theory IT-29, 35-41. MR84i:94044. *9.5*

Gross K.B.
(1975) On the maximal number of pairwise orthogonal Steiner triple systems. J. Combinatorial Theory A19, 256-263. MR52(1977)#2917. *6.5*

Gross K.B., Mullin R.C. and Wallis W.D.
(1973)$ The number of pairwise orthogonal symmetric latin squares. Utilitas Math. 4, 239-251. MR48(1974)#10852. (See also next entry.) *6.6*
(1974) Corrigenda to "The number of pairwise orthogonal symmetric latin squares". Utilitas Math. 6, 349. MR50(1975)#9620.

Grundy P.M. and Healy M.J.R.
(1950) Restricted randomization and quasi-latin squares. J. Royal Statist. Soc. B12, 286-291. Not reviewed in MR. *10.8*

Guérin R.
(1966a) Existence et propriétés des carrés latins orthogonaux I. Publ. Inst. Statist. Univ. Paris 15, 113-213. MR35(1968)#73.
(1966b) Existence et propriétés des carrés latins orthogonaux II. Publ. Inst. Statist. Univ. Paris 15, 215-293. MR35(1968)#4118. *5.6*

Gumm A.P.
(1985) A new class of check digit methods for arbitrary number systems. IEEE Trans. on Information Theory IT-31, 182-185. Not reviewed in MR. *9A*

Guy R.K.
(1989) Unsolved problems come of age. Amer. Math. Monthly 96, 903-909. Not reviewed in MR. *2A*

Häggkvist T.
(1978) A solution of the Evans conjecture for latin squares of large size. In "Combinatorics", Colloq. Math. Soc. János Bólyai, Vol.18; edited by A. Hajnal and V.T. Sos, pp. 495-514. North Holland, Amsterdam. MR80j:05023. *8.10*

Hall M.
(1945) An existence theorem for latin squares. Bull. Amer. Math. Soc. 51, 387-388. MR7(1946), page 106.   *8.2*
(1948) Distinct representatives of subsets. Bull. Amer. Math. Soc. 54, 922-926. MR10(1949), page 238.   *2.1*
(1953) Uniqueness of the projective plane with 57 points. Proc. Amer. Math. Soc. 4, 912-916. MR15(1954), page 460.   *11A*
(1954) Correction to "Uniqueness of the projective plane with 57 points". Proc. Amer. Math.Soc. 5, 994-997. MR16(1955), page 395.   *11A*
(1959) The Theory of Groups. Macmillan, New York. MR21(1960)#1996.   *6.4, 6.5*
(1980) Configurations in a plane of order 10. Annals of Discrete Math. 6, 154-174. MR82a:94087.   *11.2, 11A*

Hall M., Killgrove R. and Swift J.D.
(1959) On projective planes of order 9. Math. Tables Aids Comput. 13, 233-246. MR21(1960)#5933.   *11A*

Hall M. and Paige L.J.
(1955)$ Complete mappings of finite groups. Pacific J. Math. 5, 541-549. MR18(1957), page 109.   *2.1*

Hall M. and Roth R.
(1984) On a conjecture of R.H. Bruck. J. Combinatorial Theory A37, 22-31. MR85k:05029.   *11.2*

Hall M., Swift J.D. and Walker R.J.
(1956) Uniqueness of the projective plane of order eight. Math. Tables Aids Comput. 10, 186-194. MR18(1957), page 816.   *11A*

Hall P.
(1935) On representatives of subsets. J. London Math. Soc. 10, 26-30.   *8.2*

Hanani H.
(1970) On the number of orthogonal latin squares. J. Combinatorial Theory 8, 247-271. MR40(1970)#5466.   *5.6*
(1975a) On transversal designs. In "Combinatorics": Proc. of Advanced Study Inst. on Combinatorics, Breukelen, 1974, pp.42-52. Math. Centre Tracts No. 55, Math. Centrum, Amsterdam. MR58(1979)#5291.   *5.2*
(1975b) Balanced incomplete block designs and related designs. Discrete Math. 11, 255-269. MR52(1976)#2918.   *5.2*

Hansen N.A.
(1915) Rudefordeling og Fejl ved Markforsøg. Tidsskrift for Planteavl. 22, 493-552.   *10.3*

Haranen L.J.
(1968) Construction of projective planes over the latin squares which occur in the description of the Hughes plane of order 9. (Russian) Scientific Memoirs of Perm State Pedagogical Inst., Mathematics 61, 47-62. MR 41(1971)#2522.   *11A*

Hedayat A.
(1969) On the theory of the existence, non-existence, and the construction of mutually orthogonal F-squares and latin squares. Ph.D. thesis, Biometrics Unit, Cornell University, Ithica, New York. *12.0*
(1977) A complete solution to the existence and non-existence of Knut Vik designs and orthogonal Knut Vik designs. J. Combinatorial Theory A22, 331-337. MR55(1978)#12548. *5A*
(1978) A generalization of sum composition: a family of self-orthogonal latin square designs with sub-self-orthogonal latin square designs. J. Combinatorial Theory A24, 202-210. MR58(1979)#283. *4.4*

Hedayat A., Raghavarao D. and Seiden E.
(1975) Further contributions to the theory of F-squares design. Ann. Statist. 3, 712-716. MR53(1977)#165. *12.0, 12.1*

Hedayat A. and Seiden E.
(1970) F-square and orthogonal F-square designs: a generalization of latin square and orthogonal latin square designs. Ann. Math. Statist. 41, 2035-2044. MR42(1971)#2604. *12.0, 12.1*
(1971) On a method of sum composition of orthogonal latin squares. Proc. Conf. on Combinatorial Geometry and its Applications, Perugia, 1970, pp. 239-256. Atti del Convegno di Geometria Combinatoria e sue Applicazioni, Ist. Math. Univ. Perugia. MR49(1975)#4804. *4.4*
(1974) On the theory and application of sum composition of latin squares and orthogonal latin squares. Pacific J. Math. 54, 85-113. MR51(1976) #10125. *4.4*

Heinrich K.
(1977a) Self-orthogonal latin squares with self-orthogonal subsquares. Ars Combinatoria 3, 251-266. MR58(1979)#284. *4.4*
(1977b) Near-orthogonal latin squares. Utilitas Math. 12, 145-155. MR 58(1979)#16342. *4.4, 6.1, 6.5*
(1977c) Latin squares composed of four disjoint subsquares. In "Combinatorial Math. V" (Proc. Fifth Austral. Conf., Melbourne, 1976), pp. 118-127. Lecture Notes in Math. No.622, Springer Verlag, Berlin. MR57(1979) #12257. *4.3*
(1977d) Subsquares in latin squares. Proc. Eighth S.E. Conf. on Combinatorics, Graph Theory and Computing, 1977. Congressus Numerantium 19, 329-344. MR58(1979)#10503. *4.3*
(1977e) Approximation to a self-orthogonal latin square of order 6. Ars Combinatoria 4, 17-24. MR58(1979)#285. *11A*
(1979) Pairwise orthogonal row-complete latin squares. Proc. Eighth S.E. Conf. on Combinatorics, Graph Theory and Computing, 1979. Congressuss Numerantium 23, 505-510. MR81m:05033. *3.1*
(1980) Latin squares with no proper subsquares. J. Combinatorial Theory A29, 346-353. MR82b:05032. *4.1*
(1982) Disjoint sub-quasigroups. Proc. London Math. Soc. (3)45, 547-563. MR84b:05029. *4.3*

Heinrich K. and Wallis W.D.
(1980) Almost Graeco-Latin squares. Ars Combinatoria 10, 55-63. MR 82i:05015. *11A*
(1981) The maximum number of intercalates in a latin square. In

"Combinatorial Math. VIII" (Proc. Eighth Austral. Conf., Geelong, 1980), pp. 221-233. Lecture Notes in Math. No.884, Springer Verlag, Berlin. MR 84g:05034. *4.3*

Heinrich K.; Wu L. and Zhu L.
(199α) Incomplete self-orthogonal latin squares ISOLS(6m+6,2m) exist for all m. Discrete Math. To appear. *4A*

Heinrich K. and Zhu L
(1986) Existence of orthogonal latin squares with aligned subsquares. Discrete Math. 59, 69-78. MR88a:05026. *4.4*
(1987) Incomplete self-orthogonal latin squares. J. Austral. Math. Soc. A42, 365-384. MR88a:05027. *4.4*

Heise W. and Quattrocchi P.
(1981) Variable-order codes. Atti Sem. Mat. Fis. Univ. Modena 30, 176-184. MR84b:94026. *9.1, 9.2*
(1983) Informations und Codierungstheorie. Springer Verlag, Berlin. Not reviewed in MR. (2nd Ed., 1989). *9.2*

Hilton A.J.W.
(1973) Embedding an incomplete diagonal latin square in a complete diagonal latin square. J. Combinatorial Theory A15, 121-128. MR47(1974) #8327. *8.5*
(1974) A note on embedding latin rectangles. In "Combinatorics" (Proc. Fourth British Combinatorial Conf. Aberystwyth, 1973), pp. 69-74. London Math. Soc. Lecture Note Series 13, Camb. Univ. Press. MR53(1977)#2710. *8A*
(1975) Embedding incomplete double diagonal latin squares. Discrete Math. 12, 257-268. MR51(1976)#12565. *8A*
(1980) The reconstruction of latin squares with applications to school timetabling and to experimental design. Mathematical Programming Study 13, 68-77. MR81j:90066. *8A*
(1981) School timetables. Annals of Discrete Math. 11, 177-188. MR 83g:90061. *8A*
(1982) Embedding incomplete latin rectangles and extending the edge colourings of graphs. Annals of Discrete Math. 13, 121-138. MR84b:05030. *8A*

Hilton A.J.W. and Rodger C.A.
(1982) Latin squares with prescribed diagonals. Canad. J. Math. 34, 1251-1254. MR84f:05026. *8A*

Ho C.Y.
(1986a) Characterization of projective planes of small prime orders. J. Combinatorial Theory A41, 189-220. MR87g:51011. *11.2, 11.3*
(1986b) Involutary collineations of finite planes. Math. Z. 193, 235-241. MR87k:51022. *11.2*

Ho C.Y. and Goncalves A.
(1986) On collineation groups of a projective plane of prime order. Geom. Dedicata 3, 357-366. MR87h:51017. *11.3*

Ho C.Y. and Moorhouse G.E.
  (1985) A new characterization of the Desarguesian plane of order 11. Proc. Conf. on Groups and Geometry, Part B (Madison, Wisconsin, 1985). Algebras, Groups and Geometries 2, 428-435. MR87j:51017.   *11.3*

Hobbs A.M. and Kotzig A.
  (1983) Groups and homogeneous latin squares. Congressus Numerantium 40, 35-44. MR85i:05051.   *4.3*

Hobbs A.M., Kotzig A. and Zaks J.
  (1982) Latin squares with high homogeneity. Proc. Thirteenth S.E. Conf. on Combinatorics, Graph Theory and Computing, 1982. Congressus Numerantium 35, 333-345. MR85e:05035.   *4.3*

Hobby C., Rumsey H. and Weichsel P.M.
  (1960) Finite groups having elements of every possible order. J. Washington Acad. Sci. 50, 11-12. MR26(1963)#1356.   *4.3*

Hoblyn T.N.
  (1930) The relationship between the experimental and the demonstration plot and their relative value to the investigator, the country officer and the fruit grower. Annual Report 1929 (Part I; General), East Malling Research Station, pp. 40-55.   *10.4*

Hoffman A.J. and Kuhn H.W.
  (1956) Systems of distinct representatives and linear programming. Amer. Math. Monthly 63, 455-460. MR18(1957), page 370.   *8.2*

Hoffman D.G.
  (1983) Completing incomplete commutative latin squares with prescibed diagonals. European J. Combinatorics 4, 33-35. MR84h:05024.   *8A*

Hoffman D.G. and Lindner C.C.
  (1981) Embeddings of Mendelsohn triple systems. Ars Combinatoria 11, 265-269. MR83c:05034.   *8.8, 8.9*

Hoffman D.G. and Rodger C.A.
  (1987) Embedding totally symmetric quasigroups. Annals of Discrete Math. 34, 249-257. MR88m:20131.   *8A*

Hoghton G.B. and Keedwell A.D.
  (1982)$ On the sequenceability of dihedral groups. Annals of Discrete Math. 15, 253-258. Not reviewed in MR.   *3.1*

Holsztyński W. and Strube R.F.E.
  (1978) Paths and circuits in finite groups. Discrete Math. 22, 263-272. MR 80e:20052.   *3.1*

Horton J.D.
  (1971) Some recursive constructions of combinatorial designs. Ph.D. Thesis, University of Waterloo.   *4.4, 6.5*
  (1974) Sub-latin squares and incomplete orthogonal arrays. J. Combinatorial Theory A16, 23-33. MR50(1975)#143.   *4.4, 5.5, 6.1, 6.2, 6.5, 6.6*

Houston T.R.
   (1966) Sequential counterbalancing in latin squares. Ann. Math. Statist. 37, 741-743. MR34(1967)#905.   *3A*

Hsu D.F.
   (1980) Cyclic Neofields and Combinatorial Designs. Lecture Notes in Math. No. 824. Springer Verlag, Berlin. MR84g:05001.   *3A*
   (1990) Orthomorphisms and near orthomorphisms. To appear in Proc. Sixth Internat. Conf. on the Theory and Applications of Graphs. Wiley, New York.   *2A*

Hsu D.F. and Keedwell A.D.
   (1984)$ Generalized complete mappings, neofields, sequenceable groups and block designs I. Pacific J. Math 111, 317-332. MR85m:20031.   *2.6, 3.3*
   (1985)$ Generalized complete mappings, neofields, sequenceable groups and block designs II. Pacific J. Math. 117, 291-312. MR86k:05034.   *2.6, 3.3*

Huang C. and Rosa A.
   (1975) Another class of balanced graph designs: balanced circuit designs. Discrete Math. 12, 269-293. MR54(1977)#7292.   *3A*

Hughes D.R.
   (1957a) Collineations and generalized incidence matrices. Trans. Amer. Math. Soc. 86, 284-296. MR20(1959)#253.   *11.2*
   (1957b) Generalized incidence matrices over group algebras. Illinois J. Math. 1, 545-551. MR20(1959)#254.   *11.2*

Hughes D.R. and Piper F.C.
   (1973) Projective Planes. Springer Verlag, Berlin. MR48(1974)#12278.   *11.2, 11.6*

Hwang F.K.
   (1983) Totally symmetric complete latin squares of even order. J. Chinese Statist. Assoc. 21, 58-63. Not reviewed in MR.   *3.6*

Hwang F.K. and Lin S.
   (1977) Neighbour designs. J. Combinatorial Theory A23, 302-313. MR 57(1978)#2957.   *3A*

Hwang F.K. and Richards G.W.
   (1985) A two-stage rearrangeable broadcast switching network. IEEE Trans. on Communications COM-33, 1025-1035. Not reviewed in MR.   *9.6*

Ihrig E.C., Seah E. and Stinson D.R.
   (1987) A perfect one-factorization of K_{50}. J. Combin. Math. and Combin. Comput. 1, 217-219. MR88f:05091.   *4.2*

Imai H.
   (1984) Multivariate polynomials in coding theory. Proc. Second Internat. Conf. on Applied Algebra, Algorithmics and Error-Correcting Codes, pp. 36-60. Lecture Notes in Computer Science No. 228, Spinger Verlag, Berlin. Not reviewed in MR.   *9.4*

Iványi A.
(1987) On the d-complexity of words. Ann. Univ. Sci. Budapest. Eotvos Sect. Comput. 8, 69-90. MR?   *9.3*
(1989) Construction of infinite de Bruijn arrays. Discrete Applied Math. 22(1988/1989), 201-214. MR?   *9.3*

Jacobsen S.B.
(1988) The rearrangement process in a two-stage broadcast switching network. IEEE Trans. on Communications COM-36, 484-491. MR89g:90096. *9.6*

Janko Z.
(1984) Projective planes of order 12 with a collineation group of order 27. Rad. Jugoslav. Akad. Znan. Umjet No. 408, 1-6. MR86m:51013.   *11.2*

Janko Z. and van Trung T.
(1980a) On projective planes of order 12 which have a subplane of order three I. J. Combinatorial Theory A29, 254-256. MR82f:51005.   *11.2*
(1980b) On projective planes of order twelve and twenty. Math. Z. 173, 199-201. MR82f:51006.   *11A*
(1981a) Projective planes of order 10 do not have a collineation of order three. J. Reine Angew. Math. 325, 189-209. MR82h:51015.   *11.2*
(1981b) Projective planes of order 12 do not have a non-abelian group of order 6 as a collineation group. J. Reine Angew. Math. 326, 152-157. MR 82j:51017.   *11.2*
(1981c) Projective planes of order 12 do not possess an elation of order three. Studia Sci. Math. Hungarica 16, 115-118. MR85c:51018.   *11.2*
(1981d) On projective planes of order 12 with an automorphism of order 13. Part I. Kirkman designs of order 27. Geometriae Dedicata 11, 257-284. MR83d:05022a.   *11.2*
(1981e) Determination of projective planes of order 9 with a non-trivial perspectivity. Studia Sci. Math. Hungarica 16, 119. MR85c:51019.   *11.2*
(1982a) On projective planes of order 12 with an automorphism of order 13. Part II. Orbit matrices and conclusion. Geometriae Dedicata 12, 87-99. MR83d:05022b.   *11.2*
(1982b) The full collineation group of any projective plane of order 12 is a {2,3} group. Geometriae Dedicata 12, 101-110. MR83d:51023.   *11.2*
(1982c) A Generalization of a result of L. Baumert and M. Hall about projective planes of order 12. J. Combinatorial Theory A32, 378-385. MR 83e:51011.   *11.2*
(1982d) Projective planes of order 12 do not have a four-group as a collineation group. J. Combinatorial Theory A32, 401-404. MR83f:51016. *11.2*
(1982e) The classification of projective planes of order 9 which possess an involution. J. Combinatorial Theory A33, 65-75. MR84c:51013.   *11.2*

John J.A. and Quenouille M.H.
(1977) Experiments: Design and Analysis. 2nd Ed. Griffin, London. Not reviewed in MR.   *10.3, 10.7*

John P.W.M.
(1971) Statistical Design and Analysis of Experiments. Macmillan, New York. MR42(1971)#8625.   *10.7*

Joshi D.D.
(1958) A note on upper bounds for minimum distance codes. Information and Control 1, 289-295. MR20(1959)#5706.   *9.1*

Jungnickel D.
(1978) On regular sets of latin squares. "Problèmes Combinatoires et Théorie des Graphes" (Colloq. Internat., CNRS), pp. 255-256. CNRS, Paris. MR80j:05024.   *2.5*
(1980)$ On difference matrices and regular latin squares. Abh. Math. Sem. Univ. Hamburg. 50, 219-231. MR81m:05034.   *2.5*
(1984a) Maximal partial spreads and translation nets of small deficiency. J. Algebra 90, 119-132. MR85j:51016.   *2.5*
(1984b) Lateinische Quadrate, ihre Geometrien und ihre Gruppen. Jber. Deutsch Math. Verein 86, 69-108. MR86g:05016.   *11A*
(1988) Latin squares, their geometries and their groups. A survey. University of Waterloo Research Report CORR 88-14.   *11A*

Keedwell A.D.
(1966)$ On orthogonal latin squares and a class of neofields. Rend. Mat. e Appl. (5) 25, 519-561. MR36(1968)#3664: Erratum MR37(1969), page 1469.   *3.3*
(1974) Some problems concerning complete latin squares. Proc. British Combinatorial Conf. (Aberystwyth, 1973), pp. 89-96. London Math. Soc. Lecture Notes No.13, Cambridge Univ. Press. MR51(1976)#2943.   *3.1*
(1975) Row complete squares and a problem of A. Kotzig concerning P-qausigroups and Eulerian circuits. J. Combinatorial Theory A18, 291-304. MR51(1976)#2982.   *3.1*
(1976a) Some connections between latin squares and graphs. Colloq. Internat. Sulle Teorie Combinatorie (Roma 1973), Tomo 1, pp.321-329. Accad. Naz. Lincei, Roma. MR55(1978)#7806.   *9A*
(1976b) Recent results concerning complete latin squares. Proc. Fifth British Combinatorial Conf. (Aberdeen, 1975). Congressus Numerantium 4, 385-393. MR53(1977)#2711.   *3A*
(1976c) Latin squares, P-quasigroups and graph decompositions. (Symposium on Quasigroups and Functional Equations, Belgrade, 1974.) Recueil des Travaux de l'Institute Mathématique, Belgrade, N.S. 1(9), 41-48. MR 55(1978)#2610.   *3.1*
(1978) Uniform P-circuit designs, quasigroups and Room Squares. Utilitas Math. 14, 141-159. MR80c:05053.   *6.5*
(1980) Concerning the existence of triples of pairwise almost orthogonal 10×10 latin squares. Ars Combinatoria 9, 3-10. MR81k:05021.   *6.6, 11.1, 11A*
(1981a)$ On the sequenceability of non-abelian groups of order pq. Discrete Math. 37, 203-216. MR84h:20016.   *3.1, 3.6*
(1981b) Sequenceable groups: a survey. In "Finite geometries and designs" (Proc. Conf., Chelwood Gate, 1980), pp. 205-215. London Math. Soc. Lecture Notes No. 49. Cambridge Univ. Press. MR82i:20040.   *3A*
(1983a) Graeco-Latin squares. In Encyclopaedia of Statistical Sciences, Vol.3. Wiley, New York. MR84k:62001.   *1.2*
(1983b)$ On R-sequenceability and R_h-sequenceability of groups. In "Combinatorics '81: in honour of Beniamino Segre". Edited by A.Barlotti, P.V.Ceccherini and G.Tallini. Annals of Discrete Math. 18, 535-548. MR 84f:20025.   *3.3, 3.6*
(1983c)$ On the existence of super P-groups. J. Combinatorial Theory A35,

89-97. MR85b:20034. *2.1, 3.4*
(1983d)$ Sequenceable groups, generalized complete mappings, neofields and block designs. In "Combinatorial Mathematics X" (Proceedings of the Conference held in Adelaide, Australia, 1982), pp. 49-71. Lecture Notes in Math. No.1036. Springer Verlag, Berlin. MR85g:05035. *3.1*
(1983e)$ The existence of pathological left neofields. Ars Combinatoria 16B, 161-170. MR85k:20203. *2.6*
(1984a)$ More super P-groups. Discrete Math. 49, 205-207. MR85m:20032. *2.1, 3.4*
(1984b) Circuit designs and latin squares. Ars Combinatoria 17, 79-90. MR 85i:05036. *3A, 11A*

Kelly B.K.
(1975) Private communication, letter dated 2/12/75. *11.6*

Kempthorne O. and Wilk M.B.
(1957) Non-additivities in a latin square design. J. Amer. Statist. Assoc. 52, 218-236. MR19(1958), page 474. *10.7*

Kenjale P.S., Rajaraman V. and Sethi P.S.
(1978) An error-correcting coding scheme for alphanumeric data. Inf. Proc. Letters 7, 72-78. MR58(1979)#20798. *9A*

Kerr J.R., Pearce S.C. and Preece D.A.
(1973) Orthogonal designs for three-dimensional experiments. Biometrika 60, 349-358. MR48(1976)#1399. *10.6*

Khachatryan L.G.
(1982) Arrays with the window property. (Russian) Dokl. Akad. Nauk. Armenian SSR 74, 51-56. MR84k:94022. *9.3*

Kiefer J.C. and Wynn H.P.
(1981) Optimal balanced block and latin square designs for correlated observations. Ann. Statist. 9, 737-757. MR82h:62122. *3A*

Kiel D.I. and Killgrove R.B.
(1980) Completion of quadrangles, revisited. Proc. West Coast Conf. on Combinatorics, Graph Theory and Computing, California 1979. Congressus Numerantium 26, 187-198. MR81m:05049. *11.2*

Kiel D.I., Killgrove R.B. and Parker E.T.
(1978) Some combinatorial problems in the foundations of geometry. Proc. Ninth S.E. Conf. on Combinatorics, Graph Theory and Computing, 1978. Congressus Numerantium 21, 401-409. MR81c:05015. *4.3*

Killgrove R.B.
(1960) A note on the non-existence of certain projective planes of order nine. Math. Comput. 14, 70-71. MR22(1961)#4011. *11A*
(1964) Completions of quadrangles in projective planes. Canad. J. Math. 16, 63-76. MR28(1964)#513. *4.3*

Killgrove R.B. and Milne E.
(1974) Subsquare complete latin squares of order 12. Proc. Fifth S.E. Conf.

on Combinatorics, Graph Theory and Computing, 1974. Congressus Numerantium 10, 535-547. MR55(1978)#136.   *4.3, 11.2*

Killgrove R.B., Milne E. and Parker E.T.
(1976) Low order projective planes. Proc. Seventh S.E. Conf. on Combinatorics, Graph Theory and Computing, 1976. Congressus Numerantium 17, 365-390. MR54(1977)#12545.   *11.2*

Killgrove R.B. and Parker E.T.
(1964) A note on projective planes of order nine. Math. Comput. 18, 506-508. MR29(1965)#1573.   *11A*
(1977) Addition and multiplication in known projective planes of order nine. Proc. Eighth S.E. Conf. on Combinatorics, Graph Theory and Computing, 1977. Congressus Numerantium 19, 433-452. MR57(1979) #16106.   *11.2*
(1979) A planar latin square lacking automorphism. J. Combinatorial Theory A27, 404-406. MR81e:05046.   *11A*
(1980) Non-existence of new order-9 projective plane with order-13 collineation. J. Geometry 14, 121-153. MR81k:05029.   *11.2*

Kirkpatrick P.B. and Room T.G.
(1971) Miniquaternion Geometry. Cambridge Tracts in Math. No.60. Camb. Univ. Press. MR45(1973)#7590.   *11.2, 11.6*

Kirton H.C.
(1985) Mutually orthogonal partitions of the 6×6 latin squares. Utilitas Math. 27, 265-274. MR86m:05023.   *12.2*

Koksma K.K.
(1969) A lower bound for the order of a partial transversal in a latin square. J. Combinatorial Theory 7, 94-95. MR39(1970)#1342.   *2.2*

Kolesova G., Lam C.W.H. and Thiel L.
(1988a) On the number of 8×8 latin squares. Concordia Univ. Preprint. *11A*
(1988b) A computer search for finite projective planes of order 9. Concordia Univ. Preprint.   *11A*

Kotzig A.
(1958) Remarks on the decomposition of finite graphs with an even number of vertices into 1-factors. (Slovak). Časopis pro Pěstování Matematiky 83, 348-353. MR21(1960)#877.   *4.3*
(1964) Hamiltonian graphs and Hamiltonian circuits. In "Theory of graphs and its applications" (Proc. Sympos. Smolenice, 1963), pp.63-82. Publ. House Czechoslovak Acad. Sci., Prague. MR30(1965)#3462.   *4.2*
(1970) Circuits and edge cuts of coloured bipartite regular graphs. Publication of Centre de Récherche de Mathématiques Appliquées No. 60. Not reviewed in MR.   *4.3*

Kotzig A., Lindner C.C. and Rosa A.
(1975) Latin squares with no subsquares of order two and disjoint Steiner triple systems. Utilitas Math. 7, 287-294. MR53(1977)#5331.   *4.2*

Kotzig A. and Turgeon J.
(1976) On certain constructions for latin squares with no latin subsquares of order two. Discrete Math. 16, 263-270. MR55(1978)#137. *4.2*

Kotzig A. and Zaks J.
(1983) The three permutations theorem. Ars Combinatoria 16, 113-117. MR85h:05026. *4.3*

Kryger-Larson H.
(1913) Beretning om Lokale Markforsøg Udførte i Sommeren 1912 af de Samvirkende Landboforeninger i Fyns Stift. Odense:Trykt i Andelsbogtrykkeriet i Odense. *10.3*

Lam C.W.H.
(1989) Private Communication, letter dated 10/3/89. *11A*

Lam C.W.H., McKay J., Swiercz S. and Thiel L.
(1983) The non-existence of ovals in a projective plane of order 10. Discrete Math. 45, 319-321. MR84h:05028. *11A*

Lam C.W.H., Swiercz S. and Thiel L.
(1986) The non-existence of codewords of weight 16 in a projective plane of order 10. J. Combinatorial Theory A42, 207-214. MR88e:94029. *11A*
(1989) The non-existence of finite projective planes of order 10. Canad. J. Math. 41, 1117-1123. *11A*

Lancellotti P. and Pellegrino C.
(1982) Una extensione equidistante di codici lineari di ordine variabile. Atti. Sem. Math. Fis. Univ. Modena 31, 65-69. MR85m:94028. *9.2*
(1986) A construction of sets of pairwise orthogonal F-squares of composite order. In "Combinatorics '84", Proc. of Conf. held at Bari, Italy, 1984. Annals of Discrete Math. 30, 285-290. MR87k:05044. *12.1*

Lecointe P.
(1970) Généralisation de la notion de transversales et application à la construction de plans d'expériences. C.R. Acad. Sci. Paris A271, 804-806. Not reviewed in MR. *2.6*

Lenz H. and Stern G.
(1980) Steiner triple systems with given subspaces; another proof of the Doyen-Wilson theorem. Bol. Um. Mat. Ital. A(5)17, 109-114. MR81f:05049. *8.8*

Lewandowski J.I. and Liu C.L.
(1987) SS/TDMA satellite communications with k-permutation switching modes. SIAM J. Alg. Discrete Math. 8, 519-533. MR88m:90097. *9.2*

Lewandowski J.I., Liu C.L. and Liu J.W.S.
(1983) SS/TDMA time slot assignment with restricted switching modes. IEEE Trans. on Communications COM-31, 149-154. Not reviewed in MR. *9.2*

Lidl R. and Niederreiter H.
(1983) "Finite fields". Encyclopedia of Mathematics and its Applications,

Vol. 20. Addison-Wesley Publishing Company, Reading, Massachusetts, U.S.A. MR86c:11106. See also MR88c:11073.   *9.5, 9.6*

Lindhard E.
(1909) Om det matematiske Grundlag for Dyrkningsforsøg paa Agermark. Tidsskr. Landbrugets Planteavl. 16, 337-358.   *10.3*

Lindner C.C.
(1970) On completing latin rectangles. Canad. Math. Bull. 13, 65-68. MR 41(1971)#6702.   *8.10*
(1971a) Embedding partial idempotent latin squares. J. Combinatorial Theory 10, 240-245. MR43(1972)#1862.   *8.5*
(1971b) Extending mutually orthogonal partial latin squares. Acta Sci. Math. (Szeged) 32, 283-285. MR48(1974)#126.   *8A*
(1972) Finite embedding theorems for partial latin squares, quasigroups, and loops. J. Combinatorial Theory A13, 339-345. MR47(1974)#3200.   *8.4*
(1975a) A brief up-to-date survey of finite embedding theorems for partial quasigroups. Proc. of the Conf. on Algebraic Aspects of Combinatorics (Univ. of Toronto 1975), Congressus Numerantium 13, 53-78. MR52(1976)#7929.   *8A*
(1975b) A partial Steiner triple system of order n can be embedded in a Steiner triple system of order 6n+3. J. Combinatorial Theory A18, 349-351. MR52(1976)#129.   *8.8*
(1976) Embedding orthogonal partial latin squares. Proc. Amer. Math. Soc. 58, 184-186. MR53(1977)#12987.   *8A*
(1980) A survey of embedding theorems for Steiner systems. Annals of Discrete Math. 7, 175-202. MR84d:05039.   *8.8*

Lindner C.C., Mendelsohn E., Mendelsohn N.S. and Wolk B.
(1979) Orthogonal latin square graphs. J. Graph Theory 3, 325-338. MR 80k:05022.   *2A, 4.4*

Lindner C.C. and Mendelsohn N.S.
(1973) Construction of perpendicular Steiner quasigroups. Aequationes Math. 9, 150-156. MR48(1974)#6305.   *6.5*

Lindner C.C., Mullin R.C. and Stinson D.R.
(1983) On the spectrum of resolvable orthogonal arrays invariant under the Klein group K_4. Aequationes Math. 26, 176-183. MR85f:05027.   *4.4*

Lindner C.C. and Stinson D.R.
(1984) Steiner pentagon systems. Discrete Math. 52, 67-74. MR85m:05014.   *4.4*

van Lint J.H.
(1974) "Combinatorial Theory Seminar" (Eindhoven Univ. of Tech., 1974) Lecture Notes in Mathematics, Vol. 382. Springer Verlag, Berlin. MR 50(1975)#4311.   *5A*

van Lint J.H., MacWilliams F.J. and Sloane N.J.A.
(1979) On pseudo-random arrays. SIAM J. Appl. Math. 36, 62-72. MR 80b:05011.   *9.3*

Lu M.G.
(1985) The maximum number of mutually orthogonal latin squares. Kaxue Tongbas (English Ed.) 30, 154-159. MR87a:05039. *5A*

Lumpov A.D.
(1972) Construction of a projective plane of order 9 over latin squares of the selfsame order which consist of subsquares of order 3 and are of various structures. (Russian) Kombinatornyi Anal. 2, 93-98. MR56(1978) #11824. *11A*
(1974) Latin squares of order 9 that contain 5 latin subsquares of order 3. (Russian) Kombinatornyi Anal. 3, 107-116. MR57(1979)#145. *11A*

Lyamzin A.I.
(1963) An example of a pair of orthogonal latin squares of order ten. (Russian) Uspehi Mat. Nauk. 18, 173-174. MR28(1964)#2979. *11A*

Ma S.L.
(1984) A note on binary arrays with a certain window property. IEEE Trans. on Information Theory IT-30 , 774-775. MR86m:94005. *9.3*

MacInnes C.R.
(1907) Finite planes with less than eight points on a line. Amer. Math. Monthly 14, 171-174. *11A*

MacMahon P.A.
(1898) A new method in combinatory analysis, with application to latin squares and associated questions. Trans. Camb. Phil. Soc. 16, 262-290. *12A*

Macneish H.F.
(1922) Euler squares. Ann. of Math. 23, 221-227. *5.3, 11A*

MacWilliams F.J. and Sloane N.J.A.
(1976) Pseudo-random sequences and arrays. Proc. IEEE 64, 1715-1729. MR55(1978)#12295. *9.3*

MacWilliams F.J., Sloane N.J.A. and Thompson W.
(1973) On the existence of a projective plane of order 10. J. Combinatorial Theory A14, 66-78. MR47(1974)#1644. *11.2, 11A*

Maillet E.
(1894) Sur les carrés latins d'Euler. C. R. Assoc. France Av. Sci. 223, part 2, 244-252. *2.4*

Main V.R and Tippett L.H.C.
(1941) Statistical methods in textile research, Part 4: The design of weaving experiments. Shirley Inst. Mem. 18, 109-120. Not reviewed in MR. *10.3*

Malara N.A. and Pellegrino C.
(1984) Groups and variable order codes. Proc. Conf. on Combinatorial and Incidence Geometry: Principles and Applications, pp. 325-337. Rend. Sem. Mat. Brescia 7, Milan. MR86e:94023. *9.2*
(1986) On the maximal number of mutually orthogonal F-squares. In

"Combinatorics '84", Proc. of Conf. held at Bari, Italy, 1984. Annals of Discrete Math. 30, 335-338. MR87k:05045. *12.1*

Malih A.E.
(1972) A description of projective planes of order n by latin squares of order n-1. (Russian) Kombinatornyi Anal. 2, 86-92. MR56(1978)#11825. *11A*
(1976) Latin squares of order 8 derived from the description of Hughes planes of order 9. (Russian) Kombinatornyi Anal. 4, 40-47. MR56(1978)#11827. *11A*

Mandl R.
(1985) Orthogonal latin squares : an application of experimental design to compiler testing. Comm. ACM 28, 1054-1058. Not reviewed in MR. *9.2*

Mann H.B.
(1942)$ The construction of orthogonal latin squares. Ann. Math. Statist. 13, 418-423. MR4(1943), pages 184, 340. *2.1, 3.3, 11.5*
(1944) On orthogonal latin squares. Bull. Amer. Math. Soc. 50, 249-257. MR 6(1945), page 14. *2.2, 2.5*

Martin G.E.
(1968) Planar ternary rings and latin squares. Matematiche (Catania) 23, 305-318. MR40(1970)#3424. *11.4*

Maskell E.J.
(1925) The technique of plot experiments. Proc. Imperial Botanical Conf., London, July 7-16, 1924, Edited F. T. Brooks, pp. 12-13 and 373-381. Cambridge Univ. Press. *10.3*

Massey J.L., Maurer U. and Wang M.
(1988) Non-expanding key-minimal, robustly-perfect, linear and bilinear ciphers. Proc. Eurocrypt 87, pp. 237-247. In "Advances in cryptology", Lecture Notes in Computer Science No.?. Springer Verlag, Berlin. MR? *9.6*

Matulić-Bedenić C.
(1986) Projective planes of order 11 with a collineation group of order 5. Rad. Jugoslav. Akad. Znan. Umjet. 413, 39-43. MR87e:51011. *11.3*

McCarthy D.
(1976) Transversals in latin squares of order 6 and (7,1)-designs. Ars Combinatoria 1, 261-265. MR54(1977)#5007. *2.3*

McCarthy D. and Vanstone S.A.
(1977) Embedding (r,1) designs in finite projective planes. Discrete Math. 19, 67-76. MR57(1979)#2935. *2.5*

McEliece R. and Sarwarte D.V.
(1981) On sharing secrets and Reed Solomon codes. Comm. ACM 24, 583-584. MR82h:94024. *9.5*

McLeish M.
(1975) On the existence of latin squares with no subsquares of order two. Utilitas Math. 8, 41-53. MR53(1977)#166. *4.2*
(1980) A direct construction of latin squares with no subsquares of order two. Ars Combinatoria 10, 179-186. MR82f:05016. *4.2*

Mendelsohn N.S.
(1968) Hamiltonian decomposition of the complete directed n-graph. Theory of Graphs (Proc. Colloq., Tihany, 1966), 237-241. Academic Press, New York. MR38(1969)#4361. *3.5*
(1969) Combinatorial designs as models of universal algebras. In "Recent Progress in Combinatorics" (Proc. Third Waterloo Conf. on Combinatorics, 1968), pp. 123-132. Academic Press, New York. MR41(1971)#85. *5.1, 7.5*
(1970) Orthogonal Steiner Systems. Aequationes Math. 5, 268-272. MR 44(1972)#1587. *6.5*
(1975) Algebraic construction of combinatorial designs. Proc. Conf. on Algebraic Aspects of Combinatorics (Univ. Toronto, Ontario, 1975). Congressus Numerantium 13, 157-168. MR52(1976)#5439. *5.1, 7.5*

Mendelsohn N.S. and Wolk B.
(1985)$ A search for a non-Desarguesian plane of prime order. In "Finite Geometries", Edited by C.A. Baker and L.M. Batten. Lecture Notes in Pure and Applied Math., Vol. 103, pp. 199-208. Marcel Dekker, New York. MR 87g:51014. *11.3*

Miller G.A.
(1903) A new proof of the generalized Wilson's theorem. Annals of Math. 4, 188-190. Also in Collected Works of G. A. Miller Vol.II, pp. 247-250. *2.1, 3.3, 3.4*

Mills W.H.
(1977)$ Some mutually orthogonal latin squares. Proc. Eighth S.E. Conf. on Combinatorics, Graph Theory and Computing, 1977. Congressus Numerantium 19, 473-487. MR58(1979)#286. *5A*

Mirsky L.
(1971) Transversal Theory. (Mathematics, Science and Engineering, Vol. 75.) Academic Press, New York. MR44(1972)#87. *12.3*

Montaron B.
(1985) On incidence matrices of finite projective planes. Discrete Math. 56, 227-237. MR86m:05028. *11A*

Moore E.H.
(1896) Tactical Memoranda I-III. Amer. J. Math. 18, 264-303. *11A*

Morgan J.P.
(1988a) Terrace constructions for bordered two-dimensional neighbour designs. Ars Combinatoria 26,123-140. MR? *3A*
(1988b) Balanced polycross designs. J. Royal Statist. Soc. B50, 93-104. MR 89k:62104. *3A*

Mullen G.L.
(1981) Local permutation polynomials over a finite field. Det Kongelige Norske Videnskabers Selskab 1, 1-4. Not reviewed in MR. *9.5*
(1988) Polynomial representation of complete sets of mutually orthogonal frequency squares of prime power order. Discrete Math. 69, 79-84. MR89a:05038. *12.1, 12.5*
(1989) Permutation polynomials and nonsingular feedback shift registers over finite fields. IEEE Trans. on Information Theory IT-35, 900-902. MR? *9.6*

Mullen G.L. and Niederreiter H.
(1987) Dickson polynomials over finite fields and complete mappings. Canad. Math. Bull. 30, 19-27. MR88c:11074. *2.6*

Mullin R.C. and Nemeth E.
(1969)$ On furnishing Room squares. J. Combinatorial Theory 7, 266-272. MR41(1971)#5228. *6.5*
(1970) On the nonexistence of orthogonal Steiner systems of order 9. Canad. Math. Bull. 13, 131-134. MR41(1971)#3297. *6.5*

Mullin R.C., Schellenberg P.J., Stinson D.R. and Vanstone S.A.
(1978) On the existence of 7 and 8 mutually orthogonal latin squares. Dept. of Combinatorics and Optimization, Research Report CORR 78-14, Univ. of Waterloo. Not reviewed in MR. *5A*
(1980) Some results on the existence of squares. Annals of Discrete Math. 6, 257-274. MR81m:05036. *5A*

Mullin R.C. and Stinson D.R.
(1984) Holey SOLSSOM's. Utilitas Math. 25, 159-169. MR85h:05028. *4.4*

Nelder J.A.
(1965a) The analysis of randomized experiments with orthogonal block structure. I. Block structure and the null analysis of variance. Proc. Royal Soc. A283, 147-162. MR31(1966)#848. *10.8*
(1965b) The analysis of randomized experiments with orthogonal block structure. II. Treatment structure and the general analysis of variance. Proc. Royal Soc. A283, 163-178. MR30(1965)#4363. *10.8*
(1977) A reformulation of linear models (with Discussion). J. Royal Statist. Soc. A40, 48-76. MR56(1978)#16943. *10.7*

Niederreiter H.
(1987) Point sets and sequences with small discrepancy. Monatshefte für Mathematik 104, 273-337. MR89c:11120. *9.2*

Niederreiter H. and Robinson K.H.
(1981) Bol loops of order pq. Math. Proc. Camb. Phil. Soc. 89, 241-256. MR 82c:20121. *2.6*
(1982)$ Complete mappings of finite fields. J. Austral. Math. Soc. A33, 197-212. MR83j:12015. *2.6*

Nilrat C.K. and Praeger C.E.
(1988) Complete latin squares: terraces for groups. Ars Combinatoria 25, 17-29. MR89i:05060. *3.6*

Bibliography

Norton D.A.
(1952) Groups of orthogonal row-latin squares. Pacific J. Math. 2, 335-341. MR14(1953), page 235.   *3.5*

Norton H.W.
(1939) The 7×7 squares. Ann. Eugenics 9, 269-307. MR1(1940), page 199.   *2.5, 4.3, 10.4*

Olderogge G.B.
(1963) On some special correcting codes of matrix type. (Russian) Radiotehnika 18, No.7, 14-19. Not reviewed in MR.   *9.2*

Olesen K.
(1976) A completely balanced polycross design. Euphytica 25, 485-488. Not reviewed in MR.   *10.9*

Olesen K. and Olesen O.J.
(1973) A polycross pattern formula. Euphytica 22, 500-502. Not reviewed in MR.   *10.9*

Ostrom T.G.
(1964) Nets with critical deficiency. Pacific J. Math. 14, 1381-1387. MR 30(1965)#1446.   *2.5*
(1966) Replaceable nets, net collineations, and net extensions. Canad. J. Math. 18, 666-672. MR35(1968)#4809.   *2.5*

Owens P.J.
(1976) Solutions to two problems of Dénes and Keedwell. J. Combinatorial Theory A21, 299-308. MR 54(1977)#7285.   *3.1, 3.2*
(1983) Private Communication.   *3.1*
(1987) Knight's move squares. Discrete Math. 63, 39-51. MR88b:05031.   *9.4*
(199α) Complete sets of POLS and the corresponding projective planes. Preprint.   *11.6*

Paige L.J.
(1947) A note on finite abelian groups. Bull. Amer. Math. Soc. 53, 590-593. MR9(1948), page 6.   *2.1, 3.3, 6.3*
(1951)$ Complete mappings of finite groups. Pacific J. Math. 1, 111-116. MR13(1952), page 203.   *2.1, 3.3*

Paige L.J. and Wexler C.
(1953) A canonical form for incidence matrices of finite projective planes and their associated latin squares. Portugaliae Math. 12, 105-112. MR15(1954), page 671.   *11.4, 11.6*

Panteleeva L.I.
(1968) The construction of a projective plane over a latin square which occurs in the description of the Hughes plane of order 9. (Russian) Perm. Gos. Res. Inst. Učen. Zap. 61, 63-67. MR41(1971)#2523.   *11A*

Parker E.T.
(1959a) Construction of some sets of mutually orthogonal latin squares. Proc. Amer. Math. Soc. 10, 946-949. MR22(1961)#674.   *5.3*

(1959b) Orthogonal latin squares. Proc. Nat. Acad. Sci. USA 45, 859-862. MR21(1960)#3344.   *4.4, 5.3*
(1962) Nonextendibility conditions on mutually orthogonal latin squares. Proc. Amer. Math. Soc. 13, 219-221. MR25(1963)#2968.   *4.4*
(1978) A collapsed image of a completion of a "turn-square". J. Combinatorial Theory. A24, 128-129. MR57(1979)#287.   *2.3*
(1988) The almost orthogonal 6-squares. Congressus Numerantium 62, 85-86. MR89k:05019.   *11A*

Parker E.T. and Somer L.
(1988) A partial generalization of Mann's theorem concerning orthogonal latin squares. Canad. Math. Bull. 31, 409-413. MR?   *2A*

Patterson H.D.
(1971) Unbiassed estimation of error in the analysis of change-over trials. Bull. Internat. Statist. Inst. 44, Book 2, 320-324. Not reviewed in MR.   *10.8*
(1982) Changeover designs. In Encyclopaedia of Statistical Sciences, Vol. 3. Wiley, New York. MR84k:62001.   *3.0*

Pearce S.C.
(1975) Row-and-column designs. Applied Statist. 24, 60-74. MR52(1976) #12247.   *10.8*

Pearson K.
(1896) Mathematical contributions to the theory of evolution. III Regression, heredity and panmixia. Phil. Trans. Royal Soc. A187, 253-318.   *10.7*

Peters M. and Roth R.
(1987) Four pairwise orthogonal latin squares of order 24. J. Combinatorial Theory A44, 152-153. MR88b:05032.   *5A*

Pierce W.A.
(1953) The impossibility of Fano's configuration in a projective plane with eight points per line. Proc. Amer. Math. Soc. 4, 908-912. MR15(1954), page 460.   *11A*

Plotkin M.
(1960) Binary codes with specified mimimum distance. IEEE Trans. on Information Theory IT-6, 445-450. MR22(1961)#13361.   *9.1*

Preece D.A.
(1975) Bibliography of designs for experiments in 3 dimensions. Australian J. Statist. 17, 51-55. MR52(1976)#9525.   *10.6*
(1979) Supplementary bibliography of designs for experiments in three dimensions. Australian J. Statist. 21, 170-172. MR80b:62090.   *10.6*

Quattrocchi G.
(1984) Some constructions of sets of mutually kn-orthogonal latin squares of order n. J. Inform. Optim. Sci. 5, 49-64. MR85h:05030.   *6.7, 9.2*

Quattrocchi P.
(1968) S-spazi e sistemi di rettangoli latini. Atti Sem. Mat. Fiz. Univ.

Modena 17, 61-71. MR38(1969)#3759.   *9.2*

Raghavrao D.
(1971) Constructions and Combinatorial Problems in Design of Experiments. Wiley, New York. MR51(1976)#2187.   *5.2*

Ramanathan K.G.
(1947) On the product of the elements in a finite abelian group. J. Indian Math. Soc. 11, 44-48. MR9(1948), page 408.   *2.1*

Rédei L.
(1946) Über eindeutig umkehrbare Polynome in eindlichen Körpern. Acta Sci. Math. Szeged 11, 85-92. MR8(1947), page 138.   *9.5*

Reed I.S. and Stewart I.M.
(1962) Note on the existence of perfect maps. IEEE Trans. on Information Theory IT-8, 10-12. MR24(1962)#B1260.   *9.3*

Repphun K.
(1965) Geometrische Eingenschaften vollständiger Orthomorphismensysteme von Gruppen. Math. Z. 89, 206-212. MR33(1967)#631.   *11A*

Rhemtulla A.R.
(1969) On a problem of L. Fuchs. Studia Sci. Math. Hungar. 4, 195-200. MR 40(1970)#1468.   *2.1*

Ringel G.
(1974a) Cyclic arrangements of the elements of a group. AMS Notices 21, A95-96. Not reviewed in MR.   *3.3*
(1974b) Map Colour Theorem. Spinger Verlag, Berlin. MR50(1975)#1955. See also translation to Russian: MR57(1979)#5809.   *3.3*

Ringel G. and Youngs J.W.T.
(1969) Solution of the Heawood Map Colouring Problem: Case 11. J. Combinatorial Theory 7, 71-93. MR39(1970)#1360.   *3.3*

Robinson D.A.
(1966) Bol loops. Trans. Amer. Math. Soc. 123, 341-354. MR33(1967)#2755.   *2.1*

Rodger C.A.
(1983) Embedding incomplete idempotent latin squares. In "Combinatorial Math. X" (Proceedings of the Conference held in Adelaide, Australia, 1982), pp. 355-366. Lecture Notes in Math. No.1036. Springer Verlag, Berlin. MR85h:05031.   *8A*
(1984) Embedding an incomplete latin square in a latin square with a prescribed diagonal. Discrete Math. 51, 73-89. MR86a:05025.   *4.3, 8A*

Rogers D.G.
(199α) Turgeon's bound for the length of additive sequences of permutations. Preprint.   *5A*

Rogers K.
(1964) A note on orthogonal latin squares. Pacific J. Math. 14, 1395-1397. MR33(1967)#5501.   *5.6*

Rohrbach H.
(1953) Mathematische und maschinelle Methoden beim Chiffrieren und Dechiffrieren. In "Naturforschung und Medizin in Deutschland, 1939-1946." Band 3. Angewandte Mathematik, Teil 1, pp. 233-257. Verlag Chemie, Weinheim. MR15(1954), page 747. [English Translation: in Cryptologia 2(1978), 21-37 and 101-121. MR81e:94027a/b.]   *9.0*

Rosa A.
(1974) On the falsity of a conjecture on orthogonal Steiner triple systems. J. Combinatorial Theory A16, 126-128. MR48(1974)#10845.   *6.5*

Ruiz F. and Seiden E.
(1974) Some results on construction of orthogonal latin squares by the method of sum composition. J. Combinatorial Theory A16, 230-240. MR 48(1974)#8267.   *4.4*

Rumov B.T.
(1982) On perpendicularly separable quasigroups. (Russian) Mat. Zametki 31, 527-538. MR84c:20085. [English Translation: Math. Notes Acad. Sci. USSR 31(1982), 267-273.]   *6.5*

Rybnikov A.R. and Rybnikova N.M.
(1966) A new proof of the non-existence of a projective plane of order 6. (Russian) Vestnik. Moscow Univ. Ser. I, Mat. Meh. 21, No.6, 20-24. MR 34(1967)#4982.   *1.2*

Ryser H.J.
(1951) A combinatorial theorem with an application to latin rectangles. Proc. Amer. Math. Soc. 2, 550-552. MR13(1952), page 98.   *8.2, 8.3*
(1967) Neuere Probleme der Kombinatorik. Vortrage über Kombinatorik Oberwolfach, 24-29 Juli 1967. Mathematisches Forschungsinstitut Oberwolfach. Not reviewed in MR.   *2.2*

Sade A.
(1959) Quasigroupes parastrophiques. Expressions et identités. Math. Nachr. 20, 73-106. MR22(1961)#5688.   *1.3*
(1960) Produit direct-singulier de quasigroupes orthogonaux et anti-abéliens. Ann. Soc. Sci. Bruxelles, Sér.I, 74, 91-99. MR25(1963)#4017.   *2.1, 4.1*

Safford F.H.
(1907) Solution of a problem proposed by O.Veblen. Amer. Math. Monthly 14, 84-86.   *11A*

Sane S.S.
(1985) On extendible planes of order ten. J. Combinatorial Theory A38, 91-93. MR86f:51012.   *11.2*

Sarvate D.G. and Seberry J.
(1986) Encryption methods based on combinatorial designs. Ars Combinatoria 21A, 237-246. MR? *9.2*

Schauffler R.
(1956) Über die Bildung von Codewörtern. Arch. Elek. Übertragung 10, 303-314. MR18(1957), page 368. *9.0, 9.6*

Schellenberg P.J., Van Rees G.H.T. and Vanstone S.A.
(1978)$ Four pairwise orthogonal latin squares of order 15. Ars Combinatoria 6, 141-150. MR80c:05038. *5.0, 11.2*

Schellenberg P.J., Stinson D.R., Vanstone S.A. and Yates J.W.
(1981) The existence of Howell designs of side n+1 and order 2n. Combinatorica 1, 289-301. MR82m:05028. *4.4*

Schulz R.-H.
(1989) Some check digit systems over non-abelian groups. Preprint No. 89-10, Freie Universitat, Berlin. *9A*
(199α) A note on check character systems using latin squares. Preprint. *9A*

Scott R.A.
(1932) Potato Manurial Trials, Details of Experimental Work, Seasons 1930-32. Supplement to Tasmanian J. Agric. August 1st, 1932. *10.6*

Seah E. and Stinson D.R.
(1987) Some perfect one-factorizations of K_{14}. Annals of Discrete Math. 34, 419-436. MR89b:05099. *3.6, 4.2*

Seberry J.
(1980)$ A construction for generalized Hadamard matrices. J. Statist. Planning and Inference 4, 365-368. MR81m:05047. *12.1*

Sciden E. and Wu C.J.
(1976) On construction of three mutually orthogonal latin squares by the method of sum composition. Essays in Probability and Statistics, pp. 57-64. Shinko Tsusho, Tokio. MR82j:05031. *4.4*

Shamir A.
(1979) How to share a secret. Comm. ACM 22, 612-613. MR80g:94070. *9.5*

Shannon C.E.
(1949) Communication theory of secrecy systems. Bell System Tech. J. 28, 656-715. MR11(1950), page 728. *9.0*

Shor P.W.
(1982) A lower bound for the length of a partial transversal in a latin square. J. Combinatorial Theory A33, 1-8. MR83j:05017. *2.2*

Shrikhande S.S.
(1961)A note on mutually orthogonal latin squares. Sankhya, A23, 115-116. MR25(1963)#703. *2.5*

Shull R.
(1984) Collineations of projective planes of order nine. J. Combinatorial Theory A37, 99-120. MR85k:51014. *11.2*
(1985) The classification of projective planes of order 9 possessing a collineation of order 5. Proc. Conf. on Groups and Geometry, Part A (Madison, Wisconsin, 1985). Algebras, Groups and Geometries 2, 365-379. MR87i:51026. *11.2*

Singer J.
(1960) A class of groups associated with latin squares. Amer. Math. Monthly 67, 235-240. MR23(1962)#A1542. *2.3*

Singleton R.C.
(1964) Maximum distance q-ary codes. IEEE Trans. on Information Theory IT-10, 116-118. MR29(1965)#2118. *9.1, 9.2*

Skolem Th.
(1958) Some remarks on the triple systems of Steiner. Math. Scand. 6, 273-280. MR21(1960)#5582. *8.8*

Smetaniuk B.
(1981) A new construction on latin squares I. A proof of the Evans conjecture. Ars Combinatoria 11, 155-172. MR83b:05028. *8.10*
(1983) The completion of partial F-squares. In "Combinatorial Mathematics X" (Proceedings of the Conference held in Adelaide, Australia, 1982), pp. 367-374. Lecture notes in Math. No.1036. Springer Verlag, Berlin. MR85g:05037. *12.3*

Song H.Y.
(199α) On the relation between circular Florentine arrays and power latin sets. Preprint. *5 A*

Steedley D
(1974) Separable quasigroups. Aequationes Math. 11, 189-195. MR53(1977) #8310. *6.5*
(1976) This is a paper by C. C. LINDNER and N. S. MENDELSOHN. Constructions of n-cyclic quasigroups and applications. Aequationes Math. 14, 111-121. MR58(1979)#22370. *6.5*

Stein S.K.
(1956) Foundations of quasigroups. Proc. Nat. Acad. Sci. USA. 42, 545-546. MR18(1957), page 111. *1.3*
(1957) On the foundations of quasigroups. Trans. Amer. Math. Soc. 85, 228-256. MR20(1959)#922. *1.3*
(1975) Transversals of latin squares and their generalizations, Pacific J. Math. 59, 565-575. MR52(1976)#7930. *2.2, 2.6, 6.3*

Steven H.M.
(1928) Nursery Investigations. Forestry Commission Bullletin No. 11. London: His Majesty's Stationery Office. *10.3*

Stevens W.L.
(1939) The completely orthogonalized latin square. Ann. Eugenics 9, 82-

93. Not reviewed in MR. *9.5, 11A*

Stinson D.R.
 (1978) A note on the existence of 7 and 8 mutually orthogonal latin squares. Ars Combinatoria 6, 113-115. MR80c:05040. *5A*
 (1979a) On the existence of thirty mutually orthogonal latin squares. Ars Combinatoria 7, 153-170. MR81a:05020. *5A*
 (1979b) A generalization of Wilson's construction for mutually orthogonal latin squares. Ars Combinatoria 8, 95-105. MR81f:05032. *5.5*
 (1981) A general construction for group-divisible designs. Discrete Math. 33, 89-94. MR81m:05039. *4.1*
 (1984) A short proof of the nonexistence of a pair of orthogonal latin squares of order six. J. Combinatorial Theory A36, 373-376. MR85g:05039. *1.2*
 (1986) The equivalence of certain incomplete transversal designs and frames. Ars Combinatoria 22, 81-87. MR88c:05032. *4.4*

Stojaković Z. and Ušan J.
 (1978) Orthogonal systems of partial operations. Univ. u Novom. Sadu Zb. Rad. Prirod.-Mat. Fak. 8, 47-51. MR81f:20107. *9.2*

Straley T.H.
 (1972) Construction of Steiner quasigroups containing a specified number of subquasigroups of a given order. J. Combinatorial Theory A13, 374-382. MR46(1973)#3672. *4.3*

Strambach K.
 (1981) Geometry and loops. In "Geometries and Groups". Lecture Notes in Math. No. 893, pp. 111-147. Springer Verlag, Berlin. MR83i:51009. *11A*

Street D.J.
 (1986) Unbordered two-dimensional nearest neighbour designs. Ars Combinatoria 22, 51-57. MR88b:05034. *3.6*

Street D.J. and Street A.P.
 (1985) Designs with partial nearest neighbour balance. J. Statist. Planning and Inference 12, 47-59. MR87a:05046. *3A*

Surla A., Tošić R. and Ušan J.
 (1979) A method for constructing a system of orthogonal latin squares, codes and k-seminets. (Russian.) Univ. u Novom. Sadu Zb. Rad. Prirod.-Mat. Fak. 9, 191-197(1980). MR82d:05036. *9.2*

Szajowski K.
 (1976) The number of orthogonal latin squares. Applicationes Mathematicae 15, 85-102. MR54(1977)#5013. *5A*

Tarry G.
 (1899) Sur le problème d'Euler des n officiers. Interméd. Math. 6, 251-252. *11A*
 (1900a) Le problème des 36 officiers. C.R. Assoc. France Av. Sci. 29, part 2, 170-203. *1.2, 4.4, 11A*
 (1900b) Sur le problème d'Euler des 36 officiers. Interméd. Math. 7, 14-16. *11A*

Taylor M.A.
(1978) A generalization of a theorem of Belousov. Bull. Lond. Math. Soc. 10, 285-286. MR80j:20079. *1.3*

Tillson T.W.
(1980) A Hamiltonian decomposition of K_{2m}^*, $2m \geq 8$. J. Combinatorial Theory B29, 68-74. MR82e:05075. *3.5*

Todorov D.T.
(1985) Three mutually orthogonal latin squares of order 14. Ars Combinatoria 20, 45-48. MR87d:05045. *2.4, 4.1, 5.0, 5.6, 11.6*
(1989) Four mutually orthogonal latin squares of order 20. Ars Combinatoria 27A, 63-65. MR? *4A, 5A*

Tosić R.
(1980) On the reconstruction of latin squares. Publ. Inst. Math. (Beograd) N.S. 28(42), 209-213. MR84d:05048. *8A*

Tukey J.W.
(1955) Query 113 and Answer [Additivity in a latin square]. Biometrics 11, 111-113. Not reviewed in MR. *10.7*

Treash C.
(1971) The completion of finite incomplete Steiner triple systems with applications to loop theory. J. Combinatorial Theory A10, 259-265. MR 43(1972)#397. *8.8*

Ušan J.
(1978) Orthogonal systems of n-ary operations and codes. Mat. Vestnik 2(15) (30), No. 1, 91-93. MR81a:94041. *9.2*
(1979) Kvazigrupe. (Slovenian) Institut za Mathematiku, Novi Sad. MR 81g:20133. *9.2*

Van Rees G.H.J.
(1978) A corollary to a theorem of Wilson. Dept. of Combinatorics and Optimization Research Report, CORR 78-15, Univ. of Waterloo. Not reviewed in MR. *5.4**
(199α) Subsquares and transversals in latin squares. Preprint. *2 A*

Veblen O. and Wedderburn J.H.M.
(1907) Non-desarguesian and non-pascalian geometries. Trans. Amer. Math. Soc. 8, 379-388. *11.7, 11A*

Vedder K.
(1983) Affine planes and latin squares. Annals of Discrete Math. 18, 761-768. MR85b:05042. *11.5*

Verhoeff J.
(1969) Error-detecting decimal codes. Math. Centre Tracts No. 29, Math. Centrum, Amsterdam. MR41(1970)#1426. *9A*

Vik K.
(1924) Bedømmelse av feilene på forsøksfelter med og uten målestokk.

Meldinger fra Norges Landbrukshøiskole 4, 129-181. *10.3*

Wall D.W.
(1957) Sub-quasigroups of finite quasigroups. Pacific J. Math. 7, 1711-1714. MR19(1958), page 1159. *4.3*

Wallis W.D.
(1984) Three orthogonal latin squares. Proc. of Thirteenth Manitoba Conference on Numerical Math. and Computing (Winnipeg, 1983.) Congressus Numerantium 42, 69-86. MR86g:05017. (See also next entry.) *4.1*
(1986) Three orthogonal latin squares. Advances in Math. (Beijing) 15, 269-281. MR88e:05022.

Wallis W.D. and Zhu L.
(1983) Orthogonal latin squares with small subsquares. In "Combinatorial Mathematics X" (Proceedings of the Conference held in Adelaide, Australia, 1982), pp. 398-409. Lecture notes in Math. No.1036. Springer Verlag, Berlin. MR85k:05024. (See also next entry.) *4.1, 4.4*
(1984)The existence of orthogonal latin squares with small subsquares. J. Combin. Inform. System Sci. 9, 1-13. MR86i:05035.

Wang S.M.P.
(1978) On self-orthogonal latin squares and partial transversals of latin squares, Ph.D. thesis, Ohio State University. (Page 798 in Vol.39/02-B of Dissertation Abstracts International). *2.2, 2.4, 4.1*

Wang S.M.P. and Wilson R.M.
(1978) A few more squares II (Abstract). Proc. Ninth S.E. Conf. on Combinatorics, Graph Theory and Computing, 1978. Congressus Numerantium 21, 688. Not reviewed in MR. *4.1, 5.6*

Wang Y.
(1964) A note on the maximal number of pairwise orthogonal latin squares of a given order. Sci. Sinica 13, 841-843. MR30(1965)#3866. *5A*
(1966) On the maximal number of pairwise orthogonal latin squares of order s; an application of the sieve method. Acta Math. Sinica 16, 400-410. (Chinese) Translated as Chinese Math. Acta 8(1966), 422-432. MR34(1967) #2483. *5.6*

Weisner L.
(1963) Special orthogonal latin squares of order 10. Canad. Math. Bull. 6, 61-63. MR26(1963)#3621. *11.1, 11A*

Wells M.B.
(1967) The number of latin squares of order eight. J. Combinatorial Theory 3, 98-99. MR35(1968)#5343. *11A*

Whitesides S.H.
(1976) Projective planes of order 10 have no collineations of order 11. Proc. Seventh S.E. Conf. on Combinatorics, Graph Theory and Computing, 1976. Congressus Numerantium 17, 515-526. MR55(1978)#2617. *11.2*
(1979a) Collineations of projective planes of order 10, (I). J. Combinatorial Theory A26, 249-268. MR81b:51010. *11.2*

(1979b) Collineations of projective planes of order 10, (II). J. Combinatorial Theory A26, 269-277. MR81b:51010. *11.2*
(1985) Fixed-point-free collineations of order 7 in projective planes of order 9. Proc. Conf. on Groups and Geometry, Part B (Madison, Wisconsin, 1985). Algebras, Groups and Geometries 2, 564-578. MR87h:51020. *11.2*

Williams E.J.
(1949) Experimental designs balanced for the estimation of residual effects of treatments. Austral. J. Sci. Research A2, 149-168. MR11(1950), page 449. *3.0, 3.2, 10.5*
(1950) Experimental designs balanced for pairs of residual effects. Austral. J. Sci. Research A3, 351-363. MR12(1951), page 449. *3.0, 10.5*

Wilson R.M.
(1974a) Concerning the number of mutually orthogonal latin squares. Discrete Math. 9, 181-198. MR49(1975)#10575. *4.1, 5.4, 5.4*, 5.6, 5A*
(1974b) A few more squares. Proc. Fifth S.E. Conf. on Combinatorics, Graph Theory and Computing, 1974. Congressus Numerantium 10, 675-680. MR50(1975)#6887. *5A*

Wojtas M.
(1977) On seven mutually orthogonal latin squares. Discrete Math. 20, 198-201. MR58(1979)#21690. *5.4*
(1980a) New Wilson-type constructions of mutually orthogonal latin squares. Discrete Math. 32, 191-199. MR82b:05037. *5.5*
(1980b) A note on mutually orthogonal latin squares. Prace Nauk. Inst. Mat. Politech. Wroclaw. Ser. Stud. Materialy No.14. Analiza Dyskretna, pp. 11-14. MR81k:05022. *5.4*, 5.6*
(1981) Letter to A. E. Brouwer dated 3/5/81. *5A*

Wood R.F.
(1974) Fifty Years of Forestry Research: A Review of Work conducted and supported by the Forestry Commission, 1920-1970. Forestry Commission Bulletin No. 50. London: Her Majesty's Stationery Office. *10.4*

Woodcock C.F.
(1986)$ On orthogonal latin squares. J. Combinatorial Theory A43, 146-148. MR87j:05044. *2A*

Woolbright D.E.
(1978) An n×n latin square has a transversal with at least $n-\sqrt{n}$ distinct symbols. J. Combinatorial Theory A24, 235-237. MR57(1979)#12259. *2.2*

Wright C.E.
(1962) A systematic polycross design. Res. Exp. Rec. Min. Agric., Northern Ireland, 1961, Vol. 11, Part I, 7-8. Not reviewed in MR. *10.9*
(1965) Field plans for a systematically designed polycross. Record Agric. Res. 14, Part I, 31-41. Not reviewed in MR. *10.9*

Wu W.W.
(1985) Elements of Digital Satellite Communication, Vol.2. Computer Science Press, New York. Not reviewed in MR. *9.5*

Yamamoto K.
(1954) Euler squares and incomplete Euler squares of even degrees. Mem. Fac. Sci. Kyushu Univ. A8, 161-180. MR16(1955), page 325.   *1.2, 4.4, 6.1, 6.5*
(1961) Generation principles of latin squares. Bull. Inst. Internat. Statist. 38, 73-76. MR26(1963)#4933.   *4.4*

Yates F.
(1933) The formation of latin squares for use in field experiments. Empire J. Exper. Agric. 1, 235-244. [Reprinted in Experimental Design: Selected Papers, pp. 57-68. Griffin, London, 1970]   *10.8*
(1935) Complex experiments (with discussion). J. Royal Statist. Soc. Suppl. 2, 181-247. [Reprinted without Discussion in Experimental Design: Selected Papers, pp. 69-117. Griffin, London, 1970]   *10.6*
(1937) The Design and Analysis of Factorial Experiments. Technical Communication No. 35 of the Imperial Bureau of Soil Science, Harpenden. (Now Commonwealth Bureau of Soils.)   *10.2, 10.6*

Youden W.J.
(1937) Use of incomplete block replications in estimating tobacco-mosaic virus. Contrib. Boyce Thompson Inst. 9, 41-48.   *10.6*
(1940) Experimental designs to increase accuracy of greenhouse studies. Contrib. Boyce Thompson Inst. 11, 219-228. Not reviewed in MR.   *10.6*

Zassenhaus H.J.
(1958) The Theory of Groups. Chelsea Publishing Company, New York, 2nd edition. MR19(1958), page 939.   *2.6*

Zhang L.
(1963) On the maximum number of orthogonal latin squares I. Shuxue Jinzhan 6, 201-204. (Chinese) MR32(1966)#4027.   *5A*

Zhu L.
(1977) On a method of sum composition to construct orthogonal latin squares. Acta Math. Applicatae Sinica 1, 55-61. (Chinese.) Not reviewed in MR. [English translation in Ars Combinatoria 14(1982), 47-55. See below.] *1.2*
(1982) A short disproof of Euler's conjecture concerning orthogonal latin squares. (With editorial comment by A.D. Keedwell). Ars Combinatoria 14, 47-55. MR85e:05036.   *1.2, 4.1, 4.4*
(1983) Pairwise orthogonal latin squares with orthogonal small subsquares. Research report CORR No.83-19, Univ. of Waterloo.   *4.4*
(1984a) Orthogonal diagonal latin squares of order fourteen. J. Austral. Math. Soc. A36, 1-3. MR84k:05021.   *4.4*
(1984b) Orthogonal latin squares with subsquares. Discrete Math. 48, 315-321. MR85f:05030.   *4.4*
(1984c) Some results on orthogonal latin squares with orthogonal subsquares. Utilitas Math. 25, 241-248. MR85k:05025.   *4.4*
(1984d) Existence for holey SOLSSOMS of type 2^n. Congressus Numerantium 45, 295-304. MR86j:05042.   *4.4*
(1984e) Six pairwise orthogonal latin squares of order 69. J. Austral. Math. Soc. A37, 1-3. MR86d:05022.   *5.3**

SUBJECT INDEX

A
adjugate class (of latin squares) 1:5
admissible group 2:15-16, 6:9
admissible quasigroup 2:12-15, 6:10-11,23
affine geometry 12:18
affine plane 2:19, 5:3, 11:1-2,16, 12:18
alarm codes 9:11-12
algebra, see universal algebra
alphabet of a code 9:2,3,44
alphanumeric code 9:42
alternating group 2:2, 3:27,55
analysis of variance 10:15-22
authentication of messages 9:47-48
automorphism 2:17, 6:26, 7:4
autotopism 2:11,12

B
Bachelor square 2:35
Baer subplane 5:10
balanced identity 1:6
balanced latin square (see also quasi-complete latin square) 3:19
binary operation, definition of 7:2
blocks of a design 2:9, 3:34, 5:3-4, 7:9
blocking set of a projective plane 11:4,32
Bol loop 2:1,33
Bol-Moufang type, identity of 2:1
Bruck loop 2:1
Bruck-Moufang identity 2:2

C
canonical form 2:29
Cayley table 1:1,2, 2:17, 6:5
chain 6:9,11-15
change-over design 3:1, 10:12,18,24
check digits 9:1,42
ciphering 9:1,2,12-13,48
circular Florentine array 3:39-40
circular Tuscan-k array 3:39-40, 9:33
circular Vatican array 9:33
clear set 5:3,7,10
coding of information 6:32, 9:1
collineation group of a projective plane 11:4-9
column complete latin square 3:1,57, 9:17, 10:12,18
column method 3:30-31

column orthogonal latin squares 11:16-18
comma-free code 9:19
commutative latin rectangle, definition of 8:9
commutative quasigroup 2:1, 8:1-3
commutator subgroup 2:2, 3:33,34
comparative experiments 10:1,2
complete automorphism 2:16
complete canonical set of squares 11:24-30
complete graph 3:37, 4:16
complete latin square 3:1,56, 10:2,25
complete mapping 2:1-33,34, 3:25-26,30,32-33, 6:9, 9:44-45
complete mapping polynomial 2:33
complete set of mutually column orthogonal latin squares 11:16-18
complete set (or system) of orthogonal latin squares 1:14, 2:25,35, 6:30, 11:1,11,37, 12:18
complete set of orthogonal F-squares 12:4-5,16-18
complete set of orthogonal F-rectangles 12:14,15
complete transversal 2:8
conjectures (and open questions),
    of Bailey 3:24
    of Belyavskaya 6:11,22,23
    of Brualdi 2:8,34,35
    of Dénes and Keedwell 2:3,28, 3:36,55, 9:16, 11:3,10
    of Evans 8:1,38-49
    of Euler 1:3-4, 7:1,9-10
    of Ryser 2:8
    of Stein 2:27
conjugate identities 1:5, 8:20
conjugate invariant subgroup (of an orthogonal array) 8:21-22
conjugate latin squares 7:5
conjugate operations 1:5
conjugate orthogonal arrays 6:21
conjugate permutations 2:10
conjugate quasigroups, see parastrophic quasigroups
constraint in an experimental design 10:9
contraction of a latin square 4:21
Costas array 3:41, 9:28-37
crossed product 6:19-21
Cruse's theorem 8:3,5,10,12,23,26
cutting method (for Tuscan squares) 3:41
cyclic groups 3:21,34-35, 4:15,17,25, 6:11,21

              D
de Bruijn sequences 9:18
decimal code 9:42-47
defect of orthogonality 6:13
deficiency 2:23
derived group, see commutator subgroup
derived partial latin square 12:10-11

desarguesian projective plane 11:3
diagonal of a latin square 1:3
diagonally placed latin rectangle, definition of 8:14
Dickson polynomial 2:33
dicyclic groups 3:47,50-54
difference matrix 2:26
digraph complete set of latin squares 4:20, 11:11
dihedral groups 3:27,35-36,50-51,53,54,55
directed terrace (see also sequenceable group) 3:20-25
direct product (of latin squares) 4:3
directrix of a latin square 1:3
doubly diagonal latin square 5:20
dual translation plane 11:3,29
duplex 2:28

E

embedding a latin square in a complete set 2:24, 11:31-33
embedding a latin rectangle 8:1,4-5,6-7,10-11,14-16
embedding a partial F-square 12:9-12
embedding a partial latin square 8:1-3,8-20,37-38
embedding a partial projective plane 2:22-23
embedding partial idempotent latin squares 8:11-20
embedding Mendelsohn triple systems 8:27,33-36
embedding semi-symmetric quasigroups 8:24-27
embedding Steiner triple systems 8:27-33
embedding totally symmetric quasigroups 8:24-27
encoding a message 9:1
enumeration of F-squares 12:8
equidistant code 9:4
equi-n-square 2:27-28,35
equivalent latin squares 1:4
error-correcting codes 9:2-6
error-detecting codes 9:2-6,42-47,48,49
Euler's conjecture 1:3-4, 7:1,9-10
Evans' conjecture 8:1,38-39,43,45-47
Evans' theorem 8:6-7,9,25
even starter 3:42
exdomain element 2:30
experimental design 9:11, 10:4
experimental error 10:4
extent of a latin square 11:33-34

F

factorial experiment 10:3
Fano subplane 5:13
F-hyper-rectangle 12:14-15
Florentine square 3:38, 9;23
F-rectangle 10:9, 12:12-13
free algebra 7:2,10-14

frequency cube 12:15-16
frequency hopping 9:22-23
frequency square 12:1
frequency vector 12:1
F-square, see frequency square

G
generalized complete mapping 2:27,35
generalized direct product 4:6,14,35
generalized transversal 2:27
geometric net, see net
graeco-latin square 1:2, 10:13
graph, see complete graph
group-divisible block design 5:4,5
groups of order pq 3:27,34,56

H
Hadamard matrix 12:7
half-idempotent latin square (or quasigroup) 8:12,29,30,32
Hamiltonian circuit of a graph 3:44-46
Hamiltonian path of a graph 3:37
Hamming distance 9:2
homomorphism of an algebra 7:3
hyper-oval 5:9,10
holes in a design 5:14-16
holes in a latin square 4:7
homogeneous latin squares 4:24
Howell design 3:16
Hughes plane 11:3,18,19,29-30

I
idempotent quasigroup 1:2, 8:23,27,29,30,35
idempotent latin square 7:8, 8:19
idempotent latin rectangle, definition of 8:19
incidence matrix of a design 5:7
incidence matrix of a projective plane 11:10-14,35-36,38
incomplete array 5:15
incomplete latin square 6:2, 8:19
incomplete orthogonal arrays 6:3,28-29
incomplete pairwise orthogonal latin squares 4:7
index of a design 5:14-16
infix notation 7:2
information digits 9:1,5
intercalate 4:22-24
International Standard Book Number 9:43
isomorphic latin squares 1:5
isomorphic sets of orthogonal frequency squares 12:7
isomorphism of an algebra 7:4
isoplanar latin squares 3:56

isotopic frequency squares 12:8,18
isotopic latin squares 1:4, 3:2-3,6
isotopic quasigroups 1:5, 2:15-16
isotopism, isotopy 1:4,5, 11:25, 12:8,16
isotopy class (of latin squares) 1:5, 2:9, 11:23,35
Italian square 3:38

J
Joshi-Singleton bound 9:3,6

K
(k,λ)-complete mapping 2:32,36, 3:33
(k,λ) near-complete mapping 2:32,36
$(k,1)$-complete mapping 3:33
knight's move square 9:25, 10:6
Knut Vik design 5:20, 9:25, 10:6
k-regular complete mapping 3:32
(k,s)-invertibility 6:6-7,15-16
k-transversal 2:17-18

L
latin Costas array 9:35
latin cube 10:13,14, 12:14
latin power set 2:35, 5:20
latin rectangle 8:4-6,10,15,18, 9:8,48, 10:13-14
latin rectangle, definitions of 8:4,6
latin square, definition of 1:1
latin subsquare 2:25, 4:1-47, 6:4,8,22
latin transversal 2:27,35
left even starter 3:42
left neofield 2:30-31, 3:33
left nucleus of a loop 2:10,14
left nucleus of a quasigroup 2:11,14
left regular permutation 2:10
left starter 3:55-56
left translation 2:10
lifting (of a 2-sequencing) 3:48-50
linear space 5:3
line extendible net 5:3
local permutation polynomial 9:42
loop, definition of 1:1

M
MacNeish's theorem 2:10, 5:11,16, 7:8,10, 11:37, 12:6
main class (of latin squares) 1:5, 11:23,35
main effect in an experimental design 10:10
main diagonal, see main left-to-right diagonal
main left-to-right diagonal 8:10,46-47
major numbers 2:4-5,7

map colouring 3:26
maximal partial transversal 2:4
maximal set of mutually orthogonal latin squares 2:25
maximal set of mutually perpendicular symmetric latin squares 6:31-32
maximum distance block code 9:6,48
Mendelsohn quasigroup 8:27
Mendelsohn triple system 8:27-30,33-36
middle nucleus of a loop 2:14
middle nucleus of a quasigroup 2:11,14
middle regular permutation 2:10,13
minor numbers 2:4-5,8
(m,n)-rectangle 2:27
mobile radio telephone system 9:1,24
modified latin square 10:13
Moufang loop 2:1
mutually orthogonal latin squares, definition of 2:2

N

near complete mapping 2:28-31,36
nearfield 2:31, 11:18
near orthogonal latin squares 6:1,2,22-23,28
near sequenceable, see R-sequenceable
neofield 2:30-31, 3:33,57
net 2:18,21,22,26,37, 5:5, 9:48, 11:1-2
(n,k,σ) set of permutations 6:33
non-binary codes 9:2
non-desarguesian projective plane 11:9-10,31
non-orthogonal analysis 10:20
non-uniform direct product 4:4,15,18,21
normal form of a matrix 2:26
n-orthogonal 5:12, 6:5
n-square 2:27
nucleus 2:10-11,14
nullary operation, definition of 7:2

O

one-factorization of a graph 3:46, 4:16-17,22
order of a net 11:1
orthogonal array 5:2,4,5, 6:21,27-28, 8:21
orthogonal F-rectangles 12:13-14
orthogonal F-squares 9:48, 12:2-4,6-7,17
orthogonal latin square graph 2:36
orthogonal latin squares 1:2, 2:2,20,26, 3:8-10,29, 4:2, 6:1,4, 7:1-2,6-8,10-12, 9:6, 12:13
orthogonal latin squares of orders six and ten 1:3, 2:20, 4:2,7,11,33,36, 5:1, 6:2,31, 7:9, 11:3-5,33,35-38
orthogonal mate 1:3, 2:8-9,16, 3:30, 6:12
orthogonal matrices 5:2,6
orthogonal permutation polynomials 12:16-18

orthogonal quasi-complete latin squares 3:18-19
orthogonal quasigroups 4:35, 6:4
orthogonal row-complete latin squares 3:8-10
orthogonal vectors 5:2
orthomorphism (see also complete mapping) 2:36, 11:9
orthomorphism graph 2:33,36
overall balanced latin square 3:19

P

pairwise balanced block design 4:12,45, 5:3-4,9-11, 7:9
pairwise orthogonal latin squares, see mutually orthogonal latin squares
pandiagonal latin square, see Knut Vik design
parallel class (of a net) 5:3, 11:1,16
parastrophic operations 1:5
parastrophic quasigroups (or latin squares) 1:5, 7:5, 8:20-21
partial admissibility 6:1,9
partial commutative latin square, definition of 8:9
partial F-square 12:9-12
partial idempotent latin square, definition of 8:9
partial latin rectangle 4:30, 8:1-2
partial latin square 4:30, 8:1,8-11,16,38-45, 12:10-12
partially balanced incomplete block design 5:4
partial projective plane 2:22-23
partial quasigroup 8:23,26
partial symmetric quasigroup 8:24,26
partial transversal of a latin square 2:3-8,27, 6:10-11
perfect one-factorization 3:46, 4:16-17
perfect map 9:18
permutation of a quasigroup 2:10
permutation polynomial 2:33, 9:42,48, 12:16-18
perpendicularly separable quasigroup 6:24-26
perpendicular quasigroups (or latin squares) 6:1,3,22,24-26
P-group 2:2,34, 3:33
planar ternary ring 11:11,13,32
plane-filling Costas array, see latin Costas array
Plotkin bound 9:3-5
polycross design 10:2,25-26
polygonal path construction 3:41, 9:31,48
positively represented 2:18,21
projective plane 4:20, 5:3,4,5,6,8,9, 7:10, 11:1-10,16,18-38
prolongation 4:3,14,21,35,36,46
pseudo-characteristic of a neofield 3:33
pseudo-random array 9:18
pseudo-random permutations 9:13-16
pseudo-random sequence 9:24

Q

q-ary code 9:2,7,17
quadrangle criterion 1:2,6

Subject index

quasi-column-complete latin square 3:16-25
quasi-complete latin square 3:16-25
quasi-latin square 10:14, 12:18
quasigroup, definition of 1:1, 7:4
quasi-row-complete latin square 3:16-25
quasi-sequenceable group 3:16-25
quaternion group 11:21
q-step type 2:3,18,20
quotient sequencings 3:12-16

R

randomization of a design 3:24, 10:4-5,22-24
randomized complete block design 10:5,24
rank of a chain 6:9-11
rank sum 6:12
raster-graphics display 9:37
Rédei polynomial, see permutation polynomial
Rédei type, blocking set of 11:32
reduced degree of a polynomial 2:33
reduced frequency square 12:8,18
reduced latin square 1:1,12, 2:10, 11:35
reducible identity 1:6
Reed-Solomon code 9:37,38,40-41
regular set of pairwise orthogonal latin squares 2:26
replication in an experiment 10:4-5
representational array (of a projective plane) 11:25
residual effects of treatments 10:12
resolvable block design 5:3,9
right even starter 3:42
right nucleus of a quasigroup 2:11,14
right nucleus of a loop 2:10,14
right quasigroup 6:15
right product of quasigroups 6:5
right regular permutation 2:10
right translation 2:10
Roman square (see also row-complete latin square) 3:38
Room design 3:16
(r,1)-design 2:9
r-orthogonal latin squares 4:45, 6:1,3-4,6-7,11-14,20,25,27,29,31-34, 9:11
row-complete latin square 3:1-16,37,57, 9:17-24
row latin square 3:37, 9:18
R-sequenceable 2:2,29, 3:25-30,34,54,57
R_h-sequenceable 3:31
R-sequencings for cyclic groups 3:34,57
Ryser's theorem 8:3,5,6-9

S

secret-sharing system 9:37-41
self-orthogonal latin square 4:7, 11:3

semi-latin square 10:13
semi-symmetric quasigroup 8:22,24,26
separable block code 9:6
separable block design 5:7
separable quasigroup 6:24
separated elements 1:6
separating automorphism 6:26
sequenceable group (see also directed terrace) 2:29-30, 3:1-16,25,34,42-56
sequencings for cyclic groups 3:34
σ-symmetry 3:17
signature of a message 9:47
similar algebras, definition of 7:2
Singer difference set 5:11
Singleton bound, see Joshi-Singleton bound
singly-generated projective plane 4:20, 11:7
singly-periodic Costas array 9:28
soluble group 2:2
space communication 9:9
special orthogonal mappings 11:9
special symmetric sequencing 3:53-54
special 2-sequencing 3:53-54
spectrum of Mendelsohn triple systems 8:29
spectrum of near orthogonality 6:22-23
spectrum of partial admissibility 6:11
spectrum of partial orthogonality 6:1,18-21
spectrum of Steiner triple systems 8:29
standard form, see reduced latin square
starter-translate 2-sequencing 3:55
Steiner loop 7:6
Steiner pentagon system 4:46
Steiner quasigroup 4:33, 6:25-26, 7:5,12, 8:27
Steiner triple system 4:2, 6:25, 7:5-6,14,27-32, 8:12,14,27-33
Stein quasigroup 4:45
strength of an array 5:2
subquasigroup 4:5,26,33,45
subsquare complete 4:19-22,24
sub-variety of algebras, definition of 7:3
super P-group 2:2, 3:33-37
switching mode matrix 9:10
Sylow 2-subgroup 2:2, 3:27,34
Sylow 3-subgroup 3:26
symmetric d-sequencing 3:54
symmetric group 2:2, 3:55, 6:31
symmetric latin square 1:1, 6:3,24-25
symmetric 1-design 5:7,8,10
symmetric sequencing 3:42-56
system of distinct representatives 8:3,5,11,15-18
systematic block code, definition of 9:5

T

t-admissibility 6:11,14-16,18
terrace (see also 2-sequencing) 3:20-25,47,51
terraces for cyclic groups 3:21
three-concurrent blocking set 11:33-34
totally diagonal latin square, see Knut Vik design
totally symmetric quasigroup 8:23-27
transal 2:24, 6:9
transformation set (of latin squares) 1:5
translation plane 11:3,19,31
transposition errors in a code 9:43-46
transversal 1:3, 2:1-33, 4:2, 6:9,10,22, 11:32
transversal design 5:3-17
transversals, number of 6:23
treatment contrast 10:22
triangular blocking set 11:33
Tuscan square 3:37-42, 9:18-23
Tuscan-k square 3:38-42, 9:23
two-dimensional codes 9:24,26
2-sequencing (see also terrace) 3:25,42-56
type of an algebra 7:2
type r (for quasigroups) 6:5

U

unary operation, definition of 7:2
unipotent latin square 1:2
unipotent quasigroup 1:2
unital 5:9
universal algebra 7:1-2,4,5,7,9,14

V

valence 2:22-23
variable order code 9:3
variate 10:4
variety of algebras 7:1,4,7-8
variety of latin squares 7:4-6
variety of orthogonal latin squares 7:6-9
Vatican Costas array 9:36
Vatican square 3:38,41, 9:30,31-32

W

weighted points 5:14
Wilson bound 5:12
window property of an array 9:17-22
Wojtas bound 5:12

Y

Youden square 10:14

LATE CORRECTIONS AND COMMENTS ADDED IN PROOF.

<u>Corrections to "Additional Remarks" at the end of Chapter 2.</u>

(1) In the report on P. Yff's work on P-groups, the edition of American Mathematical Monthly referred to is that of June-July <u>1972</u>. Also, we stated that Yff had a proof for groups of even order as well as for all groups of odd order. This is not correct. His proof, which he recently submitted to Europ. J. Math., covers only a restricted class of groups of even order.

(2) We reported that E. T. Parker and L. Somer had generalized Theorem 12.3.2(a) of [DK] but failed to observe that a result of J. Bierbrauer, stated as Theorem 7.2 of Chapter 11, is stronger. (See MR90f:05025.)

<u>Comment on Lemma 1.6 of Chapter 3.</u>

A simplification of the proof is as follows:
Let ord $G = n$. Since r and n are relatively prime, there exist integers h and k such that $hr+kn = 1$. If $g_1^r = g_2^r$, then $(g_1^r)^h = (g_2^r)^h$. That is, $g_1^{1-kn} = g_2^{1-kn}$, whence $g_1 = g_2$.

<u>Further remark concerning Chapter 5.</u>

Very recently, D. Bedford has shown that the constructions which give the best values for $N(v)$ when v is small can nearly all be regarded as special cases of one single construction which uses orthogonal left-neofields. Indeed, for values of v less than 21, only the constructions which give $N(12) \geq 5$, $N(15) \geq 4$ and $N(18) \geq 3$ do not fit into this pattern. (The value $N(10) \geq 2$ has been obtained by several methods. Not all of these fit the Bedford pattern.)

www.ingramcontent.com/pod-product-compliance
Ingram Content Group UK Ltd.
Pitfield, Milton Keynes, MK11 3LW, UK
UKHW050440060526
12271UKWH00033B/22